Physical and Chemical Weathering in Geochemical Cycles

NATO ASI Series

Advanced Science Institutes Series

A Series presenting the results of activities sponsored by the NATO Science Committee, which aims at the dissemination of advanced scientific and technological knowledge, with a view to strengthening links between scientific communities.

The Series is published by an international board of publishers in conjunction with the NATO Scientific Affairs Division

A Life Sciences **B Physics**	Plenum Publishing Corporation London and New York
C Mathematical and Physical Sciences **D Behavioural and Social Sciences** **E Applied Sciences**	Kluwer Academic Publishers Dordrecht, Boston and London
F Computer and Systems Sciences **G Ecological Sciences** **H Cell Biology**	Springer-Verlag Berlin, Heidelberg, New York, London, Paris and Tokyo

Series C: Mathematical and Physical Sciences - Vol. 251

Physical and Chemical Weathering in Geochemical Cycles

edited by

A. Lerman

Department of Geological Sciences,
Northwestern University, Evanston, Illinois, U.S.A.

and

M. Meybeck

Institut de Biogéochimie Marine,
Ecole Normale Supérieure, Montrouge, France

Kluwer Academic Publishers

Dordrecht / Boston / London

Published in cooperation with NATO Scientific Affairs Division

Proceedings of the NATO Advanced Study Institute on
Physical and Chemical Weathering in Geochemical Cycles
Aussois, France
4–15 September 1985

Library of Congress Cataloging in Publication Data

Physical and chemical weathering in geochemical cycles.

(NATO ASI series. Series C, Mathematical and physical sciences ; no. 251)
Proceedings of the NATO Advanced Study Institute on Physical and Chemical Weathering in Geochemical Cycles, held at Aussois, France, September 1985.
"Published in cooperation with NATO Scientific Affairs Division."
Includes index.
1. Weathering--Congresses. I. Lerman, Abraham. II. Meybeck, M. (Michel) III. NATO Advanced Study Institute on Physical and Chemical Weathering in Geochemical Cycles (1985 : Aussois, France) IV. North Atlantic Treaty Organization. Scientific Affiars Division. V. Series.
QE570.P47 1988 551.3 88-13529

ISBN 90-277-2821-6

Published by Kluwer Academic Publishers,
P.O. Box 17, 3300 AA Dordrecht, The Netherlands.

Kluwer Academic Publishers incorporates the publishing programmes of
D. Reidel, Martinus Nijhoff, Dr W. Junk, and MTP Press.

Sold and distributed in the U.S.A. and Canada
by Kluwer Academic Publishers,
101 Philip Drive, Norwell, MA 02061, U.S.A.

In all other countries, sold and distributed
by Kluwer Academic Publishers Group,
P.O. Box 322, 3300 AH Dordrecht, The Netherlands.

Printed in The Netherlands

TABLE OF CONTENTS

INTRODUCTORY NOTE

The Organizing Committee of the NATO Advanced Study Institute on **Physical and Chemical Weathering in Geochemical Cycles**, *Abraham Lerman, Michel Meybeck, and James I. Drever*, acknowledge with thanks the financial support received from *NATO Scientific Affairs Division* (Brussels), and additional support from *Centre National de Recherches Scientifiques* (Paris).

It is also a pleasure to express our thanks to the administration and personnel of *Centre "Paul Langevin"* at Aussois (Savoie, France) for their help and arrangements made for the Advanced Study Institute. At Aussois, invaluable assistance was provided by *Mme. Jacqueline Fraisse*, Assistant to the Organizing Committee. Last but far from least, the laborious word processing of the entire book typescript at Northwestern University was done with devotion and care by *Ms. Cheril Cheverton*, without whom the typescript would not have seen the light of day for submission to the publisher.

Participants, Lecturers and Organizers of NATO ASI on *Physical and Chemical Weathering in Geochemical Cycles*, 4–15 September, 1985, Aussois (Savoie, France).

LIST OF PARTICIPANTS NATO-ASI

Aussois, Savoie

BELGIUM

L. Chou
Laboratoire d'Océanographie, Université Libre, 50, avenue F.D. Roosevelt, 1050 Bruxelles

R. Nijs
Geologisch Instituut (R.U.G.), Krijgslaan 281, 9000 Gent

R. Wollast
Laboratoire d'Océanographie, Université Libre, 50 avenue F.D. Roosevelt, 1050 Bruxelles

CANADA

D.J. Bottomley
Department of Geology, University of Ottawa, Ottawa, Ontario

A.H. MacDonald
Department of Geology, University of Western Ontario, London, Ontario

O. Slaymaker
Department of Geography, University of British Columbia, Vancouver

S. Talman
Department of Geology, University of Western Ontario, London, Ontario

R.L. Thomas
International Joint Commission, Great Lakes Regional Office, 100 Ouelette Avenue, Windsor, Ontario

J. Veizer
Department of Geology, University of Ottawa, Ottawa, Ontario

P.A. Yeats
Bedford Institute of Oceanography, P.O. Box 1006, Dartmouth, Nova Scotia B2Y 4A2

DENMARK

D. Postma
Institute of General Geology, Østervoldgade 10, 1350 Copenhagen

FRANCE

J. Berthelin
Centre de Pédologie Biologique du CNRS, B.P. 5 Vandoeuvre-les-Nancy, Cedex

S. Duchaffaut
Institut de Géologie, Université de Grenoble, Grenoble, Cedex

N. Dufrien
Centre de Sédimentologie et Géochimie de la Surface, 1 rue Blessig, 67084 Strasbourg, Cedex

D. Etchanchu
Laboratoire de Pédologie et Géochimie, Université de Toulouse, 38, rue des 38 Ponts, 31062 Toulouse, Cedex

B. Francou
Centre de Géomorphologie, Rue des Tilleuls, 14000 Caen

Z. Kattan
Centre National de la Recherche Scientifique, Centre de Sédimentologie et de Géochimie de la Surface, rue Blessig, 67084 Strasbourg, Cedex

F. Lelong
Résidence Foch, 184 rue de l'Aisne, 45160 Olivet

M. Loubet
Université Paul Sabatier, Laboratoire de Minéralogie, 38 rue des 36 Ponts, 31062 Toulouse

M.-D. Loye-Pilot
Laboratoire de Biologie-Géologie, E.N.S.J.F., 1 rue M. Arnoux, 92120 Montrouge

R.-R. Nkounkou
Université Louis Pasteur de Strasbourg, Institut de Géologie, 1 rue Blessig, 67084 Strasbourg, Cedex

J.-L. Probst
Institut de Géologie, 1 rue Blessig, 67084 Strasbourg, Cedex

P. Seyler
Laboratoire de Géologie, Ecole Normale Supérieure, 46 rue d'Ulm, 75230 Paris, Cedex

GERMANY

B. Fitzner
Geologisches Institut, RWTH Aachen, Wuellnerstrasse 2, 51 Aachen

R. Sobanski
Geol.-Palaeont.-Institut, Universitaet Hamburg, Bundesstrasse 55, 2000 Hamburg 13

GREECE

F. Sakellariadou
A.G.R.G., Department of Geology, R.S.M., Imperial College, London SW7 2BP

ICELAND

G. Gudbergsson
Agricultural Res. Institute, Keldnaholt, 110 Reykjavik

ISRAEL

D.H. Yaalon
Institute of Earth Science, Geology Department, The Hebrew University, Jerusalem 91904

ITALY

A. Prestininzi
Dipartimento Science della Terra, P. le A. Moror, 5 – Roma

NETHERLANDS

D. Eisma
Netherlands Institute for Sea Research, P.O. Box 59, 1790 AB Den Burg, Texel

H.W.J. Pos
Waranda 21, 3705 ZB Zeist

A. Rajendran
Netherlands Institute for Sea Research, P.O. Box 59, 1790 AB Den Burg, Texel

C.H. van der Weijden
Rijksuniversiteit Utrecht, Institute of Earth Sciences, Department of Geochemistry, Budapestlaan 4, P.O. Box 80.021, 3508 TA Utrecht

NORWAY

T. Frogner
Department of Geology, University of Oslo, Munkegt. 9, 0656 Oslo 6

T. Gjelsvik
Norsk Polarinstitutt, Rolfstangvn, 1330 OSLO Lufthavn

O.M. Saether
Norwegian Geological Survey, Box 3006, N-7001 Trondheim

R.F. Wright
Norwegian Institute for Water Research, P.O. Box 333, 0314 Oslo 3

PORTUGAL

F.J. Dos Santos Ramalhal
Al. D. Afonso Henriques 41–4°.D, 1000 Lisboa

M.R. De Sousa Avila Duarte Ramalhal
Al. D. Afonso Henriques 41–4°.D, 1000 Lisboa

SPAIN

J. Roca
Universitat de Barcelona, Dept. de Geomorfologia i Tectonica, Gran Via 585, 08007 Barcelona

TURKEY

M.E. Atabey
Middle East Technical University, Dept. of Geological Engineering, Ankara

UNITED KINGDOM

J.D. Burton
Dept. of Oceanography, The University, Southampton SO9 5NH

J.A. Catt
Rothamsted Experimental Station, Harpenden, Herts AL5 2JQ

C. Moon
Geology Dept., Leicester University, University Road, Leicester

J. Rae
Dept. of Geology, Whiteknights, Reading, Berks RG6 2AB

D.W.H. Walton
British Antarctic Survey, Madingley Road, Cambridge CB3 OET

U.S.A.

P. Bond
Bureau of Geology, 903 West Tennessee Street, Tallahassee, Fla. 32304

C.J. Bowser
Dept. of Geology and Geophysics, University of Wisconsin, 1215 West Dayton St., Madison, Wisc. 53706

O.P. Bricker
Water Resources Division, U.S. Geological Survey, MS 432, Reston, Va. 22092

J.I. Drever
Dept. of Geology and Geophysics, University of Wyoming, Laramie, Wyo. 82071

R.M. Garrels
Dept. of Marine Science, Unversity of South Florida, 830 First Street South, St. Petersburg, Fla. 33701

R. Hellmann
Dept. of Geological and Geophysical Sciences, Princeton University, N.J. 08544

L.M. Jones
Research and Development Department, 314 RE, Conoco Inc., Ponca City, Oklahoma 74603

R.J. Murnane
Dept. of Geology and Geophysics, Princeton University, Guyot Hall, Princeton, N.J. 08544

A.J. Ramirez
228 Deike Building, Geoscience Dept., Pennsylvania State University, University Park, PA 16802

R.F. Stallard
Dept. of Geology and Geophysics, Princeton University, Guyot Hall, Princeton, N.J. 08544

W. Ullman
Chemistry Division, Argonne National Laboratory, 9700 South Cass Avenue, Argonne, Ill. 60439

E. Zbinden
Dept. of Geological Sciences, Harvard University, Cambridge, Mass. 02138

LIST OF CONTRIBUTORS

AUTHORS' ADDRESSES FOR NATO ASI PROCEEDINGS

"Physical and Chemical Weathering in Geochemical Cycles"

Dr. J. Berthelin
Centre de Pédologie Biologique
Boîte Postale 5
54501 Vandœuvre-les-Nancy Cedex
France

Professor J. D. Burton
Department of Oceanography
The University
Southampton SO9 5NH
United Kingdom

Dr. J. A. Catt
Rothamstead Experimental Station
Harpenden, Herts. AL5 2JQ
United Kingdom

Dr. Lei Chou
(*see* R. Wollast)

Dr. D. Eisma
Netherlands Institute for Sea Research
Postbus 59
1790 AB Den Burg, Texel
The Netherlands

Professor R. M. Garrels *)
Department of Marine Science
University of South Florida
St. Petersburg, Florida 33701
U.S.A.

Professor H. D. Holland
Department of Geological Sciences
Harvard University
Cambridge, Massachusetts 02138
U.S.A.

*) Deceased 8 March, 1988

Dr. S. Kempe
Geologisch-Paläontologisches Institut
der Universität
Bundesstraße 55
2000 Hamburg 13
Federal Republic of Germany

Professor A. Lerman
Department of Geological Sciences
Northwestern University
Evanston, Illinois 60208
U.S.A.

Dr. R. H. Meade
U. S. Geological Survey
Mail Stop 413
Denver, Colorado 80225
U.S.A.

Dr. M. Meybeck
Institut de Biogéochimie Marine
Ecole Normale Supérieure
1, rue Maurice-Arnoux
92120 Montrouge
France

Professor S. O. Schlanger
Department of Geological Sciences
Northwestern University
Evanston, Illinois 60208
U.S.A.

Professor O. Slaymaker
Department of Geography
University of British Columbia
Vancouver, B.C. V6T 1W5
Canada

Dr. R. F. Stallard
U. S. Geological Survey
Mail Stop 413
Denver, Colorado 80225
U.S.A.

Dr. R. L. Thomas
Great Lakes Fisheries Research Board
Canada Centre for Inland Waters
P. O. Box 5050
Burlington, Ontario L7R 4A7
Canada

Professor Ján Veizer
Derry Laboratory
Department of Geology
University of Ottawa
Ottawa, Ontario K1N 6N5
Canada

Professor R. Wollast
Laboratoire d'Océanographie
Université Libre de Bruxelles
Campus de la Plaine, C.P. 208
1050 Bruxelles
Belgium

Dr. R. F. Wright
Norwegian Institute for Water Research
Box 333, Blindern
0314 Oslo
Norway

Ms. E. A. Zbinden
(*see* H. D. Holland)

WEATHERING RATES AND MAJOR TRANSPORT PROCESSES AN INTRODUCTION

Abraham Lerman
Department of Geological Sciences
Northwestern University
Evanston, Illinois 60208
U.S.A.

GLOBAL TRANSPORT AND WEATHERING PROCESSES

A global biogeochemical cycle is a conceptual model of the pathways and flows of individual chemical elements or their compounds in the surface environment of the Earth. The flows of materials on the Earth's surface are driven by processes of diverse nature, such as: the major physical and tectonic forces that control the shapes and elevation of the continental plates; the major climatic processes that are responsible for the global water flows and temperature distribution over the Earth's surface; the chemical and microbiological processes that result in dissolution of some crustal rock minerals and formation of other sedimentary minerals; the transport of the continental crustal materials by wind, rivers and ground-waters, and their ultimate delivery to the oceans.

The primary goals of study of the global biogeochemical cycles are understanding of the global transport mechanisms in the present, past and future. The present gives us a fleeting time-frame of reference, but it is the geologic record that enables us to see long-term secular trends in evolution of the surface environment. Changes in a number of flow paths and their magnitudes have been effected by human industrial and agricultural activity. These changes include transfer of living and fossil carbon to the atmosphere; accumulation of metals and other substances in alloys or elemental states in the environment (iron and its alloys, precious metals, organometallic compounds); acidification of the atmospheric precipitation and soils; local concentrations of high levels of radioactivity; massive fertilization of soils and natural waters; and changes in the residence times of carbon in the biosphere owing to changes in the proportions of forest and grass land. Trying to predict the longer-term consequences of any of these changes may be not less difficult than a hypothetical attempt to predict how the Earth's environment would have evolved after the appearance of the first "atmospheric polluters" —the first photosynthetic organisms that began at the time to add a new substance, free oxygen, to the atmosphere.

The global surface environment comprises a number of major reservoirs that can, depending on the scope of a cycle model and problem at hand, be further subdivided into smaller units. Thus, for example, the major reservoirs of the atmosphere, hydrosphere, continental crust, and oceanic crust can be subdivided into smaller units suitable for the scale of analysis of a particular model. Understanding of the driving forces within the geochemical cycles is fundamental to our correct perceptions of the global environmental change. The inherent flow and accumulation characteristics of the biogeochemical reservoirs also control migration of anthropogenic substances,

A. Lerman and M. Meybeck (eds.), Physical and Chemical Weathering in Geochemical Cycles, 1–10.

thereby enabling humankind to manage, at least to a limited extent, the global transport in the future.

Physical and chemical weathering are two main sets of processes responsible for release of solids and dissolved materials from crustal rocks. The products of crustal weathering are in part left *in situ* as a weathered regolith and soils, and in part transported by winds and running water. On the way from the continental source areas to the oceans, some of the material is at least temporarily stored in lakes, floodplains of rivers, and estuaries. Some of the dissolved materials may chemically react with solids in the new environments of their transport paths, or make minerals through inorganic or biological processes. In the ocean, accumulation of detrital sediments and removal of dissolved constituents from ocean water to sediments and oceanic crust are the last stages of the *exogenic* cycle, before a long return back to the continental crust through such processes as sedimentary accretion, subduction and reincorporation in the continental crust. The latter group of pathways represents a much longer *endogenic* cycle that is outside the scope of this book.

Among the sixteen chapters in this volume, some cover individual processes within the sedimentary cycle and some deal with the global picture at time scales ranging from the present to 10^9 years. A brief outline of the subject-matter of the individual chapters in the contenxt of the global cycles is given below.

The early stages of weathering involve chemical reactions between aluminosilicate minerals and waters (chapter by **Wollast and Chou**), and bacterially mediated leaching of mineral components (**Berthelin**). Acid rain is a weathering solution of major importance in the industrialized countries of the Northern Hemisphere (**Wright**). Acid rains react with minerals in soils and in the crust, causing their dissolution or alteration. Estimates of the fluxes of acidic and oxidizing substances to soils in recent and historical times (**Wright**), and in the distant past of about 2 billion years before present (**Holland and Zbinden**) provide measures of the rates of crustal weathering and formation of soils.

Mass movements of soils and bedrock erosion in the headwater parts of stream watersheds represent the initial stages of material transport that is made possible by physical weathering (**Slaymaker, Meade**). Another type of erosion —glacial —is at least in part responsible for the wide occurrence of the extensive loess deposits throughout the Northern Hemisphere (**Catt**). Historical changes in the rates of soil erosion due to activities of settlers in river and lake drainage basins are reflected in the sediment records of lakes (**Thomas**) and sediment storage rates in the floodplains of rivers (**Meade**).

The material balances of inorganic and organic carbon in a number of river basins produce a clear picture of how the carbon budgets can be affected by human practices, such as deforestation, organic pollution, fertilization of waters, and damming of rivers (**Kempe**).

Relationships between the physical and chemical weathering on a continental scale emerge through analysis of tectonic evolution of crustal plates, their geomorphological development, and geochemical reactivity of their constituent rock minerals with surface waters (**Stallard**). Continental and global budgets of riverine transport of dissolved and suspended materials, presented on an element by element basis, produce a pattern of chemical and physical denudation rates that strongly reflect the major climates, geomorphic features and lithologic structure of the continents (**Meybeck**).

Among the 39 major rivers of the world—those of greatest length and drainage basin area—seven have well developed estuarine mouths (Coleman, 1968): Amazon, Congo, Parana, Orinoco, Hwang-Ho, Rhine, and Rhone. The development at river mouths of such pronounced structures as estuaries, deltas or fjords is a result of interactions between sediment transport, tidal amplitudes, and sea-level changes (**Eisma**). Sediments in estuaries are derived from land as well as from the sea-bed. The sediment mass balance and storage in an estuary are the sum total of transport processes from land and from or to the ocean. Net transports into or out of an estuary vary considerably from one river to another depending on the water flow and sediment discharge conditions (**Eisma**). For the dissolved components of river water, storage and recycling in estuaries are important at least for such species as silica, iron, and manganese: that is, for those species involved in biological productivity and/or redox reactions in sediments and water (**Burton**).

In the ocean, imbalances in the budgets of dissolved and sedimentary materials are generally more difficult to study, because of the great physical scales. For the element strontium, global budgetary imbalances on time scales of 10^6 to 10^7 years are revealed by its higher rates of storage in shallow-water and pelagic carbonates that exceed its mean rate of delivery by rivers. Over longer term, the balance is restored by release of strontium from weathering of limestone reefs and shallow-water carbonates during the lower-level stands of the global sea level, and dissolution of Sr-rich skeletal carbonates in pelagic sediments (**Schlanger**).

The exogenic cycle of sediments is represented by the atmosphere-ocean-sediment system. The long-term working of this cycle, on time scales of 10^8 to 10^9 years of the Phanerozoic Eon, can be modeled as a closed system where the sediment mass is constant and the rates of chemical weathering are directly related to the masses of the major lithologic reservoirs (evaporites, iron sulfide in sediments, organic matter, and carbonate rocks) (**Garrels**). Weathering in the geologic past and formation of paleosoils provide evidence of the oxygen and carbon dioxide levels of the paleoatmosphere (**Holland and Zbinden**). Half-lives of the major sedimentary and tectonic units—such as the oceanic crust, oceanic sediments, continental crust and platforms, orogenic belts, and continental margins—have been estimated on the basis of a crust and sediment recycling model (**Veizer**). The results lead to estimates of the half-lives or destruction rates of the global reserves of coal, petroleum, and metallic ores on time scales of up to 10^9 years.

GLOBAL WATER FLOWS AND WEATHERING RATES

The importance of rivers and continental runoff in denudation of land and transport of the weathering products are repeatedly emphasized in the individual chapters of this volume. As a summary of the importance of the global water flows to the core questions of material transport, and physical and chemical weathering, the distribution of the global flows from the continents to the oceans is discussed below, with reference to Figures 1–3.

The water runoff from the continents to the ocean as a function of the geographic latitude and mean land elevation is shown in Figure 1. The data are given for six latitudinal belts of 30° each. The ice-covered Antarctic has the highest elevation, but its runoff to the ocean is only 1/4 to 1/2 of the runoff from the lands farther north. The runoff is the highest (greater than 500 mm/yr) in the equatorial belt from 30° S to 30° N, and it generally declines to the higher latitudes. A potential for higher rates of denudation and transport is met in the zones of the higher runoff, but greater land elevation alone is not necessarily a condition for a strong land-to-ocean transport.

Runoff from any latitudinal zone is a measure of its relative dryness or wetness, as shown in Figure 2. The plot shows the runoff or specific discharge against the land drainage areas, divided in latitudinal belts of 5° each.

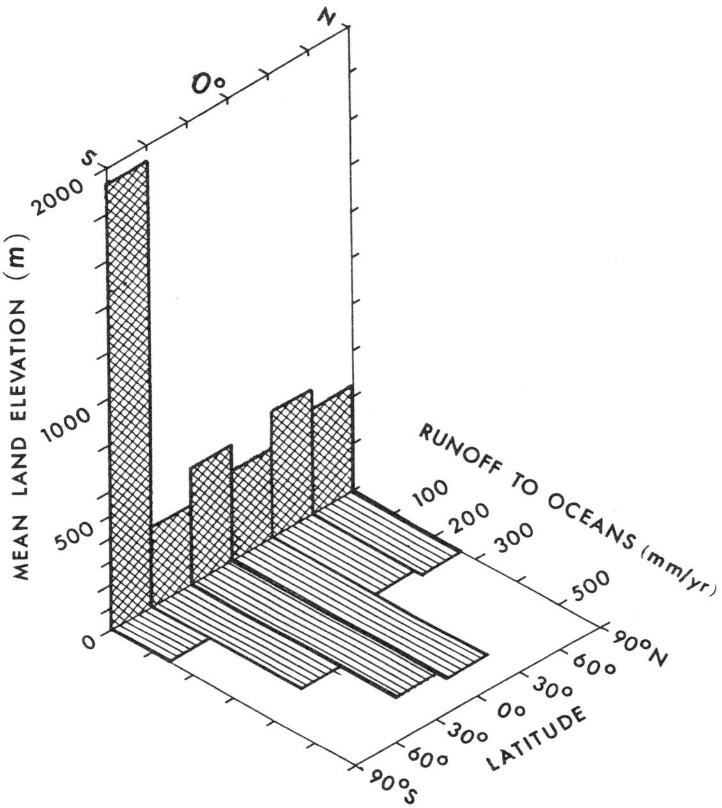

Figure 1. Mean land elevation (meters) and runoff to the oceans (mm/yr) in latitudinal belts of 30 ° each. The high elevation of the Antarctic is due to its ice cover. From data in Baumgartner and Reichel (1975), and Gates and Nelson (1975).

Within the zone from 90 ° S to 30 ° S, total discharge of water from different land areas to the ocean is bracketed by values lying within a factor of 10. However, the specific discharge or runoff within this belt varies from a low of 55 mm/yr near the South Pole to 1300 mm/yr in the wetter parts of Patagonia and New Zealand. In the equatorial belt, 30 ° S–30 ° N, runoff collects from a relatively narrow spread of drainage basin areas, about 2×10^6 to 5.5×10^6 km^2. This suggests that the differences in runoff between the individual sections of this zone are primarily related to the amounts of precipitation and evaporation.

In the Northern Hemisphere, between 30 ° N and 90 ° N, where most of the land areas of Asia, Europe and North America lie, runoff increases with an increasing drainage basin area. As a whole, however, the Northern Hemisphere is somewhat drier than the Southern Hemisphere, owing to its smaller ocean-to-land surface area ratio.

Many of the major rivers of the world and their drainage basins cut across the 5 °-wide latitudinal belts, and this makes the relationships between the water runoff

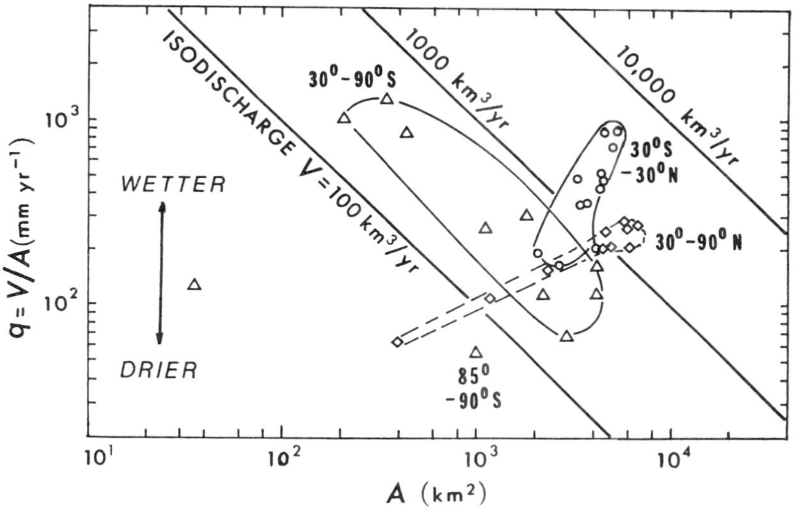

Figure 2. Runoff to the oceans and land drainage area in 5°-latitude bands, grouped into three zones, 60° each. Runoff q (mm yr^{-1}), discharge V (km^3 yr^{-1}), and land drainage area A (km^2) from Baumgartner and Reichel (1975).

and drainage areas of individual rivers (Figure 3) different from the mean latitudinal values shown in Figure 2. The total of the points in Figure 3 represents the rivers on six continents and adjacent land areas. The relationships between the runoff and the drainage area are approximated by straight lines of slopes -1, -1/2, and -1/3 in a power-law relationship of the type

$$q = K A^{-b}$$

A steeper drop in runoff (q) with an increasing drainage basin area (A) indicates a greater variance in runoff or, consequently, greater differences between the drier and wetter parts of a continent: these are the cases of Africa and Oceania. A less steep slope, or a greater scatter of the points, suggests a continent of smaller hydrological variability and, possibly, more uniform distribution of precipitation and evaporation. The Americas and Europe fall in the latter group.

For those surficial processes that are intermediated by the water cycle, their characteristic rates and data sources to the chapters in this volume are summarized in Table I. The characteristic rates are given for such diverse processes as: inorganic and bacterial mineral dissolution, dissolution and precipitation of carbonates, cation leaching from rocks and soils, rates of soil formation, soil downslope movements, eolian and riverine mass transport of solids, watershed and global denudation rates, storage of dissolved and suspended materials in estuaries and oceans, and geologically long-term recycling of sediments.

The rates of continental denudation and soil erosion in different environments and conditions cluster within an order of magnitude above and below the global mean (0.067 mm/yr), estimated by Meybeck from suspended material transport by rivers. Similar spread of values applies to the rates of formation of different soils. This poses a question, to what extent does the formation of soils on a bedrock keep pace with their erosion, and at what time scales can (if at all) a steady state be maintained?

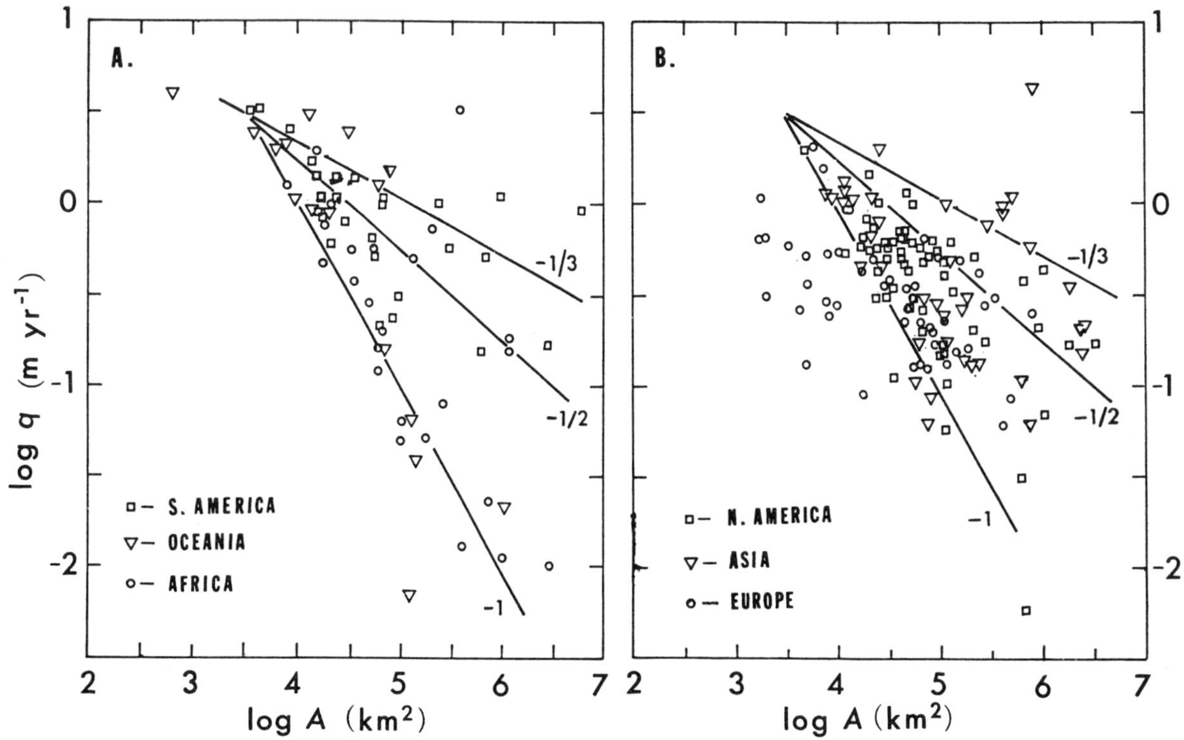

Figure 3. Plot of mean river runoff (q, in m yr^{-1}) against river basin drainage area (A, in km^2). Data from Baumgartner and Reichel (1975) and Meybeck (1984). Straight lines of slopes -1, -1/2 and -1/3 are shown as approximately bracketing the scatter of points. **A**: rivers of South America, Oceania and Africa. **B**: rivers of North America, Asia and Europe.

Entrapment of the transported sediments in lakes (**Thomas**), on river floodplains (**Meade**), and near-shore ocean floor (**Eisma**) indicates that net accumulation of the products of weathering takes place on, or near, continents at variable time scales.

A broad time spectrum of the rates of the surficial processes, summarized in Table I, provides a measure of the background information needed for insight into, and understanding of, global environmental change from a perspective that is not exclusively limited to observations of only one or two human generations.

REFERENCES

Baumgartner, A. and Reichel, E., 1975, *The World Water Balance*, Elsevier, Amsterdam and New York, 179 p.

Coleman, J. M., 1968, Deltaic evolution, *in* Fairbridge, R. W. (ed.) *Encyclopedia of Geomorphology*, pp. 255-260, Reinhold, New York.

Fairbridge, R. W., 1968, Continents and oceans—statistics of area, volume and relief, *in* Fairbridge, R. W. (ed.) *Encyclopedia of Geomorphology*, pp. 177-186, Reinhold, New York.

Gates, W. L. and Nelson, A. B., 1975, *A new (revised) tabulation of the Scripps topography on a 1° global grid. Part I: Terrain heights*, **R-1276-1-ARPA**, Rand Corp., Santa Monica, Calif.

Meybeck, M., 1984, *Les fleuves et le cycle géochimique des éléments*, Thèse de Doctorat d'Etat, Université Pierre et Marie Curie, Paris VI.

Table I. Summary of processes and their rates in physical and chemical weathering, as dealt with in this volume.

System and Process	*Process Rate or Other Measure*	*Reference to Chapter in this Volume and Explanatory Remarks*
SILICATE MINERALS		
Dissolution: albite	3×10^{-6} to 3×10^{-4} mm/yr or $3\times10^{-8\pm1}$ mol SiO_2 cm^{-2} yr^{-1}	**Wollast and Chou** High rates in acidic and alkaline solutions.
glass	0.15 g cm^{-2} yr^{-1}	**Slaymaker** Glass decomposition in soil.
Microbial solubilization	Rates 5 to 113 times higher than under sterile conditions	**Berthelin** Granitic sand.
SOILS		
Formation	1,000 to 10,000 yr	**Catt** Time of soil development on loess.
Various soils	0.06 to 0.45 mm/yr	**Slaymaker**.
Duricrusts on granite	9×10^{-3} mm/yr	**Slaymaker**.
Soil on bauxite	9×10^{-4} mm/yr	**Stallard**.
Dust deposition (storms)	up to 500 g/m^2 per event	**Catt**.
Weathering		
Oxidant/acid supply to soils	Flux ratio O_2/CO_2 = 0.03 ± 0.01	**Holland and Zbinden** For paleo-soils 1.8-2.5 billion years b.p.
Chemical weathering	0.02-0.03 and 0.01-0.15 eq m^{-2} yr^{-1}	**Wright** Cation (Ca, Mg, K, Na) leaching in soil columns.
	0.04–0.10 eq m^{-2} yr^{-1} 0.1–0.2 eq m^{-2} yr^{-1} 0.93 eq m^{-2} yr^{-1}	**Wright** Loess soil (Germany) New England Cascade Mountains.
	0.006 to 0.14 mm/yr	Estimates for preceding values, for four cations: $\times1.5$ mol/eq $\times100$ cm^3/ mol$\times10^{-3}$.
Movement and Erosion		
Landsliding	0.25–10 mm/yr	**Slaymaker**.
Soil creep	1–10 mm/yr	**Slaymaker**.
Erosion of headwater slopes in small watersheds	0.01–0.04 mm/yr or 28–95 g m^{-2} yr^{-1}	**Slaymaker** (see also **Meade**).

CRUSTAL MATERIALS AND RIVERS		
Land elevation	0.01–5 mm/yr	**Stallard**.
Land chemical denudation		
Global mean	0.014±0.001 mm/yr or 33 to 40 g m^{-2} yr^{-1}	**Meybeck** (see also **Wollast and Chou**).
Shields and sediments	0.005–0.03 mm/yr	**Stallard**.
Andean drainage:		
silicates	0.08 mm/yr	**Stallard**.
carbonates and shales	0.2 mm/yr	**Stallard**.
evaporites	0.7 mm/yr	**Stallard**.
Land erosion		
Global mean	0.067 mm/yr or 176.2 g m^{-2} yr^{-1}	**Meybeck** Suspended materials in river flow to oceans (see also **Meade**, **Stallard**, **Veizer**).
"Local"	*ca.* 1800 g m^{-2} yr^{-1} or 0.7 mm/yr: 94% stored, 6% exported	**Meade** Small watershed erosion since 1850: 85×10^6 tons over 360 km^2.
SEDIMENT TRANSPORT AND DEPOSITION		
Industrial age effects		
River sediment load	10-fold increase 1700–1900	**Meade**.
Land erosion index (part of L. Ontario basin)	40% prior to 1811; 80% since 1811	**Thomas**.
River bed rise due to hydraulic mining discharge, Sacramento R., California	3–5 m rise in 10–20 years	**Meade**.
Sediment load reduction in dammed rivers:		
Nile (Aswan Dam)	125×10^6 t/yr prior to dam 1×10^6 to 2×10^6 after	**Kempe**.
Colorado (Hoover Dam)	similar to the Nile	**Meade**.
Lakes, recent rates of sedimentation:		
Geneva	3.7–4.7 mm/yr	**Thomas** (80% to 90% porosity included in sediment thickness).
Constance	2.2 mm/yr	
Michigan	1.8 mm/yr	
ESTUARIES AND OCEANIC COASTAL AREAS		
Sediment rctention in estuaries	*ca.* 90% of riverine input	**Burton**.
Sedimentation of near-shore muds	net 0.1–10 mm/yr	**Eisma**.
Sediment fluxes		
Dutch Waddensee: total	20×10^6 t/yr	**Eisma**.
deposited	1×10^6 t/yr	

Rhine R. estuary	3×10^6 t/yr from river 13×10^6 t/yr from sea	**Eisma**.
Chao Phya R. estuary	4.7×10^6 t/yr net to sea	**Eisma**.
CARBON ON LAND AND IN WATERS		
Carbonate minerals (predominantly $CaCO_3$)		
Dissolution in soils	significant on scale 10^2 yr	**Slaymaker**.
Tombstone weathering (limestones, Yorkshire)	0.05–0.11 mm/yr	**Slaymaker**.
Limestone removal rates in karst terrains	Rate (mm/yr) = (4 to 6.7) $\times10^{-5}\times$R	**Schlanger** (R is runoff in mm/yr).
Removal of calcareous reefs	196 g m^{-2} yr^{-1} or 0.07 mm/yr	**Schlanger**.
Global weathering fluxes	estimates	**Holland and Zbinden, Kempe, Meybeck, Schlanger, Garrels**.
Deposition:		
Oceanic reefs	6×10^{14} g $CaCO_3$/yr	**Schlanger**.
Total	18×10^{14} g $CaCO_3$/yr or 2.5×10^{-3} mm/yr	Estimate for 12×10^{14} g/yr over 50% ocean floor.
Plitvice Lakes system	5000 g $CaCO_3$ m^{-2} yr^{-1} or 2 mm/yr	**Kempe**.
Nile R. basin soils	60 g $CaCO_3$ m^{-2} yr^{-1} or 0.02 mm/yr	**Kempe**.
Organic carbon		
TOC (total = dissolved + particulate)		
Global mean in rivers	3 g m^{-2} yr^{-1}	**Meybeck**.
Polluted rivers, Europe	1.7–4 g m^{-2} yr^{-1}	**Kempe**.
Temperate zone rivers	0.8–1 g m^{-2} yr^{-1}	**Kempe**.
Tropical rivers (S. Amer.)	6.4–7.3 g m^{-2} yr^{-1}	**Kempe**.
Spagnum bog (Germany)	6.9 g m^{-2} yr^{-1}	**Kempe**.
GLOBAL FLUXES AND MAJOR RESERVOIRS		
Oxidized and reduced C, S and Fe species, and Ca, Mg and SiO_2	Fluxes on time scales 10^4 to 10^8 years	**Garrels** (see also **Holland and Zbinden**).
Masses of major tectonic and sedimentary units, recycling fluxes and half-lives	Fluxes on times scales 10^7 to 10^9 years	**Veizer**.

RATE CONTROL OF WEATHERING OF SILICATE MINERALS AT ROOM TEMPERATURE AND PRESSURE

Roland Wollast and Lei Chou[1]
Oceanography Laboratory
University of Brussels
1050 Brussels
Belgium

1. INTRODUCTION

The present paper deals with the weathering mechanisms of silicate minerals from a fundamental point of view. The goal of such a study is to understand the main factors controlling the rates of this process and to predict its end products. The chemical weathering of silicates is one of the major processes responsible for the transfer of dissolved and eventually particulate components from land to sea.

The relative importance of the contribution of silicate rocks to the total dissolved load in rivers can be estimated by considering the mean global composition of river waters and by deducing the origin of the chemical constituents present in solution from various water-rock interactions (Garrels and Mackenzie, 1971). The calculation presented in Table I is a recent reevaluation of the mass balance of dissolved constituents carried annually by river waters, taking into account the recent data of river flux of Meybeck (1979) and the relative contribution to weathering of various rock types, as suggested by Holland (1978) and Meybeck (1979). The details of these calculations and further discussion are given in Wollast and Mackenzie (1983). Although this budget is well balanced and seems consistent with current ideas, it is not a unique solution of the mass-balance problem. Nevertheless, it gives a rough but fairly good approximation of the relative contribution of the various rock types to the total dissolved load in world's rivers: 17% originates from evaporites (halite, gypsum and anhydrite), 38% from carbonates (limestone and dolomite) and 45% from silicates. This points out the significant role of silicates in the overall chemical weathering on the earth surface.

From a chemical point of view, the mechanism and the rate of dissolution of silicates differ considerably from those of evaporites and carbonates: on a geological time scale, the latter are very soluble and dissolve rapidly and, in general, congruently. However as the rates of dissolution of silicates are very slow, it is essential to understand the kinetic aspects of the chemical dissolution, to be able to predict the influences of various factors on the rate of silicate weathering.

This chapter first considers some theory of the rate-controlling mechanisms of mineral dissolution reactions and the derivation of the corresponding rate laws. Next it discusses some strategies needed for elucidation of the different possible reaction steps, and it reviews briefly experimental designs used to study the dissolution rates

[1]also at: Dept. of Marine Science, Univ. of South Florida, St. Petersburg, Florida 33701 U.S.A.

A. Lerman and M. Meybeck (eds.), Physical and Chemical Weathering in Geochemical Cycles, 11–32.

Table I. Origin of the dissolved components in world average river water. The mean composition of river waters is expressed in μeq l^{-1} or μmol l^{-1} (SiO_2) (from Wollast and Mackenzie, 1983).

	SiO_2 173	Ca^{2+} 671	Mg^{2+} 281	Na^+ 223	K^+ 33	Cl^- 166	SO_4^{2-} 175	HCO_3^- 867	Total 2589
Salt recycling		3	16	64	1	76	8		
Evaporite NaCl				90		90			
(Ca,Mg)SO_4		91	30			121			
Carbonate weathering		456	107					563	
Silicate weathering									
1	111			56				56	
2	37			37				37	
3	11	80	106					186	
Sulfide oxidation	14	41	22	5	3		46	25	
	0	0	0	0	0	0	0	0	0

1 $$2NaAlSi_3O_8 + 2CO_2 + 11H_2O = \underset{\text{kaolinite}}{Al_2Si_2O_5(OH)_4} + 2Na^+ + 2HCO_3^- + 4H_4SiO_4$$

2 $$2NaAlSi_3O_8 + 2CO_2 + 6H_2O = \underset{\text{montmorillonite}}{Al_2Si_4O_{10}(OH)_2} + 2Na^+ + 2HCO_3^- + 2H_4SiO_4$$

3 $$CaAl_2Si_2O_8 + 2CO_2 + 3H_2O = \underset{\text{kaolinite}}{Al_2Si_2O_5(OH)_4} + Ca^{2+} + 2HCO_3^-$$

$$Mg_5Al_2Si_3O_{10}(OH)_8 + 10CO_2 + 5H_2O = \underset{\text{kaolinite}}{Al_2Si_2O_5(OH)_4} + 5Mg^{2+} + 10HCO_3^- + H_4SiO_4$$

of minerals. An example of such an approach will be given in the case study of the weathering of albite.

2. THEORETICAL CONSIDERATIONS

The rate of dissolution of solids in aqueous solutions may be controlled either by chemical reactions at the solid-water interface or by transport of reactants and products in the reacting system. We will not consider here very fast dissolution reactions where the rate may be controlled by transport by advection or diffusion in the aqueous solution of the dissolved components , as in the case of NaCl or of $CaCO_3$ at low pH. Under natural conditions, the rates of dissolution of silicates are too slow to be dependent on mass transfer in the aqueous phase.

2.1. Diffusion-Controlled Mechanism

The weathering of most silicates exhibits overall an incongruent character. This may be due either to the non-stoichiometric nature of the dissolution reaction or to the occurrence of secondary precipitates. If the dissolution step itself is non-stoichiometric or if precipitation occurs at the surface of the original silicate, then a protective layer could be built on the surface of the solid. Thus rate of the reaction in this case may be controlled by diffusion of reactants and products through this protective layer. This layer will usually grow as the reaction proceeds and its increasing thickness will consequently slow down the diffusion process and thus the rate of dissolution. The rate equation corresponding to this mechanism is known as the parabolic law and will be derived here in a simplified case.

Let us assume that a solid AB is dissolving non-stoichiometrically leaving a residual layer enriched in A:

$$(AB)_{solid} = (A)_{solid} + (B)_{aqueous} \tag{1}$$

The rate of dissolution is then dependent on the diffusion of B across a layer depleted in B, according to Fick's first law:

$$\frac{dq_B}{dt} = - D_B \frac{dC_B}{dX} \tag{2}$$

where dq/dt represents the mass flux of B per unit area of solid across the surface layer, dC_B/dX is the gradient of B and D_B denotes the diffusion coefficient of B in A. We will consider here the simple case where the aqueous phase is infinite or is continuously renewed. The concentration of B in solution remains constant implying that the concentration of B in the leached layer at the solid-solution interface is also maintained at a constant value C_S. Let C_o be the concentration of B in the initial phase AB and l the thickness of the residual layer after a given time t. If the values of C_o and C_B are independent of time and if D_B is constant then the gradient of B must be constant across the layer if B is neither produced nor consumed in the residual layer. The rate of dissolution at that time is then given by:

$$\frac{dq_B}{dt} = D_B \frac{(C_o - C_S)}{l} \tag{3}$$

The thickness of layer l is, however, proportional to the amount of B (q_B) released since the beginning of the reaction and thus

$$l = \alpha q_B \tag{4}$$

where α is a proportionality factor. The rate equation may be rewritten

$$q_B dq_B = \frac{D_B(C_o - C_S)}{\alpha} dt \tag{5}$$

and by integration, for $q_B = l = O$ at $t = O$

$$q_B = \left[\frac{2D_B(C_o - C_S)}{\alpha} t \right]^{1/2} \tag{6}$$

Note that substitution of equation (6) into equation (4) shows that the thickness of the layer also grows parabolically with time. Parabolic dissolution kinetics of silicate minerals have been widely used in the literature with more or less sophisticated modifications. For extended review of this subject see Petrovic (1976) and Wollast and Chou (1985). Results from laboratory experiments are often well described by parabolic kinetics. However, this agreement may be purely fortuitous since preferential dissolution of fine particles or of highly active sites at the surface of the mineral produced by grinding may exhibit similar patterns (Holdren and Berner, 1979). It is imperative to examine carefully this possibility before concluding that the reaction kinetics follows the parabolic rate law.

One of the most interesting improvements of the parabolic law of dissolution of silicates has been introduced by Paces (1973). In the previous simple model, it is obvious that after a relatively short time, the dissolution rate of compound AB will be approaching zero due to the increasing thickness of the protective layer. In reality, this is not the case and even if component A enriched in the layer is poorly soluble, the release of A to the solution will eventually become comparable to the release of component B by diffusion across the residual layer. When the rate of dissolution of component A at the solid-solution interface is equal to the rate of diffusion of B, then the dissolution becomes congruent:

$$\frac{dq_B}{dt} = \frac{dq_A}{dt} = k_A \tag{7}$$

We have assumed in this simplified model that the dissolution rate of A is constant and equal to k_A. The thickness of the layer at any time t is now given by

$$l = \alpha' (q_B - q_A) \tag{8}$$

and by differentiation, the rate of growth of the layer is

$$\frac{dl}{dt} = \alpha' \left(\frac{dq_B}{dt} - \frac{dq_A}{dt} \right) \tag{9}$$

Thus when the rates of release of A and B are equal, then $dl/dt=0$ and the thickness of the layer remains constant. The limiting thickness of this layer can be obtained by combining equations (3) and (7).

$$l^* = \frac{D_B(C_o - C_S)}{k_A} \tag{10}$$

Such a weathering mechanism leads to a steady state stage where the rate of dissolution is given by the rate of retreat of the solid-solution interface. This in turn equals to the constant rate k_A of release of component A from the leached layer according to equation (7). The layer thickness as a function of time can be calculated by substituting the rates of dissolution of A and B in equation (9) by their expressions in equations (7) and (3). The result of such a calculation is shown as an example using a set of appropriate constants in Figure 1(a). The rate of dissolution of A and B as a function of time for the same example is given in Figure 1(b). According to the model of Paces (1973), two reaction stages can be identified for dissolution of silicate minerals: (1) during the initial stage, the growth of the protective layer is the prevailing process and the kinetics are of parabolic type, and (2) during the later stage, the rate of reaction reaches a steady state and dissolution becomes congruent.

2.2. Surface-Controlled Mechanism

The rate of release of component A from the residual layer, as postulated in Paces (1973) model is, in fact, an example of a surface-controlled reaction. More generally, the stoichiometric dissolution is controlled by chemical reactions occurring at the solid-liquid interface.

In the simplest case, the rate of dissolution is only proportional to the surface area of the solid in contact with a unit volume of solution. However, the dissolution reaction is, in general, counterbalanced by the back-precipitation reaction that depends on the concentrations of the dissolved species involved in the reaction. Let us consider the example of the dissolution of quartz represented by:

$$SiO_{2(s)} \underset{k_b}{\overset{k_f}{\rightleftarrows}} SiO_{2(aq)} \tag{11}$$

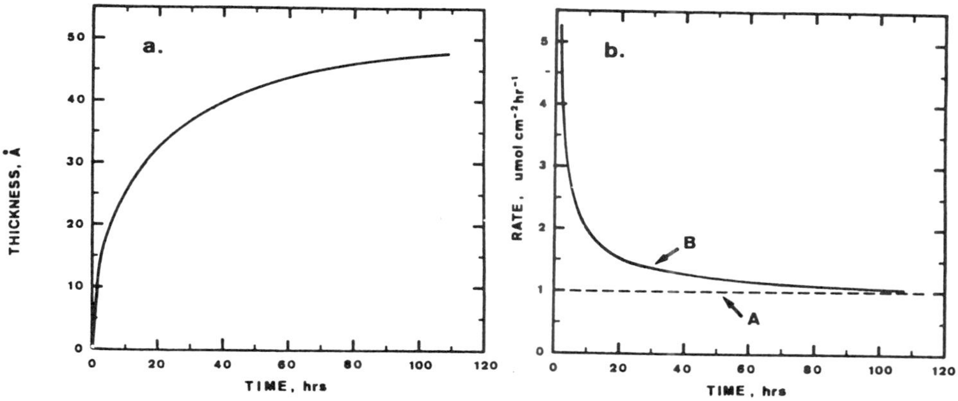

Figure 1. The evolution with time of (a) the thickness of a hypothetical residual layer and (b) the rate of dissolution of components A and B (see text).

Under fixed conditions, the forward and backward reaction rates are given by

$$\left(\frac{dq}{dt}\right)_f = k_f \tag{12}$$

and

$$\left(\frac{dq}{dt}\right)_b = k_b \, a_{SiO_{2(aq)}} \tag{13}$$

where dq/dt represents, as previously, the flux of the considered component per unit surface area of reacting solid. The net reaction rate is thus:

$$\frac{dq}{dt} = \left(\frac{dq}{dt}\right)_f - \left(\frac{dq}{dt}\right)_b = \left(k_f - k_b \, a_{SiO_{2(aq)}}\right) \tag{14}$$

Since at equilibrium dq/dt=O

$$a^*_{SiO_{2(aq)}} = \frac{k_f}{k_b} \, K_{eq} \tag{15}$$

where $a^*_{SiO_{2(aq)}}$ is the solubility of quartz expressed in terms of activity, and equation (12) becomes

$$\frac{dq}{dt} = k_b \left[a^*_{SiO_{2(aq)}} - a_{SiO_{2(aq)}}\right] \tag{16}$$

It is interesting to note the similarity of this equation to equation (3). Both mechanisms give a rate dependence proportional to the departure of the concentration of the component in solution from equilibrium, with respect to a fixed reference concentration. In other words, if l is not changing significantly during the course of an experiment, it may not be easy to distinguish between the two mechanisms from the experimental results.

According to equation (16), the rate of an elementary surface reaction should be simply proportional to the degree of departure of the system from equilibrium. However, even for a simple mineral like quartz the dissolution kinetics was found to be much more complicated. First, as shown in Figure 2, the rate of dissolution of quartz is influenced by pH, which is not predicted by the rate equation (16) nor implied by reaction (11). The results presented in Figure 2 were obtained with a fluidized bed reactor and correspond to very early stages of dissolution of quartz, when the system is far from equilibrium. Under these conditions, the backward reaction rate is negligible. Therefore, the influence of pH on the rate of reaction concerns only the forward dissolution reaction.

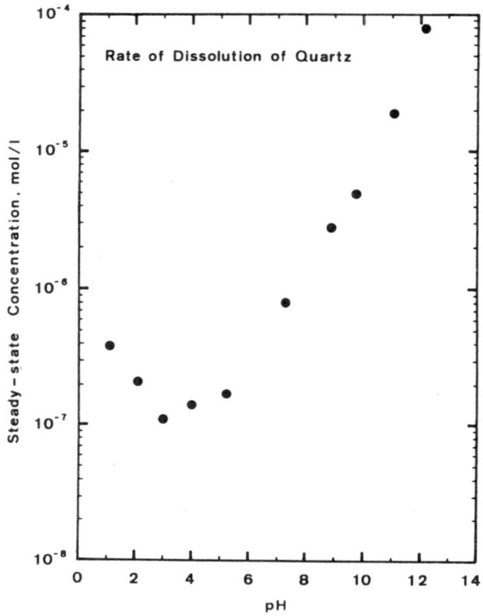

Figure 2. Rate of dissolution of quartz as a function of pH. Grain size is 50-100 μm.

Coordination chemistry of solid surfaces and its application to dissolution kinetics, mainly developed by W. Stumm and his coworkers, allows one to understand the effect of hydrogen ion on the rate of dissolution of oxide minerals. (For an introduction to this concept see Stumm and Morgan, 1981.) The theory has been further extended to the influence of foreign inorganic ions and organic acids on the reaction rate (for example, Stumm et al., 1985). Considering the fact that oxide surfaces are hydrated, the coordination chemistry model takes into account the possible chemical reactions of the hydrous oxide surfaces with protons and other cations or anions or even organic ligands. These surface coordination-reactions include complex formation and ligand exchange resulting in the formation of surface complexes. These surface species should be treated as chemical entities entering explicitly into mass-action equations and rate laws (Stumm et al., 1985). The amphoteric behavior of quartz in aqueous solutions is schematically explained in Figure 3. If the rate of dissolution depends on the concentration of the charged surface complexes, then this rate will

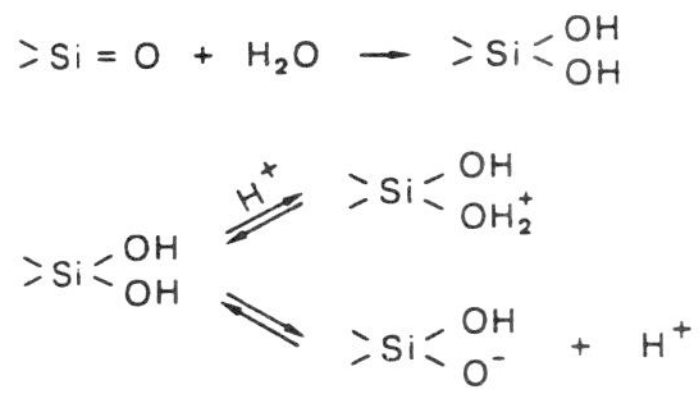

$$K_A = \frac{a_- \, a_{H^+}}{a_o}$$

$$K_B = \frac{a_+}{a_o \, a_{H^+}}$$

Figure 3. Schematic representation of the amphoteric behavior of quartz in aqueous solutions.

depend on the pH since it affects the distribution of, and the surface coverage by, these charged complexes. For example, the rate of dissolution of quartz may be enhanced by the presence of both the negatively and the positively charged surface complexes, as considered in Figure 3. The rate of dissolution is then given by

$$\frac{dq}{dt} = k_1C_+ + k_2C_- + k_3C_o \tag{17}$$

where k_1, k_2 and k_3 correspond to the rate constants of the decomposition of positive, negative and neutral surface sites respectively resulting in the release of dissolved silica to solution. In equation (17) C represents concentrations of the various surface species. Taking into account the equilibrium relations between the surface complexes and neglecting their activity coefficients, the rate equation (17) may be written as

$$\frac{dq}{dt} = (k_1K_Ba_{H^+} + k_2K_A/a_{H^+} + k_3)C_o \tag{18}$$

One may assume that C_o is almost constant since the number of charged sites remains generally small with respect to the total number of possible sites. The rate dependence on pH should then exhibit three ranges corresponding to the three terms of equation (18). Or, if k_3 is very small, there are two ranges of the reaction order with respect to a_{H^+} of +1 and -1, under acidic and basic conditions, respectively. This last case corresponds to the trend shown in Figure 2. According to equation (18), the minimum rate is found at a pH defined by

$$a_{H^+} = \left(\frac{k_2K_A}{k_1K_B}\right)^{1/2} \tag{19}$$

It is interesting to point out that this minimum nearly corresponds to the zero point of charge (ZPC) of quartz, or in other words, to the pH where the numbers of positive and of negative charges governed by the equilibrium relation of Figure 3 are equal. This pH is given by

$$pH_{ZPC} = -\log\left(\frac{K_A}{K_B}\right)^{1/2} \tag{20}$$

Comparison of equations (19) and (20) shows that if the pH_{ZPC} is close to the pH corresponding to the minimum rate, then k_1 is also close to k_2.

The above approach allows a satisfactory explanation of the qualitative role of the pH on the rate of dissolution of silica. However the fractional order dependence of the rate on the pH observed experimentally is not accounted for. A fractional order with respect to a_{H^+} has been reported in many cases for oxide and silicate minerals, as shown in Table II for experiments carried out under acidic conditions. This table shows furthermore that the reaction order is less than, or equal to, 1. There exists today no unifying theory on the origin of such a complex dependence. Stumm and his coworkers have shown in an extensive study of the dissolution of various oxides, and of Al_2O_3 in particular, that the rate of dissolution is strictly proportional to the concentration of protonated surface groups as determined from acidimetric titration curves. However they found no simple stoichiometric relation between the concentration of the protonated groups and pH during these separate titration measurements (Stumm et al., 1985). We will consider here a number of hypotheses that have been proposed to explain the fractional order and will limit the examples to the dissolution reactions under acidic conditions.

Table II. Order of dependence on a_{H^+} of dissolution rate of various silicate and oxide minerals in acid solutions reported in the literature

Mineral	Reaction Order	Source
K-feldspar	0.33	Wollast (1967)
Amelia albite	0.49	Wollast and Chou (1985)
Anorthite	0.46	Fleer (1982)
Sr-feldspar	1.0	Fleer (1982)
Diopside	0.7	Schott et al. (1981)
Enstatite	0.8	Schott et al. (1981)
Bronzite	0.5	Grandstaff (1977)
Iron oxide	0.48	Sigg and Stumm (1981)
Aluminum oxide	0.4	Furrer and Stumm (1983)

A simple explanation of the fractional order of the reactions may be the competition of various surface complexes for a limited number of active sites at the solid surface (Stumm et al., 1983). Let us consider that in the case of quartz, the number of possible sites is limited and that C_o is not constant. This problem can be solved if a Langmuir adsorption isotherm is assumed. If C_T is the concentration of the total possible sites, then

$$C_T = C_o + C_+ \tag{21}$$

and by substitution of the equilibrium condition for the formation of the positively charged complex

$$C_T = C_o(1+K_B a_{H^+}) \tag{22}$$

After neglecting the contribution of the positively charged complex in the acidic range, the rate equation (18) becomes

$$\frac{dq}{dt} = \left[\frac{k_1 K_B a_{H^+} + k_3}{1+K_B a_{H^+}}\right] C_T \tag{23}$$

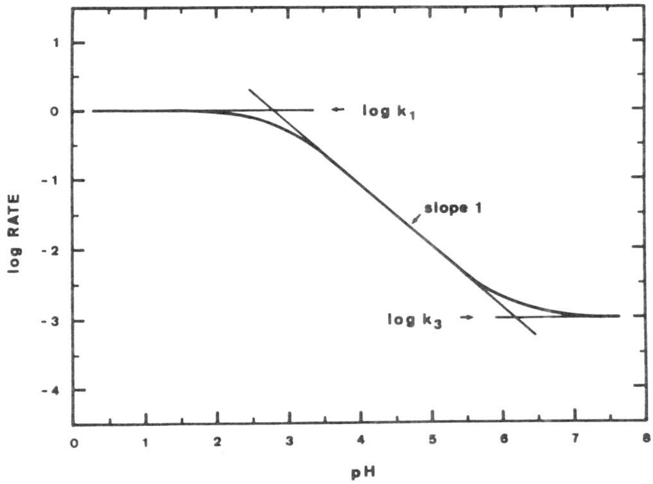

Figure 4. Relative rates of dissolution of quartz as a function of pH. Model based on competition of various surface complexes for limited active sites, and rate constants discussed in the text.

We have represented in Figure 4 the rate calculated according to equation (23) as a function of pH, for a set of appropriate values of constants. In this figure, one can see that the range over which the rate exhibits a fractional order of dependence on pH is limited to less than one pH unit. Therefore, this theory is insufficient to explain the experimental results observed over a much wider pH range. Further protonation to produce doubly charged surface complex ($>SiO_2H_4^{++}$) is possible, but it leads to an increase of the order of reaction with respect to pH. A possible explanation of the fractional order dependence over a fairly wide pH range following this theory is the occurrence of protonation of monovalent sites exhibiting a large spectrum of affinities for H^+. We have no direct proof of this hypothesis but it is not unlikely.

We have considered until now that the surface species behave ideally and their activity coefficients can be approximated to unity. Little is known about the behavior of the complexes formed at the surface of the solid. By analogy with the exchange reaction of minerals in aqueous solutions, we can assume that the surface complexes formed a system far from ideality. The activity coefficients in this case may strongly depend on the concentrations of various surface complexes. For instance, if one assumes that the surface species act as regular solid solutions, their activity coefficients γ_i are classically, given for a two-components system, expressed as:

$$\gamma_i = e^{\frac{\omega}{RT}(1-N_i)^2} \tag{24}$$

where ω is the excess enthalpy of mixing, R denotes the gas constant, T designates the absolute temperature, and N_i represents the mole fraction of the subscript surface species. Departure from ideality complicates the interpretation of the experimental data. However, introduction of equation (24) into the equilibrium conditions given in Figure 3 and expansion of the exponential into a series yield as a first approximation:

$$K_B = \left(\frac{N_+}{N_o}\right)^n \frac{1}{a_{H^+}} \tag{25}$$

where $n=1-(\omega/2RT)$. We have shown that this approximation predicts satisfactorily the exchange reaction of H^+ with Na^+ at the surface of albite and could also be used to justify the fractional order of dependence of the rate of dissolution of this mineral with respect to a_{H^+} (Wollast and Chou, 1985; Chou and Wollast, 1985). To demonstrate this approach, we will consider again the simplified case of dissolution of quartz. Under highly acidic conditions the second term of equation (17) is negligible and the rate of dissolution of silica may be rewritten as:

$$\frac{dq}{dt} = (k_1N_+ + k_3N_o)C_T \tag{26}$$

For a binary solid solution, one has furthermore

$$N_o = 1 - N_+ \tag{27}$$

and according to equation (25)

$$N_+ = (K_B a_{H^+})^{1/n}\, N_o \tag{28}$$

The rate equation (26) then becomes

$$\frac{dq}{dt} = [k_1(K_B a_{H^+})^{1/n} + k_3]C_T\, N_o \tag{29}$$

Thus under highly acidic conditions the rate becomes proportional to $(a_{H^+})^{1/n}$ if k_3 remains relatively small and if N_o remains almost constant. Since n may be greater than 1, the reaction order with respect to a_{H^+} may be a fractional number less than one as found experimentally (see Figure 3 for quartz and Table II for silicates and oxides). The weakness of this hypothesis is that the approximation implied by the expansion into series is valid only if N_+ is not too small nor too large with respect to N_o. If N_+ is comparable to N_o, then from equations (27) and (28), we have

$$N_o = \frac{1}{1+(K_B a_{H^+})^{1/n}} \tag{30}$$

A more general expression can thus be written by introducing equation (30) into equation (29).

$$\frac{dq}{dt} = \left[\frac{k_1(K_B a_{H^+})^{1/n} + k_3}{1 + (K_B a_{H^+})^{1/n}}\right]C_T \tag{31}$$

One may strongly suspect that departure from ideality of the solid surface species may contribute to the fractional order. But here again we have no independent measurement of the activity coefficient to justify the application of the model.

An alternative explanation for the fractional order of dependence related to the formation of surface complexes has been proposed by Lasaga (1981) when the formation of the complexes is related to an exchange reaction. This concept is based on the fact that one may expect that if the adsorption of H^+ on a given site is highly favorable, it may also be difficult to desorb the product of the reaction from the same site. Therefore this author assumes, somewhat arbitrarily, that the free energy of the release of the product is -m times the free energy gained by the adsorption, m being a positive constant. Under these conditions the overall rate will be controlled by the reaction at sites with optimal values of the free energy of adsorption and desorption. Assuming a Langmuir-type adsorption isotherm, Lasaga (1981) showed that the final rate expression is similar to equation (31) with k_3 neglected. In his mathematical expression, the exponent of $(K_B a_{H^+})$ appearing in equation (3) is $1/(m+1)$ where m is the positive constant mentioned earlier.

Other reasons can be evoked to explain the apparent fractional order of dependence. A last one is that the composition of the aqueous phase adjacent to the solid-solution interface may differ from that of the bulk solution. This argument has been especially raised in the case of the activity of H^+ near the surface of a solid (Plummer et al., 1979). As for the previous hypothesis there is no further information concerning this point in the case of water-mineral interactions.

2.3. Non-Uniform Dissolution Processes

Under fixed conditions and far from dissolution equilibrium weathering reactions often exhibit time dependent kinetics frequently expressed by empirical equations such as

$$\frac{dC}{dt} = k\ t^{-n} \tag{32}$$

where n is usually a fractional number. Figure 5 shows one typical example taken from Petrovich (1981) concerning the quartz dissolution experiments of Clelland et al. (1952). During the initial stage, the rate of dissolution is decreasing progressively until it reaches a constant value. This nonlinear portion is easily explained by the preferential dissolution of fine particles and areas of subsurface structural damage produced during grinding. Petrovich (1981) has formulated a mathematical expression describing this phenomenon. According to his model, the apparent rate of dissolution decreases exponentially as the thickness of the disturbed layer is dissolving away. Holdren and Berner (1979) have proposed to pretreat the sample by HF etching to eliminate ultrafine particles and damaged surfaces. This pretreatment is now widely applied.

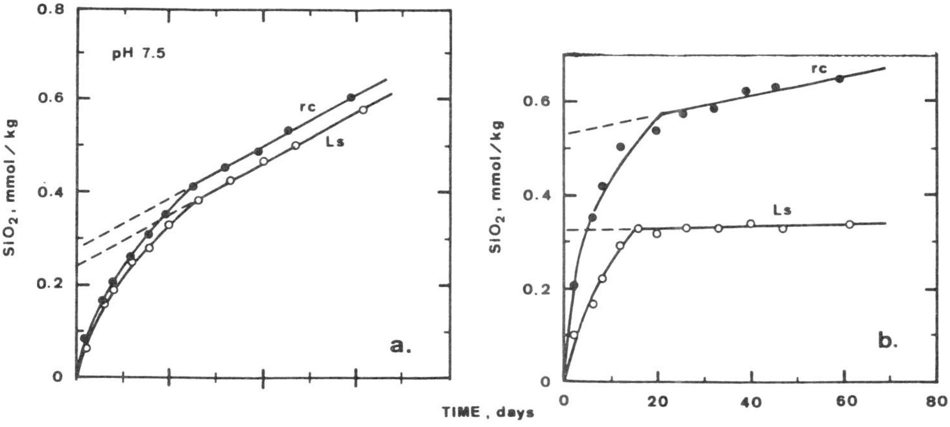

Figure 5. Cumulative release of (a) dissolved (molybdate-reactive) silica and (b) suspended particulate silica (defined as the difference between the total silica in the supernatant solution and the molybdate-reactive silica), as a function of time in two series of stepwise dissolution experiments of quartz, conducted by Clelland et al. (1952) (taken from Petrovich, 1981). rc refers to rock crystal and Ls designates Lochaline quartz. Mean grain size is 100-400 μm.

Examination of weathered surfaces of silicate minerals reveals formation of etch pits. This indicates that dissolution occurs preferentially at points of excess energy, such as dislocations and crystal defects. This nonuniform process may complicate the kinetic expressions, as shown by Dibble and Tiller (1981). According to these authors, the dissolution process involves first the formation of activated complexes followed by interface detachment reactions. The transfer of species from the solid at the interface to the bulk solution consists of several successive steps and occurs at preferential sites, such as kinks on the crystal surface. The distribution of these preferential sites and the mechanism of their formation govern finally the overall rate of dissolution observed. Dibble and Tiller (1981) have shown that linear (n=0 in equation (32)) and logarithmic (n=1 in equation (32)) rate laws can result from a variety of rate-determining surface processes under low as well as high interfacial undersaturation. Uniform detachment or detachment generated by screw dislocations can lead to linear rate laws. A parabolic time dependence (n=1/2 in equation (32)) can develop when one of the detachment parameters is a parabolic function of time, such as, for example, surface diffusion processes associated with kink motion. As pointed out by these authors, the actual interface attachment and detachment processes may be quite complex, and much more experimental work on specific silicate-water systems is needed to elucidate these mechanisms in detail.

3. EXPERIMENTAL STRATEGY

The goals of studies of mineral dissolution kinetics is either to be able to predict the rates of dissolution under various natural conditions, or to understand the details of the fundamental mechanism of the reaction process. Although these two goals seem to be very distant, they call for common approaches and steps because predictions require a good understanding of the mechanisms. Weathering predictions based on empirical rate expressions determined from laboratory studies may be misleading, as very often influencing factors encountered in nature are not taken into account in the experiments. We will consider step by step a possible strategy for such kinetic studies.

The first step is to identify the main rate-controlling process: for example, either diffusion or surface-controlled reactions. A good way to approach this is to investigate the congruency of the dissolution reaction. Since incongruency may be due to precipitation of secondary phases, it is imperative to conduct the preliminary experiments under low concentrations where no precipitation is likely to occur and where the backward reaction is negligible. As pointed out earlier, stoichiometric dissolution implies surface-controlled reactions. However, as depicted in Paces' (1973) model presented in the previous section, diffusional control of the release of some of the components can coexist with a surface-controlled mechanism.

The second step is to identify the parameters affecting the rate controlling process. If the rate-controlling step is diffusion, the possibility of extending the study further is restricted since our fundamental knowledge today concerning the diffusion mechanism of species in the solid phase at room temperature is limited, except maybe for the diffusion of H_2O or H^+ in glasses. It seems, however, that the weathering of silicates is mainly controlled by surface reactions occurring at the solid-solution interface (Berner, 1980). According to transition state theory, the rate is then governed by the formation and decomposition of an activated complex at the surface of the reacting mineral. (For an introduction of transition state theory applied to geochemical processes, see Lasaga, 1981; Aagaard and Helgeson, 1982; Helgeson et al., 1984.) Far from equilibrium, the rate of dissolution is simply proportional to the concentration of the activated surface complex C_{C^*}.

$$\frac{dq}{dt} = k\, C_{C^*} \tag{33}$$

where k is a rate constant. It is very tempting to relate the activated complexes at the solid surface to the charged surface complexes described by the coordination-chemistry theory. However, activated complexes are very short-lived species, and they never reach measurable concentrations in contrast to charged surface complexes that may be titrated easily in many cases. Nevertheless, one can consider that the charged surface complexes are the direct precursors of the activated complexes and that activation may consist of simple processes such as hydration or isomerization of the surface complexes. It is also assumed in theory that all the reaction steps involved in the formation of the activated surface complex are close to equilibrium. If, for example, the activated complex C^* results from the reaction of A and B according to:

$$2A + B = C^* \tag{34}$$

then the equilibrium condition is

$$K_{eq} = \frac{a_{C^*}}{a_A^2\, a_B} \tag{35}$$

Substitution of equation (35) into equation (33) gives

$$\frac{dq}{dt} = \frac{k\, K_{eq} a_A^2\, a_B}{\gamma_{C^*}} \tag{36}$$

where γ represents the activity coefficient of the subscript species. The study of the influence of the concentrations of the reactants and products on the reaction rate constitutes thus a powerful tool for the identification of the nature of the activated surface complexes. The determination of the reaction order with respect to the individual components allows one to fix the stoichiometry of the reaction for the formation of the activated surface complex (equation (34)). In the literature concerning the silicate dissolution studies, only the influence of pH on the rate has been systematically investigated. However, it is clear that this is insufficient to define the nature of the activated complex and to predict the effects of various factors controlling its formation and thus the rate of dissolution of the silicate investigated. Systematic studies of the influence of the concentrations of dissolved species on the dissolution rates have been conducted by Plummer and his coworkers for carbonate minerals (Plummer and Wigley, 1976; Plummer et al., 1978; Busenberg and Plummer, 1982). Their remarkable work helped considerably the understanding of the mechanisms of dissolution of these minerals. Finally, in advanced studies, it may be interesting to investigate the effects of foreign ions on the rate of reaction. The rate of surface reactions may be enhanced by ligand formation or inhibited by the adsorption of the poisonous species on the active sites. The usefulness of this approach has already been demonstrated by the work performed on the dissolution of single oxides (Stumm et al., 1985).

4. EXPERIMENTAL DESIGN

Most of the earlier attempts to simulate experimentally the weathering of minerals were conducted by simply percolating water through a column packed with the mineral grains (e.g. Correns and von Engelhardt, 1938). From an engineering point of view, this reactor is similar to a plug flow reactor where ideally it is assumed that there is no longitudinal mixing. From a practical and theoretical point of view, this kind of reactor is not recommended. First, it is not possible to avoid longitudinal mixing of the aqueous phase, and the evaluation of the fluid-flow parameters and of the mixing coefficient, is not easy. The interpretation of the chemical kinetics is also complicated by the fact that compositions of both the solid and the liquid phases are usually changing along the column. In later studies, batch reactors were used to follow evolution with time of the concentrations of the dissolved components released by the weathering reaction. In this case, a good mixing is provided by stirring and the control of some master variables, such as pH or pCO_2, can be easily realized. However, other variables, such as the concentrations of the dissolved species that may affect the reaction rate, are changing as the dissolution progresses. Furthermore, the dissolution rate is not directly measured and has to be evaluated from the rate of increase of the concentration of the products of the weathering reaction. This procedure involves usually large errors. Batch reactors on the other hand have the advantage of covering a very broad range of experimental conditions starting eventually from pure water to a system close to dissolution equilibrium. With this reactor, one can detect under favorable conditions the occurrence of changes in the reaction mechanisms such as secondary precipitation.

Most recently, geochemists started the use of mixed flow reactors in silicate dissolution studies. In this kind of reactor, the aqueous solution is continuously renewed and well mixed with the reacting solid, simply by stirring (Holdren and Speyer, 1985) or by fluidization of the solid particles (Chou and Wollast, 1984). These reactors have several other advantages. The concentrations of the dissolved components can be easily maintained at a level well below saturation with respect to any possible secondary precipitates by renewing rapidly the reacting solution. In addition, one can study the influence of various chemical parameters on the same sample of solid without manipulating it by changing the composition of the input solution or its renewal rate. Finally, the rate of dissolution of the individual components can be obtained easily by multiplying the difference in concentrations of the input and output solutions by the renewal rate of the solution, and normalizing it with respect to the surface area.

5. CASE STUDY: DISSOLUTION MECHANISMS OF ALBITE

We will present in this section, as an example, the weathering of albite that we have recently studied in detail (Chou and Wollast, 1984; Wollast and Chou, 1985; Chou and Wollast, 1985). Most of these studies were conducted using the fluidized bed reactor except the initial exchange experiments which were carried out in batch reactors. It is known that when fresh feldspars are added to pure water, they exchange rapidly their alkali ions with protons in solution resulting in a rise in pH and in the formation of hydrogen feldspar on the surface (Garrels and Howard, 1959). In the case of albite, this reaction is represented by

$$NaAlSi_3O_8 + H^+ = HAlSi_3O_8 + Na^+ \tag{37}$$

It is also known that this exchange reaction is reversible. The equilibrium condition corresponding to the above reaction is

$$K_{eq} = \frac{a_{HAlSi_3O_8} \cdot a_{Na^+}}{a_{NaAlSi_3O_8} \cdot a_{H^+}} \tag{38}$$

We have shown that it is possible to use a regular solid solution model to describe this exchange (Wollast and Chou, 1985). According to the approximation discussed in Section 2, the equilibrium condition can be rewritten as:

$$K_{eq} = \left(\frac{N_{HAlSi_3O_8}}{N_{NaAlSi_3O_8}}\right)^n \left(\frac{a_{Na^+}}{a_{H^+}}\right) \tag{39}$$

As shown in Figure 6, a plot of log (a_{Na^+}/a_{H^+}) *vs.* log $(N_{NaAlSi_3O_8}/N_{HAlSi_3O_8})$ gives a straight line over a large range of mole fractions of the surface species, indicating that the above value of n (n=0.65) corresponds to an excess enthalpy of mixing equal to 1.8 Kj/mol. The equilibrium constant K_{eq} is equal to $10^{5.76}$, demonstrating the high stability of hydrogen feldspar with respect to fresh albite in the presence of water. Further weathering steps of this mineral in water will thus involve reactions of H-feldspar at the solid surface with the aqueous phase.

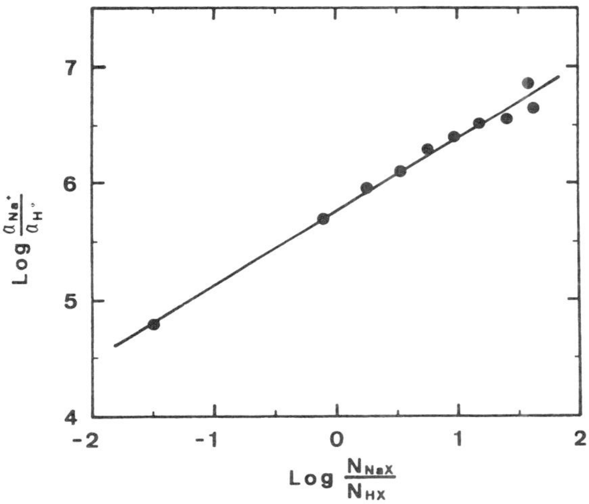

Figure 6. Exchange properties of albite surface in pure water (from Wollast and Chou, 1985).

Following the strategy outlined earlier, let us first investigate whether the dissolution of albite is congruent or not. Figure 7 shows the total amount of the three components released from fresh feldspar to solution as a function of time. These results were obtained using a fluidized bed reactor and fresh water as the input solution. During this experiment, the concentrations of the dissolved elements were always below saturation with respect to possible secondary phases such as amorphous $Al(OH)_3$ or halloysite. The very initial dissolution stage is characterized mainly by a high rate of release of all three components but also by the relatively large amount of Na dissolved with respect to Al and Si. Enhanced dissolution of fine particles or of disturbed surfaces is not sufficient to explain this observation and certainly not the highly non-stoichiometric nature of the reaction. The rate of dissolution of all three components decreases rapidly during the next stage reaching finally a steady-state stage where the dissolution becomes congruent. This behavior is in agreement with the model of Paces (1973). The first stage corresponds to the formation of a residual layer enriched in Si and to a lesser extent in Al. The mean composition of this layer is strongly pH dependent. It can be estimated in the absence of secondary precipitations by simple mass balance calculations (Chou and Wollast, 1984).

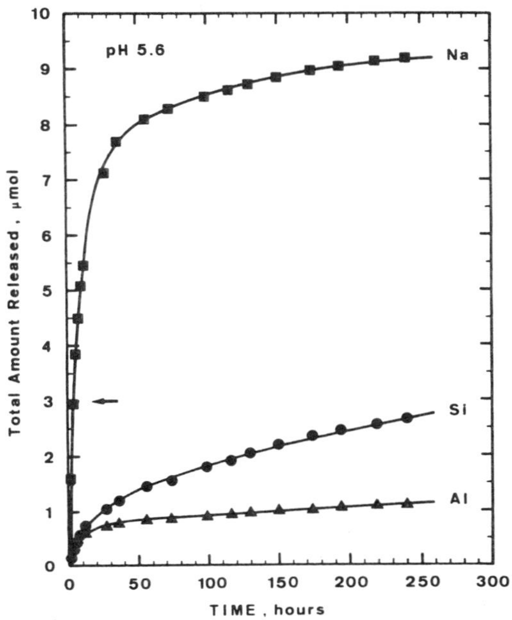

Figure 7. Total amount of dissolved components released as a function of time during the dissolution of fresh albite in pure water using a fluidized bed reactor. The arrow indicates the expected amount of Na released due to the exchange with H^+ (from Wollast and Chou, 1985).

The result of these calculations for the experiment presented in Figure 7 is shown in Figure 8. The rate of the second stage is controlled by the dissolution of the residual layer at the solid-solution interface accompanied by diffusion of Na from the fresh bulk solid across the residual layer. This leads to an overall congruent dissolution reaction of albite. It is demonstrated from the ratio of the slopes of the three curves of Figure 7, which approaches the stoichiometric ratio of Si:Al:Na in fresh albite after a sufficient long time. Under these circumstances, the steady-state rate is finally controlled by surface reactions occurring at the mineral-water boundary.

The next step is to investigate the nature of the activated complexes by examining the influence of pH and of the concentrations of the individual dissolved components on the rate of dissolution of albite. In the case of the continuous flow-through fluidized bed reactor, these studies could easily be performed by changing the composition of the input solution, or the rate of renewal of the reacting solution. Our experimental results indicate that dissolved silica and sodium have no significant effects on the reaction rate. To the contrary, dissolved Al and pH influence markedly the weathering rate of albite (Figs. 9 and 10).

The effect of pH can be described by the following empirical equation of a best-fit polynomial to our experimental results:

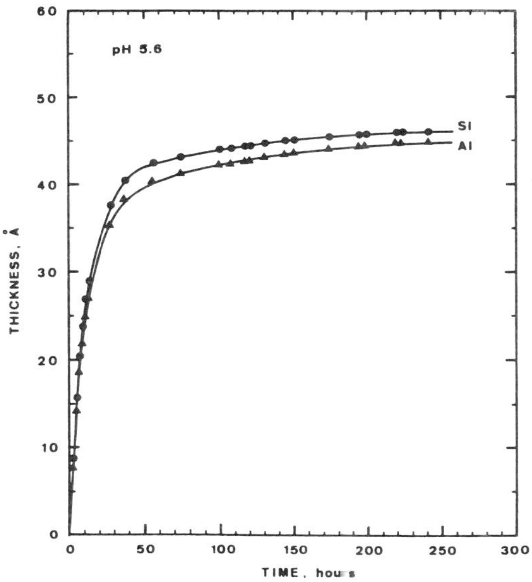

Figure 8. Thickness of a hypothetical residual layer with respect to Si and Al as a function of time for the experiment presented in Figure 7.

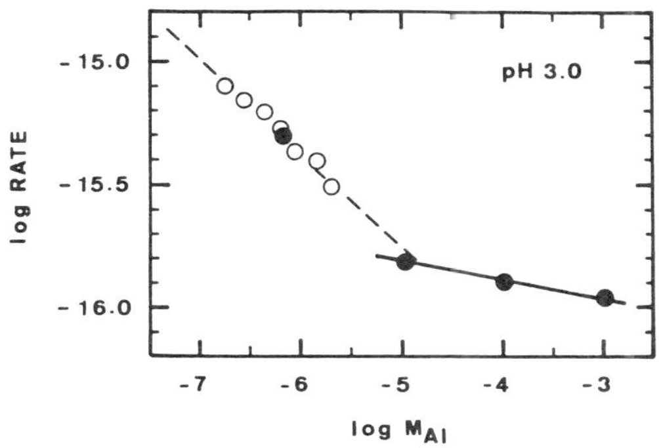

Figure 9. Influence of dissolved Al on the rate of dissolution of albite.

$$r = 10^{-13.69} a_{H^+}^{0.49} + 10^{-16.15} + 10^{-18.15} a_{H^+}^{-0.30} \qquad (40)$$

Helgeson et al. (1984) made the first attempt to interpret the influence of pH in terms of the stoichiometry of the activated complexes. They suggested the possible existence of a positively charged, a neutral, and a negatively charged surface complexes, each one prevailing under acidic, neutral and alkaline conditions. Our study (Wollast and Chou, 1985), however, was the first to show that besides H^+, dissolved Al also plays an important role in the formation of the surface complexes which are the precursors of the related activated complexes. One should also point out that the rate *vs.* pH curve exhibits a striking similarity to the solubility curve of $Al(OH)_3$. This again suggests that dissolved Al is affecting the dissolution rate of albite over the entire pH

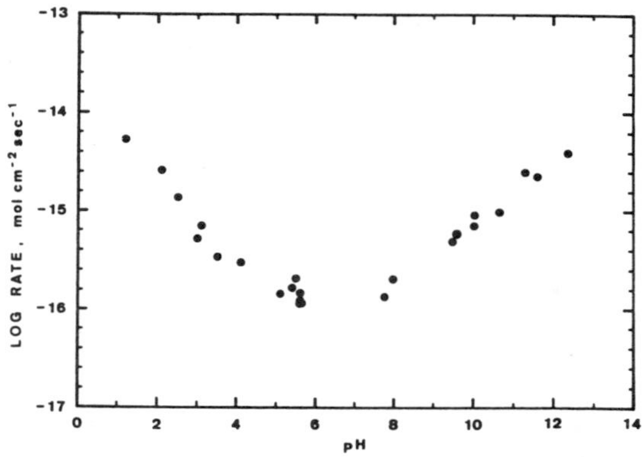

Figure 10. Influence of pH on the rate of dissolution of albite.

range. The composition of the surface complex can be identified by defining a reaction for the formation of this complex which takes into account all the species affecting the rate. The stoichiometric factors appearing in this reaction must be compatible with the reaction order determined experimentally. Consequently, in the case of albite, the formation of the surface complex under acidic conditions can be written in a general form:

$$HAlSi_3O_8 + p\,H^+ \;=\; q\,Al^{3+} + H_{1+p}\,Al_{1-q}\,Si_3O_8^{(3q-p)-} \tag{41}$$

with the equilibrium condition

$$K \;=\; \frac{a^{q}_{Al^{3+}} \cdot a_C}{a_{HAlSi_3O_8} \cdot a^{p}_{H^+}} \tag{42}$$

where C represents the surface complex $H_{1+p}\,Al_{1-q}\,Si_3O_8^{(3q-p)-}$. Under alkaline conditions, the prevailing dissolved Al species is $Al(OH)_4^-$ rather than Al^{3+}. A similar reaction to equation (41) can be written for the formation of the surface complex:

$$HAlSi_3O_8 + (p-4q)H^+ + 4H_2O \;=\; q\,Al(OH)_4^- + H_{1+p}\,Al_{1-q}\,Si_3O_8^{(3q-p)-} \tag{43}$$

with the equilibrium condition

$$K' \;=\; \frac{a^{q}_{Al(OH)_4^-} - a_c}{a_{HAlSi_3O_8} \cdot a^{p-4q}_{H^+}} \;=\; K\,{}^*\beta_4 \tag{44}$$

where ${}^*\beta_4$ is the equilibrium constant for the formation of $Al(OH)_4^-$ from Al^{3+}. Comparison of equations (42) and (44) shows that the same activities of the surface complex a_C, corresponding to the same rate of weathering, can indeed be obtained for a set of acidic and alkaline pH's as observed in Figure 10. Let us further recall that in a continuous flow reactor, the rate of dissolution is simply proportional to the concentration of dissolved products of the reaction. In this case, equal rates at low and high pHs imply that $m_{Al^{3+}} = m_{Al(OH)_4^-}$, where m represents the molality of the subscript species. From equations (42) and (44), this condition leads to

$$\frac{(a_{H^+})^p_{acid}}{(a_{H^+})^{p-4q}_{alk}} = \frac{\gamma^q_{Al^{3+}}}{\gamma^q_{Al(OH)_4^-}} {}^*\beta_4 \tag{45}$$

Taking the logarithm of equation (45) and rearranging the equation results in:

$$\frac{p}{q} = \frac{4pH_2 + \log\left({}^*\beta_4\gamma^q_{Al^{3+}}/\gamma^q_{Al(OH)_4^-}\right)}{pH_2 - pH_1} \tag{46}$$

where pH_1 and pH_2 refer to the coupled acid and alkaline pHs respectively. This equation allows the determination of the stoichiometric ratio p/q. According to our experimental results this ratio p/q was found close to 2.5. Imposing small whole numbers for p and q ($5H^+$ and $2Al^{3+}$), this requires 2 H-feldspar as the reacting solid. Equation (41) becomes thus:

$$2HAlSi_3O_8 + 5H^+ = 2Al^{3+} + H_7Si_6O_{16}^- \tag{47}$$

The reaction is only a rough representation of the overall stoichiometry. The mechanisms of the formation of the surface complex and thus the activated complex involve certainly several steps. It seems for instance more reasonable to suggest the two following steps:

$$H_2Al_2Si_6O_{16} + 6H^+ = 2Al^{3+} + H_2Si_6O_{13} + 3H_2O \tag{48}$$

$$H_2Si_6O_{13} = H^+ + HSi_6O_{13}^- \tag{49}$$

$$H_2Al_2Si_6O_{16} + 5H^+ = 2Al^{3+} + HSi_6O_{13}^- + 3H_2O \tag{50}$$

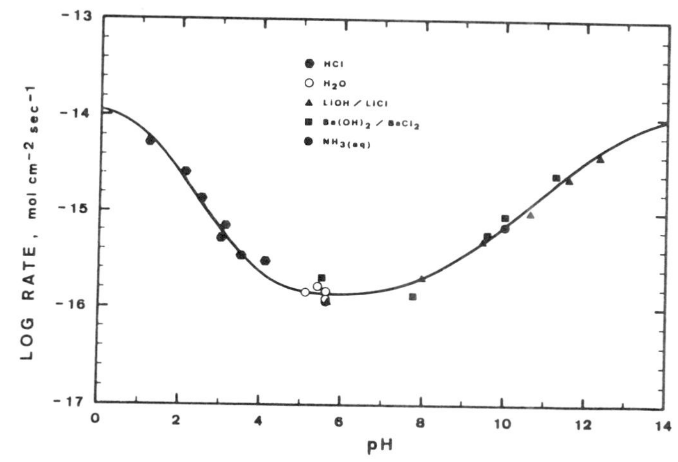

Figure 11. Comparison of the rates of dissolution of albite predicted (the curve) with those observed in various media as a function of pH (from Chou and Wollast, 1985).

The reaction scheme, however, is often a matter of speculation, as, for example, the degree of dehydration in the first reaction (48). On the other hand, the occurrence of two molecules of feldspar in the first equation may lead to the hypothesis that the activated complex is formed by the activation of two adjacent surface complexes. The second reaction equation (49) represents dissociation of the charged complex and takes into account a fundamental property of the hydrous silicate surfaces. According to the theory of coordination chemistry, these hydrous surfaces act as fairly strong acids, as demonstrated by the zero point of charge of albite, which is situated around pH 2 (Parks, 1967). The last step of the interpretation of our experimental data consists of the use of the regular solid solution model to approximate the activity of the surface complexes as discussed in the previous section. The value of n defined in equation (25) and the rate constants were adjusted to fit our experimental results (Chou and Wollast, 1985).

A unique set of kinetic and mass balance constants allowed us to describe very satisfactorily the influence of parameters such as pH, concentration of dissolved Al and renewal rate of the input solution, over a wide range of experimental conditions. Figure 11 shows good agreement between the rates predicted by our kinetic model and those observed in laboratory experiments.

6. SUMMARY

Recently, there has been an increasing effort in developing kinetic studies of water-rock interactions. The purpose of these studies is to be able to predict the weathering rates of minerals under natural conditions. However, one of the difficulties involved is that dissolution and precipitation reactions of minerals are complex processes depending on transport phenomena and surface reactions. Only little is known about the surface properties of solids in general and of the solid-liquid interface in particular. This constitutes a strong barrier to a rational progress in our understanding of the basic mechanisms of surface reactions. On the other hand, fundamental kinetic studies of dissolution and precipitation reactions are probably one of the fruitful ways to obtain valuable information on the properties of solid surfaces.

Most of the geochemical studies have been concerned with the prediction of weathering rates and, unfortunately, many geochemists have been trying to reproduce natural weathering processes in laboratory studies instead of focusing their attention on the understanding of the fundamental mechanisms. These attempts have not been proven successful and the extrapolation of their results to natural systems can be misleading.

Our principal purpose in this study was to show in some detail a possible way to deal with the chemical kinetics of weathering of silicates. Here we will point out two of the most fundamental principles in the experimental approach to kinetic studies. The first is that in each experimental run a maximum of parameters must be controlled and a minimum allowed to change. The second is that from run to run a maximum of factors should be varied but, if possible, one at a time. From this point of view, batch reactors should be avoided and well-mixed continuous flow-through reactors preferred. Finally, in the interpretation of the data it should be remembered that, even if the mechanisms of weathering reactions are complicated, nature is simple.

7. ACKNOWLEDGEMENTS

We would like to thank A. Lerman for his editorial comments. This research was supported in part by funds from the St. Petersburg Progress Chair of Robert M. Garrels at the University of South Florida, St. Petersburg, Florida.

8. REFERENCES

Aagaard P. and Helgeson H. C. (1982) Thermodynamic and kinetic constraints on reaction rates among mineral and aqueous solutions. I. Theoretical considerations. *Am. Jour. Sci.* **282,** 237-285.

Berner R. A. (1980) *Early Diagenesis.* Princeton University Press, Princeton, N. J., 241 pp.

Busenberg E. and Plummer L. N. (1982) The kinetics of dissolution of dolomite in CO_2–H_2O systems at 1.5 to 65 °C and 0 to 1 atm P_{CO_2}. *Amer. Jour. Sci.* **282,** 45-78.

Chou L. and Wollast R. (1984) Study of the weathering of albite at room temperature and pressure with a fluidized bed reactor. *Geochim. Cosmochim. Acta* **48,** 2205-2217.

Chou L. and Wollast R. (1985) Steady-state kinetics and dissolution mechanisms of albite. *Amer. Jour. Sci.* **285,** 963-993.

Clelland D. W., Cumming W. M. and Ritchie P. D. (1952) Physico-chemical studies on dusts. I. A high-solubility layer on siliceous dust surfaces. *J. Appl. Chem.* **2,** 31-41.

Correns C. W. and von Engelhardt W. (1938) Neue Untersuchungen uber die Verwitterung des Kalifeldspates. *Chemie der Erde* **12,** 1-22.

Dibble W. E. Jr. and Tiller W. A. (1981) Non-equilibrium water/rock interactions. I. Model for interface-controlled reactions. *Geochim. Cosmochim. Acta* **45,** 79-92.

Fleer V. N. (1982) The dissolution kinetics of anorthite and synthetic strontium feldspar in aqueous solutions at temperature below 1OO °C with applications to the geologic disposal of radioactive nuclear wastes. Ph.D. Thesis, The Pennsylvania State University, 197 pp.

Furrer G. and Stumm. W. (1983) The role of surface coordination in the dissolution of δ–Al_2O_3 in dilute acids. *Chimia* **37,** 338-341.

Garrels R. M. and Howard, P. (1959) Reactions of feldspar and mica with water at low temperature and pressure, *in* Proc. Sixth National Conference on Clays and Clay Minerals, Oxford, England, Pergamon Press, pp. 68-88.

Garrels R. M. and Mackenzie F. T. (1971) *Evolution of Sedimentary Rocks.* Norton and Co., New York, 397 pp.

Grandstaff D. E. (1977) Some kinetics of bronzite orthopyroxene dissolution. *Geochim. Cosmochim. Acta* **41,** 1097-1103.

Helgeson H. C., Murphy W. M. and Aagaard P. (1984) Thermodynamic and kinetic constraints on reaction rates among minerals and aqueous and solution. II. Rate constants, effective surface area and the hydrolysis of feldspar. *Geochim Cosmochim. Acta* **48,** 2405-2432.

Holdren G. R. Jr. and Berner R. A. (1979) Mechanism of feldspar weathering. I. Experimental studies. *Geochim. Cosmochim. Acta* **43,** 1161-1171.

Holdren G. R. Jr. and Speyer P. M. (1985) pH dependent changes in the rates and stoichiometry of dissolution of an alkali feldspar at room temperature. *Amer. Jour. Sci.* **285,** 994-1026.

Holland H. D. (1978) *The Chemistry of the Atmosphere and Oceans.* Wiley-Interscience, New York, 351 pp.

Lasaga A. C. (1981) Transition state theory, *in* Lasaga A. C. and Kirkpatrick R. J. eds., Kinetics of Geochemical Processes, Mineralogical Society of America, *Reviews in Mineralogy* **8,** 35-169.

Meybeck M. (1979) Pathways of major elements from land to ocean through rivers, *in* Review and Workshop on River Inputs to Ocean Systems. 26-30 March, 1979. FAO, Rome.

Paces T. (1973) Steady-state kinetics and equilibrium between ground water and granitic rock. *Geochim. Cosmochim. Acta* 2641-2663.

Geochim. Cosmochim. Acta 2641-2663.

Parks G. A. (1967) Aqueous surface chemistry of oxides and complex oxide minerals. *Adv. Chem. Ser.* **67,** 120-160.

Petrović R. (1976) Rate control in feldspar dissolution. II. The protective effect of precipitates. *Geochim. Cosmochim. Acta* **40,** 1509-1522.

Petrović R. (1981) Kinetics of dissolution of mechanically comminuted rock-forming oxides and silicates. I. Deformation and dissolution of quartz under laboratory conditions. *Geochim. Cosmochim. Acta* **45,** 1665-1674.

Plummer L. N. and Wigley T. M. L. (1976) The dissolution of calcite in CO_2-saturated solutions at 25 °C and 1 atmosphere total pressure. *Geochim. Cosmochim. Acta* **40,** 191-202.

Plummer L. N., Wigley T. M. L. and Parkhurst D. L. (1978) The kinetics of calcite dissolution in CO_2-water systems at 5° to 6O °C and 0.0 to 1.0 atm CO_2. *Amer. Jour. Sci.* **278,** 179-216.

Schott J., Berner R. A. and Sjoberg, E. L. (1981) Mechanism of pyroxene and amphibole weathering. I. Experimental studies of iron-free minerals. *Geochim. Cosmochim. Acta* **45,** 2123-2135.

Sigg L. and Stumm W. (1981) The role of surface weak acids with the hydrous goethite surface. *Colloids and Surfaces* **2,** 101-117.

Stumm W., Furrer G. and Kunz B. (1983) The role of surface coordination in precipitation and dissolution of mineral phase. *Croat. Chem. Acta* **58,** 593-611.

Stumm W. and Morgan J. J. (1981) *Aquatic Chemistry.* 2nd ed., John Wiley & Sons, New York, 780 pp.

Wollast R. (1967) Kinetics of the alteration of K-feldspar in buffered solutions at low temperature. *Geochim. Cosmochim. Acta* **31,** 635-648.

Wollast R. and Chou L. (1985) Kinetic study of the dissolution of albite with a continuous flow-through fluidized bed reactor, *in* Drever J. I. ed., *The Chemistry of Weathering,* Dordrecht/Boston/Lancaster, Reidel, NATO ASI ser. C149, pp. 75-96.

Wollast R. and Mackenzie F. T. (1983) The global cycle of silica, *in* Aston S. R. ed., *Silicon Geochemistry and Biogeochemistry,* Academic Press, pp. 39-76.

MICROBIAL WEATHERING PROCESSES IN NATURAL ENVIRONMENTS

J. Berthelin
Centre de Pédologie Biologique
B.P. 5
54501 Vandœuvre-les-Nancy Cedex
France

ABSTRACT

In order to generate energy to form new cell structures, microorganisms (primarily bacteria and fungi) that are present in large number and diversity in natural environments (soils, waters, sediments) break down organic and inorganic materials and form large amounts of gaseous, liquid and solid metabolic compounds that they excrete into their environment.

Such microbial processes are involved either in the solubilization or the insolubilization of inorganic elements, in the weathering of minerals (silicates, phosphates, carbonates, sulfides, oxides) and in the formation of deposits (carbonates, oxides, sulfides, phosphates). They depend on the environmental conditions and have been shown to occur in the cycling of elements, the formation and evolution of different soils and the formation of different deposits. But the relative importance of environmental conditions on biotic and abiotic processes is still unknown.

1. MICROORGANISMS IN NATURAL ENVIRONMENTS

Microorganisms or microbes are at the boundary line between plants and animals. They are considered as a third kingdom of organisms which can be distinguished from typical plants and animals by the simple level of their biological organization. The members of this kingdom, named protists, are either unicellular or multicellular, but the latter lack the extensive differentiation into organs and tissues characteristic of plants and animals. The protists comprise two distinct groups: lower protists and higher protists. The lower protists are procaryotic organisms that include bacteria (unicellular eubacteria of different shapes, filamentous and coryneform bacteria) and cyanobacteria (photosynthetic organisms formerly named blue-green algae). These are characterized by a lack of a proper nucleus and the absence of membrane-bound organelles (or membrane-enclosed intracellular structures) such as mitochondria. Either bacteria or cyanobacteria are the oldest living organisms found as microfossils 3,400 to 3,500 million years old.

The higher protists are eucaryotic organisms and have cell structures similar to those of plants and animals. They include the fungi, the protozoa and the algae (Dommergues and Mangenot, 1970; Richards, 1974; Alexander, 1977; Brock et al., 1984). Higher protists are more recent and their geological age has been estimated at 900 to 1,400 millions years (Cloud, 1978, 1983; Ehrlich, 1981; Margulis, 1981).

A. Lerman and M. Meybeck (eds.), Physical and Chemical Weathering in Geochemical Cycles, 33–59.

The microorganisms are widespread in all the natural environments: the lithosphere, the hydrosphere and the atmosphere (Kutznetsov et al., 1963; Zajic, 1969; Ehrlich, 1981). Microorganisms have been collected to depths of about 2,000 m in the earth's crust and to greater depths (11,000 m) in the oceans and at heights of 48 to 77 km above the earth's surface. Even if such occurrences represent living microorganisms, they are not always capable of normal activity in such habitats because of a lack of sufficient nutrients and favorable environmental conditions (Ehrlich, 1981).

In the upper layers of soils, bacteria can have a biomass of 1,000 to 4,000 kg/ha and an average number of 10^6 to 10^9 per g of dry soil. Fungi have approximately the same to slightly higher biomass. Figure 1 shows bacteria and fungi on soil aggregates and on rock fragments in the upper layers of forest acid soils. Algae are not well known in soils, where their biomass ranges from one to several hundred kg/ha. They are more abundant in desert (warm or cold) areas where they are considered as endolithic "pioneer" organisms (or first living organisms) that can be involved in rock weathering (Friedman and Ocampo-Friedman, 1984). Lichens, a symbiotic association between algae and fungi, are also interesting as colonizing organisms and weathering agents of rock surfaces (Robert and Berthelin, 1986; Wilson et al., 1981).

In soils, the number of microorganisms decreases with depth, because for most of them growth and the reproduction depend on the presence of available organic matter as a source of carbon and energy (Alexander, 1977). Large populations of microorganisms also occur in sedimentary rocks (Weirich and Schweisfurth, 1985; Kutznetsov et al., 1963). Reporting on the fossil bacteria and their role in the cycling of various substances in petroleum, Kutznetsov et al. (1963) mentioned results of Eksertsev, who found at depths from 500 to 2,000 m in Permian and Carboniferous sediments 35 to 117 million bacteria per gram of rock in samples with clear indications of oil occurrence, whereas no bacteria were observed in the overlying or underlying strata. In red sandstones, Weirich and Schweisfurth (1985) found from 10^5 to 10^3 viable bacterial cells per gram of rock in samples from 2 to 450 m depth.

2. GENERAL MICROBIAL ACTIVITY IN NATURE

2.1. The Cell Machinery

For their growth and reproduction, microorganisms can oxidize organic or mineral compounds (electron or proton donors) that they use as a source of energy (Fig. 2 and equation 1). They must also utilize different nutrients (inorganic and organic) as building blocks for the formation of their cell constituents. Carbon, nitrogen and mineral requirements can be provided as inorganic (CO_2, HCO_3^-, CO_3^{2-}, NH_4^+, NO_3^-, N_2, SO_4^{2-}, PO_4^{3-}, metal ions) and organic compounds, depending on the microbial nutritional type. A general reaction can be written as:

$$AH_2 + B \rightarrow BH_2 + A + \Delta G \qquad (1)$$

Microbes need inorganic or organic compounds that can be oxidized (AH_2, electron or proton donors) to provide utilizable energy (ΔG). They also need inorganic or organic electron or proton acceptors (B) that are reduced, as summarized in equation (1).

During such metabolic processes, microorganisms also form gaseous, liquid, solid, alkaline, acidic, complexing, and reducing compounds that are excreted from the cell as metabolic end products or intermediary metabolites. Figure 2 summarizes these different aspects of the microbial activity and shows that microbes, owing to reactions

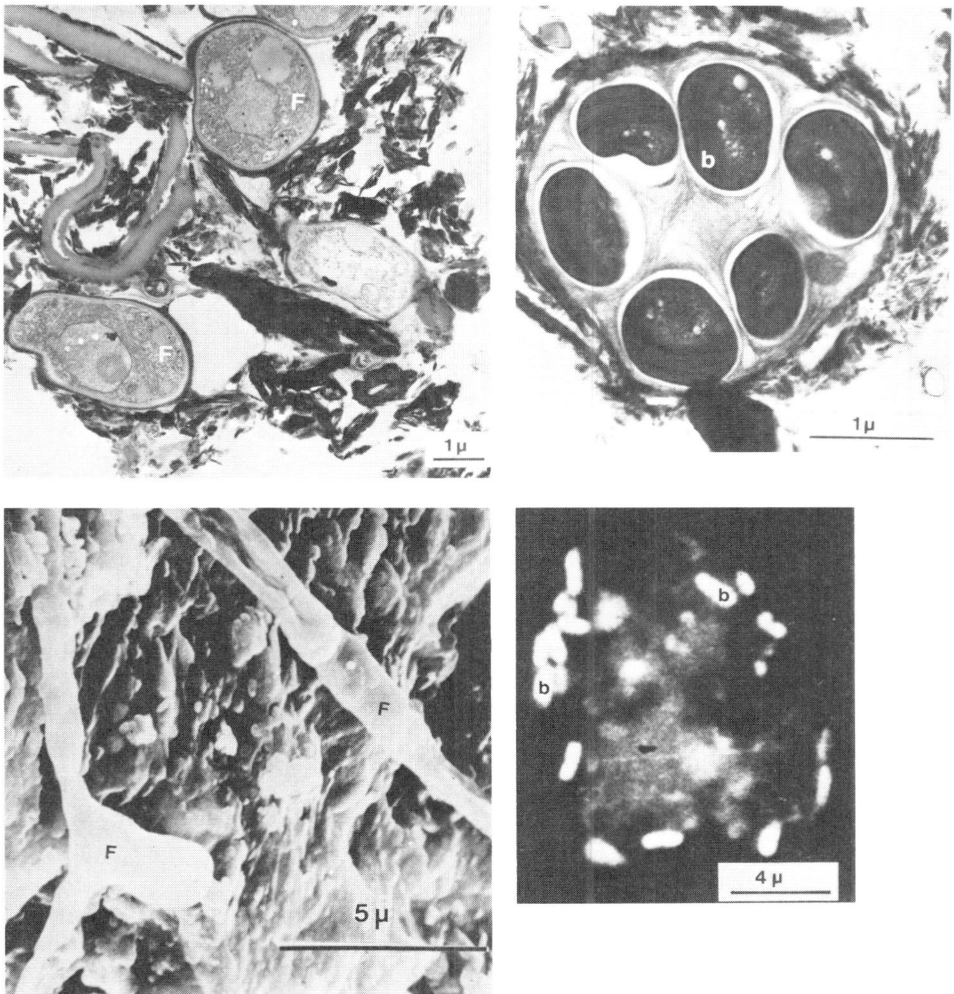

Figure 1. Bacteria (b) and fungi (F) on and in soil aggregates or on rock fragments, observed by scanning electron microscopy and transmission electron microscopy (photographs by G. Villemin, F. Toutain, C. Munier-Lamy and P. Villecourt).

of oxidation and reduction and processes of uptake and release of inorganic and organic compounds, can be efficient agents of transformation in their organic and inorganic environments.

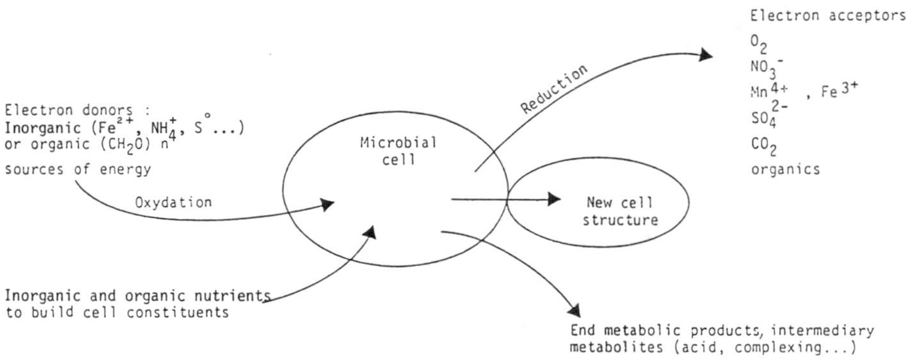

Figure 2. Diagram of general microbial activity.

2.2. Microbes and Biogeochemical Cycles

Considering the microbial cell machinery and the occurrence of microbes in three main continental and marine reservoirs (soils, waters, sediments), it appears that microorganisms play a major and fundamental role in the cycling of carbon, nitrogen and sulfur; are very active in the cycling of phosphorus, iron, manganese; and are somewhat involved in the transformation of other major (silica, aluminum, magnesium, calcium, potassium, etc.) or trace elements (copper, uranium, mercury, etc.). In addition to the transformation of inorganic and organic matter, microorganisms are also essential to the transfer of energy in the environment (Dommergues and Mangenot, 1970; Alexander, 1977; Ehrlich, 1981; Brock et al., 1984).

2.3. Influence of Environmental Factors on Microbial Activity

The chemical activity of microorganisms depends on environmental conditions: sources of carbon and energy (electron donors), type and nature of electron acceptors (O_2, NO_3^-, SO_4^{2-}, organics) corresponding either to aerobic or anaerobic conditions, nutrients to supply the cell constituents, temperature, acidity and alkalinity, salinity and water content. Favorable energy conditions, summarized by Figure 2 and equation (1), require mineral or organic compounds (AH_2, electron or proton donors) capable of being oxidized by microbes to provide utilizable energy (ΔG) and mineral or organic electron or proton acceptors (B).

Depending on the nature of the carbon and energy sources (electron donors), microorganisms are divided into four main nutritional types (Table I). Photolithotrophs use electromagnetic energy (photons), CO_2 as carbon source and H_2O as electron donor. Photoorganotrophs use electromagnetic energy (photons) and organic compounds as carbon source. Chemolithotrophs utilize oxidation of inorganic components (Fe^{2+}, S^o, NH_4^+) as chemical energy and CO_2 as carbon source. Chemoorganotrophs or heterotrophs use oxidation of organic compounds as chemical energy and also use organic compounds as source of carbon. Schematically, heterotrophic microorganisms or chemoorganotrophs are considered capable of growth only with organic compounds. They include all the fungi and most of the bacteria. Autotrophic microorganisms grow only in an inorganic medium. They include photolithotrophic organisms (algae) and chemolithotrophic bacteria. Chemolithotrophic bacteria are less numerous than heterotrophic bacteria but they play an important role in different biogeochemical cycles (nitrogen, sulfur, iron) where they are responsible for different fundamental reactions (nitrification, sulfides and sulfur oxidation, ferrous oxidation). Some bacteria, such as *Thiobacillus novellus*, can grow either in inorganic ($S_2O_3^{2-}$ and CO_2) or organic (peptone) nutrient media and are therefore named mixotrophs.

Table I. Nutritional classification of microorganisms according to the carbon and energy sources.

	Electron Donors and Carbon Sources	
	Inorganic	Organic
Electromagnetic energy (photons)	H_2O (e^- donors) CO_2 Photolithotrophs	$(CH_2O)_n$ Photoorganotrophs
Chemical energy	NH_4^+, S^{2-}, S^o, Fe^{2+} CO_2 Chemolithotrophs	$(CH_2O)_n$ Chemoorganotrophs

A second important energy consideration relates to the nature of the electron acceptors that determine the type of respiration. If the electron acceptor is oxygen, microbes have aerobic respiration. If it is an inorganic compound other than oxygen (NO_3^-, SO_4^{2-}, CO_2), they have anaerobic respiration. Finally, if it is an organic compound, they use a fermentation process (Brock et al., 1984). Such processes correspond to the different aerobic and anaerobic conditions present in natural environments.

Energy yields depend on the free energy of formation of the reactants and products, or in other words, on the free energy of the reaction. Depending on the type of respiration process, the yield in available energy (ΔG) can be modified strongly for the degradation of the same molecule. One mole of glucose can produce from 16 to 399 Kcal of available energy during alcoholic fermentation or total oxidation, respectively. The energy yields are, respectively, 28 and 67% (Senez, 1968).

The nature of the available carbon sources, electron donors, and electron acceptors promotes different microbial populations having specific nutritional requirements that are designated as physiological groups: cellulolytic, chitinolytic, ligninolytic, aerobic, anaerobic, sulfur oxidizing, nitrifying, denitrifying, and other bacteria (Dommergues and Mangenot, 1970; Alexander, 1977).

The source, the nature and the content of nutrients acting as building blocks (nitrogen compounds, soluble mineral elements, growth factors-coenzymes) are important, as they can modify the growth and the functioning of the cell machinery and the development of different microbial communities. Some organic and inorganic compounds can also act as antimicrobial agents (Dommergues and Mangenot, 1970; Berthelin et al., 1974).

Temperature is one of the most important environmental factors. Microbes are able to grow at low ($<$20°C), intermediate (between 20° and 45°C), and high ($>$45°C) temperature optima.

Acidity and alkalinity of the medium are also important environmental factors and allow development of acidophile, neutrophile and basophile microbes. Most microbes grow well at pH values between 5 and 9. Few species can grow at pH values less than 2 or greater than 10.

The humidity is also very important in terrestrial environments. In soils, microbial activity stops when water can be extracted only at pressures above 400 bars or 40 MPa (Dommergues and Mangenot, 1970). Most of bacteria isolated from soils or superficial waters are inhibited or killed by hydrostatic pressures of 200 to 600 atmospheres (Brock et al., 1984). Despite this effect of hydrostatic pressure, viable bacteria have been isolated from deep sediments and the deepest oceans (Ehrlich, 1981; Brock et al., 1984).

Salinity has an osmotic effect on microbes that also have specific requirements for salts, such as NaCl, KCl and $MgCl_2$. Most of the microbes have low salt requirements, except for halophilic microorganisms (bacteria, fungi, algae) that grow in salt environments, such as the Great Salt Lake, and the Dead Sea.

The presence of clay minerals in soils can modify microbial activity and play an important role in the survival of microbes and the maintenance of their activity (Dommergues and Mangenot, 1970; Stotzky 1980).

Although microorganisms are greatly affected by their environments, they are present and active over a wide range of redox, temperature, acidity, alkalinity, humidity, pressure and salinity conditions, and this applies even to single species. For example, as reported by Brock et al. (1984), the sulfate-reducing bacteria *Desulfovibrio*, *Desulfatomaculum*, and *Desulfomonas*, which are obligate anaerobes that use sulfate as electron acceptor in anaerobic respiration (dissimilative but not assimilative sulfate reduction), are active in the absence of O_2, NO_3^- or Fe^{3+} when redox is between +350 and -500 mV, pH is 4.2 to 10.4, pressure 1 to 1,000 atmospheres, temperature 0 to 100°C and salinity fresh water to 5 percent NaCl. For such bacteria, the electron donors generally used are H_2, lactate and pyruvate. Some others (formate, acetate, fumarate, ethanol) are of more restricted use.

In natural environments, the distribution of eucaryotic organisms (algae and fungi) is more restricted than that of procaryotic organisms (bacteria) that can grow in more diverse and extreme conditions.

3. SOLUBILIZATION PROCESSES

Solubilization, dissolution or lixiviation mean release of mineral elements from rocks or minerals by production of acid, complexing, reducing alkaline compounds or by enzymatic process (e.g., bacterial reduction of iron from oxyhydroxides). Different reviews have considered the microbial processes of solubilization of inorganic elements from rocks and minerals (Silverman and Ehrlich, 1964; Zajic, 1969; Krumbein, 1972; Berthelin, 1977, 1983, 1985; Silverman, 1979; Dacey et al., 1981; Avakyan, 1985), but only a very few papers and reviews (Berthelin and Toutain, 1982; Berthelin, 1983; Robert and Berthelin, 1986) give data on the microbial transformation and degradation of minerals or the mineralytic effects of microorganisms.

Microbial solubilization processes involve mainly the production of acid and/or complexing agents in aerobic and anaerobic environments by either heterotrophic microbes (chemoorganotrophic microbes occurring in the carbon cycle) or autotrophic microbes (chemolithotrophic microbes occurring in the nitrogen and sulfur cycles). Another important process concerns the enzymatic bacterial reduction of iron and manganese in anaerobic soils and sediments. Some secondary processes involve changes in pH by either production or assimilation of alkaline agents. These processes are presented in Table II.

Table II. Microbial solubilization and insolubilization processes of mineral elements in natural environments.

Solubilization

1. Release of metal ions by acid and complexing compounds produced by autotrophic and heterotrophic microorganisms in a variety of environmental conditions.
2. Release of Fe^{2+} and Mn^{2+} mainly from oxides by enzymatic reduction involving anaerobic or faculative anaerobic heterotrophic bacteria in non-aerated (without oxygen) environments.
3. Acidification or alkalinization by uptake or release of alkaline compounds (e.g., ammonium) in different environments

Insolubilization

1. Oxydation of Fe^{2+} and Mn^{2+} by autotrophic or heterotrophic mocroorganisms in micro-aerophilic or aerobic environments.
2. Reduction of sulfates and sulfide formation by sulfate reducing bacteria in strictly anaerobic environments.
3. Biodegradation of organo-metallic complexes by heterotrophic microorganisms in different environments.
4. Uptake of elements by bacteria and fungi in different environments.

3.1. Production of Acid and/or Complexing Compounds

3.1.1. Autotrophic (chemolithotroph) microorganisms of the nitrogen and sulfur cycles. The strictly aerobic and autotrophic nitrifying bacteria (*Nitrosomonas* and *Nitrobacter*) produce nitric acid during oxidation of ammonium according to equations (2) and (3). They occur both in neutral environments and in acidic soils (Berthelin et al., 1985):

$$NH_4^+ + 3/2\ O_2 \rightarrow NO_2^- + H_2O + 2H^+ \quad (2)$$

$$NO_2^- + 1/2\ O_2 \rightarrow NO_3^- \quad (3)$$

The sulfur oxidizing bacteria (*Thiobacilli*) produce sulfuric acid during the oxidation of sulfide, sulfur, sulfite, and thiosulfate (Ehrlich, 1981) according to equations (4), (5), (6):

$$S^{2-} \rightarrow \text{unknown compounds} \rightarrow SO_3^{2-} \rightarrow SO_4^{2-} \quad (4)$$

$$2S^o + O_2 + OH^- \rightarrow S_2O_3^{2-} + H^+ \quad (5)$$

$$S_2O_3^{2-} + 2O_2 + OH^- \rightarrow 2SO_4^{2-} + H^+ \quad (6)$$

Thiobacilli are present in neutral or slightly acidic habitats (*Th. thioparus, Th. denitrificans, Th. novellus*) and in strongly acidic habitats (*Th. thiooxidans, Th. ferrooxidans*). They are aerobic or sometimes facultative aerobic organisms. Some of them (*Th. thioparus, Th. denitrificans*) can grow in relatively anaerobic conditions in the absence of oxygen, using nitrate as electron acceptor in anaerobic respiration. One species can also use organic carbon sources (*Th. novellus*) (Dommergues and

Mangenot, 1970; Ehrlich, 1981).

The acidity (H^+) from the nitric acid produced by nitrifying bacteria is considered to be one of the main agents of solubilization of carbonates in calcimagnesian soils (rendolls, eutrochrepts) (Dommergues and Mangenot, 1970; Duchaufour, 1982). In fact, nitrifying bacteria efficiently solubilize calcium carbonates in soils (Alexander, 1977; Faurie, 1977) and in the weathering and decay of limestone of buildings and monuments (Kauffmann, 1960; Jaton, 1972). Nitrifying bacteria participate in weathering of serpentinized ultrabasic rocks (Lebedeva et al., 1978) and play a major role in the weathering of minerals of brown acid forest soils (Berthelin et al., 1985a). Table III shows the presence of relatively large amounts of nitrate in soil solutions of brown acid forest soils.

Table III. Nitrate and organic carbon contents of soil leachates and soil solutions extracted at -2.5 MPa from the A_1 horizon of a podzol and of a brown acid forest soil: average of four different periods of the year (mmol/L).

	Soil Leachates		Soil Solution extracted at -2.5 MPa	
	Podzol	Brown acid forest soil	Podzol	Brown acid forest soil
NO_3^-	0.0	0.9	0.0	2.2
Organic Carbon	4.8	1.8	9.2	2.0
pH	3.5	4.1	4.0	4.2

Production of sulfuric acid during oxidation of sulfide and sulfur by *Thiobacilli* was mainly utilized in the leaching of low grade ores to extract copper or uranium (Brierley, 1978; Murr, 1980; Ehrlich, 1981). It was shown that basic ferric sulfates (jarosite, natrojarosite, ammonium jarosite) are formed by *Thiobacillus ferrooxidans* when K, Na and NH_4 are supplied in the nutrient medium. But when K is supplied by glauconite, illite or microcline, and Na by albite instead of salts (and in absence of NH_4 which blocks the release of K from these minerals), acidification of the medium by formation of sulfuric acid leads to a release of K and Na from these minerals (Ivarson et al., 1978). Jarosite and natrojarosite were formed and simultaneously glauconite, illite and microcline were weathered. Glauconite weathers at a faster rate than illite or microcline, and albite weathers more easily than microcline. A slight expansion was observed in illite and a noticeable expansion in glauconite.

In terrestrial environments, *Thiobacilli* that can be involved in weathering of minerals (Dommergues and Mangenot, 1970) have not yet been studied in sufficient detail.

3.1.2. Heterotrophic (chemoorganotroph) microorganisms of the carbon cycle. During degradation of organic matter, heterotrophic microorganisms (bacteria and fungi) can produce, according to equation (7), carbonic (H_2CO_3), aliphatic (R-COOH), phenolic, and amino acids that are eventually involved in the solubilization of mineral elements and the breakdown of rocks and minerals.

$$\text{Organic matter}(CH_2O)_n \rightarrow \begin{cases} \text{Organic acids (R–COOH)} \\ CO_2 \rightarrow H_2CO_3 \end{cases} \quad (7)$$

In soil atmosphere, CO_2 concentration is higher than in open air (0.03%), and it increases at the expense of O_2 concentration. On average, it is in the range 0.3 to 3%, but may reach 5 to 10% or higher, and increases with incorporation of plant residues and with depth. Dissolved CO_2 in soil solution undergoes chemical speciation, and the different carbonate species account for the simultaneous occurrence of aerobic and anaerobic microbial reactions within the same soil. Above pH 6.3, bicarbonate and carbonate ions predominate over free CO_2 in soil solutions. Thus, in neutral or alkaline soils, increased production of CO_2 under the influence of microbial activity will cause a larger proportion of cations to be combined in the form of carbonates or bicarbonates. Waterborne transport to depth will be facilitated (Levy and Toutain, 1982).

The production of organic acids and/or complexing compounds depends on different environmental conditions, such as input and nature of organic substrates, acidity, aerobic or anaerobic conditions and microbial communities. Waterlogged soils can promote the production of poorly complexing agents (formic, acetic, butyric, and lactic acids).

Relatively stronger complexing agents are formed in more aerophilic environments. Well-aerated environments promote biodegradation of organic compounds and non-aerated or less aerated environments decrease biodegradation because they lack suitable electron acceptors (particularly oxygen) and also because they frequently contain antimicrobial compounds (Bruckert et al., 1970; Berthelin, 1982).

Stevenson and Ardakani (1972) and more recently a review of McKeague et al., (1986) have given the nature and concentration range of various substances in the soil solution: simple aliphatic acids, $1\text{–}4\times10^{-3}$ M; amino acids, $0.8\text{–}6\times10^{-4}$ M; aromatic acids, $0.5\text{–}3\times10^{-4}$ M.

Organic ligands can have either a specific or a non-specific affinity to metal ion complexation and transport. Compounds such as citrate or oxalate, even if they have different affinity for different metal ions, can be considered as non-specific chelating agents (Figure 3). But "siderophores" (that is, iron bearing substances), such as trihydroxamate or tridiorthophenolate, can be considered as specific agents of ferric iron chelation and transport (Figure 3) (Waid, 1975; Emery, 1977).

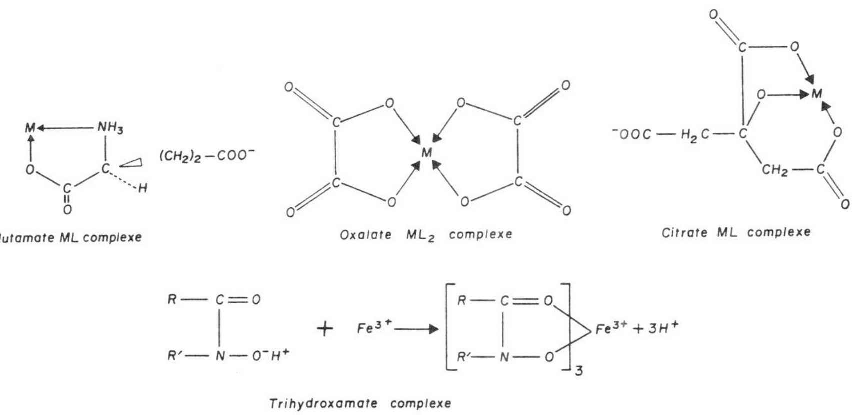

Figure 3. Examples of simple organo-metallic complexes.

Stevenson and Fitch (1986) have recently discussed, for soil environments, the formation and chemical stability of complexes of metal ions and organic compounds ranging from low-molecular-weight carboxylic acids to polymers, such as polysaccharides and fulvic acids. Stability constants of well-defined metal-organic complexes have been published (Sillen and Martell, 1964, 1971). Stability constants have been proposed for soil polymers of unknown structure (fulvic and humic acid) with a number of metal ions: Cu^{2+}, Mn^{2+}, Fe^{2+}, Al^{3+}, and others (Schnitzer, 1978; Stevenson and Fitch, 1986). These constants are valid, at best, only for specific conditions of pH, ionic strength, and temperature. Their application to sediment and soil solutions is delicate. A metal-organic complex may have a high stability constant but be unlikely to form because the metal has a greater affinity for OH^- than for the organic ligand.

The formation curves (Figure 4) of ferric and zinc citrate and ferric and zinc glutamate for a 10^{-3} M metal concentration and 3×10^{-3} M ligand concentration show that with glutamate, metal ligand (ML) complex is formed from pH 1.7 to 2.3 with Fe^{3+} and from pH 5.0 to approximately 7.0 with Zn^{2+}. Above pH 7, metal precipitates a hydroxide (dotted line). With citrate, mainly (ML) complex with Zn^{2+} and $M_2(OH)_2L_2$ with Fe^{3+} are formed from lower pH and are stable to more higher pH without precipitation. Such curves show great variation between two ligands and two metals. Formation curves of metal organic complexes (Marinsky, 1976; Linder et al., 1983) are suitable for use in natural conditions as they show the behaviour of the system as a function of the pH and concentrations of solutes.

The mechanism of dissolution by acids, summarized by equation (8), is defined as a process involving proton activity and the solubilization of mineral elements from rocks in ionized form:

$$(\text{Mineral})^-M^+ + H^+R^- \rightarrow H^+(\text{Mineral})^- + M^+R^- \quad (8)$$

$$(R^- = NO_3^-, SO_4^{2-}, HCO_3^-, R_1COO^-)$$

$$(\text{Mineral})^-M^+ = \text{silicates, phosphates, carbonates, oxides ... from rocks}$$

$$M^+ = \text{metal ions}$$

The complexation dissolution processes summarized by equation (9) may be defined as an equilibrium reaction between a metal ion and a complexing agent, characterized by the formation of more than one bond between the metal and the molecule of the complexing agent, and resulting in the formation of a ring structure incorporating the metal ion. The first step of the reaction corresponds generally to dissolution by protons and the second to the complexation that increases the stability of the metal ion in solution.

$$(\text{Mineral})^-M^+ + H^+L^- \rightarrow H^+(\text{Mineral})^- + ML \quad (9)$$

$$H^+L^- + LM \rightarrow L_2M + H^+$$

$$L = \text{organic ligands}$$

$$(\text{Mineral})^-M^+ = \text{silicates, carbonates, phosphates, oxides from rocks}$$

In weathering-simulation experiments, the formation of unidentified metalorganic complexes of Fe and Al with compounds of bacterial and fungal origins has been indicated (Tsyurupa, 1964; Arrieta and Grez, 1971; Daragan, 1971; Berthelin and Belgy, 1979). Such complexes were reported by Berthelin and Belgy (1979) to have relatively high molecular weights (2,000 to 2,900). Further analyses were interpreted as indicating that these organo-mineral complexes were composed of inorganic

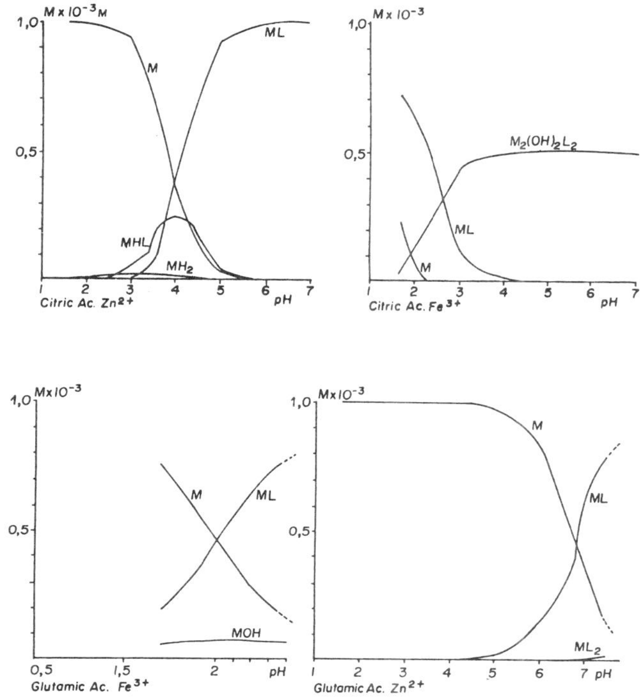

Figure 4. Formation curves for organo-metallic complexes of citrate and glutamate with Fe and Zn at different pH: M = metal (concentration 10^{-3} M); L = ligand (concentration 3.10^{-3} M) (- - - metal precipitation). (From Adrian and Berthelin, unpublished data).

elements associated with oxalic, citric, isocitric, succinic, fumaric, lactic, p-hydroxybenzoic and ferulic acids. It appeared that simple organic compounds produced by glucose metabolizing microbes were linked to form polynuclear (i.e., polycations and polyanions) complexes in which Fe and Al were bound as chelates, becoming unexchangeable with a strong cationic resin. Such studies provide information not only on complex formation but also on the transitory nature of uncombined simple organic acids in natural conditions.

In batch cultures or percolating devices the production of acid and complexing compounds by different microbial populations has increased chemical solubilization by factors of 5 to 113, compared to the release of elements obtained with nutrient media or percolating solutions without microorganisms in sterile controls. This applies to Si, Al, Fe, Mn, Mg, Ca, K, U and Cu from soils, granite sands, granites and syenite (Bertelin, 1971, 1985; Berthelin and Belgy, 1979; Berthelin et al., 1974). The general kinetics of microbial solubilization, according to cumulative curves obtained in percolation devices and shown in Figure 5a, can be given by equation (10):

$$y = a + bt - ae^{-ct} \qquad (10)$$

which is an exponential function with a linear term.

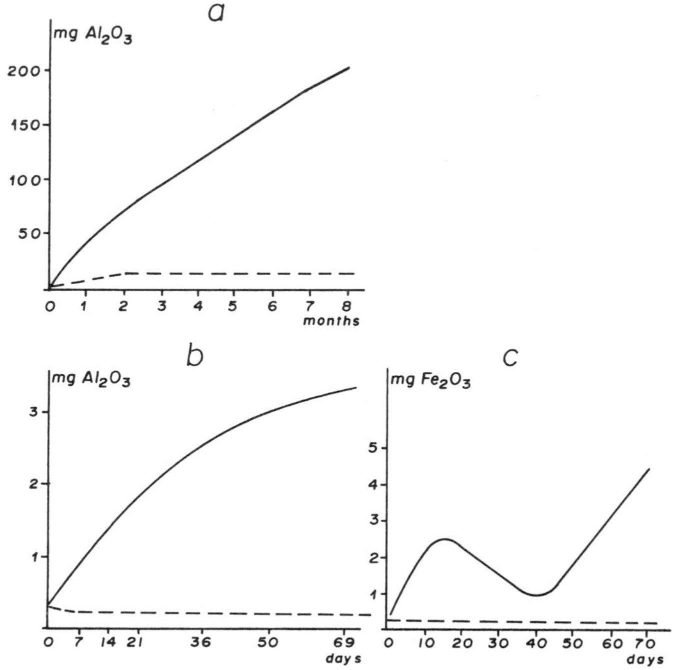

Figure 5. Cumulative curves of amounts of metals solubilized from granite sands: a) in a percolating device in the presence of a mixed microbial population; b) in a batch culture in the presence of a pure strain; c) in a batch culture, in the presence of a mixed microbial population. (——— with microorganisms; ----- without microorganisms). (After Berthelin, 1985).

In batch cultures, two main types of heterotrophic microbial solubilization curves were observed. With pure strains or simple mixed populations the kinetics of microbial solubilization is similar to exponential growth curves (Figure 5b), but with complex microflora one can observe successive phases of solubilization and insolubilization (Figure 5c) that correspond first to the production then to the biodegradation of metabolic organic compounds. Considering Figure 5c, chromatographic analysis has shown that iron microbial solubilization corresponds to the production of mainly lactic, succinic, citric acids, and that insolubilization was due essentially to citrate biodegradation (Berthelin, 1971). Similar curves have also been observed (Aristovskaya et al., 1969; Magne et al., 1974; Agbim and Doxtader, 1975) during solubilization of different elements from rocks and minerals. Such experiments simulated successive microbial processes of solubilization and insolubilization that can occur in relatively closed habitats.

Compounds involved in these microbial solubilization processes of silicates, phosphates, carbonates and oxides have been reported by Berthelin (1977, 1983, 1985), Silverman (1979), Dacey et al. (1981) and included formic, acetic, butyric, lactic, succinic, fumaric, oxalic, citric, glycolic, gluconic, 2 ketogluconic, lichenic, fulvic, humic, nitric, sulfuric and phenolic acids.

Transformation and degradation of minerals by acids and/or chelating agents of microbial origin have been observed by few workers. Anaerobic and facultative anaerobic bacteria producing mainly formic, acetic, butyric, lactic acids and also able to reduce manganese (Mn^{4+}) and iron (Fe^{3+}) have promoted the destruction of primary chlorite, vermiculite, biotite and the transformation of biotite into vermiculite, and of "illite-vermiculite" into montmorillonite (Berthelin and Boymond, 1978).

Complexing agents (citric and oxalic) of fungal origin were observed to be involved in the deterioration of sandstone monuments (Eckhardt, 1978). The production of such acids by *Aspergillus niger* leads to the removal of octahedral ions from micas, leaving a fragile matrix of amorphous material (Boyle et al., 1967). A complex microflora, obtained during partial sterilization of soils, produced complexing agents that destroyed primary chlorite, vermiculite and biotite, transformed biotite into vermiculite and into an alumino-silicate residue. An interstratified mineral (illite-vermiculite) was also transformed into vermiculite (Berthelin and Belgy, 1979).

During such microbial weathering processes, attempts have been made to determine and compare the global energy cost of microbial solubilization of elements in different environmental conditions (Table IV). A lower global energy cost of solubilization of elements from an acid brown forest soil developed on granite sand was observed during anaerobic biodegradation of grass (*Festuca*). Microbial solubilization was less efficient with plant material in aerobic conditions in this type of soil. Glucose promoted a better yield (Cheikhzadeh-Mossadegh, 1981).

Table IV. Global energy cost of microbial solubilization of major elements from a brown forest soil in different environmental conditions (carbon sources are glucose or plant materials; soils are well drained or waterlogged) expressed as mg of solubilized elements by 1 Kcal used by microorganisms.

	Microorganisms + 1 Kcal[1] (plant materials[2] or glucose)→ Solubilized Mineral Elements (in mg) From a Granitic Sand				
	Aerobic Conditions			Anaerobic Conditions	
	Glucose	Festuca	Calluna	Festuca	Calluna
SiO_2	3.2	0	0	1.7	1.8
Al_2O_3	5.0	0	0	0	0
Fe_2O_3	4.3	0	0	11.5	3.7
MnO	1.4	0.6	0	10.8	8.9
MgO	1.1	0.3	0	1.1	2.0
CaO	1.4	1.9	0	11.5	0
K_2O	0.1	1.2	0	14.9	0

[1] It is assumed that 1 kg of dry organic matter corresponds to 4,800 Kcal and 700 litres of CO_2.

[2] Plant materials are grass (*Festuca*) and shrubs (*Calluna*).

Some soil solutions, such as in spodosols (Table III), contain relatively large amounts of organic compounds of microbial and plant origin that are responsible for the acidity and are involved in the weathering of minerals. As in some hydromorphic soils, podzol soil solutions do not contain nitrate, and acidity is provided by organic compounds (Bonne et al., 1982). Vedy and Bruckert (1982) have reported that organic acids (lactic, succinic, oxalic, citric, quinic) have been found in different environmental conditions in podzolic soils, hydromorphic soils, solods, and calcareous soils. They are more abundant in podzolic and hydromorphic soils than in others.

The content of compounds of microbial origin, as shown in Table III, is generally larger in capillary soil waters extracted at -2.5 MPa than in leachates.

The organic acid content decreases rapidly from the soil surface to deeper horizons, but such gradients depend on the type of soils, as shown in Figure 6 (from Bruckert, 1970). It was observed that the organic acids extracted by maceration in water are relatively abundant in the litter (L) of brown acid soil and decrease rapidly in the A_1 horizon. In a podzol it was quite different because larger amounts of organic acids were observed in deeper horizons. Such differences between these two types of soils depend on the microbial activity and will be discussed later.

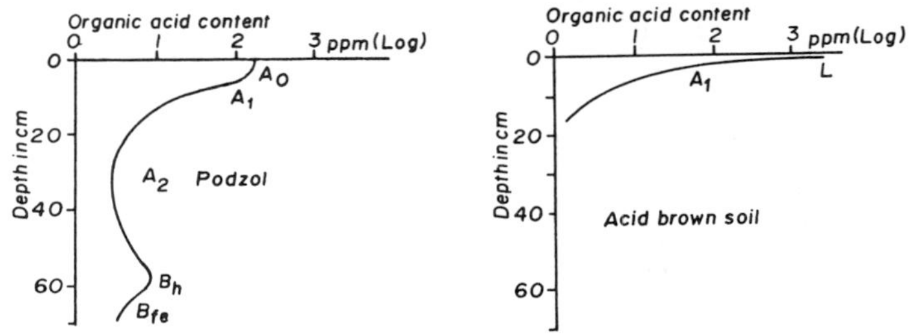

Figure 6. Distribution in soil profiles of organic acids extracted by maceration in water. Total organic acids content expressed as ppm in air dry sample on a logarithmic scale (from Bruckert, 1970).

3.2. Enzymatic Reduction of Iron and Manganese in Anaerobic Environments

Bacteria and fungi can produce organic reducing agents or promote reduction by decreasing Eh and pH. But the main reduction process involves the enzymatic reduction of ferric (Fe^{3+}) iron and manganic (Mn^{4+}) manganese by heterotrophic bacteria under anaerobic conditions (Munch and Ottow, 1977; Ehrlich, 1981). Such a bacterial reduction process promotes the dissolution of ferric oxides but requires contact between bacteria and minerals (Robert and Berthelin, 1986). The mechanisms involved are not yet understood in detail, particularly for iron. However, according to Ottow (1971), it is possible to summarize the corresponding reactions by the following equations:

$$\text{Energy source} \xrightarrow[\text{deshydrogenase}]{\text{substrate}} e^- + H^+ + \text{ATP+end products} \quad (11)$$

$$Fe(OH)_3 + 3H^+ + e^- \rightarrow Fe^{2+} + 3H_2O \quad (12)$$

Such bacterial reduction can be observed from acidic (3.5) to neutral (7.5) pH, and from relatively high (>200 mV) to low (-200 mV) Eh.

Figure 7 shows the importance and the conditions (pH, Eh) of solubilization of iron by direct enzymatic reduction of ferric iron from goethite in the presence of a *Bacillus*. In this process, two enzymatic systems may occur: one involving dissimilatory nitrate reductase*, the other one a yet unidentified ferrireductase. Both processes

*Dissimilatory nitrate reductase corresponds to an anaerobic respiration process in which nitrate is

can certainly be considered as a dissimilatory ferric reduction involved in anaerobic respiration of bacteria. As for nitrate and sulfate reduction, it may be possible to distinguish an assimilatory reduction (of nitrate, sulfate and ferric iron) associated with the microbial nutrition, and a dissimilatory reduction associated with the anaerobic respiration process in absence of oxygen. With some bacteria, such as *Bacillus*, it is possible to observe, during the degradation of organic matter, the successive reduction (or utilization as electron acceptors) of oxygen, nitrate, manganese (Mn^{4+}) and iron (Fe^{3+}). Such a process of ferric iron reduction occurs in environments and habitats where oxygen has been consumed by microbes and where organic matter is present as carbon and energy sources.

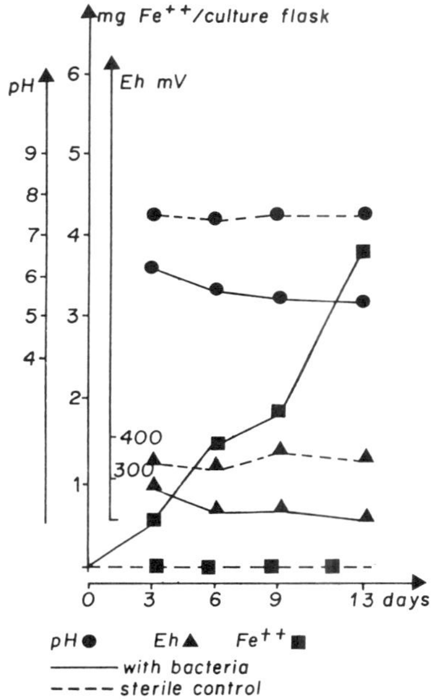

Figure 7. Cumulative curves of amounts of solubilized iron from goethite by enzymatic bacterial reduction in batch culture in the presence of a *Bacillus*. Eh and pH variations during growth. (After Cheikhzadeh-Mossadegh, 1981).

Enzymatic bacterial iron reduction can take place in hydromorphic soils and in sediments, but also in soils with short steps of anaerobic conditions during water saturation. It has been considered only under the aspect of solubilization of iron but not for their mineralytic effects (destruction of mineral structure). However, bacterial manganic and ferric reduction seems to be very important in hydromorphic and relatively anaerobic environments where organic matter promotes growth of anaerobic or facultative anaerobic bacteria able to reduce ferric iron from oxides, phosphates and silicates (Ehrlich, 1981; Berthelin, unpublished data).

an electron acceptor when oxygen is absent. Assimilatory nitrate reductase occurs in the utilization of nitrate as a nutrient.

Different experimental results have shown that iron reduction did not occur (Berthelin and Boymond, 1978) or occurred at a lower extent (Berthelin, 1982) in sterile controls (without microbial activity) in soil columns receiving soluble organic compounds or plant material (litter) even when pH and Eh were low.

Such results allow us to conclude that heterotrophic anaerobic and facultative anaerobic bacteria play a major role in the reduction of ferric iron.

3.3. Production of Alkaline Compounds

Some Russian workers (Aristovskaya and Kutuzova, 1968; Kutuzova, 1973) have considered that heterotrophic microorganisms, in particular bacteria, may form bases (ammonia during biodegradation of nitrogen compounds such as urea, sodium carbonate during biodegradation of sodium salts of organic acids) that raise the pH in alkaline environments and, eventually, promote solubilization of silica from nepheline, plagioclase, quartz and phytoliths. However, such processes are only relatively effective in limited alkaline conditions which require the presence of organic compounds producing bases in relatively large amounts.

4. INSOLUBILIZATION AND PRECONCENTRATION PROCESSES

Separately or concurrently to solubilization processes, heterotrophic and autotrophic microorganisms (essentially bacteria and fungi) are involved in insolubilization processes related mainly either to organo-mineral associations or to oxidation and reduction mechanisms. (The main insolubilization processes are presented in Table II.)

4.1. Insolubilization processes involving organo-mineral associations

Two main types of mechanisms are involved, either biodegradation of soluble organometallic complexes by heterotrophic microorganisms, or uptake and retention of metal ions by bacteria and fungi and by their constituents or their insoluble metabolites.

As discussed previously, it is well known that in natural environments (McKeague et al., 1986) metal ions are present as organo-mineral complexes. The ligands of such organo-metallic soluble complexes are used as carbon and energy sources by heterotrophic microorganisms (Berthelin and Cheikhzadeh-Mossadegh, 1977). Depending on the environmental conditions (nature of metal ions, pH, Eh, ionic strength, type of soil), biodegradation can contribute to the formation of deposits of iron, aluminum, zinc, and copper (Aristovskaya and Zykina, 1978; Berthelin and Cheikhzadeh-Mossadegh, 1977; Cheikhzadeh-Mossadegh et al., 1981; Berthelin, unpublished data; Figure 8).

Figure 8a shows the effect of metal ion (Na^+, Zn^{2+}, Cu^{2+}, Fe^{3+}, Cd^{2+}) on the rate, as CO_2 evolved, of biodegradation of citrate by a soil bacteria from maize roots. Cadmium toxicity inhibits the biodegradation of citrate. The resulting decreasing order of biodegradation rate is: $Fe < Cu < Zn < Na$.

In the experimental conditions ($7.0 < pH < 7.5$; citrate 3×10^{-3} M and metal ions 10^{-3} M) such modes of biodegradation promote the precipitation of zinc, copper and iron as hydroxides (Figure 8b).

In anaerobic conditions in waterlogged soils, and in spodosols the ligand biodegradation decreases in comparison to brown acid soils or to calcareous soils (rendzina) (Berthelin and Cheikhzadeh-Mossadegh, 1977; Cheikhzadeh-Mossadegh et al., 1981).

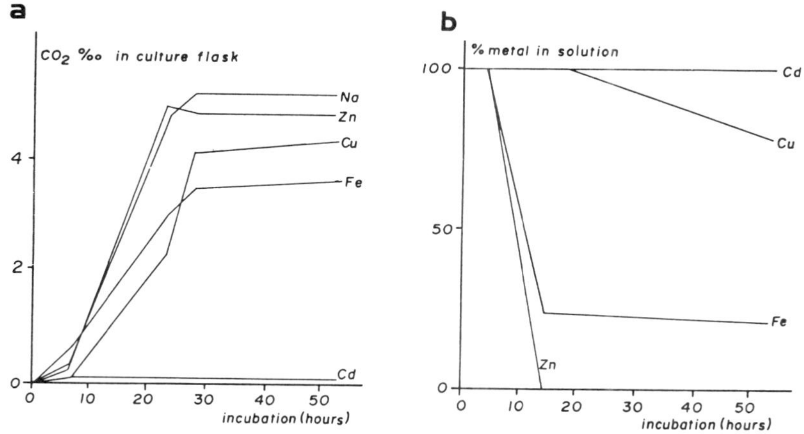

Figure 8. Biodegradation of soluble organo-metallic complexes of citrate by bacteria: a) rate of citrate biodegradation versus metal ions, b) rate of metal ions precipitation during biodegradation. (From Adrian and Berthelin, unpublished data).

Microbes can also form insoluble ligands such as mucilaginous compounds and cell constituents. Mucilaginous compounds of fulvic and humic acids seem able to play an important role in Fe immobilization, and also for other major or trace elements (Si, Al, K, U). Berthelin and Dommergues (1976) have observed that microbial production of approximately 100 mg of mucilaginous coatings that adhered firmly to mineral particles can immobilize 5 mg of Fe during the percolation of a nutrient medium containing glucose in a granite sand. Similar results were also observed in soils and granites.

The accumulation of metals by microorganisms can correspond to the absorption of elements by living organisms for their mineral nutrition but also in excess of their requirements. Such processes were mainly studied in cultures of microbes in the presence of elements in solution. Experiments performed in the presence of mineral rocks bearing phosphorus, calcium, uranium, and potassium have shown that bacteria and fungi are able to take up elements from solution, and also to displace very significantly an equilibrium between mineral and solution (Figure 9). Thus in promoting mobilization of elements (K, P, U) from rocks and minerals, they promote not only weathering but also the insolubilization and preconcentration of elements in or on their cells (Tardieux-Roche, 1966; Ennever et al., 1974; Berthelin and Munier-Lamy, 1983).

Bacterial and fungal cell walls in aqueous systems can also act as a sink for metal ions. Beveridge and Murray (1976) reported an order of uptake by cell walls of *Bacillus subtilis*: Mg^{2+}, Fe^{3+}, Cu^{2+} were strongly retained; Al^{3+}, Pb^{2+}, Ni^{2+}, Co^{2+} were not significantly retained; K^{+}, Na^{+}, Mn^{2+}, Au^{3+}, Zn^{2+} were intermediate. Berthelin et al. (1987) showed, using a bag method in the field, that in natural conditions fungal constituents are good sinks for uranium, but at a lower and certainly more realistic rate than reported by Tsezos and Volesky (1981) in laboratory conditions.

Exchange uptake of ions has been shown to promote the weathering of minerals. Weed et al. (1969) observed transformation of micas into vermiculite in a buffered nutrient solution during the growth of a fungus that absorbed interlayer K^{+}. Eventual

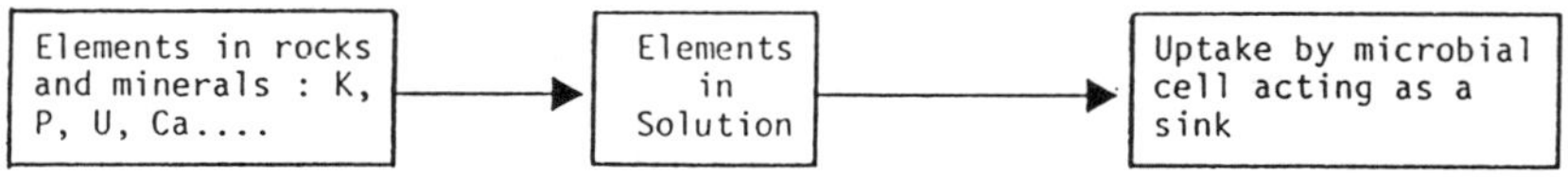

Figure 9. Exchange uptake of elements from rocks, minerals and solutions by living microbes and by microbe constituents acting as a sink.

replacement by a more hydrated ion of the nutrient medium caused a significant expansion of the layers. All the three types of processes involving organo-mineral associations need organic environment to supply carbon and energy to the microbes.

4.2. Microbial insolubilization processes involving oxidation and reduction mechanisms

Insolubilization by oxidation or reduction processes involves mainly microbial transformations of iron, manganese and sulfur. Microbes promote ferrous and manganous oxidation by affecting the environmental conditions of redox potential and pH, and by the formation of bases, biodegradation of acids, and production of O_2. The more interesting processes from an environmental and geological point of view are the enzymatic oxidation of Mn^{2+} and Fe^{2+} by autotrophic or mixotrophic bacteria and the sulfate reduction by sulfate reducing bacteria.

4.2.1 Iron and manganese oxidation. Either free Mn^{2+} or Mn^{2+} bound to Mn^{4+} oxide can be oxidized during autotrophic or mixotrophic growth of different bacteria (*Arthrobacter*, *Leptothrix*, *Oceanospirillum*, *Vibrio*) from pH 5.0 to 8.0, even at relatively high hydrostatic pressure of 476 atmospheres (Ehrlich, 1981).

Enzymatic bacterial ferrous oxidation summarized in equation (13) occurs also in different environments and involves different types of bacteria and precipitation of iron according to equation (14):

$$Fe^{2+} + 1/4\ O_2 + H^+ \rightarrow Fe^{3+} + 1/2\ H_2O \qquad (13)$$

$$Fe^{3+} + 3\ OH^- \rightarrow Fe(OH)_3 \qquad (14)$$

In acidic aerobic environments (pH 1.0 to 6.0), mesophilic (better growth around 20°C, such as *Thiobacillus ferrooxidans*) and thermophilic (growth to 85°C, *Sulfolobus acidocaldarius*) bacteria are able to oxidize Fe^{2+} (Brock et al. 1976, Ehrlich, 1981). Such bacteria that derive energy from iron and carbon from CO_2 oxidize for their growth large amounts of ferrous to ferric iron because very little energy is generated by the reaction.

Bacterial enzymatic iron oxidation occurs also at neutral pH (6.0 to 7.5). The stalked autotrophic bacteria *Gallionella* grows best under low oxygen content (0.1 to 1.0 mg of O_2 per ml). But CO_2 assimilation coupled to iron oxidation remains to be demonstrated. The sheathed bacteria (*Leptothrix*) are also associated with oxidation of ferrous iron, but their enzymatic iron oxidation remains to be shown. Figure 10 shows *Leptothrix* bacteria during the formation of a ferric iron sheath around two bacteria and empty sheaths of ferric iron observed in iron deposits in drains. Other mechanisms such as production of H_2O_2 were also suggested (Ehrlich, 1981).

Figure 10. a) *Leptothrix* during the formation of a ferric iron sheath. b) Empty sheaths of ferric iron in drains (photographs from S. Houot).

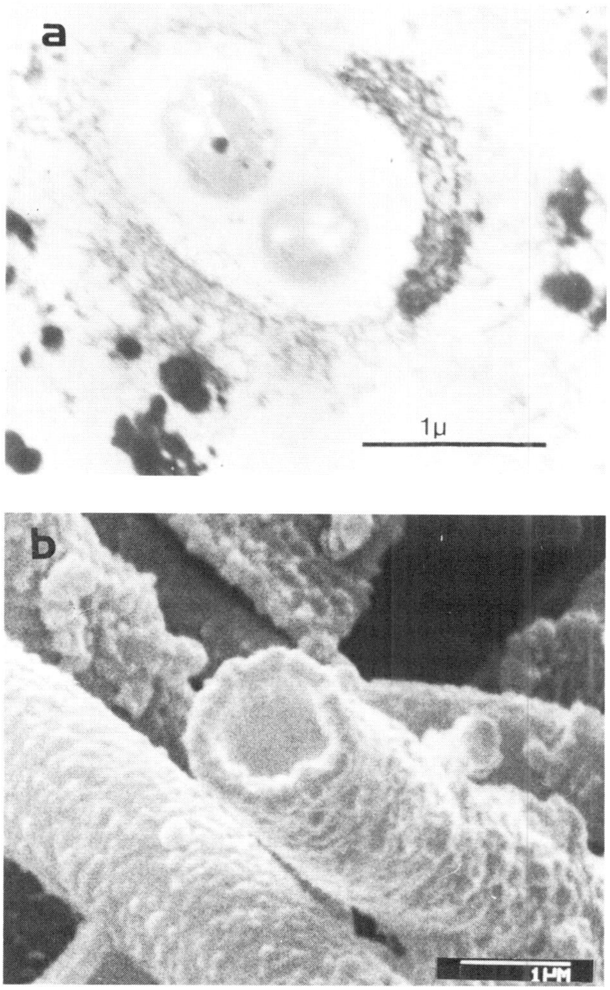

Occurrence of other sheathed bacteria is presumed for reasons of their morphological similarities.

Analyses of chemical parameters of waters in various field drains have shown (Houot and Berthelin, unpublished data) that iron ochre formation, associated with stalked and sheathed bacteria, occurs at the following annual mean values (in mg/L): Fe^{2+}, 22; organic carbon, 17; inorganic carbon, 23; total nitrogen, 2.3; $N(NH_4^+)$, 0.5; $N(NO_3^-)$, 0.03; SO_4^{2-}, 7; pH, 6.2; Eh, 226 mV. In similar fields and soils, but without iron ochre formation, mean values are: Fe^{2+}, 4; organic carbon, 10; inorganic carbon, 33; total nitrogen, 1.2; $N(NH_4^+)$, 0.02; $N(NO_3^-)$, 0.06; SO_4^{2-}, 0.8; pH, 6.7; Eh, 155 mV. Thus small differences in the chemical properties modify the microbial activity and promote the activity of iron bacteria.

4.2.2. Sulfate reduction. It is now well known that microbial reduction of inorganic sulfur compounds is a major process in biogeochemical cycles. Elemental sulfur can be reduced to H_2S in anaerobic conditions. Sulfate can be reduced either by assimilatory or dissimilatory sulfate reduction. Assimilatory sulfate reduction is associated with the transformation of inorganic sulfate into sulfur amino-acids by the microbes for their nutrition and does not require anaerobic growth conditions. But the more important and interesting process from a geochemical point of view concerns the dissimilatory sulfate reduction. A variety of bacteria in anaerobic respiration utilize sulfate as an electron acceptor in strictly anaerobic conditions, as summarized in equations (15) and (16):

$$CH_3CO\ COOH + H_2O \rightarrow CH_3\ COOH + CO_2 + 2H^+ \tag{15}$$

$$2H^+ + 1/4\ SO_4^{2-} \rightarrow 1/4\ S^{2-} + H_2O \tag{16}$$

Bacterial sulfate dissimilative reduction occurs significantly when organic electron donors or H_2 produced during fermentation are present, and thus shows the need for organic matter. There is a relationship between organic carbon and sulfate reduction. The environmental limits of bacterial sulfate reduction presented previously as the influence of environmental factors on microbial activity have shown the great range of occurrence of these bacteria.

Fractionation of sulfur stable isotopes involving selective utilization of ^{32}S instead of ^{34}S during dissimilatory sulfate reduction has shown that sulfate reducing bacteria play an important role in the formation of sulfur and sulfide deposits (Holser and Kaplan, 1966; Goldhaber and Kaplan, 1974). Hydromorphic soils and drained fields provide good examples of the occurrence of bacteria in iron and sulfur cycles, as shown in Figure 11.

5. WEATHERING PROCESSES AND SOIL FORMATION

Solubilization and insolubilization by microbial processes occur in soils where they are involved in weathering processes, soil formation, and its evolution and degradation. These are briefly presented here.

The vegetation of spodosols (*Calluna, Vaccinium*, pine, spruce) and their microbial degradation release antimicrobial compounds (pine resin and a variety of lipids) capable of modifying the soil microbial populations and their activity (Berthelin et al., 1974). The remaining microbes are able to produce large amounts of complexing agents (such as oxalic, citric, isocitric, succinic, fumaric, parahydroxybenzoic, and ferulic acids) that dissolve large amounts of mineral elements from silicates. Organometallic complexes are formed mainly with Fe and Al, and minerals are transformed

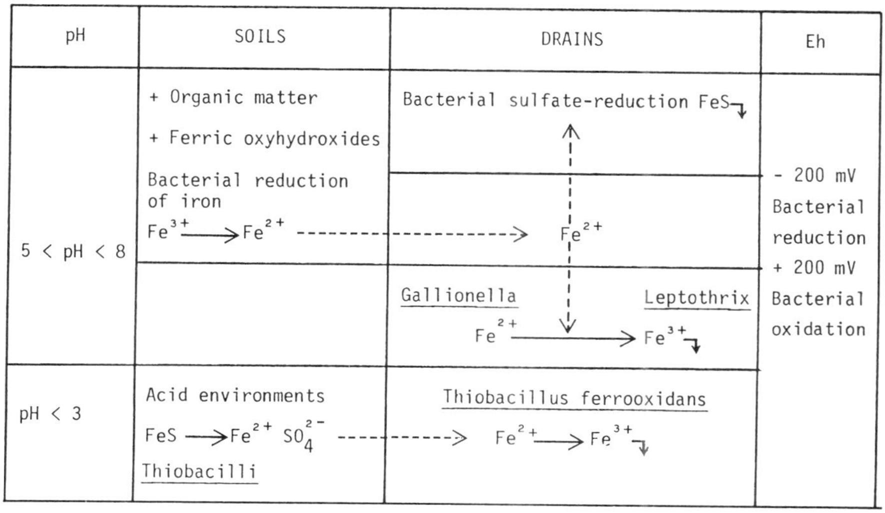

Figure 11. Diagram of occurrence of bacteria in iron and sulfur cycle in hydromorphic fields and in drains.

and/or destroyed (Berthelin and Belgy, 1979). Simultaneously, the spodosol environment decreases the activity of microorganisms that can biodegrade organic ligands (Berthelin and Cheikhzadeh-Mossadegh, 1977; Cheikhzadeh-Mossadegh et al., 1981). Both the production of complexing agents and the decrease of their biodegradation promote the strong mineral weathering occurring in such soils and the migration of elements that will precipitate later in the deeper B_h and B_s soil horizons.

In the acid brown soils of beech or fir forests (dystrochrept), it was shown that bacterial nitrification produces large amounts of nitric acid during plant litter mineralization and is therefore responsible for the solubilization of different elements. Such a process is involved in the slow weathering of minerals and in the biogeochemical cycling of Ca^{2+}, Mg^{2+}, K^+ in these soils (Berthelin et al., 1985).

The composition of the parent rock material can also play a major role in the microbial processes involved in weathering. With the same type of vegetation (beech forest) and the same climatic conditions, Toutain (1974) observed that a lower content ($<5\%$) of iron (mainly ferric oxyhydroxides) and clays in a sandstone promotes microbial activity that favours a podzolisation process (strong weathering) instead of less weathering and brown forest soil formation.

In hydromorphic environments where water tables are variable and various levels of organic matter are observed, it is possible to distinguish schematically three main types of environmental conditions, which may correspond to different aerobic/anaerobic microbial populations and activity (Berthelin, 1982).

(i) In alluvial soils, pseudo-hydromorphic soils and soils with low contents of organic matter, microbial processes occur as in aerated environments where bacterial reduction processes are not significant. In such soils, anaerobic conditions are preserved either by relatively low organic carbon content that cannot maintain a sufficient microbial activity utilizing the available oxygen or through variations in water tables that maintain aerated conditions to avoid the occurrence of bacterial reducing processes.

(ii) In hydromorphic soils *sensu stricto* (gleys, pseudogleys, stagnogleys, peats) containing available organic matter, soil waterlogging promotes the very rapid utilization of oxygen by aerobic microorganisms. Then facultative anaerobic and strict anaerobic bacteria grow. Successive bacterial reduction of nitrate, manganese, iron, and sulfate can be observed as presented in Table V. Such anaerobic respiration and fermentation processes release hydrogen, methane and acids, and become very important. Simultaneously, Eh decreases from +600 mV to less than -200 mV (Patrick, 1960; Yoshida, 1976; Berthelin, 1982). In these environments, microbial weathering involves the direct enzymatic reduction processes of manganese and iron, indirect reduction by strong reducing agents (H_2, CH_4), solubilization by acids or complexing agents. Transformation and destruction of phyllosilicates were observed (Berthelin and Boymond, 1978). Such environmental conditions also promote weathering by decreasing the biodegradation of organic ligands.

Table V. Successive microbial processes that can occur in hydromorphic soils (after Yoshida, 1975).

Observed Transformation	Initial Soil Eh (in V)	Biochemical Process
Oxygen consumption and disappearance		Aerobic respiration
Nitrate reduction	+0.6 to +0.5	Anaerobic respiration
Manganese reduction	+0.6 to +0.4	Anaerobic respiration*
Iron reduction	+0.6 to +0.3	Anaerobic respiration*
Sulfate reduction and sulfide formation	0.0 to -0.2	Anaerobic respiration
Organic acids, alcohol formation	+0.6 to +0.4	Fermentation
Formation of H_2	-0.1 to -0.2	Fermentation
Formation of CH_4	-0.1 to -0.2	Fermentation

*It is not proven that manganese and iron reduction are anaerobic respiration processes.

(iii) In relatively neutral (pH = 6.2) hydromorphic environments under more aerated conditions (microaerophilic Eh = 200-250 mV), enzymatic iron oxidation associated with stalked (*Gallionella*) and sheathed (*Leptothrix*) bacteria occurs and likely promotes precipitation of large amounts of ferric iron (Houot and Berthelin, 1985). Considering the conditions of their occurrence, it seems possible that such processes occurred in the Precambrian iron-deposit formations. Bacterial oxidation processes of manganese and iron seem also to be involved in nodule and concretion forms in aquatic, marine and terrestrial environments (Ehrlich, 1981). Sedimentary iron ores (oxides, carbonates) can also have a bacterial origin. Finally, as previously mentioned, many sulfide deposits are also of bacterial origin.

In calcareous soils, inorganic anions play an important role. Bicarbonate produced by microbial decomposition of plant material is high enough to account for up to 75% of dissolved calcium (Durand, 1978) in different soils on chalk. Organic anions are also present and active in weathering in such soils, as reported by Vedy and Bruckert (1982) from different soil analyses performed in France and Switzerland.

6. CONCLUSION

Microorganisms or microbes, essentially bacteria (procaryotic cells), and fungi (eucaryotic cells) are present in large number in different natural environments (soils, sediments and waters). In order to generate energy to form new cell structures they may either break down organic and inorganic materials by oxidation processes, or reduce inorganic (O_2, NO_3^-, Mn^{4+}, Fe^{3+}, SO_4^{2-}, CO_2) and organic compounds. They also form gaseous, liquid and solid metabolic compounds that are excreted in the environment. Microbial influence is fundamental in the biogeochemical cycles of carbon, nitrogen, and sulfur, and is a significant process in the transformation of iron, manganese, phosphorus, and other major and trace elements in soils and sediments.

Autotrophic microorganisms that live in inorganic environments and oxidize inorganic compounds (Fe^{2+}, Mn^{2+}, S^{2-}, S^{o}, NH_4^+) and heterotrophic microorganisms that oxidize organic compounds, and represent the main part of microbes are directly or indirectly involved in mineral transformations.

Weathering of minerals and solubilization of elements are realized by microorganisms which produce acids or alkaline complexing and reducing agents, during the transformation of natural compounds. They solubilize iron and manganese of oxides and silicates by direct (enzymatic) reduction of ferric iron and manganic manganese. Microorganisms can also take up inorganic elements. Such exchange and uptake processes can modify the equilibria between insoluble and soluble forms of inorganic elements in silicates, phosphates, carbonates, sulphides, and oxides.

Simultaneously or separately, microorganisms are involved in insolubilization processes by biodegradation of organo-metallic complexes, adsorption of elements, oxidation of iron and manganese, formation of insoluble complexes and reduction of sulfate.

The occurrence and intensity of the solubilization and insolubilization microbial processes depend on environmental conditions, such as nature and amount of carbon energy resources, pH, oxidation, reduction, availability of inorganic nutrients, antimicrobial compounds and grain-size and nature of minerals (Berthelin, 1977, 1982, 1983).

Observations and measurements under natural conditions, as well as in field and laboratory experiments, have shown that microorganisms are involved in the weathering and "neoformation" of soil minerals. Some processes are associated with the occurrence and activity of specific microbes in the soil environment, while others appear as the results of general biological activity. It is also possible to determine the occurrence of a specific microbial process (for example nitrification, enzymatic direct reduction of ferric iron, or uptake of elements) in laboratory or field experiments, but frequently it is only possible to relate weathering to global "plant-microbe interactions"

Microbial processes appear to be responsible for mineral transformations at different stages of weathering and of geochemical cycles. They are involved in the weathering of primary and secondary minerals in different climates and can occur from the first stages of rock weathering in mountain or arid climates (lichens, algae associated to bacteria and fungi). Microbial weathering can also occur in aerobic and anaerobic conditions and in acidic or alkaline environments. Laboratory and field data have shown the occurrence of microbial activity in the weathering of minerals in different types of soils (spodosols, acid brown soils, hydromorphic soils, calcareous soils).

The role of microorganisms in weathering and the identification of major mechanisms involved are now understood, but further research is needed to distinguish the specific effects of different organisms and to know exactly their relative impact on the total weathering processes, particularly with regard to environmental conditions. The relation between the physical and chemical parameters of different environments and microbial processes and activity are of great interest from both basic scientific and applied points of view.

7. REFERENCES

Agbim, N.N. and Doxtader, K.G., 1975, Microbial degradation of zinc silicates, *Soil Biol. Biochem.*, **7**, 275-280.

Alexander M, 1977, *Introduction to Soil Microbiology*, John Wiley & Sons Inc., New York.

Aristovskaya, T.V., Daragan, A.Yu., Zykina, L.V. and Kutuzova, R.S., 1969, Microbial factors in the movement of some mineral elements in the soil, *Soviet Soil Sci.*, **5**, 538-546.

Aristovskaya, T.V. and Kutuzova, R.S., 1968, Microbial factors in the extraction of silicon from slightly soluble natural compounds, *Soviet Soil Sci.*, **12**, 1653-1659.

Aristovskaya, T.V. and Zykina, L.V., 1978, Biological factors of aluminum migration and accumulation in soils and weathering crusts, *Problems of Soil Science*, p. 175-182, Contrib. XI Internat. Congr. Soil Sci., Nauka, Moscow.

Arrieta, L. and Grez, R., 1971, Solubilization of iron-containing minerals by soil microorganisms, *Appl. Microbiol.*, **22**, 487-490.

Avakyan, Z.A., 1985, Microflora of rock and its role in the leaching of silicate minerals, *in* G.I. Karavaiko and S.N. Groudev (eds.) *Biotechnology of Metals*, p. 175-195, Centre Internat. Projects G.K.N.T., Moscov.

Berthelin, J., 1971, Altération microbienne d'une arène granitique. Note préliminaire, *Sci. Sol*, **1**, 11-29.

Berthelin, J., 1977, Quelques aspects des mécanismes de transformation des minéraux des sols par les microorganismes hétérotrophes, *Sci. Sol*, **1**, 13-24.

Berthelin, J., 1982, Les processus microbiens intervenant dans les sols hydromorphes en régions tempérées. Incidence sur la pédogenèse, *Pédologie*, **32**, 313-328.

Berthelin, J., 1983, Microbial weathering processes, *in* W.E. Krumbein (ed.) *Microbial Geochemistry*, p. 223-262, Blackwell Scientific Publ.

Berthelin, J., 1985, Microbial leaching of metal from rocks, *in* G.I. Karavaiko and S.N. Groudev (eds.) *Biotechnology of Metals*, p. 157-174, Centre of International Projects G.K.N.T., Moscov.

Berthelin, J. and Belgy, G., 1979, Microbial degradation of phyllosilicates during simulated podzolisation, *Geoderma*, **21**, 297-310.

Berthelin, J., Bonne, M., Belgy, G. and Wedraogo, F.X., 1985, Major role of nitrification in the weathering of minerals of brown acid forest soils, *Geomicrobiol. J.*, **4**, 175-190.

Berthelin, J. and Boymond, D., 1978, Some aspects of the role of heterotrophic microorganisms in the degradation of minerals in waterlogged acid soil, *in* W.E. Krumbein (ed.) *Environmental Biogeochemistry and Geomicrobiology, Vol. 2, The Terrestrial Environment*, p. 659-673, Ann Arbor Science, Ann Arbor.

Berthelin, J. and Cheikhzadeh-Mossadegh, D., 1977, Biodégradation de complexes organoferriques dans un sol brun acide et dans un sol podzolique sur granite, *Soil Organic Matter Studies*, **Vol. I**, p. 413-424, Int. Atom. Energ. Agency, Vienna, Austria.

Berthelin, J. and Dommergues, Y., 1976, The role of heterotrophic microorganisms in the deposition of iron and carbon in soil profiles, *in* J.O. Nriagu (ed.) *Environmental Biogeochemistry*, p. 609-631, Ann Arbor Science, Ann Arbor.

Berthelin, J., Gueniot, B. and Munier-Lamy, C., 1987, An *in situ* experimental bag method to study the influence of environmental factors on uranium mobilization and preconcentration in soils, *Uranium*, in press. Environments. Int. Meeting, Nancy, France, CREGU, Vandoeuvre-les-Nancy.

Berthelin, J., Kogblevi, A. and Dommergues, Y., 1974, Microbial weathering of a brown forest soil: Influence of partial sterilization, *Soil Biol. Biochem.*, **6**, 393-399.

Berthelin, J. and Munier-Lamy, C., 1983, Microbial mobilization and preconcentration of uranium from various rock material by fungi *in* R. Hallberg (ed.) Environmental biogeochemistry, *Ecol. Bull., Stockholm,* **35**, 395-401.

Berthelin, J. and Toutain, F., 1982, Soil Biology, p. 140-183, *in* M. Bonneau and B. Souchier (eds.) *Constituents and Properties of Soils,* Academic Press, London.

Beveridge, T.J. and Murray, R.G.E., 1976, Uptake and retention of metals by cell walls of Bacillus subtilis, *J. Bact.,* **127**, 1502-1518.

Bonne, M., Andreux, F., Vedy, J.C. and Souchier, B., 1982, Etude hydrochimique saisonnière dans trois sols acides: composition des eaux gravitaires et des solutions extraites à pF 4.4, *Sci. Sol,* **4**, 275-292.

Boyle, J.R., Voigt, G.K. and Sawhney, B.L., 1967, Biotite flakes: alteration by chemical and biological treatment, *Sci.,* **155**, 193-195.

Brierley, C.L., 1978, Bacterial leaching, *CRC Crit. Rev. Microbiol.,* **6**, 207-262.

Brock, T.D., Cook, S., Petersen, S. and Messer, J.L., 1976, Biogeochemistry and bacteriology of ferrous iron oxidation in geothermal habitat, *Geochim. Cosmochim. Acta* **40**, 493-500.

Brock, T.D., Smith, D.W. and Madigan, M.T., 1984, *Biology of microorganisms,* Prentice Hall Inc., Englewood Cliffs, New Jersey.

Bruckert, S., 1970, Influence des composés organiques solubles sur la pédogénèse en milieu acide, *Ann. Agron.,* **21 (4)**, 421-452 and **21 (6)**, 725-758.

Bruckert, S. Dommergues, Y. Weinhard, P. and Boymond, D. 1970, Effet-litière. III. Influence de l'anaérobiose sur la production de composés antimicrobiens hydrosolubles, *Oecol. Plant.,* 137-146.

Cheikhzadeh-Mossadegh D, 1981, Contribution expérimentale à l'étude de la géomicrobiologie du fer dans les sols, Doctoral Thesis, University of Nancy, France.

Cheikhzadeh-Mossadegh, D., Chone, Th. and Berthelin, J., 1981, Influence de l'activité microbienne sur la migration et le devenir du fer et du carbone d'un complexe organo-ferrique dans des sols acides et un sol carbonaté, *Coll. Internat. C.N.R.S.,***303** "Migrations organo-minérales dans les sols tempérés," Nancy, 199-207.

Cloud, P., 1978, *Cosmos, Earth, and Man: A Short History of the Universe,* Yale University Press.

Cloud, P., 1983, La Biosphere. Pour la Science (French edition of Scientific American), *Numero special "La Terre, Planeté Vivante",* **73**, 138-150.

Dacey, P.W., Wakerley, D.S. and Leroux, N.W., 1981, *The biodegradation of rocks and minerals with particular reference to silicate minerals: a literature survey,* Warren Spring Laboratory, Stevenage, U.K.

Daragan, A. Yu., 1971, Decomposition of minerals containing iron by soil microorganisms, *Soviet Soil Sci.,* **3**, 567-572.

Dommergues, Y. and Mangenot, F., 1970, *Ecologie Microbienne du Sol,* Masson, Paris.

Duchaufour, Ph., 1982, *Pedology* (Pedogenesis and Classification), T.R. Paton (trans.), Allen and Unwin, London.

Durand, R., 1978, La pédogénèse en pays calcaire dans le Nord-Est de la France, Doctoral Thesis, University of Strasbourg, France, 191 pp.

Eckhardt, F.E.W., 1978, Microorganisms and weathering of a sandstone monument *in* W.E.K. Krumbein (ed.) *Environmental Biogeochemistry and Geomicrobiology, Vol. 2, the Terrestrial Environment* p. 675-686, Ann Arbor Science Publ., Ann Arbor.

Ehrlich, H.L., 1981, *Geomicrobiology,* Marcel Dekker Inc.

Emery, T., 1977, The storage and transport of iron, *in* R H. Sigel (ed.) *Metal Ions in Biological Systems, Vol. 7. Iron in Model and Natural Compounds,* p. 77-125, Marcel Dekker, New York.

Ennever, J., Vogel, J.J. and Streckfuss, J.L., 1974, Calcification by *Escherichia coli, J. Bacteriol.,* **119**, 1061-1062.

Faurie, G., 1977, Etude in vitro du rôle de la nitrification sur la lixiviation du calcium dans les sols calcaires, *Sci. Sol,* **4**, 207-218.

Friedman E.I. and Ocampo-Friedman R., 1984, Endolithic microorganisms in extreme dry environments: analysis of a lithobiotic microbial habitat, *in* M.J. Klug and C.A. Reddy (eds.) *Current perspectives in microbial ecology, Am. Soc. Microbiol. Publ.,* Washington,

D.C.
Goldhaber, M.B. and Kaplan, I.R., 1974, The Sulfur Cycle, *in* E.D. Goldberg (ed.) *The Sea*,vol. 5, p. 569-655, John Wiley, New York.
Holser, W.T. and Kaplan, I.R., 1966, Isotope geochemistry of sedimentary sulfates, *Chem. Geol.*, **1**, 93-135.
Houot, S. and Berthelin, J., 1985, Data analysis to determine the relationship between water properties and deposits associated to iron bacteria in soil drains, *Abstracts of the VIIth Int. Symp. on Environmental Biogeochemistry*, Roma, p. 55.
Ivarson, K.C., Ross, G.J. and Miles, M.M., 1978, Alterations of micas and feldspars during microbial formation of basic ferric sulfate in the laboratory, *Soil Sci. Soc. Am. J.*, **42**, 518-524.
Jaton, C, 1972, Altérations microbiologiques de l'église monolithe d'Aubeterre-sur-Dronne, *Rev. Ecol. Biol. Sol*, **9**, 471-477.
Kauffmann, J, 1960, Corrosion et protection des pierres calcaires des monuments, *Corros. Anticorros.*, **8**, 87-95.
Krumbein, W.E., 1972, Rôle des microorganismes dans la genèse, la diagenèse et la dégradation des roches en place, *Rev. Ecol. Biol. Sol*, **9**, 283-319.
Kutuzova, R.S., 1973, Possible ways of mineral weathering in alkaline soils, *Soviet Soil Sci.*, **5**, 111-116.
Kutznetsov, S.I., Ivanov, M.V. and Lyalikova, N.N., 1963, *Introduction to Geological Microbiology*, McGraw Hill, New York.
Lebedeva, E.V., Lyalikova, N.N. and Bugelsky, Y. Y., 1978, Participation of nitrifying bacteria in weathering of serpentinized ultrabasic rocks, *Mikrobiologiya*, **47**, 1101-1107.
Levy, G. and Toutain, F., 1982, Aeration and redox phenomena in soils, *in* M. Bonneau and B. Souchier (eds.) *Constituents and Properties of Soils*, p. 355-366, Academic Press, London.
Linder, P.W., Torrington, R.G. and Seeman, U.A., 1983, Formation constants for the complexes of levulinate and acetate with manganese(II): Cobalt(II), nickel(II), copper(II) and hydrogen ions, *Talanta*, **30**, 295-298.
Magne, R., Berthelin, J. and Dommergues, Y., 1974, Solubilisation et insolubilisation de l'uranium des granites par des bactéries hétérotrophes, *Formation of uranium ore deposits*, p. 73-88, Int. Atom. En. Agency, Vienna, Austria.
Marinsky, J.A., 1976, Ion binding in charged polymers, *Coordination Chem. Rev.*, **19**, 125-171.
Margulis, L., 1981, *Symbiosis in Cell Evolution: Life and its environment on the Early Earth*, W.H. Freeman and Co.
McKeague, J.A., Cheshire, M.V., Andreux, F. and Berthelin, J., 1986, Organo-mineral complexes in relation to pedogenesis, *in* P.M. Huang (ed.) *Interactions of Soil Minerals with Natural Organics and Microbes*, p. 549-592, Soil Sci. Soc. Am., Madison, U.S.A.
Munch, J.C. and Ottow, J.C.G., 1977, Modelluntersuchungen zum Mechanisms der bakterielen Eisenreduktion in hydromorphen Böden, *Z. Pfl. Bodenk.*, **140**, 549-562.
Murr, L.E., 1980, Theory and practice of copper sulphide leaching in dumps and *in situ*, *Minerals Sci. Eng.*, **12**, 121-189.
Ottow, J.C.G., 1971, Iron reduction and gley formation by nitrogen fixing clostridia, *Eocologia*, **6**, 164-175.
Patrick, W.H., 1960, Nitrate reduction rates in a submerged soil as affected by redox potential, *Proc. 7th Internat. Congr. Soil Science*, Madison, vol. II (5), 494-500.
Richards, B.N., 1974, *Introduction to the Soil Ecosystem*, Longman Group Limited, Harlow-Essex.
Robert, M. and Berthelin, J., 1986, Role of biological and biochemical factors in soil mineral weathering, *in* P.M. Huang (ed.) *Interactions of Soil Minerals with Natural Organics and Microbes*, p. 453-495, Soil Sci. Soc. Amer. Madison, U.S.A.
Schnitzer, M., 1978, Humus substances: chemistry and reactions, *in* M. Schnitzer and S.U. Khan (eds.) *Soil, Organic Matter*, p. 1-64, Elsevier.
Senez, J., 1968, *Microbiologie Générale*, Doin, Paris.
Sillen, L.G. and Martell, A.E., 1964, Stability constants of metal-ion complexes, *Chem. Soc. London Spec. Publ.*, **17**.
Sillen, L. G. and Martell, A. E., 1971, Stability constants of metal-ion complexes, *Chem. Soc.*

London Spec. Publ., **25**.

Silverman, M.P., 1979, Biological and organic chemical decomposition of silicates, *in* P.A. Trudinger and D.J. Swaine (eds.) *Biogeochemical cycling of mineral-forming elements*, p. 445-465, Elsevier Scientific Publ.

Silverman, M.P. and Ehrlich, H.L., 1964, Microbial formation and degradation of minerals, *Adv. Appl. Microbiol.*, **6**, 153-206.

Stevenson, F.J. and Ardakani, M.S., 1972, Organic matter reactions involving micronutrients in soils, *in* J.J. Mortvedt, P.M. Girodano and W.L. Lindsay (eds.) *Micronutrients in Agriculture*, p. 79-114, Soil Sci. Soc. Amer., Madison.

Stevenson, F.J. and Fitch, A., 1986, Chemistry of complexation of metal ions with soil solutions organics, *in* P.M. Huang (ed.) *Interactions of Soil Minerals with Natural Organics and Microbes*, p. 29-58, Soil Sci. Soc. Amer., Madison, U.S.A.

Stotzky, G., 1980, Surface interactions between clay minerals and microbes, viruses and soluble organics and the probable importance of these interactions to the ecology of microbes in soils, *in* R.C.W. Berkeley, J.M. Lynch, J. Melling, P.R. Rutter and B. Vincent (eds.) *Microbial Adhesion to Surfaces*, Soc. Chem. Ind. London, Horwood Ltd.

Tardieux-Roche, A., 1966, Contribution à l'étude des interactions entre phosphates naturels et microflore du sol, *Ann. Agron.*, **17**, 403-471, 479-528.

Toutain, F., 1974, Etude écologique de l'humification dans les hêtraies acidiphiles, Doctoral Thesis, University of Nancy, France.

Vedy, J.C. and Bruckert, S., 1982, Soil solution: composition and pedogenic significance, *in* M. Bonneau and B. Souchier (eds.) *Constituents and Properties of Soils*, pp. 184-213, Academic Press, London.

Tsezos, M. and Volesky, B., 1981, Biosorption of uranium and thorium, *Biotechnol. Bioeng.*, **23**, 583-604.

Tsyurupa, I.G., 1964, Some data on complex products of microbial activity and autolysis with soil minerals, *Soviet Soil Sci.*, **3**, 261-265.

Waid, J.S. 1975, Hydroxamic acids in soil systems, *in* E.A. Paul and A.D. McLaren (eds.) *Soil Biochemistry*, Vol. 4, p. 65-101, Marcel Dekker Inc., New York.

Weed, S.B., Davey, C.B. and Cook, M.G., 1969, Weathering of mica by fungi, *Soil Sci. Soc. Am. Proc.*, **33**, 702-706.

Weirich, G. and Schweisfurth, R., 1985, Extraction and culture of microorganisms from rock, *Geomicrobiol. J.*, **4**, 1-20.

Wilson, M.J., Jones, D. and McHardy, W.J., 1981, The weathering of serpentinite by *Leconora atra*, *Lichenologist*, **13**, 167-176.

Yoshida, T., 1975, Microbial metabolism of flooded soils, *in* E.A. Paul and McLaren (eds.) *Soil Biochemistry*, Vol. 3, 83-122, Marcel Dekker Inc., New York,

Zajic, J.E., 1969, *Microbial Biogeochemistry*, Academic Press, New York.

PALEOSOLS AND THE EVOLUTION OF THE ATMOSPHERE: PART I

H. D. Holland and E. A. Zbinden
Department of Earth and Planetary Sciences
Harvard University
Cambridge, Massachusetts 02138
U.S.A.

1. INTRODUCTION

As a subject of inquiry, the chemical evolution of the atmosphere has a long and distinguished history. However, few tools have been available until recently that yield more than qualitative answers to questions regarding changes in the partial pressure of O_2 and CO_2 in the atmosphere during the course of earth history. The study of paleosols (ancient soils) now promises to yield semiquantitative answers to these questions. The composition of soils today reflects not only the composition of the present-day atmosphere, but also the intense chemical effects of the biosphere. The CO_2 content of soil air is strongly enhanced by respiration and by the microbial oxidation of organic matter in soils. In soils which are poorly drained, the O_2 content of soil air is significantly depleted by the same processes. Before the advent of higher land plants the effects of the biosphere must have been much weaker and may have been negligible compared to the effects of inorganic processes. For this reason the chemistry of paleosols is potentially useful for reconstructing the pre-mid-Ordovician evolution of the atmosphere (Holland, 1984, Chapter 7).

Three problems must be solved before useful information can be extracted from compositional data for paleosols formed before the advent of higher land plants during the mid-Ordovician (Gray, 1985): 1) it must be established beyond reasonable doubt that the rock units which are being studied are indeed paleosols (Retallack et al., 1984); 2) modifications of these paleosols by later diagenesis and metamorphism must be recognized, and the original chemistry and mineralogy of paleosols must be reconstructed; and 3) a sound framework must exist for relating the properties of pre-mid-Ordovician paleosols to the composition of the ambient atmosphere.

In this paper we wish to address all three of these questions. Many of the examples are chosen from our recent experience with paleosols. The framework for interpreting the composition of paleosols is built on and extends earlier structures (Holland, 1984, Chapter 7; Kasting et al., 1985). The reader should be cautioned, however, that the entire subject is still in its adolescence, and that both the paleosol data and their interpretation are apt to change significantly during the next few years.

A. Lerman and M. Meybeck (eds.), Physical and Chemical Weathering in Geochemical Cycles, 61–82.

2. CRITERIA FOR IDENTIFYING PALEOSOLS

Paleosols tend to be ephemeral. They are preserved only rarely, because they are usually removed wholly or in part by erosional processes prior to the deposition of younger stratigraphic units. Where they are preserved, they are usually covered either by prograding sediments or by subaerial volcanic flows or ash falls. They can be confused both with sedimentary rocks and with stratigraphically controlled alteration zones. In this section we describe a variety of criteria that can be used to distinguish paleosols from other stratigraphic units. These criteria deal with three aspects of paleosols: their relationship to underlying rocks, their internal structure and composition, and their relationship to overlying rocks (for an earlier approach to this problem, see Yaalon, 1971, pp. 153-158).

2.1. The Relationship of Paleosols to Underlying Rocks

Soils, by definition, develop as a consequence of rock weathering. If the weathering products of homogeneous parent rocks have remained in place or have moved only slightly, their texture, mineralogy, and chemical composition change gradually and usually unidirectionally from the surface downward into the unaltered parent rocks. In any given geographic area, lateral changes in the nature of the bedrock are then reflected in lateral changes in soil type and in the nature of the soil profiles. These properties of soils can be used to distinguish paleosols from most sediments, since sedimentary units do not normally grade gradually into underlying rocks, particularly if these are igneous or high grade metamorphics, and the texture, mineralogy, and chemistry of sedimentary rock units do not usually reflect changes in the nature of the underlying materials.

However, distinguishing paleosols from sediments is apt to be difficult when the parent rocks of paleosols are vertically heterogeneous, and when soils have suffered significant lateral movement. Soils developed on vertically heterogeneous rocks tend not to have simple vertical profiles; soils which have moved long distances are essentially converted into sediments during this process, and lose the characteristic profiles which distinguish them from sediments prior to transport. The vertical changes downward into underlying rock units are therefore useful criteria for distinguishing paleosols from sedimentary rocks in settings where the underlying rocks are homogeneous igneous or metamorphic rocks, and where soil movement has been minimal.

Unfortunately, these criteria cannot be used to distinguish paleosols from stratigraphically controlled alteration zones. Such zones are frequently developed at the contact between impermeable and highly permeable rock units (e.g., Pavich and Obermeier, 1985). If a permeable rock unit contains a mineral assemblage that is quite different from that of a neighboring impermeable unit, the composition of ground water in the permeable unit will tend to be out of equilibrium with the mineral assemblage of the impermeable unit. A reaction zone is then apt to develop, and the texture, mineralogy, and chemistry of the impermeable unit will change gradually away from the contact. When an impermeable unit overlies a permeable unit, the alteration zone develops at the underside of the impermeable unit and hence cannot be confused with a paleosol; however, where a permeable unit overlies an impermeable unit, the alteration zone may be indistinguishable from a paleosol, particularly from the lower portions of a paleosol. Many soils were partially eroded before they were covered by layers of sediment. If such a soil developed on an impermeable rock unit such as crystalline basement, and if the overlying sediment is a sandstone, it may be impossible to distinguish the erosional remnant of the soil from an alteration zone that developed along an originally sharp sandstone-basement contact long after the deposition of the sandstone. It may be possible to make the distinction on the basis of radiometric dating, but even such measurements may be inconclusive, since most, if not all paleosols have been affected by diagenetic processes which post-date the deposition of the overlying stratigraphic units.

2.2. The Internal Structure and Composition of Paleosols

A number of textural, mineralogical, and chemical properties of soils are useful for distinguishing paleosols from sediments and from alteration zones. The presence of land plant remains that are still in their growth position is, of course, the most convincing of all criteria (e.g., Dorf, 1964; Roeschmann, 1971; Retallack, 1986). Plant remains which are not in their growth position are not particularly helpful, since they could have been transported to their present site as constituents of sediments.

Modern soils contain a variety of characteristic textures and frequently have features related to animal burrowing (e.g., Yaalon, 1971; Retallack, 1985). Similar structures have been described in paleosols, but it is often difficult to establish their origin with certainty. Many of these structures can also be produced during transport, sedimentation, and diagenesis, and are therefore of little use for distinguishing paleosols from sediments.

Soils that are of particular interest as indicators of atmospheric composition in the past predate the development of higher land plants and of terrestrial animals. Criteria for identifying paleosols based on the presence of the remains and/or on evidence for the activity of land flora and fauna are therefore useless for identifying pre-mid-Ordovician paleosols. Chemical and mineralogical criteria may, however, be helpful. As suggested by the chemistry of Precambrian soil profiles (Holland, 1984, Ch. 7), soil zonation in Precambrian soils may well have been similar to that in present day soils. Vertical changes in the composition and texture of potential paleosol horizons can therefore be useful indicators of their origin and as a means of distinguishing paleosols from sediments. They may, however, be inadequate for distinguishing paleosols from stratigraphically controlled alteration zones.

Many textural features of stratigraphic units can be used to establish their sedimentary origin. The presence of crossbedding, graded bedding, and ripple marks virtually rules out the possibility of a direct weathering origin for units overlying igneous rocks, even if chemical indicators seem favorable for such an interpretation. On the other hand, the complete absence of such sedimentary features, and the complete absence of admixtures of material from the overlying units except as infillings (see below) are encouraging signs of a direct weathering origin from the underlying igneous rocks. Paleosols developed on sedimentary rocks may, of course, contain preserved sedimentary features.

2.3. The Relationship of Paleosols to Overlying Rocks

Soils develop in contact with the atmosphere, they are soft, and they obviously predate the deposition of younger units. Overlying sedimentary units can therefore be expected to contain rip-up clasts of paleosols. Soils developed in areas of alternating wet and dry weather are apt to develop desiccation cracks, and these may be preserved at the contact between paleosols and overlying stratigraphic units. If very deep desiccation cracks develop in a soil, the overlying material, which may include algal mats, is apt to fill these cracks, and may penetrate into a soil to a depth of several meters. Such sandstone dikes tend to become intensely deformed during part of this unit and sculpted the remnants into hills, which were later covered by more recent acid volcanics. Toward lower elevations in this area, progressively lower parts of the paleosol are in contact with the overlying acid lavas. As in the case of the Sturgeon River paleosol, this geometry rules out the possibility that the unit is simply an alteration zone developed between the upper and the lower acid volcanics. The geometry does not, however, rule out the possibility that the unit is a sedimentary layer, perhaps a lake sediment, which was partially eroded prior to emplacement of the younger acid volcanics. Other criteria, such as those discussed earlier in this section, will have to be applied to decide whether the erosional remnant is a sediment or a paleosol. The chemical profiles of the unit between the Hekpoort basalt and the Dwaal Heuvel Formation in the Transvaal (Button, 1979), and of the unit between Keweenawan basalt and the Jacobsville sandstone in Michigan (Holland et al., in preparation) are consistent with the interpretation that these units as paleosols. The presently available chemical data for the Derdepoort unit are equivocal.

3. THE EFFECT OF DIAGENESIS AND METAMORPHISM ON PALEOSOLS

Although geologically young paleosols tend to look very much like modern soils (e.g., Yaalon, 1971), Precambrian paleosols rarely, if ever, do. Diagenesis and, not infrequently, metamorphism have converted these ancient soils into rocks; these processes have altered their physical properties and their chemical composition. Inferences regarding the composition of the atmosphere at the time when such paleosols were soils can be made only if those aspects of paleosol chemistry on which the inferences are based were not affected during diagenesis and metamorphism, or if the changes due to these processes can be determined, so that the chemistry of the original soil can be reconstructed.

3.1. The Evidence for Metasomatic Changes in Paleosols

Many paleosols contain more K_2O than their parent rocks. In some instances the K_2O content of paleosols is so much greater, that potassium must have been introduced during or after soil formation. This is well illustrated in Button's (1979) data for the Hekpoort paleosol at the Waterval Onder locality. These data are reproduced in Table I and in Figures 2 and 3. The K_2O content of the parent basalt is ~1%. The K_2O content of the upper part of the paleosol is between 8% and 9%. Some, but not all of this increase could be due to weathering. If none of the K_2O in the parent basalt had been lost during soil formation, the K_2O content of the original soil would have increased due to the loss of other oxides. However, the Al_2O_3 and TiO_2 content of the upper part of the paleosol is only about twice that of the Hekpoort basalt. Hence, if little or no Al_2O_3 and TiO_2 were lost during weathering, the K_2O content of this portion of the paleosol could only have risen to ~2% during weathering, even if no K_2O had been lost during this process. Most of the K_2O that is now present in the upper parts of the Hekpoort paleosol must therefore have an extraneous origin. Numerous analyses of the Hekpoort basalt and of the overlying paleosol over a strike distance of several hundred kilometers indicate that the outcrop near Waterval Onder is not exceptional, and that large scale, widespread potassium metasomatism has modified the composition of much, if not all of the original blanket of Hekpoort soil. Other Precambrian paleosols seem to have been similarly affected. Paleosols which have been spared potassium metasomatism appear to be the exception rather than the rule (e.g., Holland, 1984, Chapter 7).

compaction of the soil by the weight of the overlying stratigraphic units.

These features have been observed at and near the upper contact of the paleosol developed at the top of the 2.2 Ga Hekpoort basalt in South Africa. Button (1979) has described sandstone dikes from the outcrop of this paleosol near Waterval Onder in the Eastern Transvaal. At the same locality we have found what we believe to be rip-up clasts of this paleosol in the lower 30 cm of the overlying Dwaal Heuvel formation. We have also found abundant, well-defined desiccation cracks at the upper surface of the Hekpoort paleosol in a quarry near Koster in the Western Transvaal. These features rule out the possibility that the unit which has been interpreted as a paleosol might be an alteration zone developed at the interface between the Dwaal Heuvel formation and the underlying Hekpoort basalt.

The geometry of the upper contact of paleosols can yield similarly strong diagnostic evidence. The dissection of soil horizons prior to burial by younger stratigraphic units can leave erosional soil remnants, as well as areas where the soil has been completely eroded away. The contact between soil remnants and younger stratigraphic units is therefore apt to cut across soil horizons. Some of these features are shown in Figure 1, which indicates in diagrammatic form the major features of the remnant of a paleosol developed on Keweenawan basalt near the falls of the Sturgeon River in the Upper Peninsula of Michigan (Holland et al., in preparation). The small hill consists of a paleosol, which is enveloped by Jacobsville sandstone. Along the flanks of the hill, balls of paleosol can be seen in various stages of detachment, and remnants of paleosol "float" in Jacobsville sandstone at distances up to about 20 m from the paleosol hill.

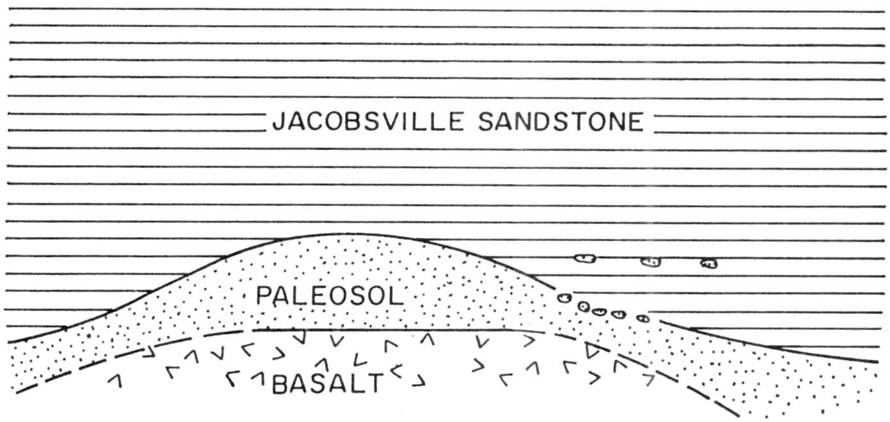

Figure 1. Diagrammatic sketch of the paleosol remnant developed on Keweenawan basalt near the falls of the Sturgeon River in the Upper Peninsula of Michigan. The paleosol hill has a horizontal dimension of about 20 m and is about 3 m in height.

A similar geometry seems to prevail in a much larger erosional remnant near Derdepoort, close to the border between Botswana and the Republic of South Africa. The area has been mapped by Tyler (1979), and the erosional remnant was mapped by us during the summer of 1985. Early Proterozoic (Crampton, 1972) acid volcanics are overlain by a unit which may be a very thick ($>$ 40 m) paleosol. Erosion removed

Table I. Chemical composition of samples from the Hekpoort paleosol (Button, 1979).

Sample No.	*1*	*2*	*3*	*4*	*5*	*6*	*7*	*8*	*9*	*10*	*11*	*12*	*13*
SiO_2	44.2	45.7	46.1	46.6	47.3	47.9	48.4	46.4	53.4	45.9	59.4	67.3	51.8
TiO_2	1.79	1.64	1.74	1.80	1.66	1.70	1.65	1.67	0.76	1.49	1.15	0.74	1.06
Al_2O_3	33.5	34.8	34.8	35.1	34.9	33.6	33.0	32.5	13.9	29.7	22.4	14.7	21.3
Fe_2O_3	3.66	2.60	1.77	0.49	0.85	1.50	2.12	3.89	8.80	4.39	3.65	3.45	10.80
FeO	2.30	0.25	0.20	0.35	0.30	0.55	0.40	1.00	16.20	5.04	2.75	5.87	3.60
MnO	0.117	0.05	0.05	<0.01	<0.01	0.01	0.02	0.07	0.24	0.13	0.12	0.15	0.20
MgO	0.5	0.5	0.3	0.3	0.3	0.5	0.6	0.6	1.8	0.9	0.8	1.2	1.4
CaO	0.13	0.09	0.09	0.02	0.11	0.08	0.04	0.05	0.03	0.13	0.17	0.06	0.37
Na_2O	0.5	0.6	0.6	0.6	0.5	0.6	0.6	0.5	0.1	0.4	0.4	0.2	0.4
K_2O	8.96	9.42	9.58	9.92	9.76	9.22	8.95	9.00	0.41	7.53	5.86	3.04	4.85
H_2O^+	5.2	4.8	4.8	4.7	4.8	4.6	5.3	4.99	6.45	5.28	3.88	3.99	5.20
H_2O^-	0.4	0.4	0.45	0.3	0.4	0.3	0.2	0.4	0.6	0.3	0.3	0.3	0.4
P_2O_5	0.11	0.05	0.04	0.02	0.07	0.05	0.04	0.05	0.02	0.06	0.14	0.06	0.06
CO_2	<0.1	<0.1	<0.1	0.1	0.1	<0.1	0.2	<0.1	0.1	<0.1	<0.1	0.2	<0.1
Total	100.37	100.90	100.52	100.30	101.05	100.61	101.52	101.12	102.81	101.25	101.02	101.26	101.44
Zr*	290	340	340	320	290	300	290	270	120	250	210	130	175
Cr*	250	200	200	160	155	145	135	180	82	135	125	110	92
Rb*	330	340	330	360	360	350	350	370	49	340	280	170	260
Sr*	76	67	60	57	56	54	56	71	7	62	58	38	75
Ba*	1160	780	590	460	340	270	220	250	98	220	210	180	460
Ni*	137	104	95	65	89	124	147	153	226	132	190	264	258
Cu*	254	28	27	14	22	86	117	74	136	7	102	102	186
Zn*	11	<5	<5	<5	<5	<5	<5	<5	72	22	27	45	77

Note: See Figure 2 for location of samples in profile.
Analysts: Bergström and Bakker, Johannesburg.
*ppm.

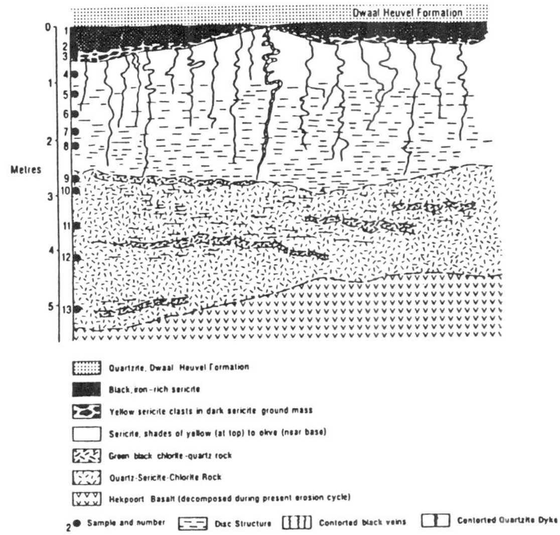

Figure 2. Stratigraphic section of the Hekpoort paleosol (Button, 1979).

The source of the excess potassium in the Hekpoort paleosol has not been determined, but there are only two likely sources, and one of these is much more likely than the other. One potential source is the water—perhaps seawater—from which the overlying Dwaal Heuvel sediments were deposited; the second potential source are later groundwaters which passed through the Dwaal Heuvel formation and which could have reacted with the paleosol at the interface with the underlying, reasonably impermeable Hekpoort basalt. We prefer the second source of potassium to the first, because the reaction of present-day seawater with marine clays is very slow indeed. The conversion of gibbsite and kaolinite to illite and/or muscovite in the oceans today is everywhere quite incomplete (e.g., Holland, 1978, Chapter 5), whereas the upper part of the Hekpoort paleosol today consists largely of muscovite. Groundwaters in the Dwaal Heuvel sandstones were almost certainly in equilibrium with quartz, muscovite, and/or potassium feldspar, and had ample time to react with clay minerals in the Hekpoort soil; these reactions probably occurred at temperatures above 100°C, as indicated by the thickness of the stratigraphic section above the Hekpoort basalt during the past 2 billion years.

It is hard to imagine that potassium was the only element which was added to the Hekpoort paleosol during metasomatism. SiO_2, Rb, and Cs were almost certainly involved as well, but Na, Ca, and Mg do not seem to have been added to the paleosol, since the vertical variation in the concentration of these elements is still typical of that in soil profiles developed on basalt today (e.g., Craig and Loughnan, 1964).

Fortunately, alkali metal and silica metasomatism does not interfere with the use of paleosols as indicators of atmospheric evolution. The critical elements in this pursuit are those that can undergo changes in oxidation state during weathering, especially those elements whose behavior in soils is strongly influenced by their valence state. There are several such elements. The list includes Fe, Mn, V, Cr, U, Eu, Ce, Cu, S, Se, Te, Sb, As, and Mo. Fe and Mn are elements of particular interest,

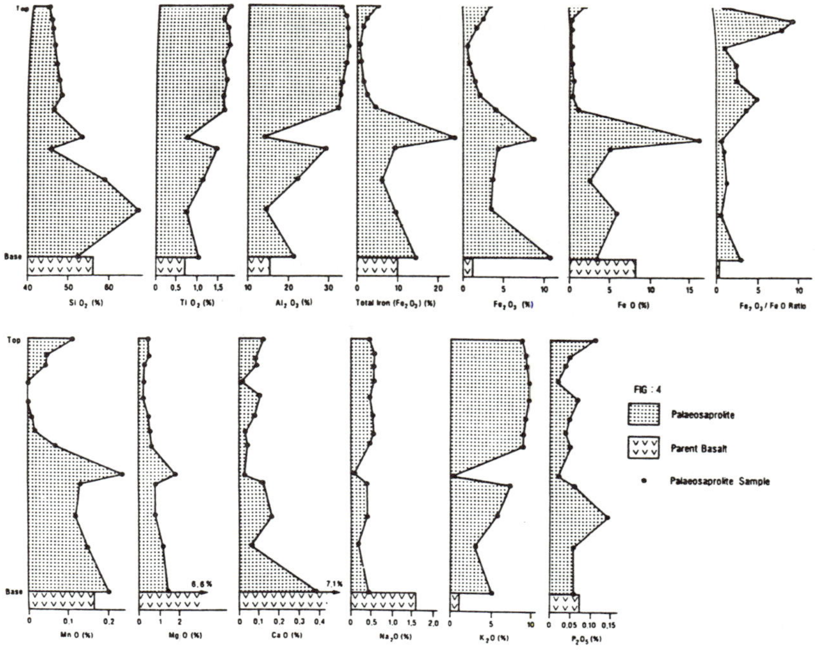

Figure 3. The distribution of the major elements in the Hekpoort paleosol at Waterval Onder, Transvaal, South Africa (Button, 1979).

iron because it is the major oxygen sink during the weathering of most igneous rocks, manganese because it is a reasonably abundant element, and because its behavior during present-day soil formation is fairly well understood. We therefore need to enquire into the behavior of these two elements during the metasomatism that has affected the chemical composition of so many paleosols.

3.2. Iron and Manganese Metasomatism in Paleosols

Iron and manganese can be transported both in groundwaters and in hydrothermal solutions. Concentrations of both elements in the range of several tens to hundreds of ppm have been observed in hydrothermal solutions at temperatures above 300°C (Von Damm et al., 1983). Even at 25°C the concentration of these elements in groundwaters can be in the ppm range. Very substantial quantities of both elements could therefore be added or subtracted during the passage of the large volumes of groundwater that are probably needed to add several percent of potassium to paleosols.

Several lines of evidence can be used to determine whether a particular paleosol has gained or lost significant quantities of either element during diagenesis and/or metamorphism. Both elements tend to be nearly immobile in oxidizing solutions, since the oxides and hydroxides of Fe^{+3} and Mn^{+4} are only very sparingly soluble. Both elements tend to be mobile in reducing solutions, although the mobility of iron is severely limited by the solubility of pyrite in reducing solutions that contain high concentrations of HS^-. If it can be established that the solutions which have had access to a particular paleosol have been in contact with Fe_2O_3 or with other, equally oxidizing compounds, then neither iron nor manganese is apt to have moved significantly since the end of weathering. The mineralogy of the overlying units suggests that this was probably the case both for the Hekpoort paleosol and for the Sturgeon River paleosol.

The distribution of iron and manganese minerals within a given paleosol may also help to settle the question whether post-weathering transport of iron and manganese was significant. If these elements occur, *inter alia,* in veinlets that were crumpled during the vertical compression of the paleosol due to the weight of the overlying stratigraphic units, if it can also be shown that these veinlets were formed during the last movement of iron and manganese in the paleosol, and if the veinlets predate the period of potassium metamorphism, then a strong case can be made for the proposition that the distribution of iron and manganese in the paleosol was not significantly affected by post-weathering processes.

Finally, arguments can be made on the basis of consistency. If it turns out that the behavior of iron and manganese in a large number of paleosols from particular parts of earth history can be explained solely on the basis of the composition of their parent rocks, then it becomes unlikely that later additions or subtractions of iron and manganese were of major import. By the same token, a paleosol in which the behavior of these elements does not conform to that in similar contemporaries, may reasonably be suspected of having suffered significant post-weathering Fe and Mn metasomatism.

4. THE RELATIONSHIP BETWEEN THE COMPOSITION OF PALEOSOLS AND THAT OF THE AMBIENT ATMOSPHERE

4.1. Introduction

The major weathering reactions in soils involve acid attack, oxidation reactions, and hydration reactions. Quantitatively, carbonic acid has always been the dominant weathering acid (Holland, 1984, Chapter 5), although organic acids may play a fairly significant role in weathering today (e.g., Eckhardt, 1985). The major oxidant today is molecular O_2. Oxidants such as H_2O_2 are produced photochemically in the atmosphere, but their concentration in rainwater is usually less than 10% of the concentration of molecular O_2. This may not have been the case during much of pre-mid-Ordovician. The production rate of H_2O_2 and of other oxidants in the atmosphere is only a weak function of P_{O_2}; hence, the ratio of the concentration of photochemical oxidants to dissolved molecular O_2 in rainwater increases rapidly with decreasing P_{O_2}, and probably begins to exceed unity at values of P_{O_2} between 0.2×10^{-2} and 0.2×10^{-3} atm (Kasting et al., 1985).

Many elements change oxidation state in soils today, among them iron and manganese. These elements are normally precipitated within soils as one or more oxides or hydroxides (e.g., Hendricks and Whittig, 1968; Holland, 1978, Ch. 2). Some downward movement of Fe^{+3} does occur in podzolized soils, presumably due to chelation of Fe^{+3} with organic compounds (e.g., Birkeland, 1974, pp. 105-106; Stumm et al., 1985). In oxygen-deficient soils, iron and manganese are not oxidized after they are released from their original host minerals, and tend to be removed from soils as dissolved constituents of groundwater (e.g., Willett, 1983). In some Precambrian rock units which are inferred to have been soils, iron and manganese have been retained; in others these elements have either been lost or transferred from upper to lower parts of the soils. The Hekpoort paleosol is probably the best example of a Precambrian paleosol in which iron and manganese were lost from the upper soil horizons.

The degree of iron and manganese loss in the Precambrian paleosols which have been studied to date is related to the ratio, R, of the oxidant demand, D_{Ox}, to the acid demand, D_{Ac}, for complete weathering of the parent rock (Holland, 1984, Chapter 7). This is not surprising. Consider an igneous rock which requires D_{Ac} moles of H^+ ions to release all of its contained Na_2O, K_2O, CaO, and MgO via the reactions

$$(Na_2O) + 2H^+ \rightarrow 2Na^+ + H_2O \tag{1}$$

$$(K_2O) + 2H^+ \rightarrow 2K^+ + H_2O \tag{2}$$

$$(CaO) + 2H^+ \rightarrow Ca^{+2} + H_2O \tag{3}$$

$$(MgO) + 2H^+ \rightarrow Mg^{+2} + H_2O \tag{4}$$

In that case

$$D_{Ac} = 2\left[M_{Na_2O} + M_{K_2O} + M_{CaO} + M_{MgO}\right] \tag{5}$$

where M is the number of moles of a particular component per kilogram of the unweathered parent rock.

The value of D_{Ox} for most igneous rocks is determined in large part by their FeO content (Holland, 1984, Chapter 7). The oxidation of this component in silicates and oxides via the reaction

$$4(FeO) + O_2 \rightarrow 2Fe_2O_3 \tag{6}$$

indicates that

$$D_{Ox} \approx \frac{1}{4}[M_{FeO}] \tag{7}$$

If, now, the ratio of the rate of oxidant supply to the rate of acid supply during weathering exceeds R, where

$$R = \frac{D_{Ox}}{D_{Ac}} = \frac{M_{FeO}}{8\left[M_{Na_2O} + M_{K_2O} + M_{CaO} + M_{MgO}\right]} \tag{8}$$

then all of the Fe^{+2} released during acid attack can be oxidized and hence reprecipitated and retained in soils developed on this parent rock. If the ratio of the rate of oxidant supply to the rate of acid supply is less than R, Fe^{+2} released during acid attack can only be partially oxidized and retained in the soil; the remainder is apt to be washed out of the upper soil horizons, and is either lost completely from the soil or is reprecipitated in the lower soil horizons, where the pH is sufficiently high, so that groundwaters there are saturated with respect to one or more ferrous silicate minerals. The value of R for igneous rocks ranges from about 0.01 to about 0.08.

Some Fe^{+2} may be washed out, particularly in acid soils, even if a small amount of free oxygen is present in soil water. At 25°C in near neutral groundwaters equilibrated with the present day atmosphere, the half life of Fe^{+2} with respect to oxidation to Fe^{+3} is ~10 minutes (Morel, 1983). The rate of Fe^{+2} oxidation is therefore sufficiently rapid under such conditions, that most of the Fe^{+2} released during acid attack is oxidized and precipitated almost in situ. In groundwaters where the concentration of O_2 is three orders of magnitude less than in rainwater today, the half life of

Fe^{+2} is ca. three orders of magnitude greater, i.e., on the order of 7 days. This is comparable to the residence time of water in the upper parts of many soils today. A significant fraction of the Fe^{+2} released during acid attack in such low-O_2 soils may therefore be flushed out of the upper soil zones during rainstorms and may be oxidized to Fe^{+3} and precipitated either as $Fe(OH)_3$ or as (Fe^{+2},Fe^{+3})-silicates in the lower parts of soil horizons. The presence of relatively highly oxidized chlorite in the lower parts of the Hekpoort paleosol could be due to this process. The oxidation of Mn^{+2} follows the same pattern as the oxidation of Fe^{+2}, but is considerably slower (Pinto and Holland (1987) give a more detailed discussion of this matter). The relative mobility of iron and manganese in Precambrian soils may therefore turn out to be a useful indicator of the concentration of dissolved oxidants in the associated soil and groundwaters.

4.2. The Rate of Oxidant Supply

The rate of supply of oxidants to Precambrian soils was almost certainly controlled in large part by inorganic processes. Algal mats may have covered the continents (Campbell, 1979), and an associated cast of bacteria may have been present within soils, but it seems likely that such algal mats were as thin as they are now, except where they were able to accumulate in topographically low areas. Since the amount of biomass per unit area was probably small, the population of soil bacteria was probably also small compared to that in present day soils. Biologic processes could well have been important in the destruction of H_2O_2 and of other products of atmospheric photochemistry, but their effect on O_2 use in soils was probably negligible (Pinto and Holland, 1987).

If these assumptions are valid, oxidants were introduced into pre-mid-Ordovician soils by three mechanisms:

(1) the addition of O_2 dissolved in rainwater,
(2) the addition of photochemically produced oxidants dissolved in rainwater, and
(3) the diffusion of O_2 into soils.

Pumping due to changes in atmospheric pressure is not important today (Joffe and Revut, 1966; Bonneau and Souchier, 1982; Buyanovsky and Wagner, 1983), and was probably similarly unimportant in pre-mid-Ordovician times. The rate of diffusion of photochemical products into soils must have been very minor compared to the flux of these products in rainwater, since they are washed out of the atmosphere very efficiently (e.g., Kasting et al., 1985).

The rate of supply to the soil of O_2 dissolved in rainwater, X_{O2}, is given by the expression

$$X_{O_2} = \frac{(m_{O_2})_{rain} \cdot r}{10^3} = \frac{B_{O_2} P^{o}_{O_2} r}{10^3} \tag{9}$$

where $(m_{O_2})_{rain}$ = concentration of O_2 in rainwater (mol/l), r = rate of rainfall that penetrates into soil (cm/yr), B_{O2} = Henry's Law constant for O_2 (mol/atm liter), P^{o}_{O2} = partial pressure of O_2 in the atmosphere (atm). The net rate of supply of photochemically produced oxidants is given by the expression

$$Y_{Ox} = \frac{(m_{POx})\, r}{10^3} \tag{10}$$

where m_{POx} is the net concentration of photochemically produced oxidants in rain water in moles per liter.

The rate of O_2 supply by diffusion from the atmosphere into soils is a complicated function of variations in the diffusion constant of O_2 with depth in soils, changes in barometric pressure, the episodicity of rainfall, and the variation with depth and in time of the rate of O_2 use by oxidation reactions within the soil (Pinto and Holland, 1987). In the simplest, but somewhat unrealistic case the diffusion constant and the O_2 gradient in the soil are constant, and the rate of O_2 supply is

$$Z_{Ox} = \frac{D^{o}_{O_2}\,\alpha}{2.2\times10^{4}}\;\frac{[P^{o}_{O_2}-P^{L}_{O_2}]}{L}\times 3.1\times10^{7}\ \mathrm{mol/cm^2yr} \qquad (11)$$

where $D^{o}_{O_2}$ = diffusion constant of O_2 in air (cm^2/sec), $\alpha \equiv D^{S}_{O_2}/D^{o}_{O_2}$ = ratio of the diffusion constant of O_2 in soil to that in air (this includes a term for porosity), $P^{L}_{O_2} = P_{O_2}$ at the groundwater table (atm), L = depth to the groundwater table (cm^3). The factor 2.2×10^4 in the number of cc/mol of gas at S.T.P., 3.1×10^7 is the number of seconds per year. The contribution of O_2 by diffusion into soil layers beyond the groundwater table is probably negligible, since the diffusion rate of gases in water is several orders of magnitude smaller than that in most soil air (see below).

Let us consider the case in which P_{O_2} goes to zero at the water table, i.e., where the supply of O_2 is exhausted just at the water table. In such a situation O_2 is available to oxidize Fe^{+2}, Mn^{+2}, and other oxidizable ions throughout the unsaturated zone, and both elements are oxidized and retained there, provided the kinetics of oxidation and precipitation are sufficiently rapid. This is also true for situations where $P^{L}_{O_2} > 0$, but it is not true when P_{O_2} becomes 0 above the water table. The condition that P_{O_2} approaches 0 at the water table therefore represents the boundary between soils in which iron and manganese can be quantitatively retained above the water table and those in which these elements are apt to be lost during weathering. Since most weathering reactions in soils take place above the water table (see below), the condition $P^{L}_{O_2} = 0$, $(P_{O_2})_{l<L} > 0$ probably represents the boundary between paleosols that have retained and those that have lost iron and manganese. When this condition obtains, the total oxidant supply to the soil is

$$X_{Ox} + Y_{Ox} + Z_{Ox} = \left[\frac{B_{O_2}\,r}{10^3} + \frac{D^{o}_{O_2}\,\alpha}{L}\times 1.4\times10^{3}\right] P^{o}_{O_2} + \frac{(m_{POx})\,r}{10^3} \qquad (12)$$

In most soils the value of the first term within the brackets is quite small compared to the second term. The Henry's Law constant of O_2 in water has a value of $(1.3 \pm 0.3)\times10^{-3}$ moles/liter atm between 0° and 50°C. The average rainfall on earth today is 100 cm/year (e.g., Holland, 1978, Chapter 3). The maximum rainfall in any 10° latitude belt is close to 200 cm/yr. It is difficult to estimate the maximum rainfall in a 10° latitude belt in the Precambrian; however, a strong upper limit on the total amount of evaporation and hence rainfall is set by the available quantity of solar energy. Today, approximately 24% of the incoming solar radiation is used to evaporate water and to drive the hydrologic cycle. It is hard to imagine that this percentage could ever have been more than double the present value. If it were doubled, if the rainfall distribution was comparable to that of the present day, and if 50% of the rainfall penetrated into Precambrian soils, the maximum value of r for a 10° belt of latitude was approximately 200 cm/yr. Locally, rates of rainwater infiltration could, of course, have been higher, but it seems likely that for most soils

$$\frac{B_{O_2}\,r}{10^3} \approx (1.3\times10^{-3})^{\times1.3}_{\div1.3}\,\frac{(50)^{\times4}_{\div4}}{10^3} = (6.5\times10^{-5})^{\times5}_{\div5}\ \mathrm{mol\ O_2/cm^2yr} \qquad (13)$$

The value of the second term within the brackets in equation 12 is potentially much more variable than the first. The diffusion constant of oxygen in air between 0° and 50°C is (0.16±0.03) cm^2/sec. The value of α is a strong function of the porosity and water content of soils. In dry, porous soils α is about 0.6 (Nielson et al., 1984; Rogers et al., 1982; Silker and Kalkwarf, 1983). In wet, non-porous soils the value of α is about 10^{-4}, and the rate of diffusion of gases in such soils is comparable to their rate of diffusion in water. This is illustrated by the data in Figure 4, in which the diffusion coefficient of radon is plotted as a function of the moisture saturation of soils. In dry soils the diffusion coefficient is only slightly less than that in air. In water-saturated soils the diffusion coefficient of radon is close to that of the gas in aqueous solutions. In most soils, the value of α lies within the limits $0.1(\div 10, \times 10)$ (e.g., Taylor, 1949; Lai et al., 1976; Richter and Grossgebauer, 1978; De Jong et al., 1983).

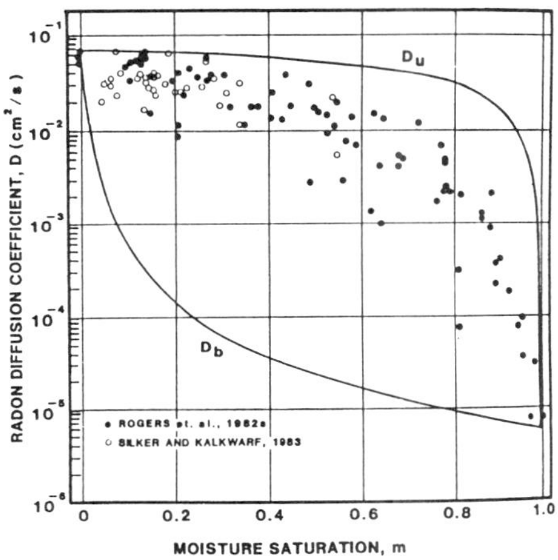

Figure 4. The diffusion coefficient of radon in soils (Nielson et al., 1984). The measured diffusion coefficients are from Rogers et al. (1982) and from Silker and Kalkwarf (1983). The solid curves represent calculated values of D for water-blocked pores (D_b) and unblocked pores (D_u).

The depth to water table in soils is highly variable, and was probably similarly variable before the development of a cover of higher land plants. It is likely that the depth to water table in many areas ranged from less than a meter to several tens of meters. As an example, the present thickness of the Hekpoort paleosol is typically about 5 m. If compression reduced the original thickness by a factor of 2 (for porosity data of modern soils, see Buchman and Brady, 1968; Lal, 1981), and if most weathering occurred then, as now, above the water table, the depth to water table in the Hekpoort soil was about 10 m. For the purposes of our rough calculation we will take L to have been $1000(\div 3, \times 3)$ cm. In soils which fall within these limits, the second term in the brackets of equation 12 therefore has a value of approximately

$$\frac{(0.16)^{\times 1.2}_{\div 1.2}\,(0.1)^{\times 10}_{\div 10}}{1000^{\times 3}_{\div 3}} \times 1.4\times 10^{3} \approx (2.2\times 10^{-2})^{\times 36}_{\div 36}$$

The most likely value of this term is ~300 times greater than the first term within the brackets in equation 12. Even the upper limit of the first term is smaller than the lower limit of the second term. It is very likely, therefore, that in the O_2 budget of most pre-mid-Ordovician soils diffusive transport of O_2 was quantitatively much more important than the delivery of O_2 dissolved in rainwater.

The magnitude of the last term in equation 12 depends on photochemical processes in the upper atmosphere; these in turn depend on the CO, CH_4 and NO_x content of the upper atmosphere, and probably on the atmospheric chemistry of sulfur as well (Kasting et al., 1985). The best estimate of the net concentration of oxidants in rainwater falling through an atmosphere containing between 0.2 and 0.2×10^{-4} atm P_{O2} is $\sim(1\times10^{-6})(\div3, \times3)$ mol/l. If total rainfall during pre-mid-Ordovician times was significantly greater or significantly less than at present, the mean concentration of photochemical oxidants in rainwater would have been proportionately smaller or greater, since the total production rate of oxidants and reductants is only a weak function of rainfall; the mean concentration of these products in rainwater at any given time is simply the total production rate divided by the annual global rainfall.

The total rate of delivery of oxidants to pre-mid-Ordovician soils was therefore approximately equal to

$$X_{Ox} + Y_{Ox} + Z_{Ox} \approx (2.2\times10^{-2})^{\times 36}_{\div 36}\, P^{o}_{O_2} + (5\times10^{-8})^{\times 3}_{\div 3}\ \mathrm{mol/cm^2yr} \qquad (14)$$

For world average soils, the flux of photochemically produced oxidants was significant only at values of P_{O2} below $\sim10^{-5}$ atm; above this pressure the oxidant flux was controlled almost entirely by the diffusion of O_2.

4.3. The Rate of Acid Supply

The supply of weathering acids to pre-mid-Ordovician soils can be expressed in terms identical to those for the supply of oxidants. However, the rate of production of weathering acids by photochemical reactions was very small compared to the supply of CO_2 dissolved in rainwater and the diffusion of CO_2 from the atmosphere into soils; hence the acid flux can be set approximately equal to

$$X_{Ac} + Z_{Ac} \approx \frac{B_{CO_2}\, P^{o}_{CO_2}\, r}{10^3} + \frac{D^{o}_{CO_2}\, \alpha}{2.2\times10^4}\ \frac{(P^{o}_{CO_2} - P^{L}_{CO_2})}{L} \times 3.1\times10^7 \qquad (15)$$

where the various symbols have the same meaning as their counterparts in equation 12. There are considerable uncertainties in the value of P^{L}_{CO2}, the CO_2 pressure in soil air at the water table. As shown by the data in Figure 5, the concentration of H_2CO_3 in groundwaters today is generally between 3×10^{-5} and 1×10^{-3} mol/l. This corresponds to a CO_2 pressure between $\sim1\times10^{-3}$ and 2×10^{-2} atm in soil air which has equilibrated with these groundwaters. Clearly, a significant fraction of the CO_2 that has dissolved in soil water is retained as carbonic acid in groundwaters.

We will attempt to set some rather broad limits on P_{CO2} in the pre-mid-Ordovician atmosphere by a two-step approximation. In the first step we will assume that P^{L}_{CO2} is zero; in the second we will estimate the effect of non-zero values of P^{L}_{CO2} on the probable range of values of atmospheric P_{CO2}

If $P^{L}_{CO2} << P^{o}_{CO2}$, equation 15 reduces to the form

$$X_{CO_2} + Z_{CO_2} \approx \left[\frac{B_{CO_2}\, r}{10^3} + \frac{1.4\times10^3\, D^{o}_{CO_2}\, \alpha}{L}\right] P^{o}_{CO_2} \qquad (16)$$

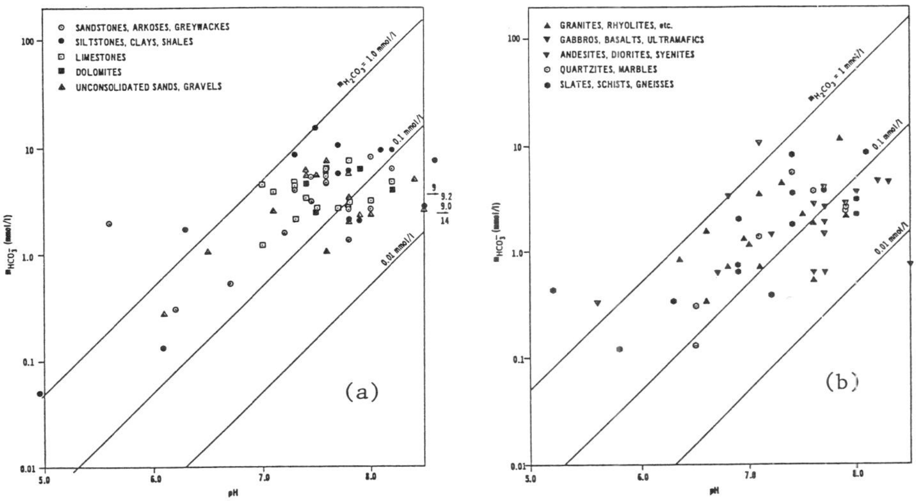

Figure 5. The pH and the concentration of HCO_3^- and H_2CO_3 in modern groundwaters; data from White et al. (1963). (a) Groundwaters in sedimentary rocks; (b) Groundwaters in igneous and metamorphic rocks.

The value of B_{CO_2} between 0° and 50° is $(3.5\times10^{-2})(\div2, \times2)$ mol/atm liter, and the value of $D^{\circ}_{CO_2}$ is $0.17(\div1.2, \times1.2)$ cm^2/sec. An approximate value of the total flux of CO_2 that is used in weathering can be obtained if we use the observation that the degree of CO_2 titration of surface rocks has remained nearly constant since the beginning of the Proterozoic (Holland, 1984, Chapter 5). On average, the rate of erosion during the Proterozoic was probably proportional to the rate of tectonism, and this in turn was probably proportional to the rate of heat generation in the earth. If these assumptions are valid, then the rate of acid use in chemical weathering has probably been within a factor of two of the present value during the past 2.5 b.y. (e.g., Turcotte, 1980). The rate of acid use at present can be estimated in several ways (e.g., Holland, 1978, Chapter 6). An upper limit is set by the total flux of HCO_3^- to the oceans. The HCO_3^- concentration of average river water is close to 1.0 x 10^{-3} mol/l (Meybeck, 1983), and the total annual flux of river water is close to 4×10^{16} l/yr (Holland, 1978). The annual river flux of HCO_3^- is therefore $\sim4\times10^{13}$ mol/yr and the mean HCO_3^- flux per cm^2 of the continents is approximately

$$\frac{4\times10^{13}\ \text{mol/yr}}{1.5\times10^{18}\ \text{cm}^2} = 3\times10^{-5}\ \text{mol/cm}^2\text{yr}$$

This figure is somewhat of an upper limit for the rate of CO_2 use in weathering today, because several tens of percent of the HCO_3^- in rivers are supplied by $CO_3^=$ released from carbonate rocks, and not by the neutralization of CO_2. If we now take the best value for r to be $50(\div2, \times2)$ cm/yr, the first term in brackets in equation 16 has the value

$$\frac{B_{CO_2}\, r}{10^3} \approx \frac{(3.5\times10^{-2})^{\times 2}_{\div 2}\,(50)^{\times 2}_{\div 2}}{10^3} \tag{17}$$
$$= (1.7\times10^{-3})^{\times 4}_{\div 4}\ \text{mol/cm}^2\text{atm yr}$$

The uncertainties in the second term within brackets in equation 16 are even larger, since neither the average value of α nor that of L are at all well known. If we simply choose half of the range of the values used earlier in the calculation of oxygen supply, the second term becomes

$$\frac{(1.4\times10^3)\, D^o_{CO_2}\, \alpha}{L} \approx 1.4\times10^3\ \frac{(0.17)^{\times 1.2}_{\div 1.2}\,(0.1)^{\times 5}_{\div 5}}{(10^3)^{\times 1.5}_{\div 1.5}} \tag{18}$$
$$= (2.4\times10^{-2})^{\times 9}_{\div 9}\ \text{mol/cm}^2\text{atm yr}$$

The most probable value of the second term is more than an order of magnitude greater than the most probable value of the first term within the brackets. It is likely, therefore, that more CO_2 was transferred into soils via diffusion than as a dissolved constituent of rainwater. However, the uncertainty in both rate terms is so large, that CO_2 dissolved in rainwater could have been an important or even the major contributor of weathering acids to pre-mid-Ordovician soils.

If we substitute both terms into equation 16, we obtain

$$P^o_{CO_2} \approx \left[\frac{3\times10^{-5}}{(1.7\times10^{-3})^{\times 4}_{\div 4} + (2.4\times10^{-2})^{\times 9}_{\div 9}}\right] \approx (1.2\times10^{-3})^{\times 8}_{\div 10}\ \text{atm} \tag{19}$$

The uncertainties in the calculated value of $P^o_{CO_2}$ are large, but the result is nevertheless interesting. The most likely value of $P^o_{CO_2}$ is 3.5 times the present value of atmospheric P_{CO_2}. The maximum value is comparable to that in many soils today. The minimum value is ~40% of present atmospheric P_{CO_2}.

The calculated maximum value of 1.0×10^{-2} atm seems to be a true maximum for the lower Paleozoic atmosphere. Today the CO_2 content of soil air is determined in large part by the rate of biologic CO_2 production within soils and by the diffusive flux of this CO_2 into the atmosphere. Less than 1% of the CO_2 that is generated within soils is used in weathering reactions. The rate of CO_2 emission from soils is highly variable (e.g., the compilation by Albertsen, 1979), but the range of emission rates straddles 120 mg/m^{2}hr; such an emission rate is to be expected if the rate is determined by the annual oxidation within soils of some 50×10^{15} g carbon, a quantity of carbon that is close to the average of recent estimates of the annual rate of carbon production via photosynthesis on land (Sundquist, 1985, Table 13).

It is certain that the supply of CO_2 to soils today greatly exceeds the rate of HCO_3^- use in weathering reactions, and that the potential supply of CO_2 to soils in pre-mid-Ordovician times would have been similarly greater than the H_2CO_3 use rate in weathering if $P^o_{CO_2}$ had been as high as 1.0×10^{-2} atm and if the diffusion rate of gases and the mean depth to the water table were comparable to those in present-day soils. The validity of these assumptions is unproven, but there is independent evidence for the proposition that $P^o_{CO_2}$ was less than 1×10^{-2} atm during late Precambrian and early Phanerozoic time. If $P^o_{CO_2}$ had been as high as this, i.e., some 30 times the present atmospheric level (P.A.L.) of CO_2, the mean surface temperature of the earth should have been significantly higher than today; if so, it then becomes difficult to explain the occurrence of glacial epochs, especially the very extensive Eocambrian glaciation, during the few hundred million years that preceded the development of vascular land plants. This argument becomes less persuasive for earlier periods of earth history; 2 billion years ago the solar luminosity may have been significantly lower than today, and a higher $P^{o}_{CO_2}$ was probably required to maintain sufficiently high surface temperatures, so that the rate of CO_2 supply to the atmosphere by volcanic and hydrothermal degassing was balanced by the rate of removal of CO_2 via chemical weathering followed by the deposition of carbonate minerals (Kasting, 1986).

From a climatic point of view, a value of 1.2×10^{-3} atm for the partial pressure of CO_2 in the early Paleozoic atmosphere is quite reasonable. However, the actual value could well have been higher, since the value of $P^{L}_{CO_2}$ was set equal to zero in the first step of the computation on which the value of 1.2×10^{-3} atm was based. In order to maintain the pH of groundwater in the same range as today, the concentration of H_2CO_3 in early Paleozoic groundwater should have been on the order of 0.2 mmol/l. This corresponds to an effective $P^{L}_{CO_2}$ of $\sim3\times10^{-3}$ atm. If this value is adopted, equation 19 becomes

$$P^{o}_{CO_2} \approx \frac{(3\times10^{-5}) + (2.4\times10^{-2})^{\times 9}_{\div 9}\,(3\times10^{-3})}{[(1.7\times10^{-3})^{\times 4}_{\div 4} + (2.4\times10^{-2})^{\times 9}_{\div 9}]} \approx (4\times10^{-3})^{\times 3}_{\div 1.3} \tag{20}$$

The minimum value of $P^{o}_{CO_2}$ is close to 3×10^{-3} atm, and the maximum value is 1.2×10^{-2} atm, i.e., only slightly higher than the value calculated for the condition $P^{L}_{CO_2} = 0$; the most likely value of $P^{o}_{CO_2}$ is some 11 times greater than its present value.

These results are reasonable, but not entirely convincing. Several of the assumptions that are implicit in the calculation of $P^{o}_{CO_2}$ before the development of the cover of higher land plants are not firmly based. They are, however, potentially testable, provided enough data become available for the chemical composition of paleosols of different thicknesses developed at a given time on a variety of rock types (see below).

4.4. The Ratio of the Rate of Oxidant to the Rate of Acid Supply

The ratio, R′, of the rate of delivery of oxidants to weathering acids is defined by the expression

$$R' = \frac{X_{Ox} + Y_{Ox} + Z_{Ox}}{X_{Ac} + Y_{Ac} + Z_{Ac}} \tag{21}$$

If we substitute the expressions developed above for the several terms in this equation, we obtain

$$R' = \frac{[(B_{O_2}P^{o}_{O_2}\,r)/10^3 + m_{POx}\,r]/10^3 + D^{o}_{O_2}\,\alpha\,(P^{o}_{O_2})/L \times 1.4\times10^3}{\dfrac{B_{CO_2}\,P^{o}_{CO_2}\,r}{10^3} + \dfrac{D^{o}_{CO_2}\,\alpha}{L}\,(P^{o}_{CO_2} - P^{L}_{CO_L}) \times 1.4\times10^3} \tag{22}$$

At any given time in earth history, the value of $P^{o}_{O_2}$ and $P^{o}_{CO_2}$ will have been constant worldwide, but the value of r and the nature of soils probably varied considerably from region to region. In general, then, the value of R′ at any given time in earth history could have been quite variable. The somewhat meager compilation of data in Figure 6 indicates, however, that the value of R′ was roughly constant for all paleosols during a rather long period of geologic time. Between ~2.5 and ~1.8 Ga iron was apparently oxidized and retained in soils developed on rocks with R values $\lesssim 3\times10^{-2}$, whereas iron was lost from soils developed on rocks with R values $\gtrsim 4\times10^{-2}$. This is best explained if during this time period the value of R′ was greater than $\sim 2\times10^{-2}$ and less than $\sim 4\times10^{-2}$. Near Flin Flon, Manitoba, a paleosol outcrops which developed between 1.7 and 1.8 Ga on FeO-rich Amisk basalt. During weathering of this basalt its FeO was apparently oxidized and largely retained (Goetz, 1980). This is the oldest known paleosol developed on unoxidized basalt in which iron behaved in this fashion. The data suggest that the value of R′ was $> 13\times10^{-2}$ between 1.7 and 1.8 Ga. The thickness of the various paleosols represented in Figure 6 varies from a few meters to several tens of meters. The depth to water table in these soils was almost certainly just as variable; nevertheless the major control on the behavior of iron in these paleosols appears to have been the R value of the parent material. This suggests that some of the terms in equation 22 are much smaller than others, and that the value of R′ at any given time depended much more on the composition of the contemporary atmosphere than on local rainfall and the properties of particular soils.

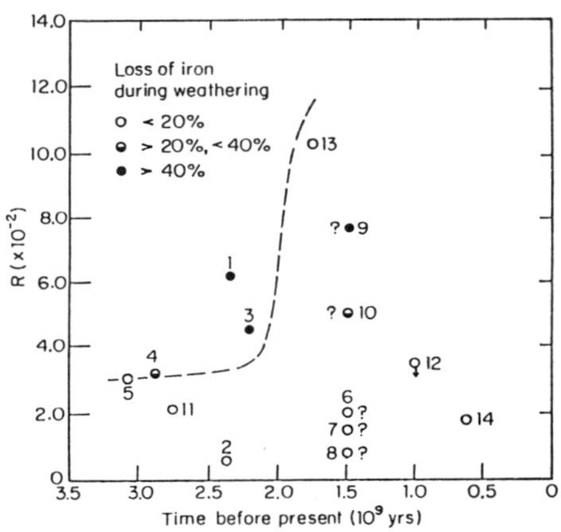

Figure 6. Summary plot of the behavior of iron in paleosols: R is the ratio of O_2 demand to the CO_2 demand during the weathering of the parent rocks. (1) Denison paleosol, Ontario, Canada; (2) Pronto paleosol, Ontario, Canada; (3) Hekpoort paleosol, Transvaal, South Africa; (4) Dominion Reef paleosol, South Africa; (5) Amsterdam paleosol, South Africa; (6) Athabaska meta-arkose paleosol, Hole 20, Saskatchewan, Canada; (7) Athabaska meta-arkose paleosol, Holes 23, 32, 51, Saskatchewan, Canada; (8) Athabaska meta-arkose paleosol, Hole 50, Saskatchewan, Canada; (9) Athabaska metapelite paleosol, Hole 81, Saskatchewan, Canada; (10) Athabaska metasemipelite paleosol, Hole 65, Saskatchewan, Canada; (11) Abitibi paleosol, Canada; (12) Cape Wrath paleosol, Northwest Scotland. (Holland, 1984); (13) Flin Flon paleosol, Manitoba, Canada (Goetz, 1980); (14) Squaw Creek paleosol, Texas (Capo, 1984).

The earlier discussion of the magnitude of these terms is in agreement with this conclusion. The dominant terms in both the numerator and the denominator of equation 22 were probably those due to the diffusive flux of O_2 and CO_2 into pre-mid-Ordovician soils, and hence that

$$R' \approx \frac{D^o_{O_2} P^o_{O_2}}{D^o_{CO_2} (P^o_{CO_2} - P^L_{CO_2})} \tag{23}$$

The ratio $D^o_{O_2}/D^o_{CO_2}$ has the value 0.94 at 25°C and is only a weak function of temperature between 0° and 50°C. If we take the value of R' to have been (0.03±0.01) between 1.8 and 2.5 Ga, the O_2 content of the atmosphere during this time interval becomes

$$P^o_{O_2} \approx (0.94)(0.03\pm0.01)(P^o_{CO_2} - P^L_{CO_2}) \tag{24}$$

In the previous section the likely limits to the difference $(P^o_{CO_2} - P^L_{CO_2})$ were estimated to have been $1.2\times10^{-3}(\div10, \times8)$; if this range of values is inserted into equation 24, we obtain

$$P^o_{O_2} \approx (3.3\times10^{-5})^{\times10}_{\div13} \text{ atm} \tag{25}$$

Kasting (1986) has shown that the CO_2 pressure during much of the Proterozoic was probably between 10^{-1} and 10^{-3} atm. If we accept a value of 10^{-2} atm for P_{CO_2}, the most likely value of $P^o_{O_2}$ was about 3×10^{-4}. Oxygen must have been a minor constituent of the early to mid-Proterozoic atmosphere, i.e., some 2.5 to 1.8 Ga to explain the behavior of iron in paleosols studied to date from this period of earth history.

5. SUMMARY AND CONCLUSIONS

Pre-mid-Ordovician paleosols are useful indicators of the composition of the atmosphere at the time of their formation. It is not always easy, however, to demonstrate that a given stratigraphic unit is indeed a paleosol; it is also difficult to reconstruct the original composition of paleosols. There are a number of criteria for distinguishing pre-mid-Ordovician paleosols from sedimentary layers and from stratigraphically controlled alteration zones. A gradual transition in composition and texture downward into igneous parent rocks distinguishes paleosols most clearly from sediments. Features such as rip-up clasts, desiccation cracks, and sandstone dikes distinguish paleosols most clearly from stratigraphically controlled alteration zones. Good three-dimensional exposures of potential paleosols are normally needed to establish their parentage beyond reasonable doubt.

Geologically old paleosols are generally rock units which have undergone diagenesis and, not infrequently, metamorphism. During their long history, potassium has been added to many paleosols, probably by groundwaters which passed through permeable overlying sediments. Other elements were almost certainly added as well, and some elements may have been subtracted during potassium metasomatism. If the distribution of iron and manganese in a paleosol is used to draw inferences regarding the composition of the atmosphere during the formation of the original soil, then it must be shown that the distribution of these elements was not affected significantly by post-weathering transport. Iron and manganese transport probably did not occur if the enclosing rocks are highly oxidized, and if the latest period of iron and manganese movement can be shown to have preceded the period of potassium metasomatism. As the number of well-studied paleosols from different geologic settings increases, the consistency of the distribution of iron and manganese in paleosols of a particular age may also become a useful criterion for deciding whether significant quantities of the two elements were gained by or lost from these paleosols during post-weathering transport.

In the absence of higher land plants, the behavior of iron and manganese during soil formation indicates whether the ratio of the supply of oxidants to weathering acids exceeded the ratio of the oxidant demand to the acid demand of the parent rocks. Oxidants and weathering acids were supplied to pre-mid-Ordovician soils by diffusion from the atmosphere and by transport as dissolved species in rain water. Except in very thick soils, diffusion was probably the more important of the two mechanisms. O_2 was probably the most important of the oxidants, and CO_2 was surely the most important of the weathering acids. The behavior of iron and manganese in Proterozoic soils indicates that the ratio of the O_2 to the CO_2 supply to soils between 1.8 and 2.5 Ga was approximately (0.03±0.01). The CO_2 pressure in the atmosphere during this period was probably between 3×10^{-3} and 1.2×10^{-2} atm. Higher values are indicated by calculations of the CO_2 pressure required to maintain present-day temperatures at the lower luminosity of the sun some 2 b.y. ago (Kasting, 1986), but these values may be in need of downward revision if the effect of cloud condensation nuclei due to the oxidation of dimethyl sulfide (Charlson et al., 1986) was less pronounced during the Proterozoic than it appears to be today. If the range of CO_2 pressures proposed in this paper is correct, the behavior of iron and manganese in the paleosols studied to date is best explained if the O_2 content of the atmosphere was very much smaller than today. It is not yet possible to define P_{O_2} during this period of earth history at all precisely; our current estimate of P_{O_2} is about 3.4×10^{-4} atm. This figure may well shift considerably under the weight of additional data.

6. ACKNOWLEDGEMENTS

The authors are very much indebted to A. Button for continuing advice and for his collection of paleosol samples, to N. Tyler for pointing out the desirability of working at Derdepoort in the western Transvaal, and to L. Robb, M. Meyer and the other members of the Economic Geology Research Unit of the University of the Witwatersrand for their unstinting support during our visit to South Africa in 1985. The manuscript benefitted greatly from discussions with many members of the NATO Conference in Aussois, and from critical reviews by C. Bowser, M.M. Kimberley, J.I. Drever, R.M. Garrels, A. Lerman, J.P. Pinto, J.F. Kasting, and T. Gardner.

The authors gratefully acknowledge financial support from NASA under Grant NAGW-599. The paper was written while the senior author was enjoying the hospitality of the Department of Geosciences at the Pennsylvania State University as a Visiting Professor during the academic year 1985-1986.

7. REFERENCES

Albertsen, M., 1979, Kohlendioxid-Haushalt in der Gasphase der ungesättigten Bodenzone, dargestellt am Beispiel eines Podsols; *Z. Pfhanzenernaehr. Bodenkd.* **142**, 39-56.

Birkeland, P. W., 1974, *Pedology, Weathering, and Geomorphological Research*; Oxford University Press, 285 pp.

Bonneau, M. and Souchier, B., 1982, *Constituents and Properties of Soils*; V. C. Farmer, Translation Editor; Academic Press, 495 pp.

Buchman, H. O. and Brady, N. C., 1968, *The Nature and Properties of Soils*; 6th Ed., The

Macmillan Co., New York, 567 pp.
Button, A., 1979, Early Proterozoic weathering profile on the 2200 m.y. old Hekpoort Basalt, Pretoria Group, South Africa: Preliminary results. *Inf. Circ. No.* **133**, Economic Geology Research Unit, University of Witwatersrand, Johannesburg, South Africa.
Buyanovsky, G. A. and Wagner, G. H., 1983, Annual cycles of carbon dioxide level in soil air; *Soil Sci. Am. J.* **47**, 1139-1145.
Capo, R. C., 1984, Petrology and geochemistry of a Cambrian paleosol developed on Precambrian granite, Llano Uplift, Texas. Unpub. M.A. thesis, University of Texas at Austin.
Campbell, S. E., 1979, Soil stabilization by a procaryotic desert crust: implications for a Precambrian land biota; *Origins of Life* **9**, 335-348.
Charlson, R. J., Warren, S. F., Lovelock, J., and Andreae, M. O., 1986, On the feedback loop of dimethylsulfide from biota, cloud condensation nuclei, and climate. *EOS*, **67**, No. 44, 869.
Craig, D. C. and Loughnan, F. C., 1964, Chemical and mineralogical transformations accompanying the weathering of basic volcanic rocks from New South Wales, *Aust. J. Soil Res.* **2**, 218-234.
Crampton, D. G. V., 1972, *The geochronology of the acid igneous rocks of the Derdepoort area, N. W. Transvaal.* Unpub. B.S. thesis, University of Witwatersrand, Johannesburg, South Africa.
De Jong, E., Douglas, J. T. and Goss, M. J., 1983, Gaseous diffusion in shrinking soils; *Soil Science* **136**, 10-18.
Dorf, E., 1964, The petrified forests of Yellowstone Park; *Scientific American,* **210**, 106-112.
Eckhardt, F. E. W., 1985, Solubilization, transport and deposition of mineral cations by microorganism-efficient rock weathering agents; pp. 161-173 *in* J. I. Drever (ed.), *The Chemistry of Weathering,* D. Reidel Publishing Co., 324 pp.
Goetz, P. A., 1980, Depositional environment of the Sherridon Group and related mineral deposits near Sherridon, Manitoba. Unpub. Ph.D. thesis, Carleton University, Ottawa, Ontario.
Gray, J., 1985, The microfossil record of early land plants, 1970-1984; *Phil. Trans. Roy. Soc. London*, Ser. **B 309**, pp. 167-192.
Hendricks, D. M. and Whittig, L. D., 1968, Andesite weathering, II. Geochemical changes from andesite to saprolite; *Jour. Soil Sci.* **19**, 147-153.
Holland, H. D., 1978, *The Chemistry of the Atmosphere and Oceans*, 351 pp., Wiley-Interscience, New York.
Holland, H. D., 1984, *The Chemical Evolution of the Atmosphere and Oceans,* Princeton University Press, Princeton, N.J., 582 pp.
Holland, H. D., Zbinden, E. A., Lazar, B. and Dobos, S., A paleosol developed on Keweenawan basalt, in preparation.
Joffe, A. F. and Revut, I. B., 1966, *Fundamentals of Agrophysics* (English Translation from the Russian).
Kasting, J. F., 1986, Theoretical constraints on atmospheric CO_2 levels during the Precambrian. *EOS*, **67**, No. 44, 867.
Kasting, J. F., Holland, H. D. and Pinto, J. P., 1985, Oxidant abundances in rainwater and the evolution of atmospheric oxygen; *Jour. Geophys. Res.*, **90**, 10,497-10,510.
Lai, S.-H., Tiedje, J. M. and Erickson, A. E., 1976, In situ measurement of gas diffusion coefficient in soils; *Soil Sci. Soc. Amer. J.* **40**, 3-6.
Lal, R., 1981, Physical properties; *in* D. J. Greenland (ed.), *Characterization of Soils*; Clarendon Press, Oxford.
Meybeck, M., 1983, Atmospheric inputs and river transport of dissolved substances, pp. 173-192, in Dissolved Loads of Rivers and Surface Water Quantity/Quality Relationships; *Proc. Hamburg Symposium,* Aug. 1983, IAHS Publ. No. **141**.
Morel, F. M. M., 1983, *Principles of Aquatic Chemistry*; Wiley-Interscience, New York, 446 pp.
Nielson, K. K., Rogers, V. C. and Gee, G. W., 1984, Diffusion of radon through soils: a pore distribution model; *Soil Sci. Soc. Am. J.* **48**, 482-487.
Pavich, M. J. and Obermeier, S. F., 1985, Saprolite formation beneath coastal Plain sediments near Washington, D.C., *Bull. Geol. Soc. Amer.* **96**, 886-900.

Pinto, J. P. and Holland, H. D., 1987, *Paleosols and the evolution of the atmosphere,* in press.

Retallack, G., 1981, Preliminary observations on fossil soils in the Clarno Formation (Eocene to early Oligocene) near Clarno, Oregon; *Oregon Geology* **43**, 147-150.

Retallack, G., 1985, *Laboratory Exercises in Paleopedology.* (unpublished laboratory manual) University of Oregon, Eugene, 74 pp.

Retallack, G., 1986, The fossil record of soils; *in,* P. V. Wright (ed.), *Paleosols, Their Recognition and Interpretation*; Blackwells, London.

Retallack, G., Grandstaff, D. and Kimberley, M., 1984, The promise and problems of Precambrian paleosols; *Episodes* **7**, 8-12.

Richter, J. and Grossgebauer, A., 1978, Untersuchungen zum Bodenlufthaushalt in einem Bodenbearbeitungsversuch. 2. Gasdiffusionskoeffizienten als Strukturmasse für Böden; *Z. Pflanzenernaehr. Bodenkd.* **141**, 181-202.

Roeschmann, G., 1971, Problems concerning investigations of paleosols in older sedimentary rocks, demonstrated by the example of Wurzelboden of the Carboniferous System; *in,* D. H. Yaalon (ed.), *Paleopedology. Origin, Nature, and Dating of Paleosols;* Halsted Press, New York, 350 pp.

Rogers, V. C., Nielsen, K. K., Rich, D. C., Sundquist, G. M. and Mauch, M. L., 1982, Radon attenuation with earthen covers; RAE-33-14. *Annual Report to U.S. Department of Energy,* UM-TRAP Project Office by Rogers Associates Engineering Corp., Salt Lake City, Utah.

Silker, W. B. and Kalkwarf, D. R., 1983, *Radon diffusion in candidate soils for covering uranium mill tailings, NUREG/CR 2924.* U.S. Nuclear Regulatory Commission, Washington, D.C.

Stumm, W., Furrer, G., Wieland, E. and Zinder, B., 1985, The effects of complex-forming ligands on the dissolution of oxides and aluminosilicates; pp. 55-74, *in* J. I. Drever (ed.), *The Chemistry of Weathering*; D. Reidel Publishing Co., 324 pp.

Sundquist, E. T., 1985, Geological perspectives on carbon dioxide and the carbon cycle; pp. 5-59 *in* E. T. Sundquist and W. S. Broecker (eds.), The Carbon Cycle and Atmospheric CO_2: Natural Variations Archean to Present; *Geophysical Monograph* **32**, American Geophysical Union, Washington, D.C.

Taylor, S. A., 1949, Oxygen diffusion in porous media as a measure of soil aeration; *Soil Science Soc. Proc.* **14**, 55-61.

Turcotte, D. L., 1980, On the thermal evolution of the earth; *Earth Planet. Sci. Lett.* **48**, 53-58.

Tyler, N., 1979, Stratigraphy, geochemistry, and correlation of the Ventersdorp Supergroup in the Derdepoort area, West-Central Transvaal, *Trans. Geol. Soc. S. Afr.* **82**, 133-147.

Von Damm, K. L., Grant, B. and Edmond, J. M., 1983, Preliminary report on the chemistry of hydrothermal solutions at 21° North, East Pacific Rise; pp. 369-389 *in* P. A. Rona, K. Bostrom, L. Laubier, and K. L. Smith, Jr. (eds.), *Hydrothermal Processes at Seafloor Spreading Centers*; Plenum Press, New York and London, 796 pp.

White, D. E., Hem, J. D. and Waring, G. A., 1963, Chemical composition of sub-surface waters; *in* M. Fleischer, Technical Editor, *Data of Geochemistry, Sixth Edition*; U.S. Geological Survey Professional Paper 440-F.

Willett, I. R., 1983, Oxidation-reduction reactions, *in Soils, An Australian Viewpoint,* Division of Soils CSIRO, CSIRO, Academic Press.

Yaalon, D. (ed.), 1971, *Paleopedology: Origin Nature, and Dating of Paleosols*; Halsted Press, New York, 350 pp.

SLOPE EROSION AND MASS MOVEMENT IN RELATION TO WEATHERING IN GEOCHEMICAL CYCLES

Olav Slaymaker
Department of Geography
University of British Columbia
Vancouver, B.C. V6T 1W5
Canada

ABSTRACT

Geomorphic models of slope development and hydrologic models of runoff generation are reviewed in order to identify the major processes of interest. Slope erosion and mass movement are then classified and typical rates of operation identified. Rates of weathering are also reviewed. The main purpose of the chapter is to identify the sediment and solute budget approach as the best way to understand the relative roles of slope erosion, mass movement and weathering in relation to geochemical cycles. Major problems that have been encountered in the quantification of sediment and solute budgets are discussed.

1. DEFINITIONS

1. *Weathering* is the process of alteration and breakdown of rock and soil materials at and near the earth's surface. The alteration of rock by weathering occurs in place and it does not directly involve removal processes (Selby 1982, p. 5).
2. *Slope erosion* is the process of detachment and downslope transport of rock and soil materials under the primary influence of water in motion.
3. *Mass movement* is the process of detachment and downslope transport of rock and soil materials under the primary influence of gravity.
4. *Fluvial erosion and sediment transport* is the process of detachment and downstream transport of rock and soil materials by open channel flow.
5. *Denudation* is the aggregate effect of the above processes. Volumes of material removed from basins and rates of downwearing are best expressed in the equivalent units 1 Bubnoff = 1 mm per 1000 years = removal of 1 m^3 km^{-2} a^{-1} (Selby, 1982, p. 226).
6. *Saturated hydraulic conductivity* is the rate (in $m{\cdot}s^{-1}$) of water movement through a saturated soil, defined as

$$K_s = BD^2 \frac{\rho_w}{\mu} \tag{1}$$

where B is a coefficient involving shape and packing, D is particle size, ρ_w is the specific weight, and μ is the absolute viscosity (Freeze and Cherry, 1979).

A. Lerman and M. Meybeck (eds.), Physical and Chemical Weathering in Geochemical Cycles, 83–111.

7. *Infiltrability* is the maximum possible rate of infiltration of water into a soil at a given time relative to onset of a rainfall event. There is an envelope curve for each soil which plots infiltrability versus time.

8. *Surface tension effects in freezing soils* control water movement. The small ice-water interfaces that occur in the pore spaces of the soil contain ice and water at different pressures such that

$$p_i - p_w = \frac{2\sigma_{iw}}{\gamma_{iw}} \tag{2}$$

where pi is ice pressure; p_w is water pressure; σ_{iw} is the surface tension between ice and water, and γ_{iw} is the radius of curvature of the interface. Thus the temperature at which freezing occurs depends partly on the interfacial curvature and partly on pore size. Moisture remains mobile in clays at lower temperatures than in silts and sands (Williams, 1982).

9. *Variable source runoff generation model* is a conceptual model of the partitioning of surface and subsurface runoff. Because precipitation intensity varies over space and time and saturated hydraulic conductivity varies over space, sources of runoff are spatially variable over time. In addition, infiltrability and surface tension variations in freezing soils lead to further variability. The best known formulation of such a model is known as the Dunne-Freeze model (Freeze, 1978). The model is sensitive to the fact that surface runoff is frequently generated at a distance from identifiable open channels.

10. *Hortonian overland flow* is surface runoff resulting from precipitation intensities greater than slope infiltrability. It is sometimes referred to as precipitation-excess overland flow (Fig. 1).

11. *Sediment budget* is a quantitative statement of the rates of production, detachment, transport and discharge of rock and soil materials from a basin as follows: slope erosion + mass movement + fluvial erosion - material stored = (input + throughput - storage) = sediment leaving the basin (output).

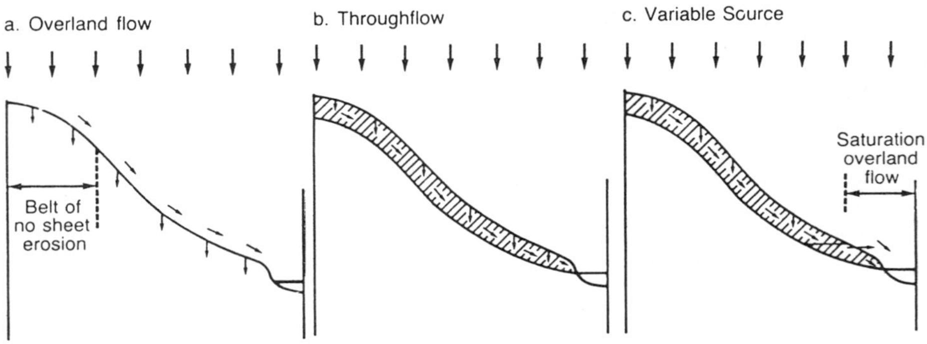

Figure 1. Schematic representations of the (A) overland flow, (B) throughflow and (C) variable source models of hillslope hydrology (Chorley, 1978).

12. *Solute budget* is a quantitative statement of solute sources, sinks and yields from a basin as follows: precipitated solutes + eroded solutes + mass moved solutes + fluvially eroded solutes = solutes stored in soil and biomass solutes leaving the

basin. Because the sediments and solutes are not tagged, it is commonly impossible to partition the sediment and solute budgets adequately. However the aggregate fluxes allow easy comparison with rates of weathering and net denudation.

2. GENERAL PROCESS-RESPONSE MODELS FOR SLOPES UNDER THE INFLUENCE OF WEATHERING, EROSION AND SEDIMENT TRANSPORT

2.1 Kirkby's Models for Mechanical and Chemical Removal

A continuity equation of the form

$$T + D = -\frac{\partial y}{\partial t} \tag{3}$$

where T is rate of mechanical lowering by slope erosion and mass movement, D is rate of chemical lowering, y is elevation above base level and t is time used as the starting point (Fig. 2). Units are in $\mu m \cdot a^{-1}$. Then, a relationship between rate of surface lowering and soil thickness is defined thus

$$\frac{\partial z}{\partial t} = -\frac{\partial y}{\partial t} - W = T + D - W \tag{4}$$

where z is soil depth and W is rate of lowering of soil-bedrock interface (Fig. 2).

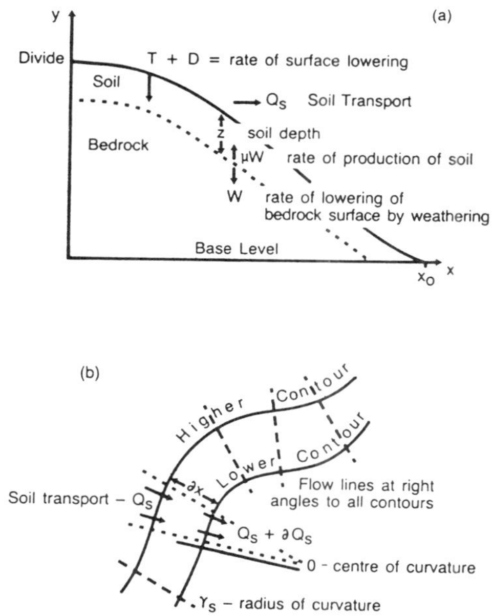

Figure 2. (a) Diagrammatic slope profile showing coordinate axes and symbols used in theoretical analysis of slope development. (b) Diagrammatic map showing symbols used in theoretical analysis of slope development where lines of steepest slope converge or diverge (i.e., where contour lines are not straight) (Kirkby, 1971).

A transport law, specified by transport capacity (c), of the form

$$c \propto a^m s^n \qquad (5)$$

where a is area drained per unit contour length and s is slope (Table I) is then introduced.

Table I. Variation of exponents m and n in empirical relationship (5)

Process	m	n
Soil creep	0	0
Rainsplash	0	1-2
Soil wash	1.3-1.7	1.3-2
Rivers	2-3	3

A relationship between transport rate (Q_s) and the transport capacity (c) can be considered for three general conditions: (i) transport limited, where $c = Q_s$, (ii) weathering limited, where $c >> Q_s$, (iii) erosion limited, where $c > Q_s$. In case (iii), erosion rate ($-\partial y/\partial t$) is

$$-\frac{\partial y}{\partial t} = k(c - Q_s) \qquad (6)$$

where k is an erosion constant.

The four relationships (3), (4), (5) and (6) can be combined to provide the components of a slope evolution model (Kirkby, 1971). A general formulation of the chemical removal process is provided by Kirkby in Carson and Kirkby (1972) as follows. Actual rates of chemical removal depend on the minerals and their amount present; their solubilities and the amount and constitution of water flowing through the soil and the way in which it makes contact with the soil particle surfaces. These controls can be approximated by considering the rate of removal of each oxide as proportional (a) to the saturated solubility of each oxide, assumed constant (b) the quantity of oxide present and (c) the flow of water through the soil.

The solution model is

$$-\frac{\partial p_s}{\partial t} = k \cdot \frac{\partial q_z}{\partial x} \cdot p_s \qquad (7)$$

where p_s is amount of oxide present as a proportion of the unweathered material, t is time elapsed, k is the solubility of the oxide, x is distance downslope and q_z is unit discharge.

2.2 Slope Profile Development with Chemical Removal Alone

From equation 7 the rate of change of total substance still undissolved ($\partial p_s/\partial t$), which expresses the changing slope profile, is a function of the unit discharge and the sum of the products of solubility and proportion of a given oxide in that profile.

In rock which is only slightly weathered ($p_s > 0.9$) solution increases porosity and q_z increases rapidly. Weathering on this model is concentrated in the zone of maximum specific discharge, which tends to be in the intermediately weathered zone. This zone is identified as a rather sharp weathering front.

2.3 Chemical-Mechanical Removal Models for Slopes

Now the rate of surface lowering ($\partial y/\partial t$) is equal to T + D (equation 3) and the rate of mechanical lowering (T) is equal to the rate of increase of sediment transport downslope ($\partial S/\partial x$) minus the ratio of sediment transport to slope radius of curvature (S/γ_s) where S is the vector sediment transport, γ_s is the radius of curvature and x is measured along the line of greatest slope. Then

$$-\frac{\partial y}{\partial t} = \frac{\partial S}{\partial x} - \frac{S}{\gamma_s} + D \tag{8}$$

If the assumption is made that the soil is in equilibrium, then

$$W = -\frac{\partial y}{\partial t} \quad \text{and} \tag{9}$$

$$W - D = \frac{\partial S}{\partial x} - \frac{S}{\gamma_s} \tag{10}$$

2.4 Implications

The above simplified process-response models tend to direct attention to the importance of hydrology, both surface and subsurface, in the understanding of the relationship between denudation and weathering. In particular, models of runoff generation need to be understood.

3. SURFACE AND SUBSURFACE EROSION ON SLOPES

3.1 Classification

Four different categories of runoff produce distinctive erosional processes:

(a) Diffuse subsurface flow produces *eluviation* of ions ("suffosion" in French).
(b) Concentrated subsurface flow produces *piping*.
(c) Diffuse surface flow (or overland flow) produces *surface wash*.
(d) Concentrated surface flow produces *rilling*.

Different combinations of the above four erosional processes can eventually generate gullies and channels. In addition, (e) *rainsplash* produces a fifth distinctive erosional effect (Fig. 3).

3.1.1 Eluviation is the vertical and lateral transport of fine particles by diffuse subsurface flow. The evidence of the process is the development of clay-rich layers in the soil profile, clay linings around vegetation roots and the removal of clay particles from the soil profile altogether. Distance of travel and volumes of material translocated are greater in hollows (convergent flow lines) than on topographic divides (divergent flow lines).

3.1.2 Piping is the vertical and lateral erosion by concentrated subsurface flow. This is a primarily mechanical process of erosion by which sediment is dislodged and entrained in water, generating tubular underground conduits. The primary landform which results from this process is a pipe but an associated suite of secondary landforms has been identified such as caves and sinkholes. The term "pseudo-karst" is sometimes used to describe this terrain.

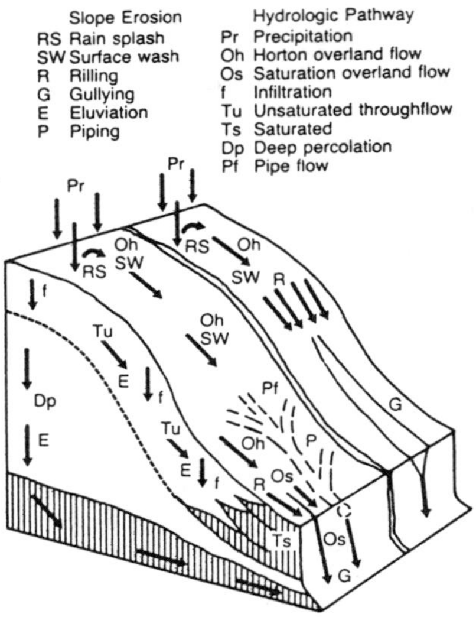

Figure 3. Slope erosion processes as a function of surface and subsurface hydrology.

3.1.3 Surface wash is erosion by diffuse surface runoff which occurs on a slope when precipitation intensity exceeds the infiltrability of the soil and its detention storage (i.e., the depth of water ponded behind individual roughness elements). A form of the Manning equation,

$$\overline{v} = \frac{1}{n} d^{2/3} s^{1/2} \tag{11}$$

where $\overline{v}$ = mean velocity (in $m \cdot s^{-1}$), n is a roughness term, d is depth of flow (in m) and s is slope, is used to define the factors that control velocity and hence the tractive force exerted on the slope. As depth of flow and slope increase, and as surface roughness decreases, so the velocity increases.

3.1.4 Rilling is produced by concentrated surface runoff and consists of small, linear rectangular channels. Erosion by rilling increases with slope. In general, concave slopes and straight slopes sustain rilling, convex slopes result from soil creep and gelifluction.

3.1.5 Rainsplash is an important slope erosional process on steep slopes not completely covered by vegetation. It is also important in reactivating surface wash on slopes where overland flow has ceased to transport sediment and raindrop impact disrupts the surface causing erosion.

3.2 Implications

Runoff generation models determine the partitioning of runoff between surface and subsurface forms. Depending on whether surface or subsurface runoff is dominant, so the appropriate combination of slope erosion processes will be emphasized (Fig. 4).

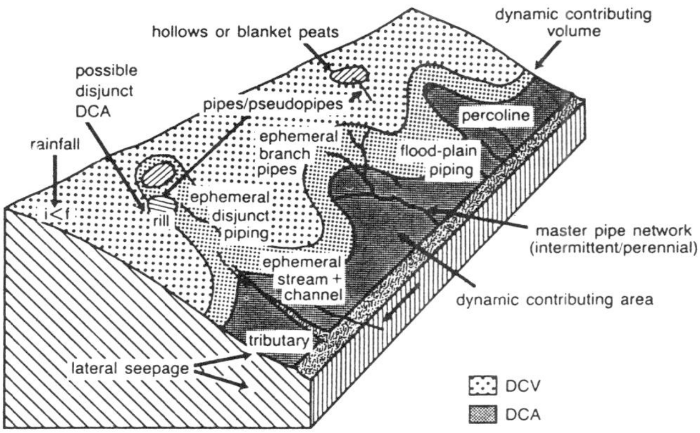

Figure 4. Proposed post-Hewlett framework for runoff generation model. Dynamic contributing volume (DCV) is mapped as horizontal projection of soil volume contributing to storm runoff by (a) pipe drainage, (b) diffuse interflow, (c) percoline flow (laterally concentrated diffuse interflow perhaps including pipeflow). Dynamic contributing area (DCA) consists of (a) saturation overland flow, (b) diffuse seepage return flow, (c) disjunct pipe discharge (Jones, 1979).

3.2.1 Summary mechanical effects in strongly piped soils. Eluviation is important for its effect on the distribution of clay size particles in the soil. The case of the strongly piped soil has been discussed by Bryan and Hodges (1982) from field work in Dinosaur Badlands, Alberta. They showed the differing behaviour of unpiped sandstones and silts by comparison with the extensively piped shales. Hortonian overland flow is initiated on sandstones and silts, whereas on shales the desiccation crack system and pipes control subsurface flow at all times except under extreme high magnitude events. When the surface approaches saturation, aggregates swell and disperse, secondary cracks disappear and primary cracks close from the base (Bryan and Yair, 1982). Pipes have important effects on the hydrologic regime and transmit as much as 40% of watershed runoff (Jones, 1984).

3.2.2 Summary chemical effects in small mountain watersheds. Slaymaker and Gallie (1984), working in first and second order mountain watersheds in the Coast Mountains of British Columbia, have reported that most water in that environment appears to bypass the soil matrix and moves either as overland flow or along discrete saturated subsurface flowpaths (Fig. 5). These preferred pathways limit primary soil and water reactions as well as ionic inputs from adjacent relatively inactive zones. Soil effluent thus contains a dilute chemical signature from the unsaturated zone regardless of the pedon of origin. Groundwater on the other hand is relatively concentrated and comparatively small additions of groundwaters in low lying sites produce calcic flow from watersheds areally dominated by sodic soils. Three fluctuating solute sinks result from this mixing pattern: unsaturated soils, transient saturated and permanently saturated soil water zones. Unsaturated zones are relatively inactive and contain concentrated capillary soil water. Transient saturated zones are active and conduct dilute mixtures of meteoric and capillary soil water during and following rainfall events. Permanently saturated zones are reservoirs of intermediate concentration, receiving additions from a range of sources in the watershed (see section 7.1).

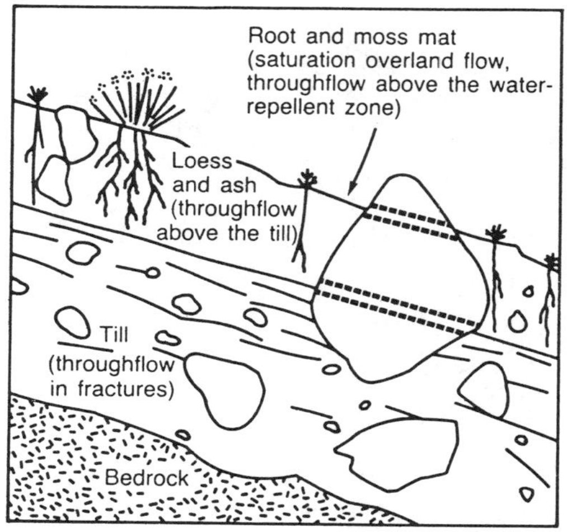

Figure 5. Water and solute flowpaths in a B.C. Coast Mountains watershed (Gallie and Slaymaker, 1984).

3.2.3 Summary effects of ice segregation and frost heaving. The process of soil freezing is complex. The processes of ice segregation and frost heaving are most relevant to present considerations. Segregation of ice in the soil, leading to formation of ice lenses and ice layers, results from the migration of pore water towards the freezing plane. As the ice lenses grow, ground surface heaving results. This is not heaving due simply to expansion of water *in situ* as it freezes, which only amounts to 9% by volume. Heaving producing expansion of over 100% is often observed so that there is a considerable increase in water content during freezing (Embleton and Thornes, 1979). Differential heaving of the surface occurs because of spatial variations in the processes of ground freezing resulting in cracking and the formation of hummocks. The amount of heave (h) in time (t) is given by

$$h = 1.09Ki\,\partial t \tag{12}$$

where K is hydraulic conductivity and i is hydraulic gradient.

In general, silts are more favourable to ice segregation than clays or sands. Finer-grained materials have a greater suction potential, but the loss of permeability means a reduction of water movement. It has often been observed that ice at a penetrating freezing-plane in clays does not advance so uniformly but tends to follow cracks or discontinuities. Ice lenses may form in places, but are often surrounded by layers of unfrozen plastic clay until the temperature falls sufficiently. Clay mineralogy also affects ice segregation (Embleton and Thornes, 1979).

4. SLOW AND RAPID MASS-MOVEMENT

The general term mass movement is applied to those processes that involve a transfer of slope material from higher to lower ground, under the influence of gravity, without the primary assistance of a fluid transporting agent. The movements may be slow or rapid, shallow or deep and include one or more of the mechanisms of creep, flow, slide or fall (Chorley et al., 1984), as shown in Fig. 6.

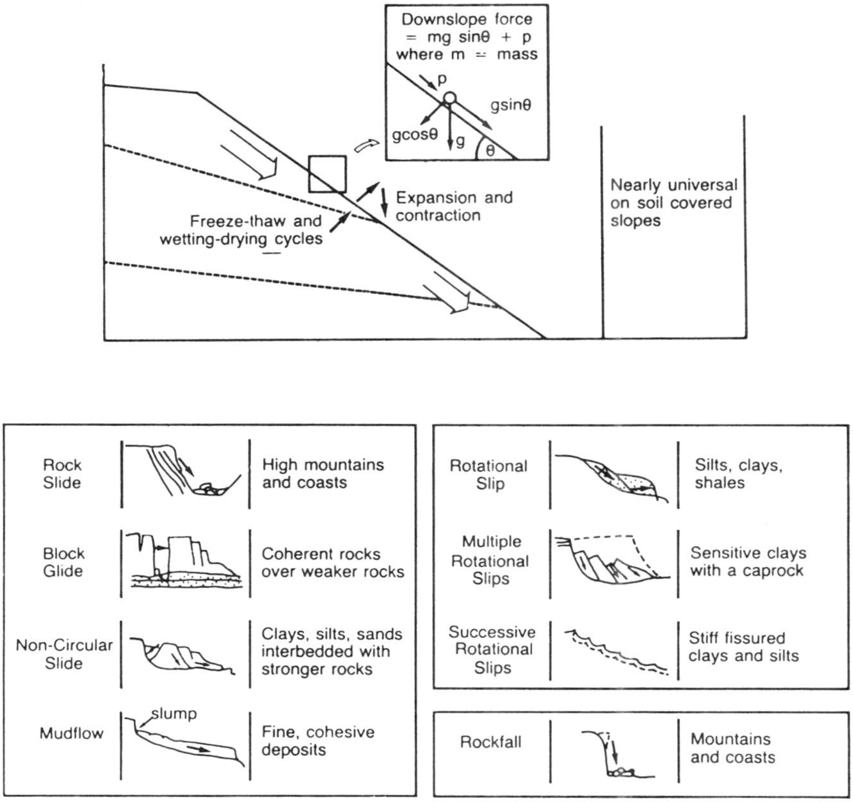

Figure 6. Some important varieties of mass movement.
a. Soil Creep. b. Slides and Flows. c. Rotational Slips. d. Falls.

4.1 The Safety Factor

The stability of a slope is assessed as the ratio of slope resistance to forces acting on the slope. Analysis is usually carried out in terms of effective stresses where the shear strength (S) under conditions of limiting equilibrium is according to Coulomb's equation:

$$S = \frac{c'}{F} + (\sigma_n - \mu)\frac{\tan\phi'}{F} \tag{13}$$

where F is the safety factor, c' is effective cohesion, ϕ' is effective angle of internal friction, σ_n is total normal stress, and μ is pore pressure at all points on the failure surface.

The influence of water pressure which fluctuates seasonally causes seasonal fluctuations of shear strength by ±10%. Changes due to weathering are no more than 1-2% per year, but they are cumulative rather than cyclic (Brunsden, 1979).

4.2 Causes of Mass-Movement

Causes of failure can be divided into two types:

(a) External causes that produce an increase in shear stress but no change in shearing resistance.
(b) Lateral changes of shearing resistance without any change in shear stresses.

Weathering leads to the following important changes in the internal strength of the regolith:

(a) Development of soil horizons that have new physical properties.
(b) Particles reduced in size and increase in clay-sized particles.
(c) Increase in water content.
(d) Possible increase in pore water pressure.
(e) Increase in the number and size of voids and fissures.
(f) Collapse of original mineral fabrics.
(g) Changes in cohesion.
(h) Solution processes that remove soluble cement, as in the case of Norwegian quick clays and the Leda clay of the St. Lawrence Lowlands, Canada.

Each of the above factors may affect the safety factor so substantially that slope failure and mass movement will occur. Freeze (1980) presents an interesting model of the interaction between ground water and mass movement.

4.3 Soil Creep and Geliflucton

Davison (1889) proposed a model for soil creep which derived from laboratory observations of a sloping soil mass under the influence of freezing and thawing. He identified that expansion-contraction cycles tend to produce downslope soil displacement.

The rates of soil creep increase in areas subject to frost heave since the heaving mechanism is aided by the growth of segregated ice lenses, needle ice, wash, mudslides and flow induced by high pore pressures created during the summer melt. Soil creep in the presence of segregated ice is termed gelifluction. The rate of operation is often an order of magnitude greater than that of soil creep in non-freezing environments (Table II) and the movements tend to be spatially separated into lobes rather than more generally spread across a hillslope.

Permanently or seasonally frozen sub-surface layers prevent downward percolation of moisture. The upper layer affected by seasonal thawing becomes soaked with water from melting snow, from melting of any temporary ice lenses or from rainfall. The effect of excess water in the active layer is to reduce its shear strength. Gelifluction may occur in areas of ground frozen permanently, seasonally or diurnally; permafrost is not essential. In areas without permafrost, deep and rapid frost penetration followed by thawing from the surface downward is required (Chorley et al., 1984).

Mackay (1981) has measured surface rates of gelifluction of 5 to 10 mm yr^{-1} on slopes flatter than 7° on Garry Island, Mackenzie Delta, and volumetric rates of 7 to 52 cm^3 cm^{-1} yr^{-1}. Uniquely in the literature, movement occurred almost exclusively in the late summer associated with thaw at the base of the active layer. Another interesting aspect of Mackay's work has been the dating of peat overridden by advancing gelifluction lobes. This has confirmed the rate of operation of gelifluction over several thousands of years. Most recently, Smith (1985) has assessed movement rates and dated an organic horizon overridden by two adjacent gelifluction lobes. He

Table II. Measured and estimated rates of seasonal soil creep (after Brunsden, 1979).

Author	Location	No. of years	Rate $cm^3 cm^{-1} yr^{-1}$	Surface rate $mm\ yr^{-1}$
Barr & Swanston, 1970	S. Alaska	-	15.0	6.0
Black, 1969	Wisconsin	-	-	5.0
Capps, 1941	Idaho	30000	-	4.5
Carson & Kirkby, 1972	Maryland	-	1.3	-
Chandler & Pook, 1971	Central England	10000	0.3	0.04
Dedkov & Duglav, 1967	Tartar USSR	-	-	1-3
Everett, 1963	Ohio	2	6.0	0.25
Eyles & Ho, 1970	Kuala Lumpur	4	12.4	5.0
Finlayson, 1976	Western England	0.75	1.12-5.09	-
Fleming & Johnson, 1975	California	1	-	5-10
Iveronova, 1964	Tien Shan	8	>0<1	1.0
Jahn, 1960	Spitsbergen	-	-	120
Kirkby, 1963	Scotland	2	2.1	-
Lewis, 1975	Puerto Rico	4.4-5.4	1.53	2.13-10.4
Leopold et al, 1964	New Mexico	3	4.9	5.0
Leopold & Emmett, 1972	Southern USA	2-9	-	0-8.6
Owens, 1969	New Zealand	0.8	3.2	-
Rapp, 1962	Lappland	-	-	300
Rudberg, 1964	Lappland	6	-	2.7-70.2
Schumm, 1956	S. Dakota	2	-	10-19
Schumm, 1964	Colorado	4	-	6-12
Slaymaker, 1972	Wales	2	2.7	-
Smith, 1960	South Georgia	-	-	50
Washburn, 1947	Arctic Canada	-	-	450
Williams, 1974	NT Australia	3	7.3,4.4	-
Young, 1960	Derbyshire	2	0.5	1-2

has also related fluctuation in the average rate of frontal advance of these lobes to the regional paleoclimatic record and the regional glacial history.

4.4 Other Mass Movement Processes

Although soil creep operates at a very slow rate, it occurs universally on soil-covered slopes. The volumetric soil transport by this process is therefore far greater than that by any other mass movement process if calculated on a regional, continental or global scale. Locally, landslides, flows, rotational slips and even falls may dominate the slope sediment cascade, but from the perspective of geochemical cycling they are not important. Rates quoted in the literature (e.g., Table III) relate to measurements only on slopes which are known to be unstable and therefore they seem to be anomalously high.

5. WEATHERING

Weathering as defined under 1.1 illustrates the following general tendencies: mineral assemblages tend towards a more stable state; irreversible changes from massive to clastic or plastic state occur; changes in volume, density, grain size, surface area, hydraulic conductivity, consolidation and strength occur; new minerals, aggregates and solutions are formed; resistance of original minerals is changed; rock is prepared for subsequent erosion and transport and new land surfaces and deposits result (Fig. 7). It is also instructive to consider the range of hydrological weathering environments as defined by Brunsden (1979) (Table IV).

Table III. Rates of ground loss through landsliding (Saunders and Young, 1983). B = Bubnoff unit of denudation (1 mm per 1000 years).

Climate	Rock	Relief	Ground Loss B	Method	Location
Temperate maritime	Sedimentary	Normal	500-1000	Volume estimate	New Zealand
Temperate maritime	London Clay	Normal	3900-10 000	Historical data	Essex, Engl.
Temperate maritime	Sand, silt, clay	Steep	1500-2000	Historical data, erosion pins	Dorset, Engl.
Temperate maritime	Greywacke sandstone and siltstone	Normal	250-1000	Volume estimate	New Zealand
Temperate maritime	Clay, mudstone, marl	Steep	130 000-2600 000	Historical data	Dorset, Engl.
Temperate maritime	Tertiary basalts	Steep	277-1617	Traps	Co. Antrim, N. Ireland
Temperate maritime	Various	Normal	3800	Volume estimate	New Zealand
Subtropical humid	-	Steep	1000	Paired photographs	Japan
Tropical savanna	Meta-igneous	Steep	270-540	Volume estimate	Tanzania
Tropical rainforest	Various	Steep	800-1000	Photographic survey, river load	Papua, New Guinea

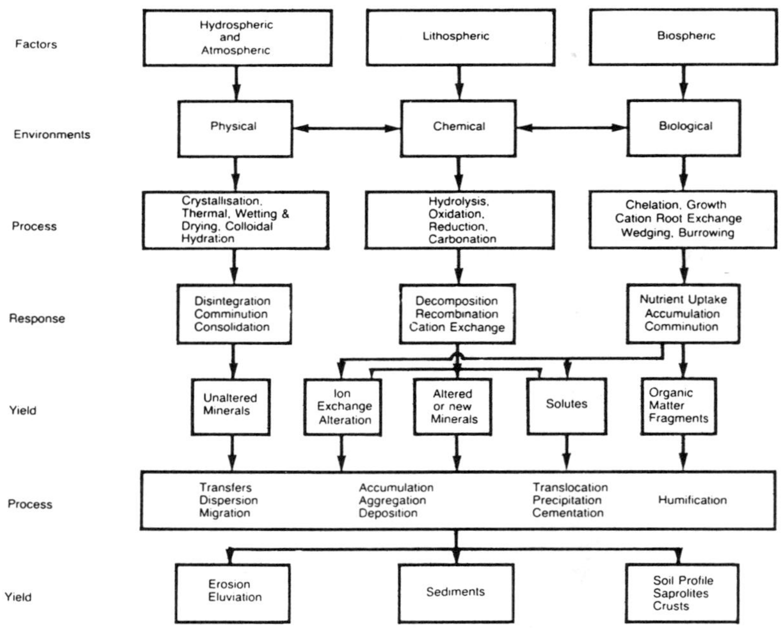

Figure 7. General diagram of the factors and processes of weathering.

Table IV. The properties of the main hydrological weathering environments (after Brunsden, 1979).

Macro-environment	Micro-environment	Properties
Atmospheric	(a) Overhead	Closely related to micro-climate of surface. Depends on rainfall constituents as primary source of cations and anions.
	(b) Near the surface	pH variable 4.9 (Av. 5.7) at 25°C. CO_2, SO_2, H_2So_4 pollutants Cations Na, K, Mg, Ca; Anions Cl, So_4.
Sub-surface	(a) Upper soil	Biologically active. Fluctuating temperature, pH, Eh, moisture. Fluctuating conc. metal ions. Al_2O_3, SiO_2 in solution. Frequent change in direction of water movement. High CO_2.
	(b) Below upper soil-solid rock	Biologically less active. Temperature and pH fluctuation less pronounced. CO_2, H_2S, H*variable. Low Eh. High conc. of M ions.
Sub-aqueous	(a) Below ground-water table	pH variable. Eh variable. Rain water renewed in permeable rocks. Stagnant conditions possible where waterlogged or impermeable.
	(b) Open fresh water	pH variable. Eh variable. Carbonate and bicarbonate anions available. Ionic conc. lower than marine. Often relatively stagnant.
	(c) Open marine water	pH usually >7. Eh variable. Chloride, sulphate, borate anions available. Ionic conc. high. Often relatively stagnant.

5.1 Rates of Rock Weathering

5.1.1 Datable geomorphic surfaces. A variety of qualitative and semi-quantitative methods is used; some of these are illustrated in Table V.

5.1.2 Qualitative measures of degree of decomposition. In the absence of datable surfaces or precise rates of breakdown, an assessment of relative change can be made. In the Pacific Northwest, tills of different age have been differentiated on this basis and relative chronologies have been established (Porter, 1975).

5.1.3 Lowering of rock surfaces. Rates of surface lowering as reported in Table VI make assumptions about the form of the initial surface and the characteristics of the material removed which cannot be checked. Data are quoted for limestone surfaces to the extent possible.

6. SEDIMENT AND SOLUTE BUDGET

6.1 Framework of Budget Studies

After reviewing some of the findings in slope erosion, mass movement and weathering

Table V. Weathering changes to datable geomorphological surfaces and mantles (after Brunsden, 1979).

Author	Method	Rate	Area	Comments
LEACHING OF $CaCo_3$				
Tamm, 1920	Leaching of $CaCo_3$	All removed (0.5%) in 124 years	L. Ragunda, Sweden	From top 10 inch soil under pine heath
Tamm, 1920	Leaching of $CaCo_3$	All removed (0.5%) in 124 years	L. Ragunda, Sweden	From top 25 inch, mixed forest
Salisbury, 1925	Leaching of $CaCo_3$	All removed (6.3%) in 280 years	Sand dunes, Southport, UK	Dunes dated by tree rings & maps
Hissinck, 1938	Leaching of $CaCo_3$	All removed (9.4%) in 300 years	Reclaimed polders, Holland	Newly exposed seafloor
LOSS OF CONSTITUENTS				
Schreckenthal, 1935	General properties	Significant in 80 years	Tyrol moraines	Soil acidity, silt & clay increased
van Baren, 1931	Soil analysis	SiO_2 decreased 6.4% Al_2O_3 increased 2.05% Fe_2O_3 increased 1.04% FeO increased 0.44% CaO increased 0.72%	Indonesia, soils developed on debris derived from eruption of Krakatoa	Pyroclastics, 45 years old 'considerable weathering'
Parsons et al., 1962	Soil analysis	Steady state org. matter in 550 years Steady state org. matter in 1000 years	Willamette Valley, Oregon, Iowa	$CaCo_3$ and pH required 10-300 years for steady state depending on initial values and amount of leaching
SOIL DEVELOPMENT				
Hay, 1960	Soil depth	6 feet soil in 4000 years Clayey soil 1.5-2 feet/1000 years Glass decomposed 15 g/cm^2/ 100 years	St Vincent, W.I.	Volcanic ash to clayey soil
Ruxton, 1966	Depth of weathering profiles	58 mm/1000 years	Hydrographers Volcano, Papua	650 000 years old
Leneuf & Aubert, 1960	Leaching rate of Si, Ca, Mg, K, Na	Ferrallitization of 1 m granite 22000-77000 years	Ivory Coast, forest zone	Granite
Haantjens & Bleeker, 1970	Nature of mantles	Skeletal soil-5000 years Immature-5000-20000 years Mature-20000 years	New Guinea	
Prendall, 1962	Duricrusts	9 m duricrust in 10^6 years	Uganda	Granites
WEATHERING OF SURFACE STONES				
Birkeland, 1974	Boulder breakdown	Weathered surface in 10^5-10^6 years Decomposition in >10^5 years Difference in mineral ratios 10^4 years	Colorado	Generalized from many studies Surface of tills of Early to Pre-Wisconsin age

Table VI. Some records of surface-weathering rates (after Brunsden, 1979).

Author	Surface	Location	Rock type	Period (years)	Result	Average mm yr^{-1}
Goodchild, 1890	Tombstones	Yorkshire	Kirkby Stephen lmst.	500	Based on 1 inch	0.051
			Tailbrig lmst.	300	of weathering	0.085
			Penrith lmst.	250	in period	0.102
			Askrigg lmst.	240		0.106
Emery, 1960	Great Pyramid	Giza	Hard grey lmst.	1000	Little	
			Soft grey lmst.	1000	1-2cm pits	0.01-0.02 in part
			Grey shale	1000	20 cm niches	0.2
			Yellow limey shale	1000	Rubble	0.2 (average over whole surface)
Sweeting, 1960	Erratic block	N. England	Carb. lmst.	12000	30-50 cm	0.025-0.042
Sweeting, 1960	Glacial striae	N. England	Carb. lmst.	13	3-5 cm	2.2-3.8
Sweeting, 1960	Runnels in glacial surface	N. England	Carb. lmst.	13	15 cm	11.5
Hodgkin, 1964	Jetty intertidal	Norfolk Is.	Limestone	16	28 mm	0.5-1.0
Emery, 1941	Inscriptions	La Jolla, CA	Limestone	-	-	0.5
Kaye, 1959	Intertidal notch	Puerto Rico	Limestone	155	-	1.0
Rapp, 1960	Glacial surface	Spitsbergen	Lmst. chert	70	-	0.02-0.2
Rapp, 1960	Glacial surface	Karkevagge	Micaschist, garnet mica, lmst.	8	-	0.04-0.15
Trudgill, 1976	Lmst. pavement	Co. Clare	Poorly cem. limestone	-	-	0.003-6.3

studies, the reader should be warned that most geomorphologists do not claim expertise over the whole range of these processes and that there exist three quite discrete literatures covering the subject matter of sections 3, 4 and 5—not to mention a fourth, that of fluvial process studies (see Meade, this volume). In an attempt to relate the findings in these disparate literatures a budgeting framework has been developed. Sections 1.1 provides definitions of sediment and solute budgets and the following discussion identifies the major problems encountered in attempting to operationalize these budgets.

6.2 Recognition and Quantification of Transport Processes

6.2.1 Sediment budget. "At present, no transport process is sufficiently well understood that field data on major controls can be used to predict transport rates" (Swanson et al., 1982). Extensive monitoring is still required but, before monitoring, it is essential to recognize and assess the order of importance of each sediment transporting process. A recognition of the dominant hillslope erosion processes and mass movements in the field is a crucial step in developing a sediment budget (see sections 3 and 4). A formal sampling design which is sensitive to the variety of process is also necessary (Figs. 8 and 9). The hierarchical sampling design (Fig. 8) is related to ten sediment budgets of small basins in mid-Wales; Figure 9 expands the spatial arrangement of sampling sites and plots within a basin.

6.2.2 Solute budget. Similarly, it is necessary to assess the order of importance of each solute transporting process in the field (see section 3.7.2).

6.3 Recognition and Quantification of Storage Elements and/or Solute Sinks

6.3.1 Sediment budget. A variety of sediment reservoirs exist in a watershed, ranging from the soil mantle on a hillslope, gravel bars in a stream channel, debris fans and "active" channel sediment (all the sediment in the channel bed down to maximum scour depth). The time scale that defines the period over which anything approximating a steady state can be assured is proportional to the residence time of sediment in each reservoir (see Table VII).

6.3.2 Solute budget. Storage of solutes in soils, bedrock and biomass varies according to ionic species. For example chloride moves through soils and biomass with minimal uptake; potassium, calcium and sulphate experience preferential uptake by biomass; the primary nutrients phosphorus and nitrogen experience differential uptake by soils. In general, field experiments and observations are required.

6.4 Identification of Linkages Among Sediment and Solute Transport Processes and Storage Elements and/or Pathways of Solute Movement

6.4.1 Sediment budget. Identification of linkages in the budget suggests effects of successive sediment transfers on the sediment moved. Dietrich and Dunne (1978) have contrasted low order and high order basins, pointing out that in low order channels sediments transported by the river should reflect the texture of the soil on tributary slopes; along the floor of higher order valleys residence time of sediment in storage is longer and particle size distribution of load carried by rivers reflects the result of weathering in the storage zone and may bear little resemblance to slope soil texture.

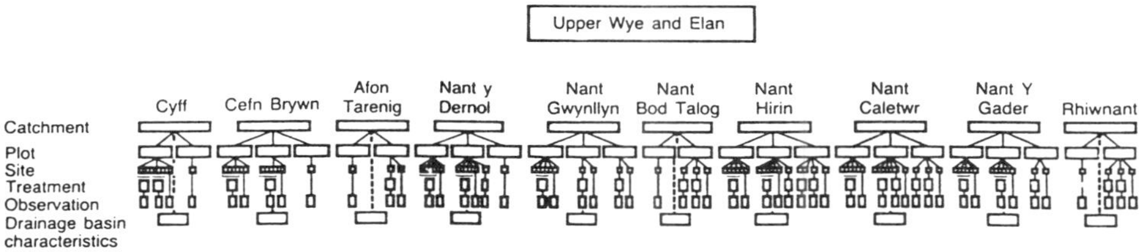

Figure 8. Hierarchical sampling design for determination of sediment budget (Slaymaker, 1968).

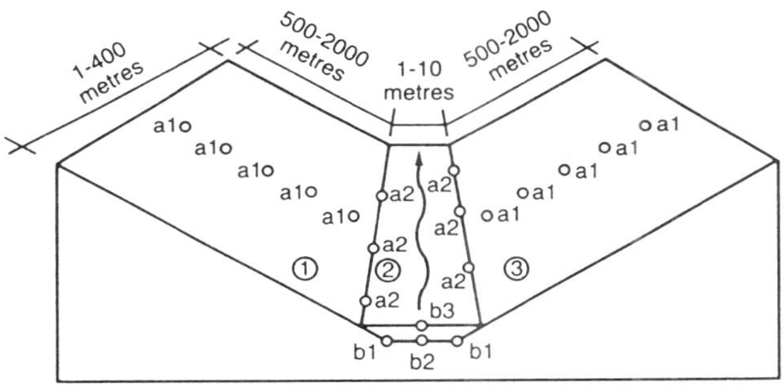

Catchment	1 + 2 + 3	Sampling Unit
Plot	1 + 3 2	Slope Plots Stream Reach Plot
Site	a1 a2 b1 b2 b3	Soil Creep Sites Bank Erosion Sites Scour and Fill Sites Bed Load Sites Suspended Load Sites

Figure 9. Schematic diagram of sampling units, sampling plots and sampling sites (Slaymaker, 1972).

Table VII. Channel storage in relation to mean annual sediment discharge (Swanson et al., 1982).

Stream (author)	Drainage area	Storage	Annual Sediment discharge	Storage-annual discharge
	Square kilometers			
Watershed 10, OR (Swanson et al., 1982)	0.10	510 t (subject to debris torrent)	6.6 t Total particulate transport including episodic debris torrents	77
Average of seven streams in Idaho batholith (Megahan, 1982)	1.27	171 m^3 "behind obstructions"	11 m^3 debris basin accumulation	15
Rock Creek, OR (Dietrich and Dunne 1978)	16.2			
First-order tributaries			Main channel, active sediment	1/19 yr
Third-order tributaries				1/231 yr
				1/619 yr
Big Beef Creek, WA (Madej, 1982)	38	49,200 t "active, to depth of scour"	1,000 t bedload	50
Redwood Creek, WA (M.A. Madej, Redwood National Park, personal communication)	720	7,200,000 main channel only	1,860,000 total sediment load	3.9
Otamarahu River, New Zealand (Mosley 1977, cited in Pearce and O'Loughlin 1978)				35-40

1/Residence time for particle entering channel of this order at midpoint of channel segment.

Dietrich and Dunne (1978) have discussed the implications of linkages through a model (Fig. 10) and this is now being extensively quoted by geomorphologists (e.g., Selby, 1982) as a significant breakthrough in relating geomorphological problems to processes. Similar models by Selby (Fig. 11) and Caine (1980) emphasize bedrock slope processes.

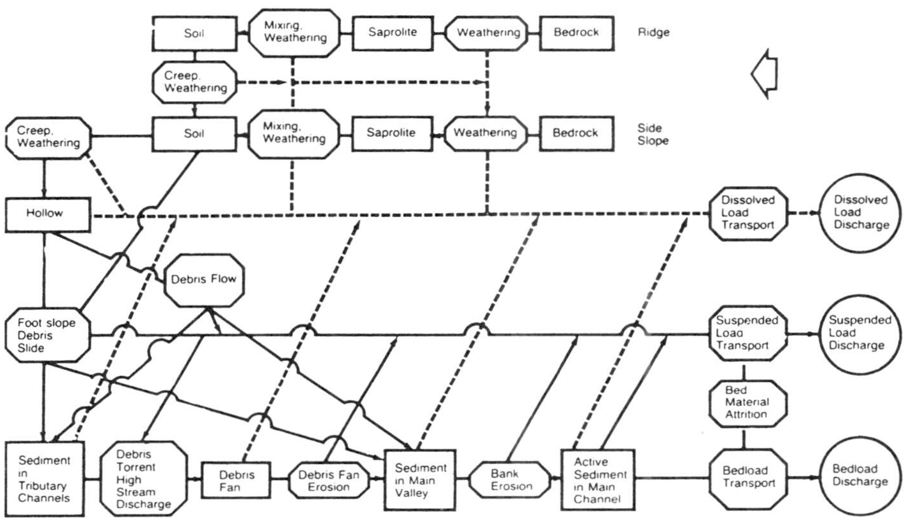

Figure 10. Sediment-budget model for the Rock Creek basin, central coastal Oregon: rectangles, storage sites; octagonals, transfer processes; circles, outputs; solid lines, transfer of sediment; and dotted lines, migration of solutes (from Dietrich and Dunne, 1978).

Identification of linkages is critical in distinguishing those erosion processes that are separate additive components of sediment discharge from those that are carrying the same material along sequential parts of its pathway.

Estimating sediment transport by episodic processes (such as debris slides) is an essential but difficult part of computing a sediment budget. A major difficulty is evaluating the interaction of successive major disturbances in a watershed (e.g., Beven, 1981). Evidence is now overwhelming that successive episodic events are not independent; for example, major floods change the quantity of sediment available for transport (Brown and Ritter, 1971) and channel conditions (Baker, 1977) encountered by subsequent events. "Geomorphic recovery" after major episodic events is one of the criteria which is used in the field to assess effect on sediment yield. Wolman and Gerson (1980) used interesting but inconsistent criteria and there is as yet no uniform standard used.

6.4.2 Solute budget. Identification of pathways of solute movement is essential to interpret the magnitude of solute flux and the precise components of that flux. The watershed can be viewed as a series of hydrologic reservoirs. Solute export is a function of: (a) the number and type of hydrologic reservoirs, (b) the net chemical reactivity of each hydrologic reservoir, and (c) the volume and flow rate of water within each hydrologic reservoir.

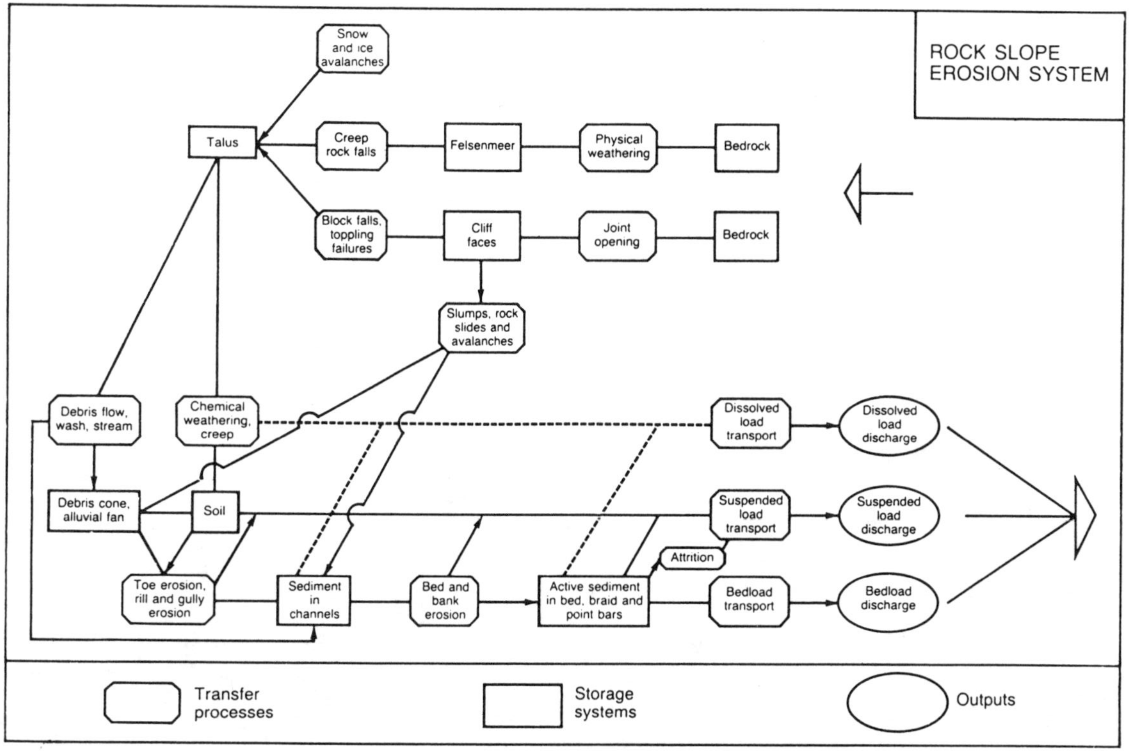

Figure 11. The system of stores and transfer processes on a rock slope (Selby, 1982).

6.5 The Role of Biotic Factors

6.5.1 Sediment budget. Organic matter may comprise a large proportion of sediment in transport, thus potentially complicating sediment sampling and confounding interpretation of the data. Interactions between dissolved and fine particulate organic and inorganic matter (Jackson et al., 1978) present problems for sampling, distinguishing, and interpreting dissolved and suspended sediment yield from a sediment-routing standpoint. These interactions affect the fate and persistence of pollutants in ecosystems.

Plants and animals also affect soil and sediment movement and temporary storage in a variety of ways. Large woody debris in streams and on hillslopes and tree roots retain sediment at temporary storage sites. The multiple, cumulative effects of vegetation on sediment routing through small (less than 100 ha) drainage basins have been demonstrated by studies in both forested and disturbed conditions. Gurnell and Gregory (1984) have discussed the general problem.

6.5.2 Solute budget. Figure 12 demonstrates the integral role of biomass with respect to nutrient cycling. If gross primary productivity, net primary productivity and living and dead biomass accretion data are not available, then the assumption that biomass solute input equals biomass solute output is commonly made for forest ecosystems. However, this is clearly an inadmissible assumption for ecosystems that are in process of change, either aggrading or degrading actively (see also Wright, this volume).

6.6 The Role of Weathering

6.6.1 Sediment budget. Study of weathering and its effect on availability of plant nutrients, soil development, and soil stratigraphy is advanced compared with knowledge of weathering as a regulator of sediment routing. Weathering affects the availability of material for transport and the types and rates of transport processes operating in an area (Dietrich and Dunne, 1978). Geological material enters the sediment-routing system by weathering of bedrock, which makes it available for transport.

Dietrich and Dunne (1978) suggest that weathering is a critical rate-limiting factor in the long-term movement of sediment in many mountain environments (section 2.1). This may be particularly true in steeplands with shallow soils over competent bedrock. Weathering is less crucial in determining the availability of erodible material where primary sediment sources are deep soils, unconsolidated sediments, or tectonically shattered rock. Weathering processes and their interaction with biota alter soil cohesion, bulk density, hydraulic conductivity and mechanical properties, consequently controlling the rates of virtually all hillslope transport processes.

6.6.2 Solute budget. Cleaves, Godfrey and Bricker (1970), in a classic paper on the geochemical balance of a small watershed (Pond Branch in Maryland), used a combination of known weathering reactions (Bricker, Godfrey and Cleaves, 1968), water chemistry and hydrologic balance to infer weathering contribution. Figure 13 shows their weathering model which divides the watershed into three parts—channel, floodplain and valley sides and upland.

THREE LEVELS OF BIOGEOCHEMICAL ANALYSIS
OF HUMID TERRESTRIAL ECOSYSTEMS

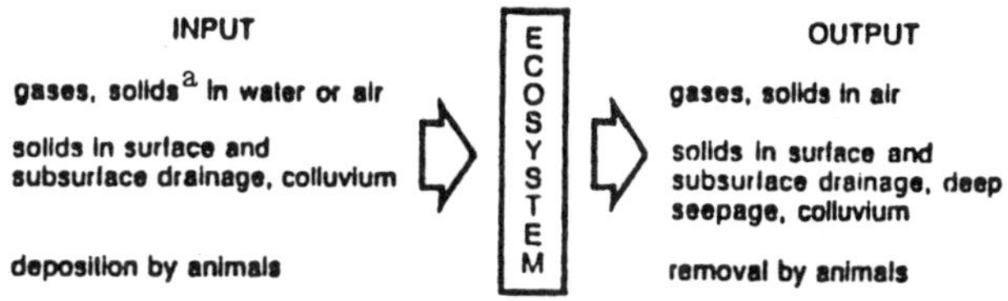

[a]Solids = dissolved chemicals or particulate matter.

1. The ecosystem is considered as a black box, and its boundaries are not coincident with those of a small watershed.

Potential for biogeochemical study:

1. Hydrologic and nutrient input in precipitation.
2. Partial estimate of geologic input.
3. Insight into aspects of ecosystem biogeochemistry.

2. The small-watershed technique where the ecosystem is considered as a black box, but its boundaries are coincident with those of a small watershed underlain by an impermeable substrate. The watershed is in a biologically homogeneous area.

Additional potential for biogeochemical study:

1. Complete hydrologic budgets.
2. Estimate of evapotranspiration.
3. Input-output budgets for sedimentary elements.
4. Partial input-out budgets for gaseous elements.
5. Net change patterns for individual elements.
6. Estimate of minimum weathering rates.
7. Relationship between streamwater flow rate and streamwater chemistry.
8. Annual and seasonal aspects of system biogeochemistry.

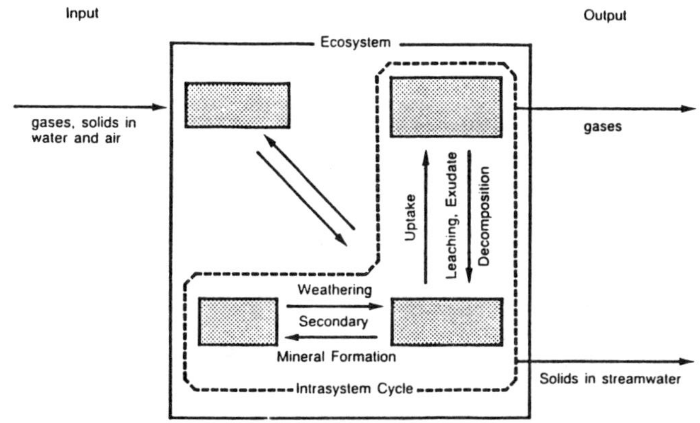

3. The small-watershed technique coupled with measurement of internal features according to the Hobb and Brook Nutrient Flux and Cycling Model.

Additional potential for biogeochemical study:

1. Annual nutrient uptake by living biomass.
2. Nutrient accretion in living and dead biomass.
3. Estimates of leaching, exudation, and decomposition.
4. Estimates of nutrient input by gaseous uptake, impaction, and nitrogen fixation.
5. More complete estimate of weathering.

Figure 12. Three levels of biogeochemical analysis of humid terrestrial ecosystems (Bormann and Likens, 1979).

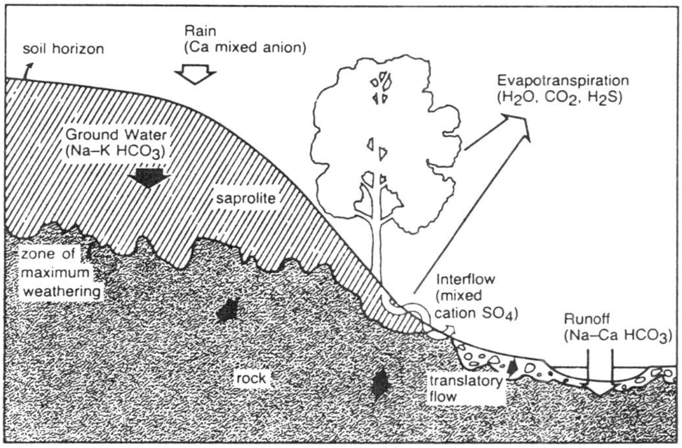

Figure 13. Pond Branch weathering model (Cleaves et al., 1970).

6.7 Interpretation of Stratigraphy (Sediment Budget)

Soil stratigraphy (the use of surface and buried soils to subdivide and to correlate sediments primarily of Quaternary age (Tonkin et al., 1981; Birkeland, 1984) may be useful in constructing sediment budgets by identifying the relative stability of different landscape elements and by placing limits on frequency and rate of erosion.

Easily reversible properties, such as pH, organic material, carbonates, and others, are probably in equilibrium with existing climatic and vegetation conditions at a site, but irreversible properties, such as plinthite, oxic horizons, duripan, and others, may persist under conditions that are drastically different from those under which they developed (Yaalon, 1971). Plinthite is mottled, iron-rich material that can harden irreversibly on exposure to repeated wetting and drying (Birkeland, 1984) and duripan is defined as a diagnostic horizon used in soil taxonomy in which strong silica cementation has occurred (Birkeland, 1984).

Soil stratigraphy may be used to identify relative contributions of different landscape units to the overall sediment yield from a drainage basin. Low erosion rates on hillslopes and ridges are indicated by soils displaying strongly developed profiles and by clay mineralogy and pollen reflecting climatic and vegetation conditions that have not existed for tens or even hundreds of thousands of years. In contrast, hillslopes bearing weakly developed soil profiles or active colluvium indicate more rapid erosion rates.

Another potential use of soil stratigraphy in studying erosion processes is an indicator of recurrence intervals of episodic erosion events. Different ages of landslide episodes can often be recognized by differences in soil depth and weathering intensity, or by buried soil horizons.

Time-dependent weathering characteristics as thickness of weathering rinds of clasts, ratios of fresh to weathered clasts, and etching of heavy minerals are diagnostic and often easier to use than soil-profile development. In some settings, differences in specific surface and subsurface rock-weathering characteristics actually give a clearer separation of Quaternary sediments than differences in soil-profile development (Burke and Birkeland, 1979; see also Catt, this volume).

6.8 Evaluation of the Hydrologic Balance (Solute Budget)

Whereas in the construction of a sediment budget a determination of surface runoff is all that is required with respect to hydrology, in the construction of solute budgets all components of the hydrologic balance must be evaluated (Zeman and Slaymaker, 1978) in the form

$$P - Q - ET - GW \pm \Delta S = 0 \tag{14}$$

Depending on the time frame and location of the study it will frequently be necessary to add a snow melt term so that

$$P + SM - Q - ET - GW \pm \Delta S = 0 \tag{15}$$

where P is precipitation, SM is snowmelt, Q is surface runoff, ET is evapotranspiration, GW is groundwater and S is soil moisture.

The formulation of equation 15 assumes a time frame long enough for groundwater to have responded to precipitation and snowmelt input but short enough that soil moisture redistribution is still proceeding.

The most difficult component to evaluate in a solute budget context is ΔS, especially in light of the variety of pathways discussed in section 3. Although P can be determined with adequate precision (except in mountainous terrain, see Zeman and Slaymaker, 1978), there are significant problems with the temporal and spatial variability of wet and dry fall out.

7. EVALUATING THE RELATIONSHIP OF SEDIMENT AND SOLUTE BUDGETS TO DENUDATION AND GEOCHEMICAL CYCLES

7.1 Watershed Scale Implications

Sediment storage and solute sinks are poorly understood components of sediment and solute budgets respectively. Total stored sediments and solutes, duration of storage at various sites, clays in volume of material stored, and changes in the properties of material while in storage have major implications for sediment and solute routing. The volume of temporarily stored alluvium in river channels is commonly more than ten times greater than the average annual export of particulate sediment (Table VII). Mean residence times of soil on hillslopes are several orders of magnitude greater. For a soil loss rate of 100 tons $km^{-2}\ yr^{-1}$ the mean residence time of a 1 m soil profile is 25,000 years.

Kelsey (1982) has calculated the implication of sediment storage for hillslope evolution and sediment movement in northern coastal California. The study was made on the 160 km^2 headwater basin of the Van Dusen River. Erosion in steep, forested first-order drainage basins is mainly by debris avalanches, any one of which can encompass a large proportion of the drainage-basin area. Residence time of debris on headwater slopes ranges from 15,000 to 50,000 years.

Using the estimates of residence time for the surficial slope mantle, an erosion rate for the headwater basin area can be calculated if we know the average thickness of the surficial layer lost by periodic avalanching. Avalanche failures in the headwater basins ranged between 1.0 and 4.5 m thick, and 2.5 m is a representative thickness. Assuming the average density of the rock and soil to be 1.92 g cm^{-3}, the five different residence times (Table VIIIa) give the following range of erosion rates (Table VIIIb).

Table VIII. (a) Determination of residence time of the surficial slope mantle for six cases in the steep headwater region of the Van Dusen drainage basin (Kelsey, 1982).

Case	Recurrence interval of major episode of avalanching	Portion of basin affected	Residence time of surficial mantle
	Years	Percent	Years
I	500	2	25,000
II	750	2	37,500
III	1,000	2	50,000
IV	500	3	16,700
V	750	3	25,000
VI	1,000	3	33,300

(b) Erosion rates calculated from residence times (Kelsey, 1982).

Residence time	Erosion rate
Years	Tons/km^2 per year
16,700	280
25,000	190
33,300	140
37,500	125
50,000	95

Changes in sediment storage can drastically affect interpretations of erosional conditions within a drainage basin based on sediment-yield data alone (Janda, 1978). The timing and magnitude of erosion of soil and its ultimate delivery as sediment to a downstream point may be very different. Trimble (1981), and others have demonstrated that storage in channels and flood plains may delay and subdue the peak of downstream delivery of sediment introduced from hillslope sources.

Residence time of sediment in storage determines the opportunity for stabilization by rooting of vegetation on the scale of years and decades and for changes in size distribution of material by weathering over centuries and millenia. Weathering of deposits in gravel bars, floodplains, and other storage sites facilitates breakdown of sediment, thus changing the relative importance of transport as dissolved, suspended, and bedload (Dietrich and Dunne, 1978).

In a study of hydrologic controls on alpine stream chemistry, Gallie and Slaymaker (1984) have used solute budgeting to elucidate the relationship between solute sources, sinks and yields. Major cation and silicon budgets from a small Coast Mountains' watershed illustrate effects of soil ice and soil water-repellency on direct runoff mechanisms and on stream chemistry. Annual runoff was similar in the study years 1979 and 1980 (904 mm and 1027 mm respectively) but net solute yields varied from 1.6 to 3.4 ton km^{-2}. The low 1979 yield was caused by segregated soil ice which reduced infiltrability and caused most snowmelt to runoff as precipitation-excess overland flow. In contrast, the dominant runoff mechanisms during 1980 were saturation overland flow above a water-repellent soil zone and saturated throughflow above strong textural contrasts among surficial materials.

In both years, net solute yields from soils were much lower than expected because soil ice, water-repellency, and pipe flow provide quick pathways to channels and reduce soil-water contact.

7.2 Regional Scale Implications

In general, it appears that rates of regional and continental denudation have been greatly overestimated from suspended sediment yield data. Not only for the reasons identified by Meade (1969), Meade and Trimble (1974) and Meade (1982) but because of the extent to which sediment yield data for basins of 1,000 km^2 or greater are an expression of redistribution of sediment stores, primarily through bank erosion (Fig. 14).

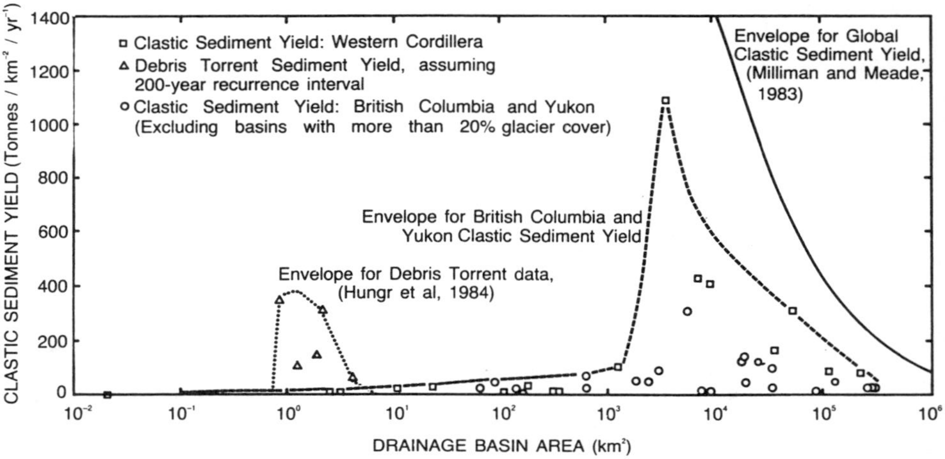

Figure 14. Effects of scale on clastic sediment yield in British Columbia and Yukon (Slaymaker, 1987).

Solute yield data (Janda, 1971) also overestimate denudation but not as severely as the suspended sediment yield data. It seems that while data on sediments and solutes contributed to ocean basins are becoming increasingly precise (e.g., Meybeck, 1979) our understanding of the links between present bedrock denudation and pathways to the ocean basins requires substantial further research.

7.3 General Conclusions

From our studies of surface and subsurface erosion, mass wasting and weathering on slopes, it seems that even in freezing soil environments chemical and mechanical weathering are of comparable orders of magnitude, that the absolute rate of denudation in mountain environments has been greatly overestimated (Slaymaker, 1977) and that, with respect to geochemical cycling, the rate of delivery of sediments and solutes to the oceans bears strictly limited relationship to the rate at which sediments and solutes are being produced in the river headwaters regions.

Use of sediment and solute budget and routing analysis in watersheds is in its infancy (Foster and Grieve, 1984). It is hoped that the link between smaller scale systems (<1,000 km^2) and the continental scale geochemical cycling will become possible through extended use of this methodology.

8. REFERENCES

Baker, V.R., 1977, Stream channel response to floods, *Geol. Soc. Amer. Bull.*, **88**, 1057-1071.

Beskow, G., 1935, *Tjalbildningen och Tjallyftningen Med Sarskild Hensyn Till Vagar, Sveriges Geol. Unders. Afh. Ser.* **C**, 375.

Beven, K, 1981, The effects of ordering on the geomorphic effectiveness of hydrologic events, *in* Erosion and Sediment Transport in Pacific Rim Steeplands, *Intern. Assoc. Hydrol. Sci. Pub.* **132**, 510-526.

Birkeland, P.W., 1984, *Soils and Geomorphology*, Oxford University Press.

Bormann, F.H. and G.E. Likens, 1979, *Pattern and Process in a Forested Ecosystem*, Springer-Verlag, New York.

Bricker, O.P., A.E. Godfrey and E.T. Cleaves, 1968, Mineral-water interaction during the chemical weathering of silicates, *Amer. Chem. Soc. Advan. in Chem. Ser.* **73**, 128-142.

Brown, W.M. and J.R. Ritter, 1971, Sediment transport and turbidity in the Eel River basin, *U.S.G.S. Water Supply Paper 1986.*

Brunsden, D., 1979, Weathering, *in* C. Embleton and J. Thornes, eds., *Process in Geomorphology*, 73-129, Annold.

Bryan, R.B. and W.K. Hodges, 1984, Runoff and sediment transport dynamics in Canadaian Badland microcatchments, *in* T.P. Burt and D.E. Walling, eds., *Catchment Experiments in Fluvial Geomorphology*, 115-131, Geobooks, Norwich.

Bryan, R.B. and A. Yair, 1982, *Badland Geomorphology and Piping*, Geobooks, Norwich.

Burke, R.M. and P.W. Birkeland, 1979, Reevaluation of multiparameter relative dating techniques, *Quat. Res.*, **11**, 21-51.

Caine T.N., 1984, Elevational contrasts in contemporary geomorphic activity, *Studia Geomorph. Carpathobalcanica*, **18**, 5-31.

Carson, M.A. and M.J. Kirkby, 1972, *Hillslope Form and Process*, Cambridge University Press.

Chorley, R.J., S.A. Schumm and D.E. Sugden, 1984, *Geomorphology*, Methuen.

Cleaves, E.T., A.E. Godfrey and O.P. Bricker, 1970, Geochemical balance of a small watershed and its geomorphic implications, *Geol. Soc. Amer. Bull.*, **81**, 3015-3032.

Conacher, A.J. and J.B. Dalrymple, 1977, The nine-unit landsurface model, *Geoderma*, **18**, 1-154.

Davison, C., 1889. On the creeping of the soil cap through the action of frost, *Geol. Mag.*, **6**, 255.

Dietrich, W.E. and T. Dunne, 1978, Sediment budget for a small catchment in mountainous terrain, *Zeit. Geomorph. Supp. Bd.* **29**, 191-206.

Embleton, C. and J. Thornes, 1979, *Process in Geomorphology*, Arnold.

Foster, I.D.L. and I.C. Grieve, 1984, Some implications of small catchment solute studies for geomorphological research *in* T.P. Burt and D.E. Walling, eds., *Catchment Experiments in Fluvial Geomorphology*, 359-378, Geobooks, Norwich.

Freeze, R.A., 1978, Mathematical models of hillslope hydrology, *in* M.J. Kirkby, ed., *Hillslope Hydrology*, 177-225.

Freeze, R.A., 1980, A stochastic conceptual analysis of the rainfall-runoff processes on a hillslope, *Water Resources Res.*, **16**, 391-408.

Freeze, R.A. and J.A. Cherry, 1979, *Groundwater*, Prentice-Hall.

Gallie, T.M. and O. Slaymaker, 1984, Variable solute sources and hydrologic pathways in a coastal subalpine environment, *in* T.P. Burt and D.E. Walling, eds., *Catchment Experiments in Fluvial Geomorphology*, 347-358, Geobooks, Norwich.

Gurnell, A.M. and K.J. Gregory, 1984, The influence of vegetation on stream channel processes, *in* T.P. Burt and D.E. Walling, eds., *Catchment Experiments in Fluvial Geomorphology*, 515-535, Geobooks, Norwich.

Jackson, K.S., I.R. Jonasson and G.B. Skippen, 1978, The nature of materials—sediment—

water interactions in freshwater bodies, *Earth Sci. Rev.*, **14**, 97-146.

Janda, R.J., 1971, An evaluation of procedures used in computing chemical denudation rates, *Geol. Soc. Amer. Bull.*, **82**, 67-80.

Janda, R.J., 1978, Summary of watershed conditions in the vicinity of Redwood National Park, *U.S.G.S. Open File Report*, 78-25.

Jones, J.A.A., 1979, Extending the Hewlett model of stream runoff generation, *Area*, **11**, 110-114.

Jones, J.A.A. and F.G. Crane, 1984, Pipeflow and pipe erosion in the Maesnant experimental catchment *in* T.P. Burt and D.E. Walling, eds., *Catchment Experiments in Fluvial Geomorphology*, 55-72, Geobooks, Norwich.

Kelsey, H.M., 1978, Earthflows in Franciscan melange, Van Dusen River basin, *Geology*, **6**, 361-364.

Kelsey, H.M., 1982, Hillslope evolution and sediment movement in a forested headwater basin, Van Dusen River, *in* F.J. Swanson et al., eds., *Sediment Budgets and Routing in Forested Drainage Basins*, 86-96, Gen. Tech. Rpt. PNW-141, U.S. Dept. Agri. Forest Service, Portland, Ore.

Kirkby, M.J., 1971, Hillslope process-response models based on the continuity equation, *Inst. Brit. Geog. Spec. Pub.* **3**, 15-30.

Likens, G.E., F.H. Bormann, R.S. Pierce, J.S. Eaton and N.M. Johnson, 1977, *Biogeochemistry of a Forested Ecosystem*, Springer-Verlag, New York.

Mackay, J.R., 1981, Active layer slope movement in a continuous permafrost environment, *Can. J. Earth Sci.*, **18**, 1666-1680.

Meade, R.H., 1969, Errors in using modern streamload data to estimate natural rates of denudation, *Geol. Soc. Amer. Bull.*, **80**, 1265-74.

Meade, R.H., 1982, Sources, sinks and storage of river sediment in the Atlantic drainage of the U.S. *J. Geol.*, **90**, 235-252.

Meade, R.H. and S.W. Trimble, 1984, Changes in sediment loads in rivers of the Atlantic drainage of the U.S. since 1900, *Intern. Assoc. Hydrol. Sci. Spec. Publ.*, **113**, 99-104.

Meybeck, M., 1977, Dissolved and suspended matter carried by rivers *in* H.L. Golterman, ed., *Interaction between Sediments and Freshwater*, 25-32, Junk and Pudoc, The Hague.

Meybeck, M., 1979, Concentrations des eaux fluviales en éléments majeurs et apports en solution aux oceans, *Rev. Géol. Dyn. Géog. Phys.*, **21**, 215-246.

Porter, S.C., 1975, Weathering sinks as a relative age criterion, *Geology*, **3**, 101-104.

Rapp, A., 1960, Recent development of mountain slopes in Karkevagge and surroundings, *Geog. Ann.*, **42A**, 65-200.

Selby, M.J., 1982, *Hillslope Materials and Processes*, Oxford University Press.

Slaymaker, O., 1968, *Patterns of Erosion in Instrumented Catchments*, Ph.D. Thesis, University of Cambridge.

Slaymaker, O., 1972, Patterns of present subaerial erosion and landforms in mid-Wales, *Trans. Inst. Brit. Geog.*, **55**, 47-67.

Slaymaker, O., 1974, Rates of operation of geomorphological processes in the Canadian Cordillera, *Ab. Akad. Wiss. Gottingen, Math. Phys. Kl. 3*, **29**, 319-331.

Slaymaker, O., 1977, Estimation of sediment yield in temperate alpine environments, *Intern. Assoc. Hydrol. Sci. Spec. Pub.* **122**, 109-117.

Slaymaker, O., 1982, Land use effects on sediment yield and quality, *Hydrobiologia*, **91**, 93-109.

Slaymaker, O., 1987, Sediment and solute yields in British Columbia and Yukon, *in* V. Gardiner, ed., *International Geomorphology 1986*, part 1, 925-945, Wiley.

Slaymaker, O., and H.J. McPherson, 1972, *Mountain Geomorphology*, B.C. Geographical Series 14, Vancouver, Tantalus Research.

Slaymaker, O. and H.J. McPherson, 1977, An overview of geomorphic processes in the

Canadian Cordillera, *Zeit. Geomorph.*, **21**, 169-186.

Smith, T.R. and T. Dunne, 1977, Watershed geochemistry, *Earth Surf. Proc.*, **2**, 421-25.

Smith, D.J., 1985, *Turf-banked Solifluction Lobe Geomorphology in the Alberta Rocky Mountains*, Ph.D. Thesis, University of Alberta.

Swanson, F.J., R.J. Janda, T. Dunne and D.N. Swanston, 1982, *Sediment Budgets and Routing in Forested Drainage Basins,* U.S. Dept. Agri. Forest Service, P.N.W. Forest and Range Experiment Station G.T.R., PNW-141.

Tonkin, P.J., J.B.J. Harrison, I.E. Whitehouse and A.S Campbell, 1981, Methods for assessing the late Pleistocene and Holocene erosion history, *Inter. Assoc. Hydrol. Sci. Spec. Pub.* **132**, 527-543.

Trimble, S.W., 1981, Changes in sediment storage in the Coon Creek basin 1853 to 1978, *Science,* **214**, 181-183.

Trudgill, S.T., 1976, Rock weathering and climate, *in* E. Derbyshire, ed., *Geomorphology and Climate*, 59-99, Wiley.

Williams, P.J., 1982, *The Surface of the Earth,* Longman.

Wolman, M.G. and R. Gerson, 1978, Relative scales of time and effectiveness of climate in watershed geomorphology, *Earth Surf. Proc.*, **3**, 189-208.

Yaalon, D.H, 1971, Soil forming processes in time and space *in* D.H. Yaalon, ed., *Paleopedology,* 29-39, Int. Soc. Soil Sci., Jerusalem.

Zeman, L.J. and O. Slaymaker, 1978, Mass balance model for calculation of ionic input loads in atmospheric fallout and discharge from a mountainous basin, *Hydrol. Sci. Bull.*, **23**, 103-117.

LOESS—ITS FORMATION, TRANSPORT AND ECONOMIC SIGNIFICANCE

John A. Catt
Soils and Plant Nutrition Department
Rothamsted Experimental Station
Harpenden, Hertfordshire AL5 2JQ
U.K.

ABSTRACT

Thick deposits of loess or windblown silt are common in many parts of the world. They accumulated mainly in areas close to extensive Pleistocene glaciers, the silt being blown principally from proglacial outwash plains. Some smaller deposits of hot or desert loess, as opposed to the cold or periglacial loess, were formed by repeated dust storms originating in hot arid regions. An alternative hypothesis, that loess originates by weathering *in situ*, does not explain all of its characteristics, but some aspects of this loessification theory may be necessary to account for the wide range of secondary carbonate forms almost universal in loess and the peculiar distribution of thin loess deposits in parts of Britain.

Thick loess sequences, subdivided by buried soils formed during warmer and wetter Quaternary stages, occur in central Europe, Tadzhikistan and China. These provide the best evidence from terrestrial deposits for the extremely complex sequence of Quaternary climatic changes originally inferred from ocean floor sediments. Soils formed in loess are very valuable agriculturally, but are prone to disaggregation, crusting and erosion under heavy rain. Some clay-deficient loesses are also problematic as metastable foundation materials in civil engineering. Dust storms, leading to the formation of modern desert loess deposits, cause several other economic problems in many parts of the world.

1. INTRODUCTION

In terms of outcrop area and probably also of total volume, loess is the single most abundant Quaternary deposit on land surfaces of the earth. Keilhack (1920) estimated that it covers over 26 million km^2 or about 20% of the earth's land surface. He and others (e.g., Grahmann 1932; Scheidig, 1934; Kriger, 1965; Pye, 1984a) have mapped the global distribution, but deposits are still often being discovered and Keilhack's estimate of total area is probably a minimum value. The thickest loess deposits (often >200 m) occur in northern China, Tadzhikistan and Uzbekistan (Soviet Central Asia), Ukraine, the Siberian steppes, Kashmir, eastern Europe and central northern U.S.A., but thin deposits (30 cm to 15 m) occur extensively in many other areas, including northwestern Europe, Sweden, Spain, Italy, Yugoslavia, north Africa, Israel, Iran, Iraq, Saudi Arabia, Pakistan, southeast India, northwestern U.S.A. (Washington, Oregon, Idaho and Alaska), Canada, Mexico, southwest

A. Lerman and M. Meybeck (eds.), Physical and Chemical Weathering in Geochemical Cycles, 113–142.

Greenland, Iceland, Argentina, western Australia, and New Zealand (Figure 1). Although a few consolidated loess deposits (loessites) have been reported from the late Precambrian of Norway and some Upper Paleozoic and Triassic beds (Edwards, 1979), loess is characteristically a Quaternary material.

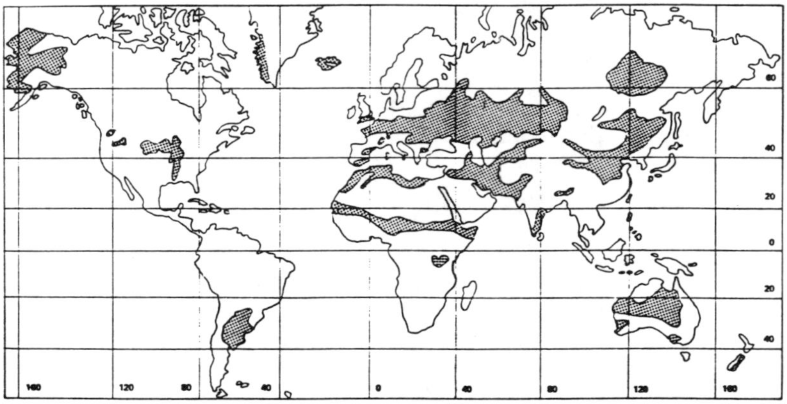

Figure 1. World distribution of loess, compiled from various sources.

Typical loess is a yellow calcareous silt, which is unconsolidated, unbedded and structureless when freshly cut, but on drying develops a coarse prismatic structure with deep vertical joints or fissures. It is also quite firm or hard and strong when dry, so that many of the vertical faces formed in natural and artificial cuttings persist indefinitely and do not degrade to a slope. The carbonate content is usually 10-15% in total, but often exceeds 40% in some Mediterranean and Middle Eastern accumulations which Smalley and Krinsley (1978) called carbonate loess. It is either dispersed through the deposit in microscopic detrital particles, or is secondarily concentrated as irregular yellowish brown nodules ("loess-dolls") or white crystalline masses lining narrow, branching channels or coating the vertical joints. Fossils such as terrestrial molluscs (mainly *Pupilla muscorum* and *Columella columella*) and the bones of steppe animals (mammoth, reindeer, voles) occur occasionally, suggesting deposition under cold dry conditions. Freshwater molluscs are also found locally, often in weakly laminated loesses which were probably deposited by streams. Typical loess has 10-20% clay (<2 μm), <10% sand (>63 μm) and a modal silt size between 20 and 60 μm, though the silt in some loesses is slightly coarser or finer than this. In chemical composition it is similar to the average upper continental crust (Taylor et al., 1983), though loess is usually enriched in Si, Zr and Hf because quartz and zircon are more abundant than in many other crustal rocks.

Loess usually occurs immediately beneath the ground surface, forming a regionally extensive blanket of fairly uniform thickness that covers all but the most recent geomorphological components of the landscape and often smooths out irregularities in the underlying surface. This suggests that if loess is a deposit, most of it was laid primarily by the wind during a fairly recent period. No other form of sediment transportation (marine, fluvial, glacial or mass movement on slopes) can account for a deposit of such uniform particle size distribution and thickness occurring in such diverse physiographic situations. Also, the surface of some loess deposits is built into elongate dune-like features known as *gredas* in slavonic countries or *pahas* in U.S.A. However, some Russian pedologists believe that a special weathering process (loessification) may convert any fine-grained rock into loess, thus explaining its blanket distribution over present land surfaces.

Other important characteristics recognized by most workers are the large porosity (35-50%) and small dry bulk density (1.25-1.65 g/cm^3) of loess, which reflect an open microstructure consistent with either of the origins suggested above. However, the open microstructure and firm consistence do not survive saturation. Loess flooded with water collapses and flows under its own weight, or slumps to a compact deposit occupying little more than half the original volume. This weakness often causes engineering problems in loess areas, such as the collapse of improperly drained structural foundations or the failure of canal walls. Together with the tendency for large blocks to break away from faces along the vertical joints, this characteristic probably accounts for the name loess (German *loesen* = to detach), which was originally used in the Rhine valley in the early 19th century.

Horizons showing some but not all of the properties already mentioned are often termed loessoid or loess-like. They may differ from typical loess in colour, absence of calcium carbonate, increased clay, sand or gravel content, increased bulk density or presence of lamination (bedding), all of which can be explained either by soil formation during or since deposition, or by various processes of reworking and partial mixing with other deposits. These processes of alteration and redeposition have resulted in the proliferation of names for loess-like deposits in several languages (Russell, 1944). They also present a problem in the definition of loess; many geologists feel that the term should be restricted to the unweathered and unmixed typical loess, but other workers, especially soil scientists, include weathered and mixed deposits. The main problems with the latter approach are, first, that the limits of alteration and admixture are as yet undefined, and second, that sediments are usually categorised according to the final and not the penultimate mode of deposition. But if the stricter definition is employed, there is no simple comprehensive term for the thin silty deposits on many land surfaces which determine soil characteristics over quite large areas. The Loess Commission of the International Union for Quaternary Research has long been aware of this problem, and has recently proposed a worldwide survey to establish what types of loess and loess-like sediments have been recognized and the criteria used to differentiate them. This should lead eventually to a much needed universal classification, though it is unlikely that it will be available until the last few years of this century.

2. LOESS STRATIGRAPHY

Before considering the problems of the origin and classification of loess in more detail, it is worth summarising the evidence from various parts of the world for its age and stratigraphic subdivision. This has shown that most of the loess was deposited during the coldest parts of the Pleistocene, when glaciers extended into many mid-latitude lowland regions close to the main sites of loess formation. During intervening warmer stages the uppermost layers of loess were modified by weathering and other processes related to growth of forests, shrubs or grassland vegetation; such layers were often preserved as buried soils usually differing from the unaltered loess above and below in having darker colours, no calcium carbonate and more clay.

2.1. Younger loesses equivalent to oceanic Stages 2-5 (Table 1)

The radiocarbon and thermoluminescence methods have been used to date the youngest loess deposits, which are usually the thickest, most extensive and least altered. They are correlated with Stage 2 (29,000-11,000 yr B.P.) of the sequence of Quaternary climatic stages based on oxygen isotope ratios in deep sea sediments (Shackleton and Opdyke, 1973, 1976), and usually date from the later part of this stage, when glaciers invaded mid-latitude lowlands on either side of the North Atlantic (Cepek, 1965; Penny et al., 1969; Willman and Frye, 1970; Fliri et al., 1971). Thus in eastern England this loess underlies till deposited between 18,5OO and 14,000 yr B.P., but is mineralogically similar to the till and probably close to it in age (Catt et al., 1974). In midwestern U.S.A., the youngest major loess is the Peoria Loess of the Woodfordian substage. This was deposited mainly to the south and west of the region invaded by the Woodfordian glaciers, but near the ice margin (e.g., in Illinois) it is divided into two, the Morton Loess beneath the Woodfordian tills and the Richland Loess above them (Figure 2). This shows that the loess was deposited in proglacial areas during the ice advance in the early Woodfordian and also during retreat in the later part of the substage. McKay (1979) estimated that the base of the Morton Loess is 25,000 yr old. A minimum age is indicated by dates for the buried Brady Soil developed on the Peoria Loess in southern Nebraska and northern Kansas; these range from 12,700 to 9,160 yr B.P., but Frye et al. (1968) suggested that the true age of the soil is about 11,000 yr. A buried soil within the Peoria Loess (the Jules Soil) was formed between 16,500 and 15,500 yr B.P. (Frye et al. 1974).

Throughout Europe and North America the youngest and most extensive loess was therefore deposited during the period of the last major ice advance after 28,000 yr B.P., perhaps mainly between about 20,000 and 14,000 yr B.P. This period is equivalent to stage 2 of the sequence of climatic stages based on oxygen isotope ratios in deep ocean sediments. In Europe there is no clear evidence for more recent loesses, even though there was a short, intensely cold period with some regrowth of glaciers between 11,000 and 10,000 yr B.P. In midwestern U.S.A., the Brady Soil is usually overlain by Bignell Loess, which must be younger than 11,000 yr, but is thin (<3 m), localised, and clearly derived from the Peoria Loess by local reworking.

The dating of loess equivalent to oceanic stages 3 and older is less precise, mainly because most of it is beyond the range of radiocarbon dating. However, it has been subdivided in many areas by the recognition of buried soils. Some of these soils (Parabraunerde) have well developed and often reddened (rubefied) argillic B horizons, and were probably formed under forest, implying a long period of warm and fairly humid interglacial climate, at least in mid- and high-latitude regions. Others are chernozems with little evidence for alteration of the mineral components but fairly deep incorporation of humus, or greyish-mottled (gleyed) soils formed in wetter situations; these probably formed under grassland vegetation during short (interstadial) periods of humid but slightly cooler climate. The youngest of the interstadial soils are often composite or polycyclic, with evidence of arctic conditions (frost-wedges, irregular frost disturbance patterns or cryoturbations) after incorporation of humus. They just fall within the range of radiocarbon dating, and are often separated from the most recent of the rubefied interglacial soils (usually correlated with oceanic stage 5e) by thin loess-like deposits. The radiocarbon dates are not very precise, but together with some thermoluminescence dates they show that weak pedogenesis occurred probably at various times during oceanic stages 3, 5a and 5c following minor periods of loess deposition during stages 3, 4, 5b and possibly 5d (Table 1). Stage 4 of the oceanic succession was short but intensely cold, and although

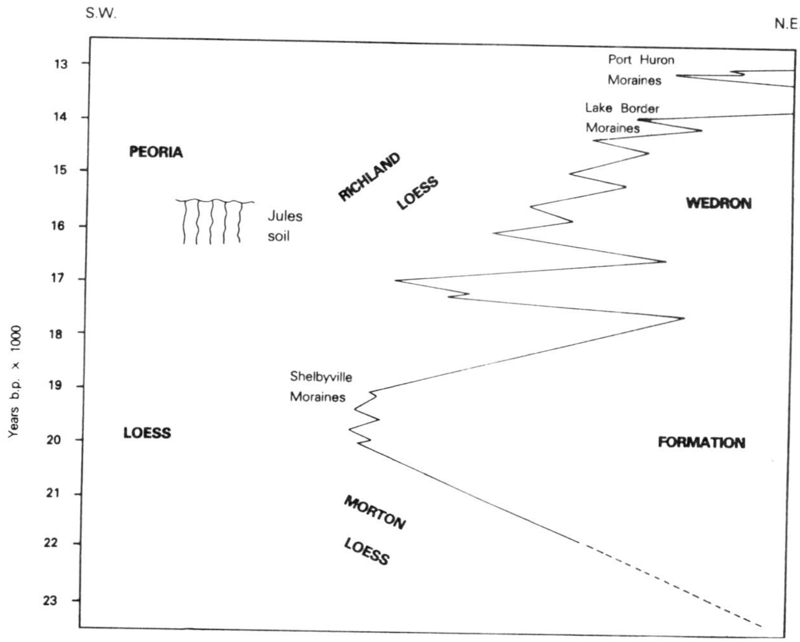

Figure 2. Relationship of the Peoria Loess to Woodfordian glacial deposits (the Wedron Formation) in Illinois and adjacent areas, U.S.A. (based on Willman and Frye, 1970).

Stages 3 and 5 were mainly warm they did include short cold spells (e.g., 5b and 5d).

2.2. Loess older than oceanic Stage 5

Rubefied interglacial soils thought to be equivalent to oceanic stage 5e, such as the Sangamon Soil in midwestern U.S.A. and the St. Romain I Soil in northern France, are developed partly in older loess (Table I) and partly in glacial, glaciofluvial, gelifluction or fluvial deposits. The best evidence for the ages of both the loess and the interglacial soil comes from Normandy, where Wintle et al. (1984) obtained thermoluminescence dates of $110 \pm 10 \times 10^3$ to $120 \pm 10 \times 10^3$ yr for the St. Romain I Soil and $125 \pm 11 \times 10^3$ to $139 \pm 12 \times 10^3$ yr for the loess. These clearly correspond with oceanic stages 5e and 6 respectively. But elsewhere the dating is less precise. In particular, the rubefied soil could often be a composite profile formed during two or more warm stages (Boardman 1985), and the underlying loess correspondingly older.

At many sites in central and eastern Europe (Kukla, 1970, 1977), Tadzhikistan (Goudie et al., 1984) and China (Heller and Liu Tungsheng 1982, 1984) the thick loess deposits contain sequences of buried soils, of both interglacial and interstadial character. Both the soils and the intervening loesses show fairly strong remanent magnetism, and this has often been used to help date older parts of the successions which are beyond the range of the radiocarbon method. The main magnetostratigraphic

Table I. Stratigraphy of younger loess deposits and buried soils on either side of the North Atlantic.

Oceanic Stages	Dating (yr B.P.)	Equivalent loess deposits and buried soils	Reference
1			
	11 000		
2		Stabroek Soil (Belgium)	Paepe and Vanhoorne (1967)
		Upper Pleniglacial Loess (Belgium)	Paepe and Vanhoorne (1967)
		Upper Pleniglacial Loess (France)	Lautridou (1981, 1982), Wintle et al. (1984)
		Late Devensian Loess (U.K.)	Wintle (1981)
		Late Würm Loess (Germany)	Semmel (1973)
		Dunaujvaros-Tapiosuly Loess (Hungary)	Butrym and Maruszczak (1984)
		Desna Loess (Russia)	Velichko (1975)
		Peoria Loess (U.S.A.)	Willman and Frye (1970), McKay (1979)
	29 000		
3		Kesselt Soil (Belgium)	Paepe and Vanhoorne (1967)
		Kesselt Soil (France)	Lautridou (1981, 1982)
		Lohner Soil (Germany)	Semmel (1973)
		Farmdale Soil (U.S.A.)	Willman and Frye (1970)
		Bruansk Soil (Russia)	Velichko (1975)
		Middle Pleniglacial Loess (Belgium)	Paepe and Vanhoorne (1967)
		Hoboken Soil	Paepe and Vanhoorne (1967)
		Poperinghe Soil	Paepe and Vanhoorne (1967)
	63 000		
4		Early Glacial Loess (France)	Wintle et al. (1984)
		Mende-Basaharc Loess (Hungary)	Butrym and Maruszczak (1984)
		?Meadow Loess (U.S.A.)	Willman and Frye (1970)
	75 000		
5a		?Warneton Soil (Belgium)	Paepe and Vanhoorne (1967)
		Basaharc Soil (Hungary)	Butrym and Maruszczak (1984)
		?Pleasant Grove Soil (U.S.A.)	Willman and Frye (1970)
	87 000		
5b		Mende-Basaharc Loess (Hungary)	Butrym and Maruszczak (1984)
		?McDonough Loess (U.S.A.)	Willman and Frye (1970)
	96 000		
5c		?Chapin Soil (U.S.A.)	Willman and Frye (1970)
	108 000		
5d		?Markham Loess (U.S.A.)	Willman and Frye (1970)
	118 000		
5e		Mende Base Soil (Hungary)	Butrym and Maruszczak (1984)
		St. Romain I Soil (France)	Wintle et al. (1984)
		Sangamon Soil (U.S.A.)	Willman and Frye (1970)
		Rocourt Soil (Belgium)	Paepe and Vanhoorne (1967)
	126 000		
6		Loveland Loess (U.S.A.)	Willman and Frye (1970)
		Saalian Loess (Belgium)	Paepe and Vanhoorne (1967)
		" " (France)	Lautridou (1981, 1982), Wintle et al. (1984)

boundary used for this dating is the Brunhes-Matuyama reversal at approximately 0.73 m.y. B.P. (Makinen and Dalrymple, 1979); this has also been recognised in the oceanic oxygen isotope record close to the boundary between Stages 19 and 20. Other useful paleomagnetic boundaries are the Jaramillo Event (0.90-0.97 m.y. B.P.) the Olduvai Event (1.61-1.90 m.y. B.P.) and shorter excursions such as the Blake Excursion (104,000-117,000 yr ago), though the excursions are detected only if the loess was deposited rapidly and is sampled at close vertical intervals. Relative dating of loess layers and buried soils is also assisted by biostratigraphical evidence, especially the occurrence of particular species of snails, voles or large mammals in certain horizons. The snails are also very useful for interpreting climatic conditions during loess deposition or soil formation.

Using these types of data, and also assuming that the boundaries between thick layers of unaltered loess and overlying interglacial soils (the "marklines" of Czechoslovakian loess successions) are equivalent to the dated periods of rapid warming ("terminations") found in the oceanic succession, Kukla (1977) was able to date many of the older loess layers in central Europe. These occur between buried soils mainly on older terraces of rivers such as the Danube, Vltava, Rhine and Dniestr, and contain snails such as *Pupilla* and *Columella* indicating deposition in cold dry conditions. The main loess layers older than oceanic stage 5, designated as subcycles C3, D3, E3, F3, G3, H3, I3, J3 and K3, correspond with cold stages 6, 8, 10, 12, 14, 16, 18, 20 and 22 respectively, the two oldest predating the Brunhes-Matuyama reversal. Thus over the last 0.9 m.y. loess was deposited in central Europe during every major cold stage known from the oceanic succession. Also, as Morrison (1978) pointed out, it is possible to match many of the interstadial chernozems as well as the interglacial Parabraunerde with warm peaks in the oceanic isotopic record (Figure 3).

Loess layers older than 0.9 m.y. occur locally in central Europe, for example in the Danube valley in Czechoslovakia (Fink and Kukla, 1977), but they are thin and often strongly altered throughout by rubefication and pedogenic clay enrichment (Pécsi, 1984). Their exact ages are uncertain, partly because the oceanic oxygen isotope stages before about 1 m.y. B.P. are less well defined, so a correlation with cold stages is less certain than with later episodes of loess deposition.

In the loesses of Tadzhikistan and Uzbekistan there are successions of nine soils above the Brunhes-Matuyama reversal (Dodonov, 1981, 1984; Lazarenko et al., 1981). This suggests similar sequences of climatic events over the last 0.73 m.y. to that recorded in the loess of central Europe, but the dating of individual loess layers and buried soils is too imprecise as yet for detailed correlation of the three successions. The Tadzhikistan succession also includes many loess layers and buried soils older than the Brunhes-Matuyama reversal. Indeed some may be Pliocene because they are older than the Olduvai Normal Event, the upper boundary of which was recommended by Bowen (1978) as a suitable date for the beginning of the Quaternary, and Penkov and Gamov (1977) recognised over 40 buried soils in Tadzhikistan dating from the period since the beginning of the Matuyama Paleomagnetic Epoch (2.48 m.y. B.P.).

Paleomagnetic dating has shown that the loess in northern China also includes layers as old as about 2.4 m.y. (Heller and Liu Tungsheng, 1984), though Liu Tungsheng and Menglin (1982) include these in the Pleistocene, because they draw the Plio-Pleistocene boundary at the beginning of the Matuyama Epoch. The maximum known age for Plio-Pleistocene loess (2.4 m.y.) in China and Tadzhikistan coincides approximately with a sharp increase in northern hemisphere ice volumes, as indicated by the oceanic oxygen isotope record (Shackleton and Opdyke, 1977), and with the first sedimentary evidence for ice-rafted debris in north Atlantic ocean floor

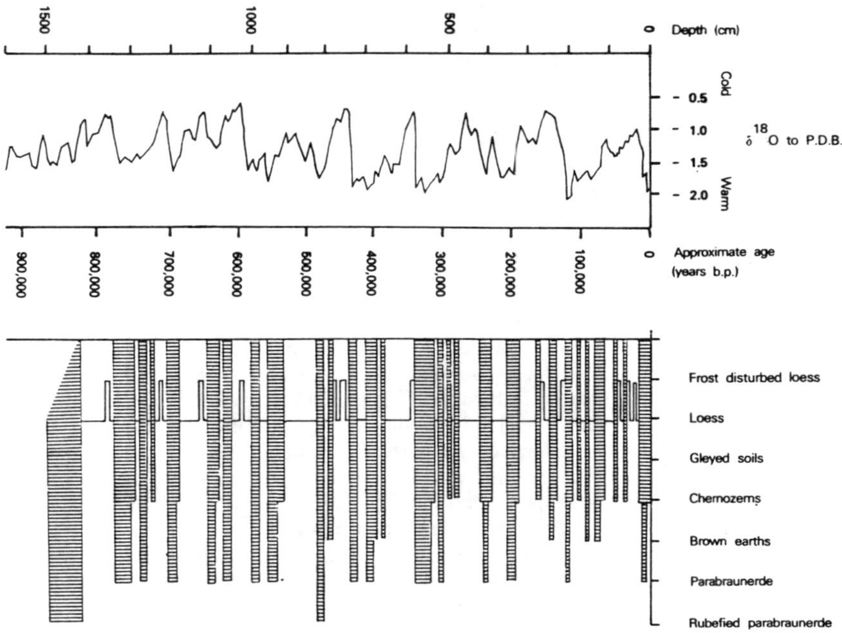

Figure 3. Correlation of buried soils in the loess succession of central Europe with paleotemperature curve based on oxygen isotope changes (relative to P.D.B., a standard belemnite from the Cretaceous Pee Dee Formation, S. Carolina) in ocean sediment core V28-238 from the Equatorial Pacific (redrawn from Morrison, 1978).

deposits (Backman, 1983). This reinforces the relationship between loess deposition and glaciation.

In a borehole at Lochuan in the Schaanxi province of China, the loess sequence post-dating the Brunhes-Matuyama geomagnetic reversal contained 8 buried soils (Parabraunerde), many of which were rubefied or contained gastropods, such as *Metodontia hueiensis, M. yantaiensis, M. beresowskii* and *Succinea erythrophana,* indicating a warm humid climate. The gastropods found in the intervening layers of unaltered loess (*Cathaica pulveratrix, Vallonia costata* and *Pupilla aeoli*) indicate in contrast a dry cold steppe environment (Heller and Liu Tungsheng, 1984). So although the Chinese loess succession has yet to be correlated in detail either with the sequences in central Europe and Tadzhikistan or with the oceanic isotope stages, it is clear that it covers approximately the same period of time as the other long loess sequences, was deposited under similar cold, arid conditions, and can be subdivided into approximately the same number of layers (probably equivalent to the cold stages of the oceanic sequence) by soils formed during warmer, wetter periods. Pye (1984a) drew attention to minor differences between the successions in Europe, China and Soviet Central Asia, such as the fact that the number of major buried soils formed since 110,000 yr B.P. range from one in China to four in Tadzhikistan. However, some cold stages may not be represented by loess in all these regions, and geographical variations in climate may mean that Parabraunerde in one area correlate with less developed soils, such as chernozems, elsewhere.

Some paleopedologists have also questioned the correlation of buried soils with the warm isotope stages of the oceanic succession; they feel it is a relationship forced partly by Kukla's assumption that the "mark-lines" are equivalent to "terminations." Boardman (1985) has pointed out that the main prerequisite for soil formation is a stable land surface with little or no erosion or deposition. Buried soils therefore may have developed simply when there was a diminished supply of loess or other sediment, and were likely to have been independent of climatic changes. Birkeland and Shroba (1974) expressed a similar opinion, based upon evidence in southwestern U.S.A. for soil development at times equivalent to glacial periods elsewhere. They consequently suggested that rain and warmth are less important as soil-forming factors than surface stability and time.

However, these opinions ignore the malacological evidence from central Europe and China for alternating cold dry conditions during loess deposition and warm wet conditions when the main buried soils were formed. The relative importance of time and climatic change in determining the types of soils formed during the Quaternary depends very much on geographic location. Reconstructions of global climate of 18,000 years ago, the time of the last glacial maximum, show that although the earth's surface overall was then slightly cooler and drier, many tropical and subtropical regions were only slightly cooler and no drier than they are today (Gates, 1976). The main differences between the present and 18,000 yr B.P. were the enlarged polar regions and subtropical desert areas (Sarnthein, 1978), and irregular compression of the intervening climatic zones. Many low latitude regions, including southwestern U.S.A., had a "glacial" climate slightly cooler, but no drier and locally slightly wetter than the present (and probably also than the interglacials). In such regions there would be little difference between rates of soil formation at different times in the Quaternary and, as Birkeland and Shroba (1974) inferred, the main factor determining the extent of development of a buried soil was length of the soil-forming period. In contrast, mid-latitude land areas close to the greatly expanded Laurentide, Scandinavian and Himalayan ice sheets, such as northern U.S.A., northwestern Europe and China, were much colder and drier at 18,000 yr B.P. (and at previous glacial maxima) than they are today (Gates, 1976, Figs. 7 and 23). Here one would consequently expect the large difference in rates of pedogenesis between warm and cold stages of the Quaternary which Kukla (1977) and others claim to have found, and length of soil-forming periods would be less important in determining how strongly developed the buried soils are.

We can therefore conclude that all the major loess deposits of the world accumulated during cold stages over about the last 2.4 m.y., that is since the time when glaciers in the northern hemisphere first increased greatly in size. Some of the loess deposits are quite close to the old glacial limits, or were even locally over-run by glaciers soon after their formation. However, many of the smaller deposits are far from any areas known to have been glaciated, so although close association with glaciers or glacial deposits was often important, it does not seem to have been essential for formation of all loesses.

2.3. Desert loess

A subdivision into periglacial ("cold") loess and desert, continental or "hot" loess has been recognised by many workers on the basis of proximity to areas previously glaciated and present hot arid regions, respectively (Obruchev, 1945). However, the origin of many deposits originally regarded as hot loess has been disputed (Butler 1956), Smalley and Vita-Finzi, 1968; Smalley and Krinsley, 1978), mainly because many desert areas could have been more humid and much closer to glaciers during cold stages of the Pleistocene. Also some desert areas have very little loess, suggesting that those which do have loess acquired it in special circumstances, perhaps unrelated to present conditions. Another problem arises over the lack of a universally accepted definition of loess; according to some workers, many of the desert deposits which have been called loess are too sandy, too clayey or too calcareous, and should be given a different name (see also 1. Introduction).

Despite these objections to the existence of hot loess, there remains enough evidence from certain areas to conclude that some of the smaller loess or loess-like deposits in or close to modern desert areas are not periglacial in origin. The best documented is in the northern Negev Desert and Beersheba Basin of Israel (Yaalon and Dan, 1974). This deposit is up to 12 m thick and contains 5-30% calcium carbonate and 10-20% clay on a carbonate-free basis. Its mineralogical composition and microfaunal content suggest that its main sources are Upper Cretaceous and Tertiary limestones in nearby southwestern parts of the desert (Ginzbourg and Yaalon, 1963). This probably explains why the deposit is slightly coarser than most loesses, with a modal particle size of 50-70 μm. Modern aeolian dust blown from the Negev is similar in particle size and mineralogical composition (Yaalon and Ginzbourg 1966), which suggests that the loess is still accumulating and contributing to Israel soils (Dan and Yaalon, 1964). Yaalon and Ganor (1979) estimated that a total of 25×10^6 t of dust is deposited annually in the East Mediterranean basin; it originates from various desert areas in Libya, Egypt, and Israel.

The main site where the stratigraphy of the Negev loess has been investigated is Netivot (Bruins and Yaalon, 1981), situated on the northern desert fringe where any past climatic changes would have had a maximum effect on rates of deposition or pedogenesis. Cyclic variations in clay and coarse silt content and the development of six calcic horizons (semi-arid soils with calcareous nodules formed by evaporation from the soil surface) resulted from minor rainfall fluctuations within a fairly narrow semi-arid range. The periodic soil development was contemporaneous with continuing loess deposition over approximately the last 140,000 years. Taken with the evidence for present-day loess deposition in the area, this shows that the relationship between loess formation and the universally synchronous major cold stages of the Pleistocene, established in other important mid-latitude areas of loess accumulation, does not hold for Israel. This is sufficient to distinguish it from the periglacial or cold loesses.

Other deposits of hot loess occur in Iran (Fookes and Knill, 1969), Iraq (Kukal and Saadallah, 1970), Australia (Gill and Segnitt, 1982), Nigeria (Smith and Whalley, 1981) and some Mediterranean areas (Brunnacker, 1973, 1974). However, the exact origin, stratigraphy and ages of these deposits are often uncertain. Many of them are carbonate loesses, which Smalley and Krinsley (1978) regarded as different in origin from glacial loess, but at present we cannot be certain that all carbonate loesses have a desert origin.

3. ORIGIN OF THE SILT IN LOESS

The abundance of silt particles, mainly in the range 20-60 μm equivalent spherical diameter, is a definitive characteristic of loess. Some other deposits can be silty in part, but none is as consistently silty as loess, and few have the predominance of coarse silt that detailed particle size analyses of most loess deposits show (Figure 4).

The abundance of coarse silt particles in loess may be attributed to one or both of the following: (a) production of large amounts of silt by some process involving weathering of coarser particles close to the site of deposition, (b) concentration of silt from material with a mixed population of particle sizes by segregation during fluid transportation, possibly over long distances. However, the uniformity of loess composition over large areas suggests that (b) was usually more important than (a).

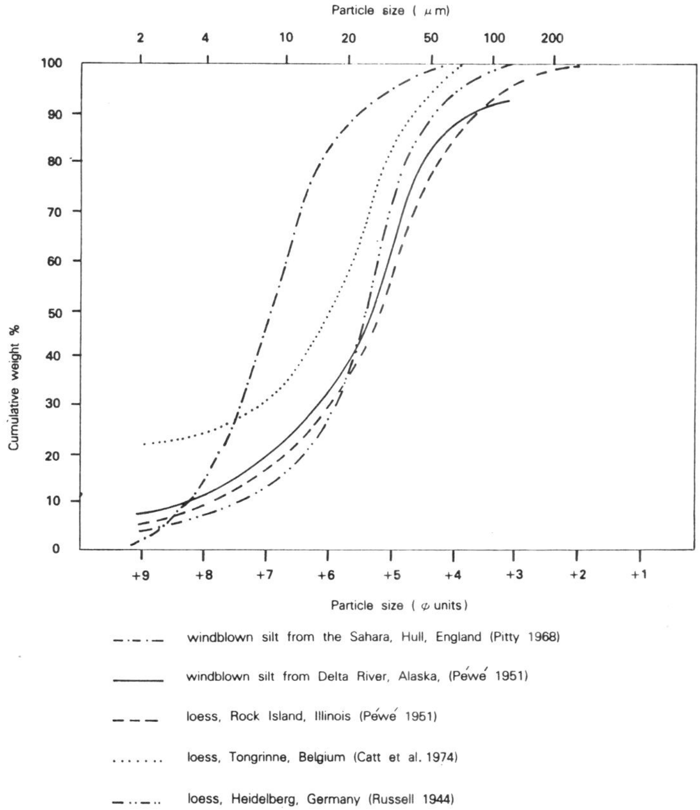

Figure 4. Particle size distributions of typical loess samples compared with windblown silt from periglacial and hot desert sources.

Most of the silt particles in loess (other than the extremely calcareous carbonate loesses) are sharply angular fragments of quartz. These were originally attributed to insolation or frost shattering (Kerr 1881, Wood 1882) of the larger (sand-sized) quartz particles usually found in other sedimentary, igneous and metamorphic rock types. However, Smalley (1966) showed that glacial grinding of granitic and other quartzose rocks can produce similar quartz silt particles, and this fitted the observation that many of the major loess deposits are associated with areas previously glaciated. Also, other workers have shown that the deposits of recent glaciers abound in morphologically similar quartz silt; for example, Manecki et al. (1980) showed that fine-grained sediments from ice caves beneath, and morainic features in front of, the Broggi Glacier in Peru contained up to 50% by weight of 20-60 μm quartz resembling that in loess. Nevertheless, Pye (1984b) has shown that quartz grains morphologically similar to those in loess can be formed by frost action, glacial grinding and other processes discussed below.

To explain the smaller loess deposits originating in hot deserts, Smalley and Vita-Finzi (1968) suggested that the other most likely process for formation of quartz silt is spalling by grain-to-grain impacts of angular sand grains during wind transportation. However, aeolian abrasion experiments by Kuenen (1960, 1969) showed that spalling produces almost no quartz particles in the 20-60 μm range, and this led Smalley and Krinsley (1978) to propose that most desert loess was reworked from the glacial deposits in mountain regions nearby. Rogers et al. (1963) suggested that some quartz silt is produced by abrasion of pebbles and coarse sand in rivers and on beaches, but they concluded that most of this silt is <25 μm, so it is unlikely that much of the silt in loess is formed in this way. Moss et al. (1973) and Riezebos and Van der Waals (1974) noted marginal cracks and other inherent defects in quartz sand grains, and suggested that these are important for the production of silt grains by abrasion in glacial, fluvial, marine and aeolian environments. In deserts the crystallisation of salts in such cracks (salt-weathering) may well produce significant amounts of silt, as shown by the experimental work of Goudie et al. (1979).

Nahon and Trompette (1982) maintain that microfractures, generated in quartz grains during the transformation of high- to low-temperature quartz on cooling of igneous rocks, are exploited by other weathering processes in tropical and Mediterranean soils to produce large amounts of quartz silt. The processes mainly involve dissolution of the quartz along the planar fractures, so as to produce several silt particles from each sand-sized grain. The silt residues are concentrated in the thick weathered mantles common in tropical areas, and further concentrated to form loess-like deposits by surface reworking. These authors quote the lack of silt particles in the late Proterozoic and Upper Ordovician glacial deposits of Africa as evidence for the ineffectiveness of glacial grinding in the production of silt particles. They explain the abundance of silt in Quaternary glacial and periglacial deposits as resulting from glacial and glaciofluvial reworking of weathered silty zones formed after "intensification of weathering processes."

Recently further observations and new experimental work have revived the idea that frost action can produce significant amounts of silt (Zeuner, 1949; Moss et al., 1981; Konishchev, 1982; Pye and Paine 1984). Also, Whalley (1979) showed that silt fragments can be spalled from sand grains during fluvial transportation, and despite Kuenen's results Whalley et al. (1982) have produced coarse silt by aeolian abrasion of angular quartz sand. So silt for formation of loess could have originated by various processes. Additionally some could have come from the few other rock types that contain primary silt-sized quartz crystals, such as schists (Kuenen, 1960; Brockie, 1973). However, the amounts that could come from primary sources or be created by aeolian or fluvial abrasion, frost shattering or chemical weathering in tropical and subtropical soils were almost certainly too small to account for the thick and extremely extensive loess deposits in Europe, Asia and U.S.A. The processes that seem most likely to have provided large amounts of silt are glacial grinding for the periglacial loesses and salt-weathering for some of the small desert loess deposits in areas far from glaciated mountains. But even these processes do not produce deposits with as much silt and as little of other size particles as loess contains, so processes of silt concentration during transport from source to depositional site must also be considered.

4. TRANSPORTATION AND DEPOSITION OF LOESS

4.1. Various early hypotheses

Early geologists such as Lyell (1834, 1847, 1852), Prestwich (1864), and Fuller and Clapp (1903) believed that loess was alluvial—that is, deposited by rivers. The largest loess deposits are indeed associated with the valleys of major rivers, such as the Rhine, Danube, Mississippi and Huang-Ho (Smalley, 1972), but this could be because during cold stages of the Quaternary such rivers carried large amounts of meltwater and sediment from major glaciers. The sediment could have been deposited in part by the rivers, but in very cold conditions with large amounts of sediment load the valleys would have been occupied by braided, multichannel streams, and the sediment would have been frequently exposed to reworking by winds blowing off the glaciers. This would explain why some of the loess both close to and far from these valleys occurs at heights above any possible river water level. Richthofen (1877, 1882) used this argument and other features of the Chinese loess to popularise the aeolian theory of loess deposition originally proposed by Virlet d'Aoust (1857). Richthofen believed that the Chinese loess was blown eastwards from the central Asian deserts and settled in prairies bordering enclosed basins at a time when the climate was much drier than today and the land extended a long way east of the present Chinese coast. Wind transportation of the Chinese and most other loess deposits is now almost universally accepted. Smalley and Vita-Finzi (1968) suggested that most of the Chinese loess was blown northeastwards from the glaciated mountains of southern China, but this has been questioned by Derbyshire (1984).

Other early workers suggested that loess is lacustrine (Pumpelly, 1866; Warren, 1878; Call, 1882), marine (Kingsmill, 1871), volcanic (Howorth, 1882), cosmic (Keilhack, 1920) or colluvial (Kayser, 1923) in origin, and Russell (1944) proposed that the loess of the Mississippi valley originated by weathering and downslope creep of fine backswamp terrace deposits. However, no theory of loess deposition has ever achieved the same universal support as the aeolian hypothesis. The final deposition of some loess or loess-like accumulations may have been lacustrine, marine, colluvial (i.e., moved downslope by runoff, creep or gelifluction) or fluvial, and others may indeed have volcanic or even cosmic components. But none of these theories accounts very readily for the abundance of silt particles in loess and the frequent presence of terrestrial gastropods, rodents etc.

4.2. Aeolian transportation and sorting

Bagnold (1941) showed that because air flow over the ground surface is usually turbulent rather than laminar, and therefore includes vertical velocity components at right angles to the main (horizontal) direction of flow, small particles can be carried up into the atmosphere and travel in suspension at the same speed as the air. At the highest frequently recorded wind speeds the maximum diameter of quartz particles that can be carried in suspension is about 200 μm, but the smaller the particle the higher and further it is carried. Particles somewhat coarser than the size which can be carried in suspension may be moved short distances close to the ground, usually about 10 cm and rarely more than 1 m above the surface. These larger particles are initially entrained when they project above the general surface level and the drag force of the air moving over them overcomes the combined effects of their mass and the friction of grains beneath. There is a square root relationship between the threshold velocity required to entrain grains and grain diameter above about 80 μm

(Figure 5), but increasing wind velocities are required to entrain particles finer than 80 μm because the drag is no longer carried by a few exposed grains but is distributed over the whole surface. After large grains (>200 μm) have been entrained by gusts of sufficiently high velocity, they quickly fall back to the surface. The path of each grain through the air is defined by its initial upward velocity and changes in wind velocity with height above the ground surface and with time, but the mean hop distance is always short. The particles moved by this saltation process consequently travel much more slowly and less far than those in suspension.

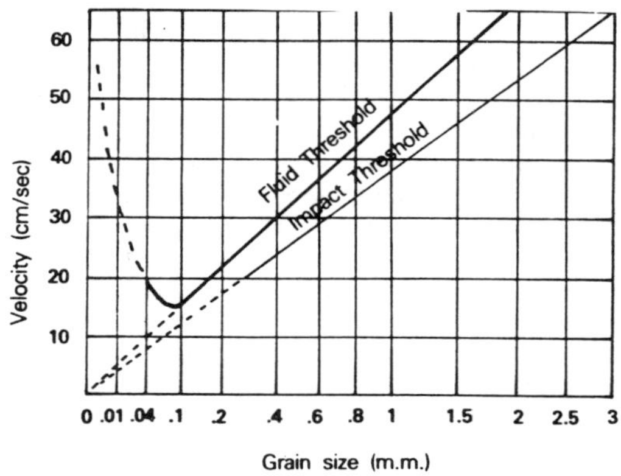

Figure 5. Wind velocities required to entrain grains of various sizes (redrawn from Bagnold, 1941).

When the saltating particles return to the ground surface, their momentum on impact can push or roll grains up to six times their size along the surface. By disturbing the surface, they also make it easier for particles to be entrained as either saltation or suspension load; this gives the slightly lower impact threshold velocity shown in Figure 5.

Although the atmospheric suspension load should have a modal size of 80 μm, because particles of that size require the least threshold velocity, in practice the modal size is usually much less; 80 μm particles are entrained most easily, but smaller particles are carried further because they fall out of suspension less readily. The frequent dust clouds in and near modern desert areas contain particles with median or modal sizes between 15 and 68 μm (Goudie, 1978), but the less frequent falls of far-travelled dust in areas such as southeastern Australia (Walker and Costin, 1971), New England (Robinson, 1936), the eastern North Atlantic (Game, 1964), eastern England (Pitty, 1968) and the California Channel Islands (Muhs, 1983) are finer, with 4-20 μm modal sizes (Figure 4).

Aeolian processes therefore seem to be quite effective in sorting particles of different sizes from an originally mixed population. Sand particles >200 μm travel only by saltation or by surface creep resulting from the impacts of saltating grains, and are therefore likely to accumulate within a short distance of their source. Particles <200 μm are likely to be carried in suspension, but the coarsest of these require high wind speeds in order to remain in suspension; particles around 80 μm are lifted most easily, but it is only particles <30 μm diameter that have settling speeds under gravity less than the minimum wind speed (15-20 cm/sec) capable of entraining grains of any size. Particles of 30-200 μm consequently have a short life as part of the suspension load, and are likely to be redeposited as soon as wind speeds slacken, though the finer ones in this range are often carried tens of kilometres. Particles <30 μm are likely to stay much longer in suspension, and may be carried to heights of several kilometres, especially if cyclonic disturbances cause strong turbulence (Pye, 1984c), so that they can travel hundreds or even thousands of kilometres around the world. In addition, all components of the suspension load >1 μm may be removed from the atmosphere by incorporation into raindrops (Twomey 1977).

So wind transportation can produce deposits with particle size distributions similar to loess; those with a majority of particles significantly finer than 30 μm are likely to have come hundreds or thousands of kilometres without encountering rain clouds, but the more common loess deposits with modal sizes >30 μm would have come from sources a few tens of kilometres distant, perhaps as much as 200 km away if there were consistently strong winds and no rain. For example, from the mineralogical composition of aeolian dust samples collected at Tempe in the Arizona desert, U.S.A., which had 75% of particles between 5 and 50 μm diameter and particle size distributions similar to many loesses, Péwé (1984) concluded that most of the particles had come from 50-200 km away.

Another aspect of wind action which is probably important in the formation of loess is the inability of wind to entrain particles finer than about 30 μm (Bagnold, 1941, p. 90) unless the surface is disturbed, for example by saltating sand grains. Once an area of land has been covered by far-travelled silt the saltation load is effectively zero, because sand particles originally entrained with the silt have been long since deposited, any local sand is no longer exposed to erosion, and although coarse silt particles in the cover can be entrained by strong winds they do not saltate. Consequently the silt cover is virtually resistant to erosion, at least while the same conditions of climate and sediment supply prevail; it is more likely to remain or continue accreting than be blown away.

4.3. Fluvial activity in the early stages of silt transportation

Although it is clear that aeolian processes can account for many aspects of loess formation, it is likely that water transportation is also important, if only in originally separating silt particles from the clay with which they are often intimately mixed in likely source materials, such as glacial tills. Dust clouds arising from the seasonally desiccated surfaces of proglacial outwash plains have been reported from Greenland (Kornerup 1881), Alaska (Tarr and Martin, 1914; Péwé, 1951, 1955), New Zealand (Cotton, 1952) and Iceland (Spethman, 1912), and in at least one of these countries (Hobbs, 1931) silty loess-like aeolian deposits are accumulating at the present day. The silt produced by glacial crushing and abrasion is spread across these broad plains by braided streams in periods of spring and summer melting, before it is taken up by the wind from the unvegetated surface when the sediments dry out in the autumn. Small amounts of silt are also blown from the surfaces of glaciers where debris accumulates by ablation, but again only after there has been local disaggregation and sorting by meltwater. Similarly in hot deserts water often plays an important preliminary role in making silt more available for entrainment by the wind. Yaalon (1969) noticed that the intermittent floods in the Negev Desert (Israel) spread silt and clay across the floors of wadis and surfaces of alluvial fans, and a thin film of clay is often deposited on the surface above siltier layers beneath. When the clay dries it forms a crust, which cracks and curls, thus exposing to wind erosion the drying silt beneath.

So although the fluviatile hypothesis for the origin of all loess deposits has long been abandoned, mainly in favour of an aeolian origin, running water was nevertheless quite important in the earlier stages of silt transportation and concentration in both periglacial and hot desert environments.

Smalley (1966), Smalley and Vita-Finzi (1968), and Smalley and Smalley (1983) have formalised the various processes thought to be involved in the formation of loess into "event sequences." These include processes for production of silt grains (P events), transportation and sorting (T events), and deposition (D events). Hypothetical sequences involving several P, T and D events have been suggested for the loess in different areas; for example, the Tashkent loess was probably produced from silt formed by frost action and glacial crushing in the Tien Shan Mountains (P1), which was originally transported by mountain streams (T1), then deposited in the foothills (D1), but reworked and carried by major rivers (T2) to floodplains in desert areas (D2), where it was eroded and transported by the wind (T3) and eventually deposited as aeolian silt at the desert fringe (D3) (Smalley and Smalley, 1983). This is a convenient way of clarifying what may often be a complex sequence of processes in the formation and concentration of silt particles to form loess.

4.4. Accumulation of loess

Silt particles carried in suspension by the wind can be deposited directly when the wind slackens, or be rained out of the atmosphere, and there are reasons of fluid dynamics why the particles once deposited in abundance are not re-entrained by the wind. However, other processes could have been involved in the trapping and stabilisation of aeolian dust, and in the modification of dust accumulations after deposition, to confer some of the special properties shown by loess.

Richthofen (1882) and Deecke (1920) suggested that grasses, mosses and other small plants helped trap the dust, a process which has been observed to occur in present-day tundra and hot arid regions, and is also indicated by the sparse pollen assemblages in many loesses (Fink and Kukla, 1977; Liu Tungsheng et al., 1982; Urban, 1984). Roots and stems of grasses may also account for the white tubular concretions of secondary carbonate abundant in many loess deposits. Tuck (1938) noted a vertical gradation in the still accumulating loess-like deposits of Alaska from fibrous living roots close to the surface, through a zone with dead carbonised root remains, into a lower zone in which the remains are replaced by tubular carbonate concretions. The carbon dioxide formed during oxidation of the dead rootlets may cause calcite to precipitate if the pore-water in freshly deposited loess is alkaline and rich in calcium ions.

It is unlikely that rain played much part in deposition of existing loesses. Once deposited, loess is very easily eroded by water, and frequent heavy rain would certainly carry silt from the atmosphere directly into rivers and the sea. Indeed the absence or scarcity of loess in areas such as the coastal regions of western Europe, eastern and western U.S.A., and Japan, that probably had an oceanic climate even at the glacial maxima, is best attributed to surface erosion by rain, because aeolian silt was probably almost as plentiful here as elsewhere during cold Pleistocene periods. In the drier areas with a more continental climate, occasional light rain showers or snowfall (Walther, 1900) may have assisted loess deposition, but the paleontological evidence for cold arid conditions suggests that the main process was dry fall-out from the atmosphere.

Cegla (1969) pointed out that the particle size and pore structure of loess favour capillary rise of water, and that surfaces moistened by this process would trap windborne particles more effectively than completely dry surfaces. This may explain the accumulation of some thick loess deposits that contain no evidence for contemporary vegetation.

4.5. Post-depositional changes

Some of the common characteristics of loess, such as the large porosity, the abundance of small secondary carbonate concentrations and the structural collapse on flooding, are not explained simply by deposition of silt by any of the processes discussed above. They can either be explained by pedogenetic or diagenetic changes occurring during and soon after deposition, or by some process of loess formation not dependent on deposition of silt, such as loessification.

When considering secondary carbonate accumulations in loess, it is important to distinguish between forms related to processes occurring during loess deposition or soon after deposition in the same (cold or hot) arid environment, and those related to pedogenesis during later periods of more humid climate. Smalley (1971) referred to these two forms as "original" and "contemporary" respectively, though the latter term is slightly misleading because it may include some deeply buried interglacial accumulations in older loesses. The "contemporary" forms occur in a limited zone beneath present or interglacial humid soils, and are derived from the "original" accumulations and from detrital (wind blown) carbonate particles, which are slowly dissolved in higher soil horizons made acidic by decaying organic matter, the carbonate then being partly reprecipitated in lower (calcareous) horizons where the pH increases. The main "contemporary" forms are (a) subspherical nodules up to 5 cm across, often hollow and with a concentric structure, and (b) thin coatings (calcans) along vertical fissures. The "original" forms of secondary carbonate include (a) the fine tubular concretions formed around decaying rootlets, (b) "loess-dolls," which are usually composed of several solid, pale brown, lenticular concretions each 5-20 cm across and cemented together at their edges to form compound concretions, and (c) submicroscopic encrustations (gnarls) on quartz and other silt grains (Smalley, 1970). Capillary rise and evaporation in semi-arid environments during loess deposition (Cegla, 1969) would give a continuing supply of calcium-rich pore water to form the "original" accumulations of secondary carbonate, and oxidation of organic matter (grass roots) incorporated in the accumulating loess would produce CO_2 to precipitate the calcium as carbonate in an alkaline environment.

However, the source of calcium ions in the rising pore water is uncertain. The calcium cannot be derived from detrital carbonate particles, because they dissolve only in acidic conditions and the lower zones of individual loess deposits are usually alkaline throughout and have obviously remained so since deposition. Other possible sources within the loess are calcium ions and compounds such as calcium oxalate in plant remains, the exchangeable calcium on clay particles, and the calcium in other detrital minerals such as plagioclase felspars, but it is unlikely that any one of these could supply enough calcium for all the secondary carbonate present in some loesses. The most calcareous loesses invariably overlie calcium-rich rocks, such as limestones or some ultrabasic igneous rocks, so some of the calcium could have been drawn ultimately from bedrock sources by weathering processes in alkaline conditions and capillary rise during loess deposition. The weathering processes and water chemistry were perhaps similar to those of alkaline springs associated with ultrabasic rocks in Oman, Yugoslavia and other areas (Barnes et al., 1978).

The Russian theory of loess formation by pedogenesis or loessification takes the idea of derivation of calcium from the bedrock much further, by suggesting that loess originates entirely by weathering of any fine-grained, permeable, calcium-rich sediment or rock. Originally proposed by Berg (1916, 1932, 1964), loessification was later supported by Gerasimov (1939), Lukashev (1970) and other Russian workers, and a modified version was offered by Russell (1944). Berg saw several problems in the aeolian theory of loess formation, and suggested instead that weathering processes in the steppe environment can produce a calcareous silty residuum by partly dissolving large particles (compare Nahon and Trompette, 1982) and aggregating fine (clay) particles in situations where carbonate-rich solutions migrate upwards because of capillary rise and evaporation from the ground surface. As Smalley (1971) pointed out, many of Berg's objections to the aeolian theory have now been explained satisfactorily, but some aspects of the loessification theory are probably necessary to explain the abundance of "original" secondary carbonate dispersed throughout the deposit. Russell and some of the Russian workers were also impressed by the fact that the deposits seen as the unloessified source materials grade vertically and laterally into loess, but this can usually be explained by progressive incorporation of windblown material into other (glacial, fluvial or colluvial) sediments.

The large porosity and small bulk density of loess are also left unexplained by simple aeolian deposition. Some macropores visible to the naked eye were formed when rootlets of grass and other plants growing in the accumulating loess were decomposed. However, most of the total porosity results from much smaller pores, which must be attributed to some other factor. According to Berg's loessification theory, the porosity of loess results from aggregation of the clay and other colloidal particles by flocculation in the calcium-rich pore-water, and then stabilisation of the aggregates by precipitation of carbonate coatings. But scanning electron micrographs of broken surfaces of loess do not show this type of microstructure. Instead they often show that micropores are formed because silt particles are propped apart by biconcave clay bridges (Grabowska-Olszewska, 1983), which must have originated on drying when clay dispersed within the voids between silt particles was carried towards the pore walls by retreating water menisci.

The micropores therefore seem to be another feature related to the evaporation of capillary water from the surface of the slowly accumulating loess. But simple drying, even by rapid evaporation in desert conditions, does not entirely explain the large porosity of loess. The surface tension of water menisci may have moved dispersed clay particles, but cannot have pushed apart the much larger silt grains, which would have been in contact with one another after aeolian deposition. The only process capable of forcing the silt grains apart is growth of ice crystals within upper layers of the freshly-deposited loess. The particle size distribution of loess provides optimum conditions for growth of ground ice crystals. When loess soils in present-day humid environments freeze downwards from the surface, ice lenticles are formed because water is supplied by capillary rise and is also attracted by suction induced by initial ice crystallization; if the temperature is low enough, water vapour also migrates to the freezing front (Van Vliet-Lanoë, 1985). The ice lenticles cause heaving of the soil and create visible pores which remain after thawing. In the more arid environment of the periglacial loess steppes, there was probably a cycle of very shallow nocturnal freezing followed by daytime evaporation and drying, which led to the creation of very small pores because the ice lenticles were never permitted to grow.

The tendency for loess to lose its very porous structure and collapse on wetting is obviously related to the clay bridges, which must disperse on saturation, allowing the silt particles to pack more tightly again in close contact with one another. The collapse shows that the hardness of dry loess and its tendency to maintain vertical surfaces is not caused primarily by carbonate cementation, as suggested by the loessification theory, for saturation would not remove or weaken a carbonate cement. Also decalcified loess layers retain the ability to stand in vertical faces (Lohnes and Handy, 1968) and the tendency to collapse on saturation.

5. DISTRIBUTION OF LOESS IN BRITAIN

Although Britain is at the western margin of the main northwest European loess belt (Figure 1), and has very thin loess deposits, 1 to 4 m thick in a few small areas only (Figure 6), it has some importance in loess studies because the deposits less than 1 m thick are quite widespread and display interesting relationships to different types of substrata. Some of the thickest loess in southeastern England is shown as "Brick-earth" on maps of the British Geological Survey, but the much more extensive distribution of all the loess-like accumulations (Figure 6), many of which are decalcified throughout and partly mixed with underlying deposits, has become clear only since the 1:250,000 soil map of the country was completed (Soil Survey of England and Wales, 1983). These thin, weathered and mixed deposits are recognised by upward increases in coarse silt in the soil profiles and by mineralogical comparisons with unweathered calcareous loess and with glacial deposits that were the source of the aeolian silt (Catt et al., 1974). The minimum identifiable thickness of loess in British soils is 30 cm.

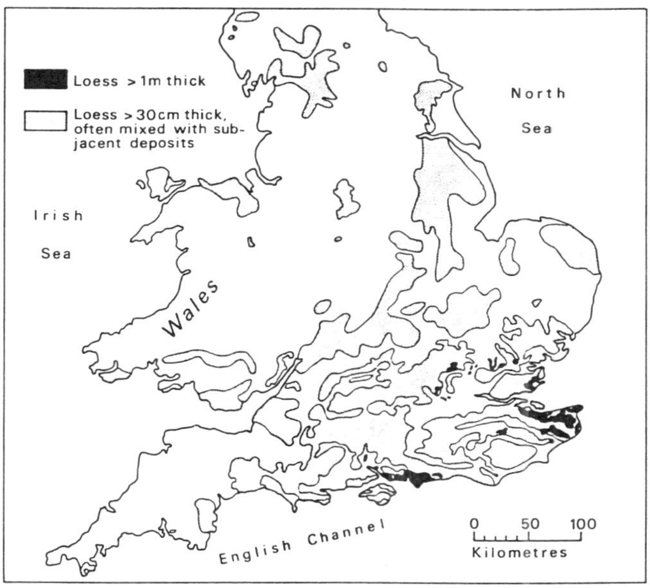

Figure 6. Distribution of loess deposits in England and Wales based upon 1:250,000 soil map (Soil Survey of England and Wales, 1983).

In southeast England the loess occurs on all types of substrata, though the thickest deposits occur mainly on Chalk, calcareous Paleocene sands (Thanet Beds) and Pleistocene terrace gravels of the Thames and other rivers, and some non-calcareous clay and sand formations have little or no loess cover. The loess here was probably blown from glacial outwash deposits in the North Sea area by east or northeast winds, and tends to become thinner, finer and richer in flaky minerals (for example, chlorite) westwards (Catt, 1978). However, in extreme southwestern England the loess coarsens and contains a different heavy silt mineral suite from that further east, probably because it was derived from glacial deposits in the Irish Sea basin (Catt and Staines, 1982).

In northern England, Wales and all western parts of the country, the loess is thin and more patchy in distribution than in southeast England. The patches in these northern and western regions occur mainly on outcrops of Carboniferous Limestone, though some smaller areas are on Devonian and Jurassic limestones, granite or ultrabasic rocks; other types of substrata are either completely devoid of loess (most sandstones) or carry only a few, very small and thin patches of doubtful loess. The restriction of loess to limestone surfaces is sometimes quite striking; for example, on parts of the North Yorkshire Moors interbedded sandstones and oolitic limestones form parallel outcrops on almost level plateau surfaces, and the limestones have a loess cover 40-120 cm thick but the sandstones give rise to sandy podzols.

Nearby on the Yorkshire Wolds the weathered loess cover contains many angular frost-shattered flints derived from nodules in the underlying Chalk, and where it passes eastwards beneath a protective cover of till the unweathered loess also contains much frost-shattered chalk. Catt et al. (1974) interpreted this chalky, flinty loess as a head (gelifluction) deposit formed by cryoturbation (disturbance and mixing by frost action) of loess and frost-shattered bedrock. They suggested that a period of widespread surface erosion in the Holocene had removed loess from most of northern England, but that the loess on the Yorkshire Wolds and other limestone outcrops in the north was preserved during this erosive episode because it was partly cemented by secondary carbonate derived from the abundant chalk or limestone fragments in the original head deposit. However, there is no known remnant of the postulated erosion-resistant, calcreted loess, and the same explanation cannot apply to the loess overlying granite or ultrabasic rocks. Also, although there was considerable soil erosion in Britain during the Holocene, following Neolithic and later periods of deforestation and cultivation, there is no evidence from colluvial and alluvial deposits in northern and western Britain that the eroded soil was especially loess-rich. In contrast, the later Holocene colluvial and alluvial deposits of southeast England are often loess-rich (Burrin, 1981).

The loessification hypothesis would explain the present distribution of loess in Wales and northern and western England in terms of selective production by weathering of certain suitable rocks, calcium-rich as opposed to carbonate-rich, rather than selective retention. However, production of British loess by weathering rather than aeolian deposition does not explain why throughout almost the whole country it contains the same heavy mineral suite as the glacial deposits in eastern England. A third possible explanation is that silt was blown fairly evenly across the country but accumulated to form a layer of loess only where it was trapped by vegetation or stabilised by a supply of groundwater rising by capillary attraction. But there is no reason to suppose either that suitable vegetation existed only on certain rock types or that capillary rise occurred on them only. The one unifying characteristic of all the loess-covered outcrops is that they are of calcium-rich rocks so that, as Berg (1916) originally suggested, loess accumulation was perhaps dependent, in some situations at least, on a weathering environment in which calcium could be mobilised upwards towards the ground surface.

At present it is impossible to be more precise about the processes involved in the accumulation and preservation of loess in England and Wales. However, we can probably learn more about the formation of loess by studying the thin and apparently unpromising deposits in areas like Britain than by examining the thicker and more spectacular loess sequences in eastern Europe, Asia and U.S.A. Also, if thin loess deposits are as widespread in some other regions thought to have little or no loess as they are now known to be in Britain, the total area of the earth's land surface affected by loess is probably much greater than the 20% envisaged by Keilhack (1920).

6. ECONOMIC SIGNIFICANCE OF LOESS

Because loess is such a widespread surface deposit in many parts of the world, its somewhat unusual characteristics have a considerable impact on human lives through agriculture, horticulture, civil engineering and atmospheric pollution. The natural fertility of loess soils has influenced the location of settlement centres in Europe and Asia since the Neolithic (Smalley, 1968). In many countries loess has also been worked extensively for making bricks, both the high temperature fired and the sun-baked mud types, though decalcified loess is usually preferred for fired bricks to avoid the problem of "blowing" when carbonate decomposes.

6.1. Importance of loess in agriculture and horticulture

The particle-size distribution, large porosity and mineralogical composition of loess make it a very suitable medium for the growth of many arable and horticultural crops. Silty soils retain larger amounts of water available to plants than either sandy or clayey soils, so crops are less likely to suffer from drought during the growing season. Loess soils are also well aerated, unless affected by a high groundwater table, and their structural units (peds) are softer than those of clay soils, so that plant roots can explore intrapedal as well as interpedal voids in the search for water and nutrients. The inherent reserves of potassium, phosphorus, calcium, magnesium and many essential trace elements are larger than in many other soils, because loess contains a wide range of minerals derived directly from primary rocks by glacial abrasion without significant chemical weathering. Loess soils usually have large earthworm populations, which help break down plant remains and mix the humus with mineral constituents; this is because earthworms prefer moist but well aerated calcareous soils. The small amounts of clay also make loess soils easy to cultivate, and because they are usually well-drained and do not compact or smear even when wet, the timing of cultivations is less critical than with more clayey soils. However, the small artificial peds (crumbs) produced during seedbed preparation are often unstable under heavy rain, and disaggregation or slaking, a process similar to the structural collapse of loess on saturation, is followed on drying by formation of a hard crust, which can be impenetrable to the growing seedlings. Disaggregation also makes the soil prone to water erosion on slopes, and in thick loess deposits this can lead to the formation of deep gullies. To overcome the problems of crust formation, poor germination and erosion under heavy rain, loess soils are often best suited to cultivation under glass (such as in Guernsey). The absence of stones in most loess soils is also beneficial in agriculture, as farm machinery frequently in contact with the soil (ploughs, tractor tires etc.) suffers less wear, and root crops can be harvested mechanically without damage.

Because of these properties, loess soils are regarded in many countries as among the best agricultural and horticultural soils (Chesworth, 1982). For example, in Britain soils in loess >1 m thick are graded as Class 1 (land capable of producing a very wide range of crops) in the Soil Survey's Land Capability Classification (Bibby et al., 1982), and give greater yields of winter wheat (often >10 t/ha) than other soils (Weir et al., 1984). Even those in loess <1 m thick over hard rock, sand or clay perform slightly better than soils in the same substrata but without superficial loess horizons, leading to the suggestion that it might be profitable to remove loess from areas where thick deposits are going to be covered by industrial development, airports etc., and spread it thinly over nearby loess-free farmland (Catt, 1978). Experiments in Iowa, U.S.A., by Drake and Ririe (1981) showed that the cost of reclaiming opencast coal mines for profitable corn-growing by spreading loess was equivalent to 4.9% of the cost of the coal as delivered to an electric utility 50 miles (80 km) distant, or 1% of the price of the electricity generated for domestic use.

6.2. Loess in civil engineering

The tendency of loess to collapse on wetting, with rapid disruption of intergranular bonds and large losses of porosity and volume, is usually termed metastability or hydro-consolidation (Holtz and Gibbs, 1951). Construction problems arising from failure to ensure adequate drainage of foundations in thick loess deposits are common in Europe (Feda, 1966), U.S.A. (Turnbull, 1948; Krinitzsky and Turnbull, 1967) and Russia (Denisov 1951; Krayev, 1969). Clevenger (1956) found that loess with a dry density <1.28 g/cm^3 collapses by a large amount, whereas the settlement is small or zero if dry density exceeds 1.44. Also, the more clay-rich a loess the less likely it is to collapse (Handy, 1973), and loess redeposited by water is not metastable. If the porosity is decreased to <40% by compaction, the collapse on saturation is small (Lehr, 1967), so a common practice in engineering is to flood and roll a loess site before starting construction work.

Although well-drained natural and artificial sections of loess tend to maintain vertical faces, those higher than about 5 m are unstable because deep tension cracks form and allow large blocks to break away along slip planes inclined at approximately 50°, which eventually build up a talus slope of about 38° (Lohnes and Handy, 1968). In U.S.A. it is therefore common practice to terrace highway and other cuttings in 5 m steps.

Another common problem with roads in loess areas is frost-heave arising from formation of segregation ice lenticles, which are larger and consequently cause greater heave than in most other foundation materials. The amount of heave is often greater if the loess is fairly thin over a less permeable horizon such as a clay or even a clay-enriched buried soil within the loess (Bollen, 1945).

6.3. Dust storms

Atmospheric transportation of silt in arid and semi-arid regions often causes damage to crops and buildings, disruption of transport, death of cattle by suffocation, the spread of disease through dispersion of pathogens, the interruption of radio communications, and the pollution of water supplies (Idso, 1976). Dust storms, defined as occasions when the visibility is less than 1000 m because of atmospheric dust, are as frequent as 60 per year in Kazakhstan (U.S.S.R.) and 68 per year in Mexico City

(Goudie, 1978). The amounts of sediment deposited by single storms can reach 500 t/km^2, and some storms carry several million tons of dust and affect very large areas; Goudie (1978) estimated that the total quantity of aeolian dust currently deposited on the earth's surface is similar in order of magnitude to worldwide denudation by fluvial erosion.

7. CONCLUSIONS

Despite their undoubted importance, both to man and as a geological process of sediment transportation, modern dust storms are only a minor remnant of the extensive loess clouds that seasonally filled the atmosphere in mid- and high-latitudes for long periods on many occasions during the Quaternary. Periglacial atmospheric pollution by loess could have been involved in a feed-back mechanism during glaciations, in which the dust clouds generated originally by glacial erosion became increasingly responsible for decreasing incoming solar radiation and thus worsening the climate. The extent to which this occurred is unknown, but could be investigated by computer-based simulations of the global ice-age climate.

The problems of finding an acceptable definition and method of subdividing loess are important to resolve, if only because the various deposits included within the term by some workers have different geotechnical and pedological properties. However, it is important to retain a general term for surface deposits which are essentially silty because they were at some stage transported and sorted by the wind. For some purposes, such as soil mapping, redeposition by fluvial or colluvial processes is of secondary importance. It has long been convenient to use "loess" in this general sense, even though many of the materials included, perhaps modified by weathering or admixture with other deposits, do not resemble typical calcareous loess at all closely in the field. The term conveys important characteristics which distinguish such materials from other sediments, notably their widespread distribution and uniformity of thickness on almost all facets of the land surface, and the preponderance of silt which is fairly uniform in size distribution and mineralogy. A stricter definition of loess based on lithological details may satisfy the petrologist, but is of little use to the soil scientist, geomorphologist or civil engineer.

Smalley and Smalley (1983) extended the definition of loess to include silty alluvial, glaciofluvial and other deposits (that is, those which had reached the P1-T1-D1 stage of events where T1 was not aeolian), because they felt that in soils such deposits would confer similar agricultural properties to aeolian silt. However, this seems an unnecessary extension of the term because these deposits do not have the same pedological significance as aeolian loess; they are unlikely to be as consistently silty as aeolian loess over large areas and are unlikely to occur on as many physiographic components of the landscape. These characteristics depend on an aeolian episode of deposition, and this should be a critical aspect of the definition of loess.

Thick loess sequences with multiple buried soils provide a terrestrial record equivalent to the extremely complex history of Quaternary climatic changes deduced from deep ocean sediments. There is also value in the study of thin loess deposits, which are often weathered throughout, more sparsely distributed and partially mixed with other materials, as in parts of Britain. These thin deposits throw light on the problems of loess formation, which obviously involves more than deposition of an atmospheric suspension load by rain or mere slackening of the wind. Few would fully concur with the loessification theories of Berg and Russell (section 4.5), but the distribution of the thin British deposits in relation to certain substrata, the unusual structure and geotechnical properties of loess, and the reorganisation of its carbonate content are all best explained by some loessification process following deposition of aeolian silt. This was essentially a pedological process, probably involving repeated diurnal cycles of freeze-drying, capillary rise of calcium-rich water and oxidation of organic matter incorporated in the loess during and after deposition, but as some workers claim, the process may have depended on environmental conditions which have no widespread modern analogue.

8. ACKNOWLEDGEMENTS

I thank Mrs. P. Ashcroft for drawing the figures and Professor D.H. Yaalon for critical comments on the text.

9. REFERENCES

Backman, J., 1983, The age of the first cycle of ice-rafted debris in the North Atlantic Ocean, *Terra Cognita* **3**, 221.

Bagnold, R.A., 1941, *The Physics of Blown Sand and Desert Dunes*, Menthuen & Co., London, xx + 265 pp.

Barnes, I., O'Neil, J.R. and Trescases, J.J., 1978, Present-day serpentinization in New Caledonia, Oman and Yugoslavia, *Geochim. Cosmochim. Acta* **42**, 144-145.

Berg, L.S., 1916, O proiskhozhdenii lessa, *Izvestiya Russkogo Geograficheskogo Obshchesta* **52**.

Berg, L.S., 1932, The original loess. Beitrage zur Geophysik, Abhandlungen aus dem Geographischen Seminar der Universität Strassburg, **35**, 130-150.

Berg, L.S., 1964, *Loess as a Product of Weathering and Soil Formation*, (Trans. A. Gourevitch), Israel Program for Scientific Translations, Jerusalem, 206 pp.

Bibby, J.S., Douglas, H.A., Thomasson, A.J. and Robertson, J.S., 1982, *Land Capability Classification for Agriculture*, Soil Surv. Scotland Monogr., Macaulay Inst. Soil Res., Aberdeen, viii + 75 pp.

Birkeland, P.W. and Shroba, R.R., 1974, The status of the concept of Quaternary soil-forming intervals in the western United States, *in* W.C. Mahaney (ed.) *Quaternary Environments*, pp. 241-276, York University, Toronto, Geograph. Monographs 5, vii + 318pp.

Boardman, J., 1985, Comparison of soils in midwestern United States and western Europe with the interglacial record, *Quat. Res.* **23**, 62-75.

Bollen, R.E., 1945, Characteristics and uses of loess in highway construction, *Am. J. Sci.* **243**, 283-293.

Bowen, D.Q., 1978, *Quaternary Geology. A Stratigraphic Framework for Multidisciplinary Work*, Pergamon Press, Oxford, xi + 221 pp.

Brockie, W.J., 1973, Experimental frost-shattering, *Proc. 7th New Zealand Geogr. Congr., Hamilton 1972*, 177-186.

Bruins, H.J. and Yaalon, D.H., 1981, Stratigraphy of the Netivot section in the desert loess of the Negev (Israel), *Acta Geol. Acad. Sci. Hungaricae* (1979) **22**, 161-169.

Brunnacker, K., 1973, Einiges über Löss-Vorkommen in Tunisien, *Eiszeitalter und Gegenwart* **23-24**, 89-99.

Brunnacker, K., 1974, Lösse und Pälaoböden der letzten Kaltzeit im mediterranen Raum. *Eiszeitalter und Gegenwart* **25**, 62-95.

Burrin, P.J., 1981, Loess in the Weald, *Proc. Geol. Assoc.* **92**, 87-92.

Butler, B.E., 1956, Parna—an aeolian clay, *Austr. J. Sci.* **18**, 145-151.

Butrym, J. and Maruszczak, H., 1984, Thermoluminescence chronology of younger and older loesses, *in* M. Pécsi (ed.) *Lithology and Stratigraphy of Loess and Paleosols*, pp. 195-199, Proc. Symp. INQUA Commission on Loess and Paleopedology, XIth INQUA Congress, Moscow, Geogr. Res. Inst., Hungarian Acad. Sci., Budapest, 325 pp.

Call, R.E., 1882, The loess of North America, *Amer. Natural.* **16**, 369-381 and 542-549.

Catt, J.A., 1978, The contribution of loess to soils in lowland Britain, *in* S. Limbrey and J.G. Evans (eds.) *The Effect of Man on the Landscape: the Lowland Zone*, pp. 12-20, Counc. Brit. Archae. Res. Rep. 21, v + 153 pp.

Catt, J.A. and Staines, S.J., 1982, Loess in Cornwall, *Proc. Ussher Soc.*, **5**, 368-375.

Catt, J.A., Weir, A.H. and Madgett, P.A., 1974, The loess of eastern Yorkshire and Lincolnshire, *Proc. Yorkshire Geol. Soc.* **40**, 23-39.

Cegla, J., 1969, Influence of capillary ground moisture on eolian accumulation of loess, *Bull. Acad. Polonaise Sci., Sér. Sci. Géol. Géogr.* **17**, 25-27.

Cepek, A.G., 1965, Geologische Ergebnisse der ersten Radiokarbondatierungen von Interstadialen im Lausitzer Ursttomtal, *Geologie* **14**, 625-657.

Chesworth, W., 1982, Late Cenozoic geology and the second oldest profession, *Geosci. Canada* **9**, 54-61.

Clevenger, W.A., 1956, Experience with loess as foundation material, *Proc. Amer. Soc. Civil Eng.* **82** (SM3), Paper 1025.

Cotton, C.A., 1952, *Geomorphology. An Introduction to the Study of Landforms*, 6th Ed., Whitcomb & Tombs, Christchurch, 505 pp.

Dan, J. and Yaalon, D.H., 1964, The application of the catena concept in studies of pedogenesis in Mediterranean and desert fringe regions, *Trans. 8th Internat. Soil Sci. Cong., Bucharest, 1964*, **V**, 751-758.

Deecke, W., 1920, Kritische Studien zu Glazialfragen Deutschlands, *Zeitschrift für Gletscherkunde* **11**, 34-84.

Denisov, N.Y., 1951, *Stroitel'nie svoista lessa i lessovidnykh suglinkov*, Gosstroiizdat, Moscow, 133 pp.

Derbyshire, E., 1984, Granulometry and fabric of the loess at Jiuzhoutai, Lanzhou, People's Republic of China, *in* M. Pécsi (ed.) *Lithology and Stratigraphy of Loess and Paleosols*, pp. 95-103, Proc. Symp. INQUA Commission on Loess and Paleopedology, XIth INQUA Congress, Moscow, Geogr. Res. Inst., Hungarian Acad. Sci., Budapest, 325 pp.

Dodonov, A.E., 1981, Stratigraphy of the Upper Pliocene-Quaternary deposits of Tajikistan (Soviet Central Asia). *Acta Geol. Acad. Sci. Hungaricae* **22** (1979), 63-73.

Dodonov, A.E., 1984, Stratigraphy and correlation of Upper Pliocene-Quaternary deposits of central Asia, *in* M. Pécsi (ed.) *Lithology and Stratigraphy of Loess and Paleosols*, pp. 210-211, Proc. Symp. INQUA Commission on Loess and Paleopedology, XIth INQUA Congress, Moscow, Geogr. Res. Inst., Hungarian Acad. Sci., Budapest, 325 pp.

Drake, L.D. and Ririe, G.T., 1981, A low-cost method of reclaiming strip-mined land in Iowa to agriculture, *Env. Geol.* **3**, 267-279.

Edwards, M.B., 1979, Late Precambrian glacial loessites from North Norway and Svalbard, *J. Sed. Petrol.* **49**, 85-92.

Feda, J., 1966, Structural stability of subsident loess soil from Praha-Dejvice, *Eng. Geol.* **1**, 201-219.

Fink, J. and Kukla, G.J., 1977, Pleistocene climates in central Europe: At least 17 interglacials after the Olduvai Event, *Quat. Res.* **7**, 363-371.

Fliri, F., Hischer, H. and Markgraf, V., 1971, Weitere Untersuchungen zur Chronologie der alpinen Vereisung (Banderton von Baumkirchen, Inntal, Tirol), *Zeit. Gletscherkunde Glazialgeologie* **7**, 5-24.

Fookes, P.G. and Knill, J.L., 1969, The application of engineering geology in the regional development of northern and central Iran, *Eng. Geol.*, **3**, 81-120.

Frye, J.C., Willman, H.B. and Glass, H.D., 1968, Correlation of midwestern loesses with the glacial succession, *in* C.B. Schultz and J.C. Frye (eds.) *Loess and Related Aeolian Deposits of the World*, pp. 3-21, Univ. Nebraska Press, Lincoln, 369 pp.

Frye, J.C., Leonard, A.B., William, H.B., Glass, H.D. and Follmer, L.R., 1974, The Late Woodfordian Jules Soil and associated molluscan faunas, *Ill. State Geol. Surv. Circ.* **486**, 11 pp.

Fuller, M.L. and Clapp, F., 1903, Marl-loess of the lower Wabash valley, *Bull. Geol. Soc. Amer.* **14**, 153-176.

Game, P.M., 1964, Observations on a dust fall in the eastern Atlantic, Feb., 1962, *J. Sed. Petrol.* **34**, 355-359.

Gates, W.L., 1976, The numerical simulation of ice-age climate with a global general circulation model, *J. Atmos. Sci.* **33**, 1844-1873.

Gerasimov, I.P., 1939, K voprosu o genezise lessov i lessovidnykh otlozhenii, *Izvest. Akad. Nauk SSSR, Ser. Geogr. Geofiz.* **1**.

Gill, E.G. and Segnitt, E.R., 1982, Strata of carbonate loess in S.E. Australia, *Loess Lett.* **8**, 3-4.

Ginzbourg, D. and Yaalon, D.H., 1963, Petrography and origin of the loess in the Be'er Sheva Basin, *Israel J. Earth Sci.* **12**, 68-70.

Goudie, A.S., 1978, Dust storms and their geomorphological implications, *J. Arid Env.* **1**, 291-310.

Goudie, A.S., Cooke, R.U. and Doornkamp, J.C., 1979, The formation of silt from quartz dune sand by salt-weathering processes in deserts, *J. Arid Env.* **2**, 105-112.

Goudie, A.S., Rendell, H.M. and Bull, P.A., 1984, The loess of Tajik SSR, *in* K. Miller (ed.) *International Karakoram Project*, pp. 399-412, Cambridge University Press.

Grabowska-Olszewska, B., 1983, Microstructural sensitivity of loesses, *Bull. Acad. Polonaise Sci., Sér. Sci. Terre* **30**, 3-4 (1982), 181-188.

Grahmann, R., 1932, Der Loess in Europa, *Mitt. Gesell. Erdkunde Leipzig*, **51**, 5-24.

Handy, R.L., 1973, Collapsible loess in Iowa, *Proc. Soil Sci. Soc. Amer.* **37**, 281-284.

Heller, F. and Liu Tungsheng, 1982, Magnetostratigraphical dating of loess deposits in China, *Nature* **300**, 431-433.

Heller, F. and Liu Tungsheng, 1984, Magnetism of Chinese loess deposits, *Geophys. J. Roy. Astr. Soc.* **77**, 125-141.

Hobbs, W.H., 1931, Loess, pebble bands, and boulders from glacial outwash of the Greenland continental glacier, *J. Geol.* **39**, 381-385.

Holtz, W.G. and Gibbs, H.J., 1951, Consolidation and related properties of loessial soils, *Amer. Soc. Testing Mater., Spec. Tech. Pub.* **126**, 9-33.

Howorth, H.H., 1882, The loess—a rejoinder. *Geol. Mag. II*, **9**, 343-356.

Idso, S.B., 1976, Dust storms, *Sci. Amer.* **235**, 108-114.

Kayser, E., 1923, *Lehrbuch der Geologie, Band 2*, 8th Ed., F. Enke, Stuttgart, xii + 625 pp.

Keilhack, K., 1920, Das Rätsel der Lössbildung, *Zeit. Deutsch. Geol. Gesell.* **72B**, 146-167.

Kerr, W.C., 1881, On the action of frost in the arrangement of superficial earthy material, *Amer. J. Sci., Ser. 3*, **21**, 345-358.

Kingsmill, T.W., 1871, The probable origin of deposits of "loess" in north China and eastern Asia. *Quart. J. Geol. Soc. London* **27**, 376-384.

Konischev, V.N., 1982, Characteristics of cryogenic weathering in the permafrost zone of the European SSR, *Arctic Alpine Res.* **14**, 261-265.

Kornerup, A.,1881, Geologiske iagttagelser fra Vestkysten af Grönland, *Meddel. Grönland* **2**, 149-194.

Krayev, V.F., 1969, On the subsidence of loess soils of the Ukraine, *Actes du Colloque de Tokyo IASH-AIHS UNESCO* **1**, 321-322.

Kriger, N.I., 1965, *Loess, its Characteristics and Relation to Geographical Environments*, Nauka, Moscow, 296 pp.

Krinitzsky, E.L. and Turnbull, W.J., 1967, Loess deposits of Mississippi, *Geol. Soc. Amer. Spec. Paper* **94**, 64 pp.

Kuenen, P.H., 1960, Experimental abrasion, 4: Eolian action, *J. Geol.* **68**, 427-449.

Kuenen, P.H., 1969, Origin of quartz silt, *J. Sed. Petrol.* **39**, 1631-1633.

Kukal, Z. and Saadallah, A., 1970, Composition and rate of deposition of the recent dust storm sediments in Iraq, *Casopis Mineralogii Geologii* **15**, 227-234.

Kukla, J., 1970, Correlation between loesses and deep-sea sediments, *Geol. Föreningens Stockholm Förhandl.* **92**, 148-180.

Kukla, G.J., 1977, Pleistocene land-sea correlations, 1: Europe, *Earth-Sci. Revs.* **13**, 307-374.

Lautridou, J.P., 1981, Lithostratigraphie et chronostratigraphie des loess de Haute Normandie, *Acta Geol. Acad. Sci. Hungaricae* **22** (1979), 125-132.

Lautridou, J.P., 1982, The Quaternary of Normandy. *Quat. Res. Assoc. Field Guide*, Cambridge, 88 pp.

Lazarenko, A.A., Bolikhovskaya, N.S. and Semenov, V.V., 1981, An attempt at a detailed stratigraphic subdivision of the loess association of the Tashkent region, *Int. Geol. Rev.* **23**, 1335-1346.

Lehr, H., 1967, Foundation engineering problems in loess soils, *Proc. 3rd Asian Regional Conf. on Soil Mechanics and Foundation Engineering*, pp. 20-24, Academic Press, Jerusalem.

Liu Tungsheng and Menglin, D., 1982, Pleistocene stratigraphy and Plio/Pleistocene boundary in China, *in* Liu Tungsheng (ed.) *Quaternary Geology and Environment of China*, pp. 1-6, China Ocean Press, Beijing, v + 224 pp.

Liu Tungsheng, Zhisheng, A., and Baoyin, Y., 1982, Aeolian processes and dust mantles (loess) in China, *in* R.J. Wasson (ed.) *Quaternary Dust Mantles of China, New Zealand and Australia*, pp. 1-17, A.M.U., Canberra.

Lohnes, R.A. and Handy, R.L., 1968, Slope angles in friable loess, *J. Geol.* **76**, 247-258.

Lukashev, K.I., 1970, *Lithology and geochemistry of the weathering crust*, Israel Program for Scientific Translations, Jerusalem, iv + 367 pp.

Lyell, C., 1834, Observations on the loamy deposit called "loess" of the basin of the Rhine, *Edinburgh New Philos. J.* **17**, 110-122.

Lyell, C., 1847, On the delta and the alluvial deposits of the Mississippi River and other points in the geology of North America, observed in the years 1845-1846, *Amer. J. Sci., Ser. 2*, **3**, 34-39 and 267-269.

Lyell, C., 1852, On the Tertiary strata of Belgium and French Flanders, Part II. The Lower Tertiaries of Belgium, *Quart. J. Geol. Soc. London* **8**, 277-370.

Makinen, E.A. and Dalrymple, G.B., 1979, Revised geomagnetic polarity time scale for the interval 0-5 m.y. B.P., *J. Geophys. Res.* **84**, 615-626.

Manecki, A., Muszynski, M. and Wrzak, J., 1980, Fine grained deposits from the bottom of Broggi Glacier and its foreland, *Polska Akad. Nauk, Oddzial Krakowie Komisja Nauk Mineralogicznych, Prace Mineral.* **64**, 27-46.

McKay, E.D., 1979, Wisconsinan loess stratigraphy of Illinois, *in* Wisconsinan, Sangamonian and Illinoian stratigraphy in central Illinois, pp. 95-108, *Ill. State Geol. Surv. Gdbk.* **13**, 139 pp.

Morrison, R.B., 1978, Quaternary soil stratigraphy—concepts, methods and problems, *in* W.C. Mahoney (ed.) *Quaternary Soils*, pp. 77-108, GeoAbstracts, Norwich, xv + 508 pp.

Moss, A.J., Walker, P.H. and Hutka, J., 1973, Fragmentation of granitic quartz in water, *Sedimentology* **20**, 498-511.

Moss, A.J., Green, P. and Hutka, J., 1981, Static breakage of granitic detritus by ice and water in comparison with breakage by flowing water, *Sedimentology* **28**, 261-272.

Muhs, D.R., 1983, Airborne dust fall on the California Channel Islands, U.S.A., *J. Arid Env.* **6**, 223-238.

Nahon, D. and Trompette, R., 1982, Origin of siltstones: glacial grinding versus weathering, *Sedimentology* **29**, 25-35.

Obruchev, W.A., 1945, Loess types and their origin, *Amer. J. Sci.* **243**, 256-262.

Paepe, R. and Vanhoorne, R., 1967, The stratigraphy and palaeobotany of the Late Pleistocene in Belgium, *Mémoires pour servir à l'explication des Cartes géologiques et minières de la Belgique* **8**, 95 pp.

Pécsi, M., 1984, Is typical loess older than one million years? *in* M. Pécsi (ed.) *Lithology and Stratigraphy of Loess and Paleosols*, pp. 213-224, Proc. Symp. INQUA Commission on Loess and Paleopedology, XIth INQUA Congr., Moscow, Geogr. Res. Inst., Hungarian Acad. Sci., Budapest, 325 pp.

Penkov, A.V. and Gamov, L.N., 1977, Palaeomagnetic datums in Pliocene-Quaternary strata of southern Tajikstan, *in* A.E. Dodonov, Y.R. Melamed and K.V. Nikiforova (eds.) *Guidebook, Intern. Symp. Neogene-Quaternary Boundary*, pp. 46-47, Moscow.

Penny, L.F., Coope, G.R. and Catt, J.A., 1969, Age and insect fauna of the Dimlington Silts, East Yorkshire, *Nature* **224**, 65-67.

Péwé, T.L., 1951, An observation on wind-blown silt, *J. Geol.* **59**, 399-401.

Péwé, T.L., 1955, Origin of the upland silt near Fairbanks, Alaska, *Bull. Geol. Soc. Amer.*, **66**, 699-724.

Péwé, T.L., 1984, Deposition of windblown dust in central Arizona, U.S.A., *in* M. Pécsi (ed.) *Lithology and Stratigraphy of Loess and Paleosols*, pp. 305-325, Proc. Symp. INQUA

Commission on Loess and Paleopedology, XIth INQUA Congr., Moscow, Geogr. Res. Inst., Hungarian Acad. Sci., Budapest, 325 pp.

Pitty, A.F., 1968, Particle size of the Saharan dust which fell in Britain in July 1968, *Nature* **220**, 364-365.

Prestwich, J., 1864, Theoretical considerations on the conditions under which the (Drift) deposits containing the remains of extinct Mammalia and flint implements were accumulated, and on their geological age. On the loess of the valleys of the south of England, and of the Somme and the Seine, *Phil. Trans. Roy. Soc.* **154**, 247-309.

Pumpelly, R., 1866, Geological researches in China, Mongolia and Japan during the years 1862 to 1865, *Smithsonian Inst. Contr.* **202**, 1-162.

Pye, K., 1984a, Loess, *Prog. Phys. Geogr.* **8**, 176-216.

Pye, K., 1984b, SEM investigations of quartz silt micro-textures in relation to the source of loess, *in* M. Pécsi (ed.) *Lithology and Stratigraphy of Loess and Paleosols,* pp. 139-151, Proc. Symp. INQUA Commission on Loess and Paleopedology, XIth INQUA Congr., Moscow, Geogr. Res. Inst., Hungarian Acad. Sci., Budapest, 325 pp.

Pye, K., 1984c, Some perspectives on loess accumulation, *Loess Lett.* **11**, 5-10.

Pye, K. and Paine, A.D.M., 1984, Nature and source of aeolian deposits near the summit of Ben Arkle, northwest Scotland, *Geol. Mijnbouw* **63**, 13-18.

Richthofen, F.F. von, 1877, *China. Ergebnisse eigener Reisen und darauf gegründeter Studien,* Band 1, Dietrich Reimer, Berlin, xliv + 758 pp.

Richthofen, F.F. von, 1882, On the mode of origin of the loess. *Geol. Mag. II,* **9**, 293-305.

Riezebos, P.A., and Van der Waals, L., 1974, Silt-sized quartz particles: a proposed source, *Sed. Geol.* **12**, 279-285.

Robinson, W.O., 1936, Composition and origin of dust in the fall of brown snow, New Hampshire and Vermont, February 24th, 1936, *Monthly Weather Rev.* **64**, 86.

Rogers, J.J.W., Krueger, W.C. and Krog, M., 1963, Sizes of naturally abraded materials, *J. Sed. Petrol.* **33**, 628-632.

Russell, R.J., 1944, Lower Mississippi Valley loess, *Bull. Geol. Soc. Amer.* **55**, 1-40.

Sarnthein, M., 1978, Sand deserts during glacial maximum and climatic optimum, *Nature* **272**, 43-46.

Scheidig, A., 1934, *Der Löss und seine geotechnischen Eigenschaften,* Dresden.

Semmel, A., 1973, Area between the Scandinavian and the Alpine Glaciation, 1. Periglacial sediments and their stratigraphy, *Eiszietalter und Gegenwart* **23-24**, 293-305.

Shackleton, N.J. and Opdyke, N.D., 1973, Oxygen-isotope and paleomagnetic stratigraphy of equatorial Pacific core V28-238: oxygen isotope temperatures and ice volumes on a 10^5 year and a 10^6 year scale, *Quat. Res.* **3**, 39-55.

Shackleton, N.J. and Opdyke, N.D., 1976, Oxygen isotope and paleomagnetic stratigraphy of Pacific core V28-239 late Pliocene to latest Pleistocene, *in* R.M. Cline and J.D. Hays (eds.) *Investigation of Late Quaternary Paleoceanography and Paleoclimatology,* pp. 449-464, Geol. Soc. Amer. Memoir 145.

Shackleton, N.J. and Opdyke, N.D., 1977, Oxygen isotope and paleomagnetic evidence for early northern hemisphere glaciation, *Nature* **270**, 216-219.

Smalley, I.J., 1966, The properties of glacial loess and the formation of loess deposits, *J. Sed. Petrol.* **36**, 669-676.

Smalley, I.J., 1968, The loess deposits and Neolithic culture of northern China, *Man* **3**, 224-241.

Smalley, I.J., 1970, Calcium carbonate encrustations on quartz grains in loess from the Karlsruhe region, *Naturwiss.* **57**, 87.

Smalley, I.J., 1971, "In-situ" theories of loess formation and the significance of the calcium-carbonate content of loess, *Earth-Sci. Rev.* **7**, 67-85.

Smalley, I.J., 1972, The interaction of great rivers and large deposits of primary loess, *Trans. New York Acad. Sci, Ser. II,* **34**, 534-542.

Smalley, I.J. and Krinsley, D.H., 1978, Loess deposits associated with deserts, *Catena* **5**, 53-66.

Smalley, I.J. and Smalley, V., 1983, Loess material and loess deposits: formation distribution and consequences, *in* M.E. Brookfield and T.S. Ahlbrandt (eds.) *Eolian Sediments and Processes,* pp. 51-68, Developments in Sedimentology 38, Elsevier, Amsterdam.

Smalley, I.J., and Vita-Finzi, C., 1968, The formation of fine particles in sandy deserts and the nature of 'desert' loess. *J. Sed. Petrol.* **38**, 766-774.
Smith, B.J. and Whalley, W.B., 1981, Late Quaternary drift deposits of north central Nigeria examined by scanning electron microscopy, *Catena* 8, 345-367.
Soil Survey of England and Wales, 1983, *Soil Map of England and Wales—Scale 1:250,000*, Soil Survey of England and Wales, Harpenden.
Spethmann, H., 1912, Forschungen am Vatnajökull auf Island, *Zeit. Gesell. Erdkunde zu Berlin* **47**, 414-433.
Tarr, R.S. and Martin, L., 1914, The glaciers and glaciation of Yakutat Bay, *Alaskan Glacier Studies*, pp. 23-231, Nat. Geograph. Soc., Washington.
Taylor, S.R., McLennan, S.M. and McCulloch, M.T., 1983, Geochemistry of loess, continental crustal composition and crustal model ages, *Geochim. Cosmochim. Acta* **47**, 1897-1905.
Tuck, R., 1938, The loess of the Matanuska valley, Alaska, *J. Geol.* **46**, 649-653.
Turnbull, W.J., 1948, Utility of loess as a construction material, *Proc. 2nd Int. Conf. Soil Mechanics and Foundation Engineering* **5**, pp. 97-103, Gehr Keesmaat, Rotterdam.
Twomey, S., 1977, *Atmospheric Aerosols*, Elsevier Scientific Pub. Co., Amsterdam, xiv + 302 pp.
Urban, B., 1984, Palynology of central European loess-soil sequences, *in* M. Pécsi (ed.) *Lithology and Stratigraphy of Loess and Paleosols*, pp. 229-248, Proc. Symp. INQUA Commission on Loess and Paleopedology XIth INQUA Congr., Moscow, Geogr. Res. Inst., Hungarian Acad. Sci., Budapest, 325 pp.
Van Vliet-Lanoë, B., 1985, Frost effects in soils, *in* J. Boardman (ed.) *Soils and Quaternary Landscape Evolution*, pp. 117-158, J. Wiley & Sons, Chichester, x + 391 pp.
Velichko, A.A., 1975, Problems of correlation of the Pleistocene events within glacial, loess-periglacial and maritime regions of East-European plain, *in* V. Sibrava (ed.) *Quaternary Glaciations in the Northern Hemisphere*, Report 2 on the session in Salzburg (Austria), Sept. 10-14, 1974, pp. 82-107, IUGS-UNESCO Int. Geol. Correlation Prog., Proj. 73/1/24, Geol. Surv., Prague, 151 pp.
Virlet-d'Aoust, P.T., 1857, Observations sur un terrain d'origine météorique ou de transport aérien qui existe au Mexique, et sur le phénomène des trombes de poussière auquel il doit principalement son origine, *Bull. Soc. Géol. France, Sér. 2*, **2**, 129-139.
Walker, P.H. and Costin, A.B., 1971, Atmospheric dust accession in south-eastern Australia, *Austr. J. Soil Res.* **9**, 1-5.
Walther, J., 1900, *Das Gesetz der Wüstenbildung in Gegenwart und Vorzeit*, Dietrich Reimer, Berlin, xiv + 175 pp.
Warren, G.K., 1878, Valley of the Minnesota River and of the Mississippi River to the junction of the Ohio; its origin considered, *Amer. J. Sci., Ser. 3*, **16**, 417-431.
Weir, A.H., Rayner, J.H., Catt, J.A., Shipley, D.G. and Hollies, J.D., 1984, Soil factors affecting the yield of winter wheat: analysis of results from I.C.I. surveys 1979-80, *J. Agric. Sci., Cambridge* **103**, 639-649.
Whalley, W.B., 1979, Quartz silt production and sand grain surface textures from fluvial and glacial environments, *Scanning Electron Microscopy* **1**, 547-554.
Whalley, W.B., Marshall, J.R. and Smith, B.J., 1982, Origin of desert loess from some experimental observations, *Nature* **300**, 433-435.
Willman, H.B. and Frye, J.C., 1970, Pleistocene stratigraphy of Illinois, *Ill. State Geol. Surv. Bull.* **94**, 204 pp.
Wintle, A.G., 1981, Thermoluminescence dating of late Devensian loesses in southern England, *Nature* **289**, 479-480.
Wintle, A.G., Shackleton, N.J. and Lautridou, J.P., 1984, Thermoluminescence dating of periods of loess deposition and soil formation in Normandy, *Nature* **310**, 491-493.
Wood, S.V., 1882, On the origin of the loess. *Geol. Mag. II*, **9**, 339-343.
Yaalon, D.H., 1969, Origin of desert loess, *in* M. Ters (ed.) *Etudes Quaternaires du Monde*, p. 755, Proc. 8th INQUA Congress, Paris, 1969, 2.
Yaalon, D.H. and Dan, J., 1974, Accumulation and distribution of loess-derived deposits in the semi-desert and desert fringe areas of Israel, *Zeit. Geomorph. Supplementband* **20**, 91-105.

Yaalon, D.H. and Ganor, E., 1979, East Mediterranean trajectories of dust-carrying storms from the Sahara and Sinai, *in* C. Morales (ed.) *Saharan Dust: Mobilization, Transport, Deposition,* pp. 187-193, J. Wiley & Sons, Chichester, xvi + 297 pp.

Yaalon, D.H. and Ginzbourg, D., 1966, Sedimentary characteristics and climatic analysis of easterly dust storms in the Negev (Israel), *Sedimentology* **6**, 315-332.

Zeuner, F.E., 1949, Frost soils on Mount Kenya, and the relation of frost soils to aeolian deposits, *J. Soil Sci.* **1**, 20-30.

LAKE SEDIMENTS AS INDICATORS OF CHANGES IN LAND EROSION RATES

R.L. Thomas
Institute F.-A. Forel
Université de Genève
10, route de Suisse,
1290 Versoix
Switzerland

ABSTRACT

The dating and analysis of lake sediment cores in unravelling the historical influence of man in his use of lake basins are rapidly becoming routine. However, interpretation of core profiles must be carefully carried out to resolve the effects of in-lake processes, such as mixing and eutrophication, on the compositional characteristics of the sediment. Both sediment texture and major element composition can be used with considerable success in qualitatively assessing changes in both temporal and spatial basin soil erosion rates. Much of the work, however, is new and considerable research is needed to establish detailed and possibly quantitative measures of soil erosion rates by examination of the sediment column in recipient lakes.

1. INTRODUCTION

Lakes represent small sedimentary basins of the shortest geologic duration. They are regarded as being geologically ephemeral and have a time span controlled either by sedimentary infilling or by emptying due to the breakdown of a formative sill or barrier. In both cases the record of the history of its life span is preserved in its sedimentary column. Lakes existing in those areas that were subject to glaciation preserve a record from the last ice retreat, on the order of 11,000 years, whereas lakes in non glaciated regions may preserve a record of many tens of thousands of years, such as Lake Patzcuaro (Watts and Bradburg, 1982).

The analysis of lake cores is a relatively recent endeavor which was seriously started in the mid-1960's. The emphasis at that time, and it remains so to this day, was an attempting to determine the effects of human use of the lake basin upon the limnology of the receiving lake and to document the sequence of lake changes that could be interpreted from the sedimentary record. Initially these attempts were hindered by the lack, in most areas, of a suitable stratigraphic marker horizon to date the core profiles. However, with the introduction of radio-isotopic dating techniques in the late 1960's studies on lake sediment cores proliferated and have led to a large literature on this subject. The techniques have now become routine and are performed as an ancillary to almost all lake sedimentary studies concerned with understanding lake processes for the purpose of effective lake management.

A. Lerman and M. Meybeck (eds.), Physical and Chemical Weathering in Geochemical Cycles, 143–164.

Lake sediments have proved to be much more effective in the analysis of human influences than marine sediments for two reasons. First, the rate of sedimentation is faster, which allows for easier sub-sampling of the core in small units of time leading to higher resolution. Second, the higher lake to drainage basin ratio amplifies the observed effects of basin change on the profiles of individual components. In addition, a lake may be considered a closed system in the sense that the greatest proportion of the deposited input will be sedimented within the lake and the lake will not receive input from other sources by lateral transport, as is the case in open marine basin systems. The lake sediment is thus not diluted by sediment from sources other than its own basin.

Changes in lake sediments are varied and may be textural, mineralogical, chemical or biological in nature. They may be primary in that they directly represent a change in external conditions such as weathering or a volcanic ash deposit, or secondary in that they represent a lake response to an external change such as eutrophication. In addition sediment may be chemically imprinted by the adsorption of an element or compound reflecting a specific use or characteristic of the delivery system. Such imprinting is of particular value in understanding the effects of modern agriculture and industry, and is diagnostically of enormous value to lake managers in determining the cause, history and source of detrimental perturbations of lake ecosystems.

This paper presents a brief and often abbreviated overview of the subject from a number of perspectives but in the final analysis attempts to indicate how lake sediments may be indicative of recent weathering or erosional processes in the associated watersheds.

2. IN-LAKE PROCESSES

Sediment delivered to lake systems is subjected to the many in-lake processes which affect transportation, sorting and deposition. Two major size-fractions may be considered to be important, namely: sand and clay. Geochemical imprinting is only important with respect to the clay fraction and, to a lesser extent, on particles in the silt-size range. The sand representing the bottom traction load of rivers is deposited as bars or deltas at the river mouths or is entrained in the coastal currents to form beach and dune deposits. The finer particles are subject to intermittent deposition with wave induced resuspension and onward transportation by lake currents to a final depositional zone in the lake. This is normally a deep water zone, but it is more complex in a shallow lake situation. The processes of fine particle movement to a depositional zone are collectively defined as "sediment focusing."

To better describe these processes some simple sedimentation models are presented in Figure 1 and summarized below:

2.1 Shallow Lake: Non-Depositional (Figure 1A)

In this model the input of fine grained sediment is approximately equal to the output of fine sediment. Coarse sediment normally forms a delta which is progressive and will ultimately result in lake infilling. Fine particles are deposited on the open lake bed but are resuspended under storm conditions. The sediment veneer remains thin and is intermittently mixed by physical processes.

Sediment texture tends to be coarse and the sediment textural characteristics appear to be in equilibrium with the shallow water conditions despite the physical reworking. This condition may be described as "dynamic textural equilibrium."

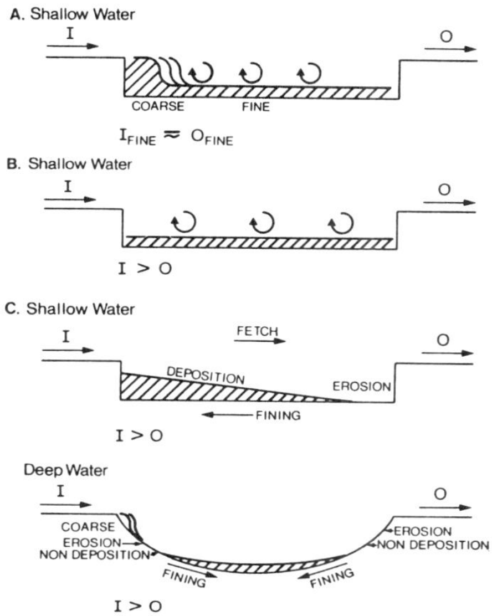

Figure 1. Diagrammatic representation of simple lake sedimentation models.

Resuspended material is in part thus transported to the outlet river and replaced by material entering the lake. There is no net accretion. A good example has been given by Thomas et al. (1977) in their description of Lake St. Clair in the North American Great Lakes system. These authors were able to show that, with the control of industrial mercury input to the lake, the half-life for sediment retention is on the order of 3.5 years.

Sediment cores taken in lakes of this type will have thin modern sediment in which the profiles of elements will be random throughout due to mixing.

2.2 Shallow Lake: Depositional (Figure 1B)

This model is essentially the same as that described above, other than the fine sediment input load is greater than the output, hence net accretion or deposition must occur. Deltas or bars may form depending on the sand input but the lake essentially will fill from the bottom up. The sediment is finer than expected from the water depth and is hence in "textural disequilibrium". Excess energy derived from storm waves resuspends and mixes the surface sediment to depths sometimes exceeding 10 cm. A good example is the western basin of Lake Erie (Thomas et al., 1976) which does not have an input delta at the mouth of the Detroit River, as Lake St. Clair serves as a sediment filter removing coarse sediment derived originally from Lake Huron.

Coring of this type of lake will give random element profiles for the top layers which have been subjected to physical mixing with smooth profiles below. These profiles reflect the upward movement of averaged concentrations in a mixing zone of constant thickness analogous to moving averages in smoothing data trends.

2.3 Shallow Water, Fetch Controlled Deposition (Figure 1C)

This occurs in moderate to large lakes in which wind fetch dominates over water depth as the controlling factor bringing about the focussed deposition of fine material. The fine material input exceeds output hence net deposition must occur. Resuspension is less significant and textural trends are clearly observed focusing the finer materials in the quieter upwind reaches relevant to the prevailing winds. The sediment texture coarsens downwind with increasing wave energy until deposition of fines ceases, and erosion and lag deposit formation occurs. The finer materials in these regions are driven across the erosion zone to the basin outlet. Textural equilibrium is established with surface mixing occurring due to bioturbation rather than physical mixing. Lakes of this type may be often deep enough to stratify which is rarely the case in the two previously described shallow water models. A good example of this type is the central basin of Lake Erie (Thomas et al., 1976). Basin infilling is from the bottom up but progresses downwind as a function of the prograding texture (see Figure 1C). Coring provides good elemental profiles.

2.4 Deep Water Model (Figure 1D)

This model describes the most common lake condition. Coarse materials occur as bars or deltas and the shallow water periphery is almost exclusively an erosional or non-depositional zone. Fine sediments occur in the deeper water and fine outwards from all sides into the deep water basins. This represents the bathymetric or morphological control of texture commonly recognized as "sediment focusing." These sediments are not subjected to physical mixing and any disruption of the surface is due exclusively to bioturbation. The sediments are in perfect textural equilibrium with the wave energy as determined by bathymetry and there appears to be constant sedimentation. Annual fluctuations, however, may occur and may be directly observed by varve deposits under some lake conditions. In some lakes disruption of sediment may occur due to turbidity flows or storm induced slumping each of which have different textural characteristics. Many good examples exist but Lake Ontario may be cited here as being typical (Thomas et al., 1972).

Coring of these sediments tends to produce elemental profiles which are readily interpreted in the context of lake and basin history. Care has to be taken to account for any bioturbation, or the occurrence of turbidities and slump deposits.

3. LAKE SEDIMENT DATING

Many techniques have been developed in attempting to produce relative and absolute dates for lake sediments. These range from long term (e.g., up to 40,000 years, Lake Patzcuaro, Mexico) to within the historical record of the lake basin. The methods selected in any one case depends on the lake basin and sediment characteristics as well as upon the availability of technical equipment and expertise. In virtually all cases, however, more than one method is needed in order to produce internal verification and confirmation.

The methods used fall under three main categories: event related, stratigraphic, and radio-chemical, many of which are summarized in Table I. Absolute dating is generally provided by radio-chemical methods, but once stratigraphy and events have been placed in a dated sequence then they in turn provide rapid absolute dates and correlation.

Table I. Methods used for dating lake sediment cores.

Events	Stratigraphic	Radiochemical
Ash bands	Magnetostratigraphy	C^{14}
Slumps	Fossil Assemblages	Pb^{210}
Turbidites	Chemical	(Cs^{137})
Hydraulic Regime	Textural	
Cs^{137}		
Faunal Change		
Anthropogenic Materials		

3.1 Event Markers

A marker horizon within the sediment column can be related to a specific event occurring either in the lake, its basin or to an external event of sufficient magnitude, to provide a record of its influence in the sediment. Marker horizons may be readily related to a known event for which the date is known or may merely serve as a means of correlation between cores. Event horizons given in Table 1 are discussed below with specific examples.

3.1.1. Ash Bands. Ash bands are the visible horizons within the sediment column related to a specific incident of volcanic activity, such as the recent Mount St. Helens eruption in the Western United States. Such horizons are being used by Slaymaker (this volume) in studies in British Columbia and were used by, for example, Watts and Bradburg (1982). These horizons only occur in regions close to the eruptions, where concentrations of ash are easily identifiable. Detailed examination of particles may increase the utility of the method and may allow identification of the horizons at large distances from the source.

3.1.2 Slumps. Slumps imply the mass downslope movement of sediment from one location to another due to some strong external stimulus such as an earthquake or storm. The material moved is generally similar to that occurring at the point of final deposition and may not be readily visible. An excellent example is given in Figure 2 (Edgington and Robbins 1976). This diagram shows the profile for excess Pb^{210} in a core from Lake Michigan in which block deposits of constant Pb^{210} concentration occur indicating instantaneous deposition. These three events can be related to slumping induced by three intense storms which occurred in 1938, 1940 and 1915.

3.1.3 Turbidites. The occurrence of turbidites has been recorded in many lakes. A good example from Lake Geneva has been provided by Vernet et al., (1984) whose information is given in Figure 3. It can be seen from Figure 3 that a change in deposition occurred which, at a depth of 12 cm, resulted in a decrease in Corg, Cd, Hg, Zn and Cs^{137}. The mean particle size showed an increase in particle size followed by a decline representing the normal fining upward composition of a turbidite unit. The basal boundary is not precise due to the sub-sampling interval and to the perturbation of the lower sediment by the turbidity flow. Coarse basal material is rarely present in lake sediment but the increase in $CaCO_3$ is indicative of movement from coastal silts enriched in $CaCO_3$. This turbidite is dated at approximately 1971 by Cs^{137}. A turbidite flow was known to have occurred in response to a major flood of

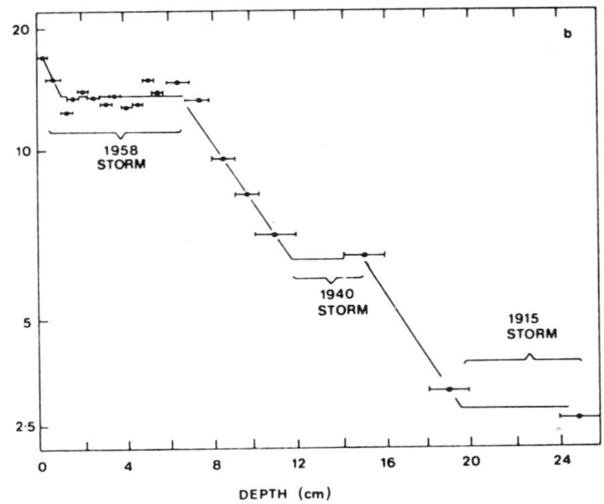

Figure 2. Modification of Pb^{210} profiles in a core from Lake Michigan showing three slumping episodes triggered by intense storms (after Edgington and Robbins, 1976).

the River Drance in 1968 and indeed it is quite possible that the sediment has recorded this event. Turbidites are recognizable from slumps by the fining upward sequence characteristic of the transport sedimentation process whereas slumps are homogenized by the mass flow of sediment downslope.

3.1.4. Hydraulic Regime. Major changes in the flow characteristics of the input to lakes may have a profound visual impact on the sediment, Changes are normally due to human influence such as daming or channel modification. An excellent example has been given by Dominik et al. (1981) from Lake Constance. In 1900 the Upper Rhine river was channelized such that its outlet to the lake was shifted some eight kilometers northeast of its previous outlet (Figure 4). A core taken at station SM-2 showed a significant change in composition at a depth of 11 cm from a coarse grain homogeneous material to a finer grain laminated sediment (Figure 4). Pb^{210} dating of this core indicated that the date 1900 (Figure 4) is close to the break in sedimentation, confirming the date of the change in sedimentation.

3.1.5. Cs^{137}. Despite the fact that Cs^{137} is a radionuclide, it is included as an event marker rather than a radio-isotope dating technique. Cs^{137} concentration in sediment cores and atmospheric fallout parallel the atmospheric testing of nuclear devises with maximum Cs^{137} activity in sediment profiles corresponding to the year 1964. Atmospheric fallout by year is shown in Figure 5 as recorded by Hemmerich (1980) in Heidelberg. For the corresponding peak in sediment Cs^{137} activity, it is assumed that incorporation into the sediment is rapid and that post-depositional migration is not significant. This migration may, however, occur (Lerman and Lietzke, 1975; Robbins et al., 1979; Albert et al., 1979; Dominik et al., 1981) but it is generally agreed that, where a well defined profile and clear peak occurs then the method can be used with

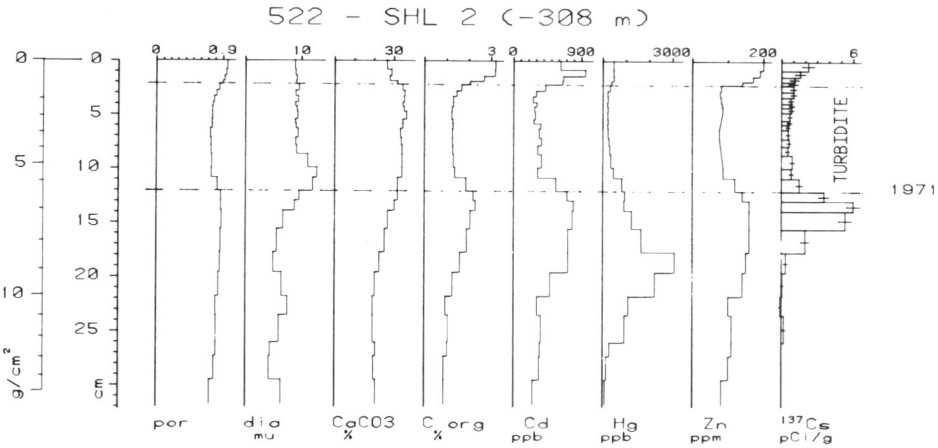

Figure 3. Changes in particle size and element profiles due to turbidite occurrence in Lake Geneva (after Vernet et al., 1984).

considerable reliability. This has been well demonstrated in Lake Geneva (Vernet et al., 1984). Four types of Cs^{137} distribution with depth were observed. In Figure 5 these distributions are compared with the fall-out history of this nuclide as measured in Heidelberg (after Hemmerich 1980). The diagram shows that a large part of the total fall-out was recorded in 1963 and 1964.

In core 533, recovered in the area close to the Rhone delta, sedimentation rate is very high and thus, the resolution is good. There is a more dispersed Cs^{137} activity with a larger portion deposited above the maximum activity horizon, in comparison to the atmospheric fall-out curve. Cs^{137} distribution in sediment core 521 presents an example of a well-developed, sharp peak activity at a moderate sedimentation rate, a condition in which the average sedimentation rate can be accurately determined. A similar distribution can be observed in core 532, but the Cs^{137} profile is more symmetrical which, under the assumption of a constant sedimentation rate, would imply post-depositional redistribution of the radionuclide with a resulting "flattening" of the curve. In such a case, the peak activity can still be used with a certain confidence for the determination of sedimentation rate, but the first appearance of the Cs^{137} activity does not correspond to the sediment layer deposited in 1954.

Finally, the distribution of the type occurring in core 524 (Figure 5) presents a major problem. It may be interpreted as indicative of a very low deposition rate coupled with some post-depositional redistribution of the nuclide (by bioturbation, migration in solution, etc.). Some sedimentological observations suggest that this type of distribution may be characteristic for areas in which erosion and redistribution of sediment are common. A similar distribution may be caused by a major delay in Cs^{137} delivery to the sediment or to Cs^{137} migration toward the sediment surface.

These distributions which are characteristic of the types of profiles from most lakes show that the data must be interpreted with caution, but that, in general, the method under moderate to rapid continuous sedimentation provides excellent results.

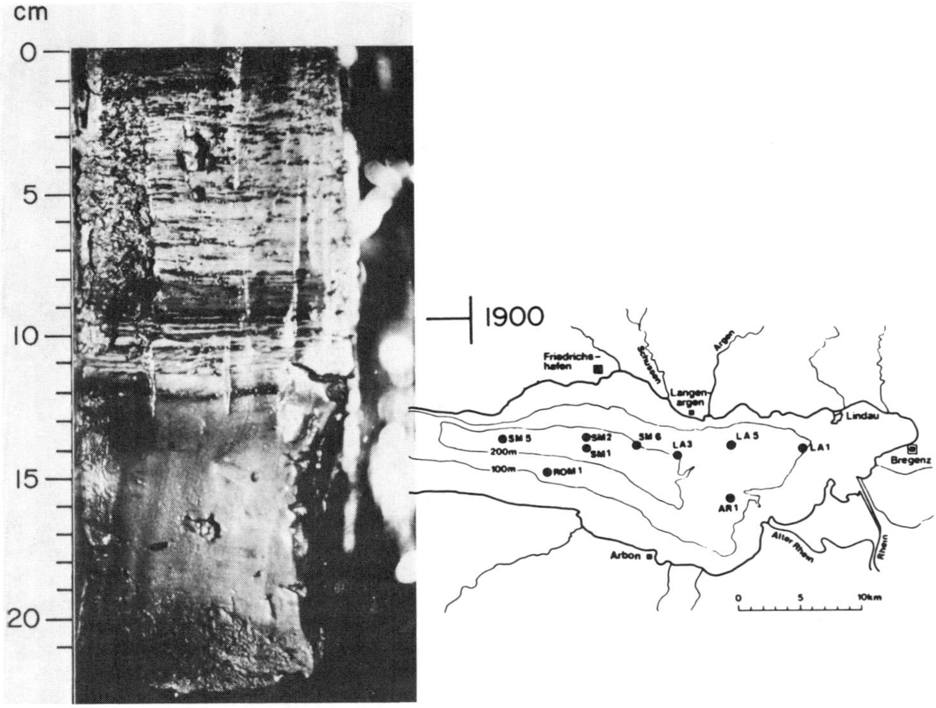

Figure 4. Visual change in core profile SM2 due to a change in the location of the Upper Rhine River mouth into Lake Constance (after Dominik et al., 1981).

3.1.6. Faunal Change. This type of event is one in which a very rapid change in fauna occurs as a result of a very specific phenomenon. However, it does not relate to the slower evolutionary changes occurring in response to climate or changing lake conditions. One of the best examples that has been used extensively is the Ambrosia pollen rise in forested areas of eastern North America. The increase in this pollen can be observed easily and is a most effective marker. *Ambrosia,* a common ragweed, increased as a result of deforestation and the increase in its open meadow habitat. Local records of land transfer provide precise dates at which deforestation occurred for example, in 1850 in Lake Ontario (Kemp and Harper, 1976; McAndrews, 1976).

3.1.7. Anthropogenic Materials. This technique has not been applied to any great extent but could be useful in certain areas. For example, the occurrence of fly ash and coal particles is very common in the sediments of the Great Lakes. This material came from coal-fired steam vessels and it is certain that if the spatial concentrations were to be mapped then the shipping tracks of the Great Lakes would be clearly seen. In cores the onset of these particle would likely mark the onset of steam shipping and the profiles would probably reflect the slower subsequent change to diesel-motor vessels. This work remains to be done, but clearly has wide applicability to other regions impacted by industrialization.

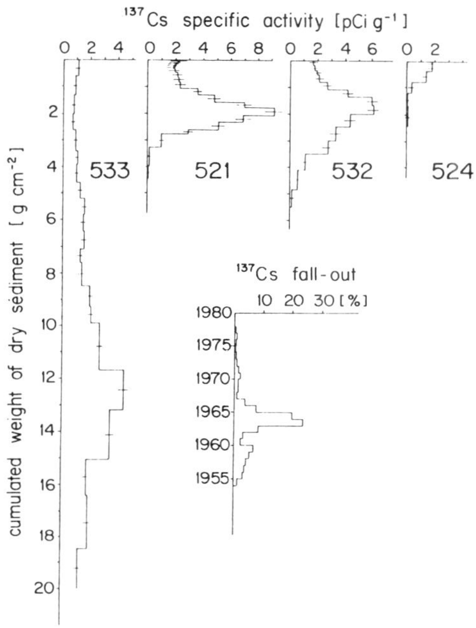

Figure 5. Atmospheric Cs^{137} fall-out history as measured in Heidelberg (Hemmerich, 1980) and typical Cs^{137} profiles from four cores from Lake Geneva (after Vernet et al., 1984).

3.2 Stratigraphic Dating

Stratigraphic methods of dating, as used here, emphasize those single or composite changes which occur as a result of natural changes (for example climatic) or broad man-induced changes, such as eutrophication, industrialization and, most recently, acidification. Some of these are described below.

3.2.1. Magnetostratigraphy. The basis for magnetostratigraphy of lake sediments has been well described by Thompson (1984). It is based on the fact that magnetic declination and inclination are continuously changing, and the residual magnetism of mineral particles in sediment cores represents a measurable record of these changes. Master curves have been prepared and dated for the entire Holocene record (about 10,000 years) against which comparison may be based. Age estimations and correlations can thus be made though it should be noted that this field requires specialized training and equipment.

3.2.3. Fossil Assemblages. The literature abounds in excellent studies of palaeolimnology based on stratigraphic evidence from plant remains. The reader need only be referred to the compendium provided by Haworth and Lund (1984), to obtain a series of good examples. Most studies involve algal and/or pollen stratigraphy: the former to obtain an insight on changing lake conditions relative to external forcing, while the latter represents the sedimentary preservation of the history of plant changes in the basin as affected by both natural and human causes.

3.2.4. Chemical. These may be divided into a number of categories relative to the process or cause of chemical change in the sediment column. These include a) eutrophication which can be verified by biologic stratigraphy, b) changes in weathering or soil erosion, c) industrialization and the onset of point sources of pollutants, which may be individually datable but collectively define the history of industrialization.

a) *Eutrophication.* Cultural eutrophication is the onset of increased and detrimental algal productivity in a lake resulting from increased population and associated land use, increasing the supply of plant nutrients, P and N to lake systems. A good example has been given for Lake Constance by Müller et al. (1979). Their data are summarized in the core profiles in Figure 6. These profiles show an increase in P and N starting around 1870 with a large increase in 1954 which is sustained to the time of collection in 1977. Parallel to the nutrient increase there is an increase in Corg directly representing the increased primary production of the lake. Carbonate also shows an associated increase to a more calcite rich sediment, indicating the precipitation of calcite in surface waters resulting from high pH values due to the increase in photosynthesis occurring in the surface waters. The grain size changes as a direct result of an increase in silt size carbonate while the Al profile decreases inversely to the increased percentages of carbonate and Corg, and is hence a product of compositional dilution.

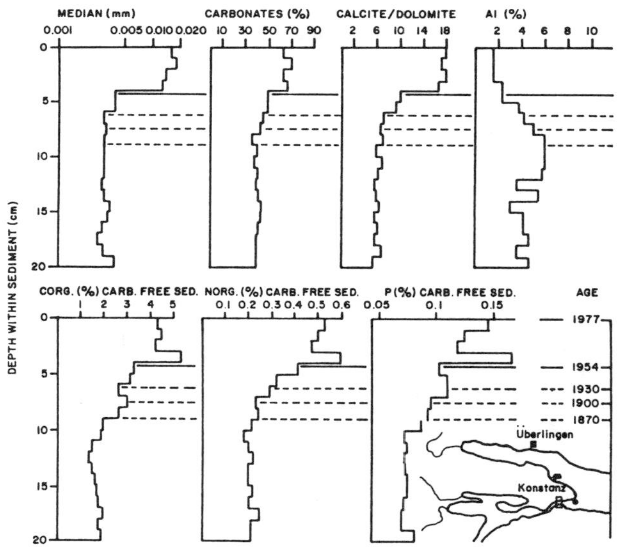

Figure 6. Changes in P, N, Corg, Al, silt and carbonate profiles as a result of increasing eutrophication in Lake Constance (after Müller et al., 1979).

b) *Weathering and Erosion.* Makereth (1966), and Engstrom and Wright (1984) have drawn attention to the fact that during periods of active erosion, the mass transport of raw unleached soils should increase levels of Na, K and Mg in lake sediments, whereas during episodes of relatively stable soils deep weathering should diminish the mineral material available for erosive removal, transport and sedimentation. Thus, elements that represent the mineral or detrital fractions should directly reflect erosional intensity in the basin. This is discussed subsequently.

c) *Industrialization.* The onset of the industrial revolution is strongly reflected in the lake sediment column. In the North American Great Lakes, trace metals start to rise immediately following the arrival of Caucasian man and deforestation of the basin. The general rise accelerates at the turn of the century and is amplified by the Second World War. Certain responses by specific elements occur over and above the general trends, for example, Pb in response to the addition of lead to gasoline, and Hg from the building of chlor-alkali mercury cell plants (Kemp et al., 1976; Walters et al., 1974). The post-Second World War era is characterized by the onset of organic compounds which show dramatic increase in use after about 1950. This increase in organochlorine compounds has been documented in the Great Lakes by Frank et al. (1977; 1979a, 1979b), and an example is shown in Figure 7 for two cores in Lake Geneva.

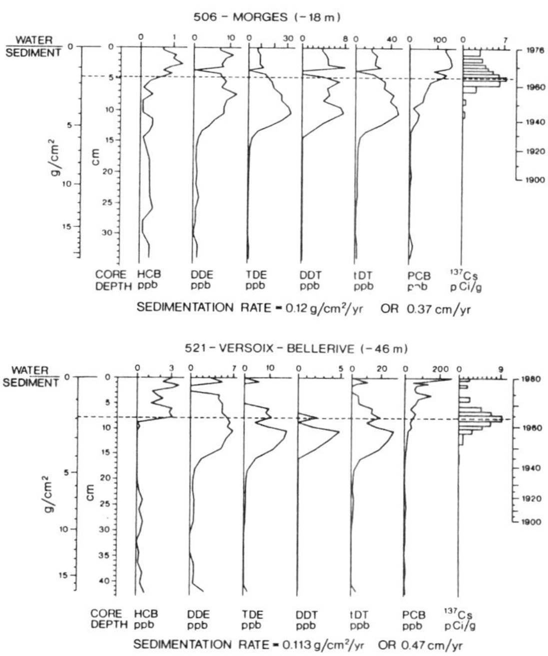

Figure 7. Profiles for organo-chlorine compounds in two cores from Lake Geneva (after Thomas et al., 1984).

It appears, in general, that the concentration of an element or compound in a dated sediment core closely follows the production or sales of the same material. This has been shown by Vernet (personal communication, 1985) for mercury in Lake

Geneva where mercury trends exactly follow mercury imports into Switzerland. A further example is given in Figure 8 for Lake Ontario in which the dated sediment profile for PCB closely follows the annual sales figures for the same compound. Such data clearly and not surprisingly suggest that environmental levels of contaminants are directly proportional to production, which translates into sales and distribution.

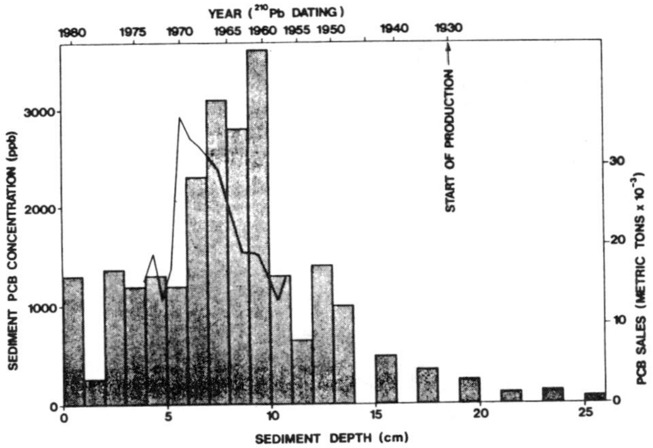

Figure 8. Profile for PCB for a core from Lake Ontario compared to annual sales for the same compound over the same period (after Durham and Oliver, 1983).

3.3 Radiochemical Methods

Two major radiochemical techniques, C^{14} and Pb^{210}, are in use for absolute dating of sediment cores. The former is generally used to establish the chronology of stratigraphic data. It is difficult to use and interpret, and it is dependent on the core intercepting suitable biological material for analysis. With a half-life of 5736 years it is useful in corroborating major events over the post-glacial period. Pb^{210}, of half-life 22 years is becoming a standard dating tool in lake sediment studies and is rapidly replacing the more time consuming stratigraphic and faunal change techniques. It is of particular value in determining the historical record from the latter part of the 19th Century to present. Many excellent papers have been written on its use (Krishnaswamy et al., 1971; Krishnaswamy and Lal, 1978; Oldfield and Appleby, 1984).

4. BASIN DELIVERY CHARACTERISTICS

It was previously noted that rapid erosion in the watershed delivering sediment to a lake was characterized geochemically by elements incorporated into the lattice of detrital minerals. These elements, as noted by Engstrom and Wright (1984), include Na, K and Mg. The relationships of K and Na to the proportion of mineral matter in the sediments of some United Kingdom Lake District lakes is given in Figure 9. The data appear to support the weathering or erosion concept but since the sediment composition may have been influenced by in-lake processes further confirmation appears to be necessary.

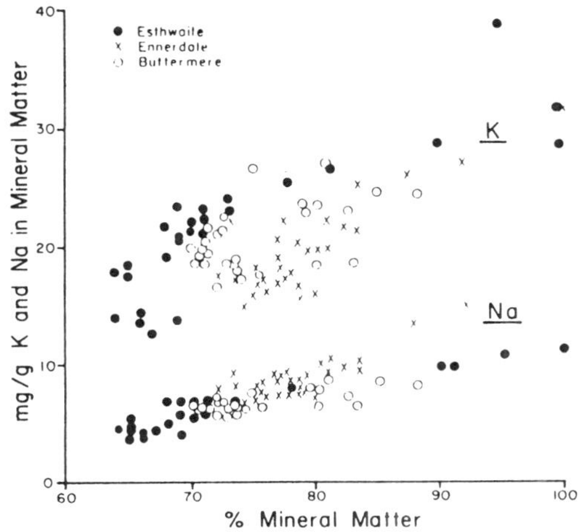

Figure 9. K and Na relationships to mineral matter in sediment of lakes from the English Lake District (after Engstrom and Wright, 1984).

The author conducted a major analysis of tributaries in the Canadian portion of the Great Lakes drainage basin during the period 1974 to 1976. Samples of suspended matter were taken at the mouths of rivers as close to the lakes as possible, under conditions of positive fluvial flow. Each river was sampled once during spring runoff and an attempt was made to sample ten rivers monthly over a full twelve month cycle. These ten monitor rivers were used to provide some insight into the annual variability in the quality of the delivered sediment in order to be able to provide some confidence to the single sampling of the remaining rivers. The composition of the suspended solids for six of the monitored rivers is given for SiO_2, Al_2O_3, Fe_2O_3, K_2O, TiO_2 and P_2O_5 in Table II. Mean values together with the coefficient of variation are given for each river. The coefficient of variation as given here is the standard deviation expressed as a percentage of the mean. The coefficient of variation for all the elements other than P_2O_5 is remarkably low. This is particularly so when the whole system of sampling, centrifugation, and analysis by X-ray fluorescense is considered. Indeed, the variation observed is often within the variation expected of the analytical technique. High variability is observed for the elements in the Thames River, which is a larger and slower flowing river than the remaining rivers, excluding the Niagara. The increased variability is likely due to seasonal variations related to instream productivity, however, this needs to be further evaluated. The Niagara is a special case because it is a direct conduit between Lake Erie and Lake Ontario, and it does not directly drain a basin as is the case with the other rivers. Previous work (Sly and Thomas, 1974) has indicated that SiO_2 in sediments is indicative of detrital quartz, whereas Al_2O_3, Fe_2O_3, K_2O and TiO_2 are indicative of clay minerals dominated by illite. These elements thus represent the detrital minerals indicative of erosion in the watershed, as postulated by Makereth (1966). The data from these rivers clearly indicate that mineral characteristics do not vary significantly within the basins and that variable seasonal erosion delivers sediment of the same characteristics throughout the year. Higher variability can be observed for phosphorus and is believed to be due to the addition of phosphorus from many anthropogenic sources within the basins, such as community sewage discharge, septic tank systems, animal husbandry, and agricultural fertilization practices.

Table II. Variability of the major element composition of the suspended solids of six monitored rivers (in %).

	SiO_2	Al_2O_3	Fe_2O_3	K_2O	TiO_2	P_2O_5
Niagara N=3						
Mean	56.25	10.98	5.97	3.23	0.73	0.24
S.D.	2.21	1.05	0.22	0.18	0.06	0.01
C.V.	3.90	0.50	3.70	5.60	7.50	2.40
Grand N=12						
Mean	46.72	9.46	5.16	2.75	0.69	0.35
S.D.	1.82	0.73	0.39	0.15	0.04	0.06
C.V.	3.90	7.70	7.60	5.60	5.80	16.80
Cedar N=11						
Mean	52.06	11.65	5.85	3.21	0.69	0.26
S.D.	6.87	1.51	1.02	0.30	0.08	0.05
C.V.	4.10	10.60	7.20	5.20	5.80	12.40
Thames N=11						
Mean	44.49	10.75	5.27	3.26	0.63	0.30
S.D.	6.87	1.51	1.02	0.30	0.08	0.05
C.V.	15.50	14.00	19.40	11.90	13.20	18.40
Saugeen N=7						
Mean	37.90	7.17	3.04	2.47	0.49	0.29
S.D.	2.70	0.39	0.35	0.22	0.04	0.09
C.V.	7.00	5.40	11.64	8.80	8.80	32.00
Nottawasaga N=7						
Mean	44.26	7.24	3.85	2.33	0.50	0.39
S.D.	3.19	1.02	0.16	0.20	0.06	0.08
C.V.	7.20	14.10	4.30	8.50	12.90	21.00

S.D. = Standard Deviation; C.V. = Coefficient of Variation

The data for major elements from the monitor streams indicate that there is a high probability that the single sampling episode from all the rivers draining into the lakes will be representative of the individual drainage basins. Values for Al_2O_3 for the single spring sampling of rivers draining the Canadian watershed in Lakes Ontario, Erie, Huron, and Georgian Bay are given in Figure 10. The vertical axis gives the Al_2O_3 concentration of the suspended solids and the horizontal axis gives the streams by number, starting at the eastern end of Lake Ontario and moving continuously along the shoreline around Lakes Erie, St. Clair, and Lake Huron to Georgian Bay. The major physical fabric is noted on Figure 10 and it can be seen that high Al_2O_3 or high percentage clay-content occurs in the clay-rich intensively farmed regions of the Niagara Peninsula and Western Lake Erie. Low concentrations occur in till areas (rivers 10-30), the Brantford Sands on Lake Erie, and the soil-deficient limestone region of the Bruce Peninsula. Clearly the distribution reflects changes in erosion as related to increased soil loss under intensive agriculture. This occurs predominantly in clay regions where intensive cash cropping occurs.

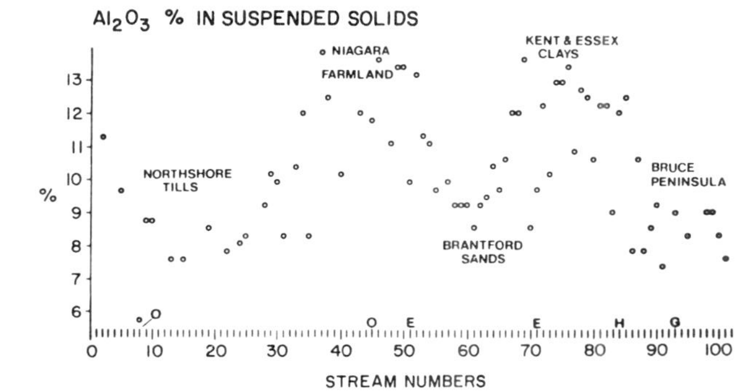

Figure 10. Al_2O_3 concentrations in suspended solids delivered to the Great Lakes from Canadian rivers (O = Lake Ontairo, E = Lake Erie, H = Lake Huron, G = Georgian Bay).

Figure 11 shows the same distribution by streams for the percentage of sediment finer than 12 phi (0.25 μm). The pattern is less well-defined than Al_2O_3 but shows that once again the intensively farmed areas of higher erosion have a higher proportion of very fine material as determined by textural analysis. This is discussed subsequently in relation to the soil erosion index utilized by Warwick (1980) to assess agricultural effects on sediment cores taken in the Bay of Quinte in Lake Ontario.

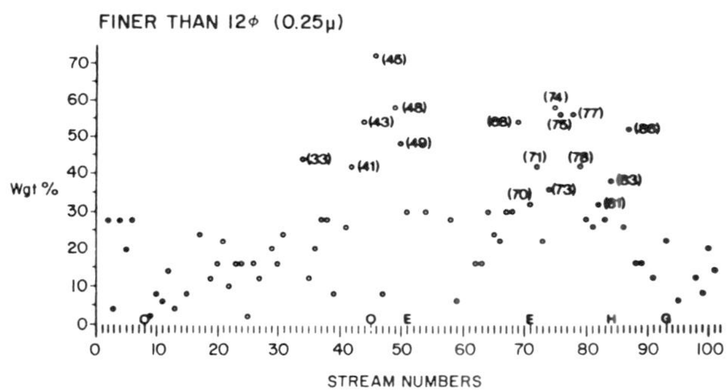

Figure 11. Percentage of material finer than 12 phi (0.25 μm) in suspended solids delivered to the Great Lakes from Canadian Rivers. Numbers in parentheses are the sample numbers allocated by river.

This somewhat perfunctory analyses of river data on the Great Lakes leads to a number of conclusions of considerable significance to fluvial basin analysis and erosion cycles which may be summarized as follows:

a) The composition of suspended sediment at the point of delivery, for naturally occurring elements associated with mineral composition is characterized by low variability reflecting the in-basin mixing and homogenization processes that occur between point of erosion and final delivery.

b) The composition is unique to individual basins and it may vary from one to the other as a function of land fabric and land use. As land use changes, the composition will change and stabilize to a new level.

c) Compositional features of anthropogenic origin as indicated by phosphorus show large in-basin variability reflecting the combined effects of human activity.

5. BAY OF QUINTE SEDIMENTS

Probably the most comprehensive analysis of lake sediment cores was undertaken by Warwick (1980) on the Bay of Quinte in Lake Ontario (Figure 12). In this study, three 2-meter long sediment cores were taken downstream from the Glenora Ferry (Figure 12) and carefully subsampled. Sedimentation chronology was established using *Ambrosia* pollen, Pb^{210}, C^{14} and an Erosion Index (to be discussed in 5.2 and 5.3). The cores were further analysed for sediment texture, total C, Organic C, total N and P and for a complete taxonomic evaluation of the chironomids. These parameters representing a chemical and biological stratigraphy were then evaluated against the core chronology and the known history of the region.

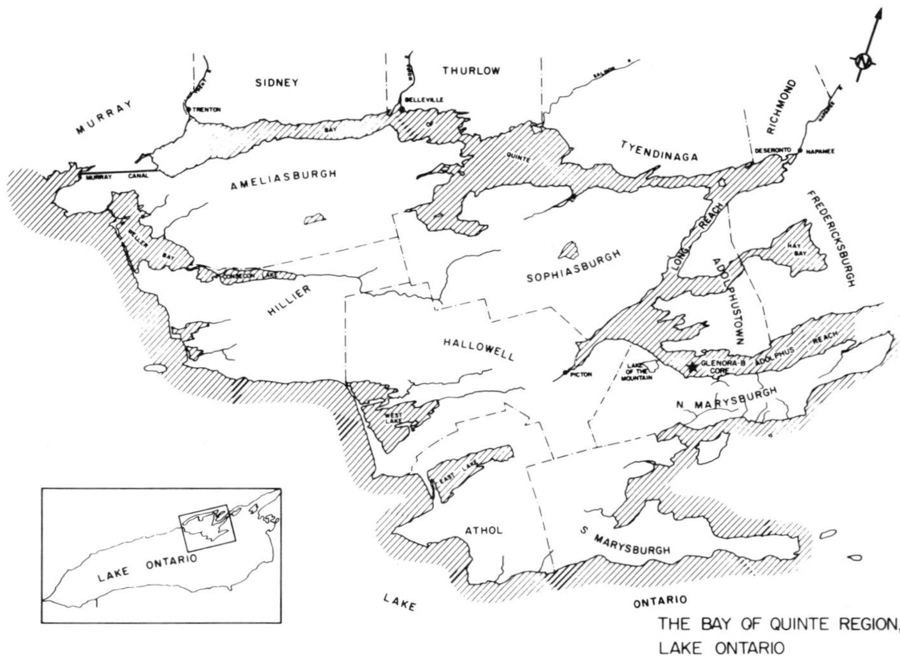

Figure 12. Coring location in the Bay of Quinte, Lake Ontario (after Warwick, 1980).

5.1 History of the Bay of Quinte

The history of the Bay of Quinte may be summarized after Warwick (1980) as follows.

5.1.1. Cultures

1. Aboriginal

a. Hopewell — flourished about 2000 years B.P. Fairly sophisticated and extensive culture

b. Algonkian — more primitive and nomadic were replaced by:

c. Iroquois — culturally more advanced, remained in area from 1000 to 1650 of the present era, evolved with increasing sophistication into the Huron, Petun, Neutral, and Erie tribes. Were destroyed by tribes from the south also of Iroquoian origin.

2. European

a. French. 1669-1682
1669 Sulpician Mission built
1682 Sulpician Mission abandoned
1682-1760 Period of warfare and desultory fur trading
1760 Fall of the French regime.

b. British. 1760-present
1784. Colonization began dramatically with the arrival of 1,814 United Empire Loyalists. Population rose slowly, then more rapidly and steadily after the War of 1812, and rapidly after 1838 to a maximum in 1860. 1860 to approximately 1900 declining population. 1900 to present. Revitalization with increasing population related to urbanization and industrialization and, more recently, tourism.

5.1.2. Agriculture

1. Practised by the Hopewell and Iroquois aboriginal cultures.
2. French conducted subsistence agriculture only.
3. British. Slow initial development of agriculture but accelerated after forest clearance. The soils used were often poor and with poor farming practice led to the decline in population between 1860 and 1900.

5.1.3. Forestry

1. Initial land clearing by the first settlers, French and British. Wood used for ash and charcoal were the first cash crops of the region.
2. 1804-1805. Start of the "square timber trade" to provide oak to meet a critical shortage in the Royal Navy for ship construction. A differential tax introduced by the British parliament made Canadian lumber economical.
3. 1808-1809. All types of naval wood required following the closure of the Baltic Sea area by Napoleon.
4. 1846. Approximate end of square lumber trade.
5. 1851. American lumber companies allowed to enter Canada.
6. 1851-1860. Like a swarm of locusts, the lumber companies stripped Southern Ontario and entered Michigan.
7. 1860-1892. Local wood manufacturing companies.
8. 1892. Pulp and paper industries started after lumber grade woods exhausted.

5.2 Core Chronology

Figure 13 summarizes the data utilized by Warwick (1980) to establish a chronology for the Bay of Quinte Cores by using Ambrosia Pollen, Pb^{210} C^{14} and the Erosion Index. The Erosion Index is based on the measurement of grain size and is defined as the fraction of clay-sized material finer than 10.50 phi divided by the total clay fraction finer than 8.5 phi (Warwick, 1980). This approximates the weight of less than 1 μm particles divided by the weight of the less than 4 μm fraction. Warwick (1980) interpreted this as representing the ratio of fine mineral sediment introduced from the watershed erosional processes to coarser materials produced in the aquatic environment. In the establishment of the chronology, Warwick (1980) used the Ambrosia Horizon as a control for the Pb^{210} chronology which was used between 1929 and 1909, and was further used to calibrate events on the Erosion Index profile. From there the Erosion Index was used to inter-calibrate specific events such as the construction of the Sulpician Mission in 1669 and the arrival of the United Empire Loyalists in 1784. Older sediments were then dated by the C-14 method.

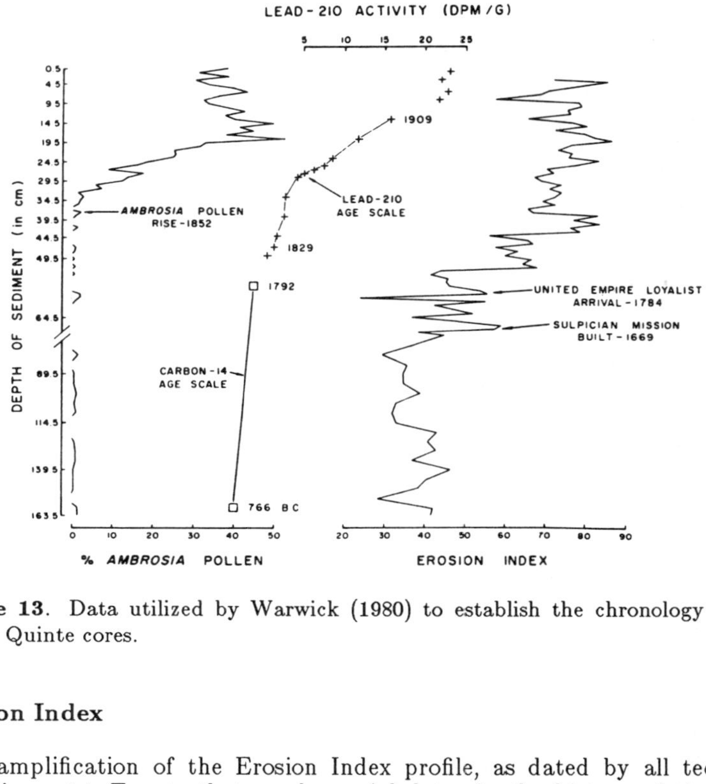

Figure 13. Data utilized by Warwick (1980) to establish the chronology of the Bay of Quinte cores.

5.3 Erosion Index

A further amplification of the Erosion Index profile, as dated by all techniques, is given in Figure 14. Events that mark special dates in the history of the region are shown as follows:.

1. A decline in the index in 1932 to a low, marks a period of low erosion during the Great Depression; and
2. An increase in erosion from 1803 to 1811, marks the beginning of the square timber trade, and the rise of 1841 shows the last mass immigration to British North America.

Following the building of the Sulpician Mission in 1669, a large decrease in erosion occurred in 1771 as an interim period between the French and British Regimes, and the arrival of the United Empire Loyalists in 1784.

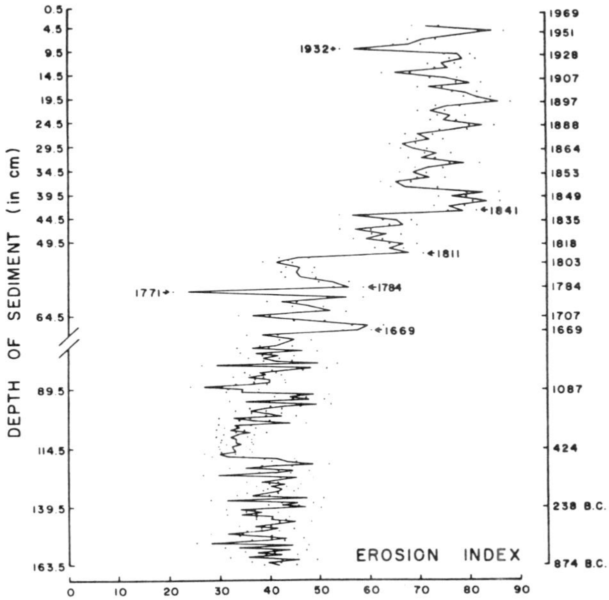

Figure 14. Vertical distribution of the Erosion Index in the Bay of Quinte core (after Warwick, 1980).

At first inspection the Erosion Profile Index clearly indicates a major shift in the ratio at 49.5 cm depth (or in 1811), with a step increase in 1841, and a slow increase to time of sampling in 1969. Clearly the trend suggests that the curve may be interpreted in a more general fashion, with increasing erosion due to the arrival of the first colonials and the onset of major deforestation. The resolution, as given by Warwick (1980), may at first sight appear to be an overly imaginative interpretation of the curves. However, two factors tend to support the interpretation as given above. First, the fit between the core chronology and the historical record of the region is too good to be entirely coincidental. Second, the interpretation of the river suspended sediment data, as given by the author earlier, would appear to strongly support the concept of changing particle size and the Erosion Index as sensitive indicators of watershed erosion.

The interpretation of the changing texture, as indicated by the Erosion Index, clearly reflects a change in particle sizes delivered from the basins, where fine-particle erosion is produced by accelerated sheet erosion from agriculturally exposed soils. Warwick (1980) also used other sediment-related properties, in particular the C/N ratio and the community structure of chironomid remains to track the onset and progress of cultural eutrophication relative to the history and development of the region.

6. CONCLUSIONS

The question as to whether variable soil erosion in a lake basin can be observed or measured in the lake sediment column can be answered with a qualified yes. First, however, an analysis of a core must be carried out and changes in textural, chemical and biological parameters must be tested on the basis of in-lake processes and a sound chronology established. Changes due to mixing, both physical and biological, eutrophication, precipitation, slumping and turbidity deposits must be accounted for before a valid interpretation may be made.

From preliminary work the following techniques may be used as indices of soil erosion in a watershed.

1. Particle texture. A fining of the particle size associated with an increase in sediment load of fine particle size. This was shown here by the changes in the particle size of suspended solids in rivers draining into the Great Lakes.
2. A Soil Erosion Index or a variable based on particle size analysis. A ratio of this type offers more resolution than direct particle size measurement and appears to be a direct result of increased mass movement of fine particles from the land surface due to tillage or natural increase in sheet erosion.
3. Variation in major element composition as noted both in the Great Lakes and the English lake district. This is a reflection of the proportional increase in loading of allochthonous mineral matter from rivers as distinct from autochthonous particle generation in the lake. In large measure this is a reflection of changes in particle size (increased clay mineral composition) as noted in the Great Lakes rivers.
4. The field of using lake sediment columns to unravel the history of human influence in the basins is as well developed as the methods of dating core sequences. The use of lake sediments to interpret basin erosion is a new and exciting application for future research. Such research might involve stable isotopes as tracers, or radionuclides of short half-lives to attempt to determine the flux of sediment particles from erosion at the field plot level to delivery at the basin mouth, to the characterization of the sediment as it accumulates on the lake bed.

7. REFERENCES

Alberts, J.J., Tilly, J., and Vigersted, T.S., 1979, Seasonal cycling of cesium-137 in a reservoir, *Science,* **203**, 649-651.

Dominik, J.A., Mangini, A., and Müller, G., 1981, Determination of recent deposition rates in Lake Constance with radioisotopic methods, *Sedimentology,* **28**, 653-677.

Durham, R.W., and Oliver, B.G., 1983, History of Lake Ontario contamination from the Niagara River by sediment radiodating and chlorinated hydrocarbon analysis, *Jour. Great Lakes Res.* **9**, 160-168.

Edgington D.N., and Robbins J.A., 1976, Records of lead deposition in Lake Michigan sediments since 1800, *Environ. Sci. Technol.,* **10**, 266-274.

Engstrom, D.R., and Wright, H.E., 1984, Chemical stratigraphy of lake sediments as a record of environmental change in sediments and environmental history, *in,* E. V. Haworth and J. W. G. Lund (eds.) *Lake Sediments and Environmental History,* pp. 11-68, Leicester University Press.

Frank, R., Thomas, R.L., Holdrinet, M., Kemp, A.L.W., and Braun, H.E., 1979a, Organochlorine insecticides and PCB in sediments of Lake Ontario, *Jour. of Great Lakes Research.* **5**, 18-27.

Frank, R., Thomas R.L., Holdrinet, M., Kemp, A.L.W., Braun, H.E. and Dawson, R., 1979b, Organochlorine insecticides and PCB in the sediments of Lake Huron (1969) and

Georgian Bay and North Channel (1973), *Sci. Total Environ.*, **13**, 101-107.

Frank., R., Thomas R.L., Holdrinet, M., Kemp, A.L.W., Braun, H.E. and Jaquet, J.-M., 1977, Organochlorine insecticides and pollutants in the sediments of Lake St. Clair and Lake Erie, *Sci. Total Environ.*, **8**, 205-227.

Haworth, E.V., and Lund, J.W.G., 1984, *Lake Sediments and Environmental History.* Haworth and Lund, eds., Leicester University Press., 411 p.

Hemmerich, R., 1980, *Bestimmung des Pb^{210} Flusses an Bodenprofilen*: Staatsexamenabeit, Universitat Heidelberg, 34 p.

Kemp, A.L.W., Anderson, T.W., Thomas, R.L., and Mudrochova, A., 1974, Sedimentation rates and recent sediment history of Lakes Ontario, Erie and Huron, *J. Sed. Petrol.* **44,** 207-218.

Kemp, A.L.W., and Harper, N.S., 1976, Sedimentation rates and a sediment budget for Lake Ontario, *J. Great Lakes Res.* **2**, 324-340.

Kemp, A.L.W., and Thomas, R.L., 1976, Cultural impact on the geochemistry of the sediments of Lakes Ontario, Erie and Huron, *Geoscience Canada* **3**, 191-207.

Krishnaswamy, S., Lal, D., Martin, J.M., and Meybeck, M., 1971, Geochronology of lake sediments, *Earth Planet Sci. Lett.* **11**, 407-414.

Krishnaswamy, S., and Lal, D., 1978, Radionuclide limnochronology, *in,* A. Lerman (ed.) *Lakes Chemistry, Geology and Physics*, pp. 153-177, New York.

Lerman A. and Lietzke T.-A., 1975, Uptake and migration of tracers in lake sediments, *Limnol. Oceanogr.* **20**, 497-510.

Makereth F.J.H., 1966, Some chemical observations on post-glacial lake sediments, *Phil. Trans. R. Soc. B.* **250**, 165-213.

McAndrews, J.H., 1976, Fossil history of man's impact on the Canadian flora; an example from southern Ontario, *Can. Bot. Assoc. Bull. Suppl.* **9**, 1-6.

Müller, G., Dominik, J., and Mangini, A., 1979, Eutrophication changes sedimentation in part of Lake Constance, *Naturwissenschaften.* **66**, 261-262.

Oldfield, F. and Appleby, P.G., 1984, Empirical testing of Pb^{210} dating models for lake sediments *in:* E. V. Haworth and J. W. G. Lund (eds.) *Sediments and Environmental History* pp. 93-124, Leicester University Press.

Robbins, J.A., Krezoski, J.R., and Morley, S.C., 1979, Radioactivity in sediments of the Great Lakes: Post depositional redistribution by deposit-feeding organisms, *Earth. Plant. Sci. Lett.* **36**, 325-333.

Sly, P.G. and Thomas, R.L., 1974, Review of geological research as it relates to an understanding of Great Lakes Limnology, *J. Fisheries Res. Board, Canada* **31**, 795-825.

Thomas, R.L., Frank, R., and Vernet, J.-P., 1984, $\sum$DDT, PCB's and HCB in the sediments of Lake Geneva and the Upper Rhone River, *Environmental Geology* **5**, 3, 103-113.

Thomas, R.L., Jaquet, J.-M., Kemp, A.L.W., and Lewis, C.F.M., 1976, The surficial sediments of Lake Erie, *J. Fisheries. Res. Board, Canada* **33**, 385-403.

Thomas, R.L., Jaquet, J.-M., and Mudroch, A., 1977, Sedimentation processes and associated changes in surface sediment trace metal concentrations in Lake St. Clair, *Proc. Int. Conf. in Heavy Metals in the Environment*, Toronto, 1975, 691-708.

Thomas, R.L., Kemp, A.L.W., and Lewis, C.F.M., 1972, Distribution, composition and characteristics of the surficial sediments of Lake Ontario, *J. Sedim. Petrol.* **42**, 66-84.

Thomas, R.L., Vernet, J.-P., and Frank, R., 1984, $\sum$DDT, PCB's and HCB in the sediments of Lake Geneva and the Upper Rhone River, *Environ. Geol.* **5**, 103-113.

Thompson, R., 1984, A global review of palaeomagnetic results from wet lake sediments, *in* E. V. Haworth and J. W. G. Lund (eds.) *Lake Sediments and Environmental History,* pp. 145-164, Leicester University Press.

Vernet, J.-P., Dominik, J., and Favarger, P.Y., 1984, Texture and sedimentation rates in Lake Geneva, *Environ. Geol.* **5**, 143-149.

Walters, L.J., Wolery, T.J and Myser, R.D., 1974, Occurrence of As, Cd ,Cr, Cu, Fe, Hg, Ni, Sb and Zn in Lake Erie sediments, *Proc. 17th Conf. Great Lakes Res.*, Internat. Assoc. Great Lakes Res., 219-234.

Watts, W.A. and Bradbury, J.P., 1982 Paleoecological studies at Lake Patzcuaro on the west-central Mexican Plateau and at Chalco in the Basin of Mexico, *Quat. Res.* **17**, 56-70.

MOVEMENT AND STORAGE OF SEDIMENT IN RIVER SYSTEMS*

Robert H. Meade
U. S. Geological Survey, MS 413
Denver, Colorado 80225
U. S. A.

1. INTRODUCTION

River sediments play two major roles in geochemical cycles. First, the river sediment fluxes represent about three-quarters of the total denudation of the continents under present-day conditions. Second, river sediments adsorb and transport a number of aqueous ionic constituents. Two recent compilations of the available data have estimated that the total discharge of river sediment to the coastal zones of the world under present-day conditions is 15 to 16×10^{12} kg/yr (Milliman and Meade, 1983; Walling and Webb, 1983). This number, however, cannot be taken at face value to compute a world-wide mean rate of long-term mechanical denudation. In the first place many of the measurements on which the estimate is based are probably inaccurate. Among the large rivers of the world, accurate records that span several decades are available only in China, Europe, and North America. Second, the present-day sediment loads of many of the world's major rivers have been increased or decreased markedly by human activities, such as deforestation, crop farming, surface mining, and construction of dams and reservoirs. Third, the direct linkage between upland soil-erosion rates and downstream river-sediment discharges is so tenuous in large river systems that the common practice of dividing a river's sediment load by its drainage area to obtain a mean rate of erosion or denudation may yield a number that has little significance. Relations between soil erosion and river-sediment discharge change so markedly in time and space, and the time lags between upland causes and downriver effects are so large relative to the rates of these changes, that any assumptions of equilibrium, or even steady state, must be carefully considered in the context of whatever time scale is relevant (Schumm, 1977; Tricart, 1962; Trimble, 1975, 1977).

This chapter is written as a guide to (rather than a comprehensive summary of) some of the more helpful and enlightening recent publications on the subject of sediment in rivers, especially to those published during the decade 1975-85. In all cases, the reader is encouraged to seek out the cited publications for more complete discussions and graphic material.

A. Lerman and M. Meybeck (eds.), Physical and Chemical Weathering in Geochemical Cycles, 165–179.

2. MOVEMENT OF SEDIMENT IN RIVERS

At least three dozen books, symposium volumes, and review articles on the sedimentology and morphology of rivers have been published within the last decade. Among the most useful review articles to students of geochemical cycles are those by Nordin (1981, 1985). Among the useful longer treatments are the recent reviews by Richards (1982) and Hadley et al. (1985), the volume of engineering discussions edited by Vanoni (1975), and especially the seminal book by Schumm (1977). This chapter will review some of the aspects of sediment movement in rivers that are important to studies of geochemical cycles.

2.1. Some Definitions

Confusion has stemmed from different usages of terms in the literature of geochemistry, on the one hand, and river engineering, on the other. The first area of confusion is related to terms describing the concentration, discharge, yield, and load of river sediment:

(i) Sediment concentration is the mass of sediment suspended in a unit volume of river water, usually expressed as grams or milligrams per liter.
(ii) Sediment discharge is the mass of sediment transported by a river past a given point or through a given cross section in a unit of time, usually expressed as kilograms per second, tons per day, or tons per year.
(iii) Sediment yield is the mass of sediment transported by a river in a unit of time, divided by the drainage area upriver of the measuring point; it usually is expressed as tons per square kilometer per year. Although it has been used as such, sediment yield is not synonymous with erosion rate or denudation rate (see the discussion in 5. Storage of Sediment in River Systems).
(iv) Sediment load is a qualitative term that is best used in descriptive discussions to denote the material being transported by a river. The term is frequently used as a synonym of sediment discharge. Geochemists tend to use the term as a synonym for sediment concentration; this latter usage is to be discouraged.

The second area of confusion includes the terms that describe the components of the sediment load. The following definitions, taken mostly from Colby (1963), are used by river engineers and geomorphologists.

(v) Bed material is the sediment on the bed of a river. The term applies regardless of whether the sediment is moving or stationary.
(vi) Bedload is the sediment that is being moved along the river bed, by rolling, sliding, or slipping, within a few grain diameters of the bed. The term bedload sometimes is used mistakenly to describe stationary bed material.
(vii) Suspended load is the material being carried in suspension by a river, and being maintained in suspension by the upward component of the turbulence of the river.

In engineering usage, suspended load is commonly divided into two categories:

(viii) Suspended bed-material load consists of suspended particles whose sizes are large enough to occur in appreciable quantities at the surface of the stream bed.
(ix) Wash load consists of suspended sediment so fine grained that it is only an inappreciable fraction of the sediment on the stream bed.

In practice, the boundary between suspended bed-material load and wash load in large rivers is commonly assumed to be a particle diameter of 0.063 mm, the conventional boundary between sand and silt. Wash load is often uniformly distributed in the water column (Colby, 1963, p. 25), but in at least one large river, particles having diameters as small as 0.020 mm are more concentrated near the river bed than near the river surface (Curtis et al., 1979).

2.2. Bedload and Suspended Load in Geochemical Cycles

Bedload has received a great deal more attention than suspended load in the recent and current sedimentological literature, mostly because of its bearing on the evolution of river channels and the genesis of coarser grained sedimentary rocks. In geochemical cycles, however, bedload is much less important than suspended load.

A consensus of river sedimentologists maintains that, in today's large rivers, bedload generally accounts for only ten percent or less of the total sediment load. However, because bedload is difficult to measure in large rivers, very few direct measurements have been made to substantiate this generalization. Measurements made in two moderately large rivers of the United States, the Tanana River of Alaska and the Clearwater-Snake River of Idaho, yielded bedload discharges that amounted respectively to one and five percent of the suspended-sediment discharges (Nordin, 1985, p. 189; Emmett, 1981, p. 14). Bedload in the biggest rivers is more frequently estimated by computation than by direct measurement: for example, recent computations have estimated that the bedload discharge of the lower Amazon River of Brazil is only one percent of the suspended-sediment discharge (Mertes, 1985).

In extrapolating present-day conditions to the understanding of geochemical cycles, therefore, we are able to ignore bedload transport and focus our attention on suspended sediment. Because the errors in most measurements of suspended-sediment discharge of large rivers are of an order of ± 10 percent, we can assume that the bedload discharges of today's large rivers lie within the range of error of measurement of suspended-sediment discharge. However, let us not lose sight of the fact that a significant proportion of the geologic column consists of sandstones, and, to the extent that these sandstones are alluvial in origin and made of particles coarser than about half a millimeter (the maximum sand size transported in suspension by today's large rivers), fluvial bedload transport must have played a role in older geochemical cycles.

3. ACCURACY OF ESTIMATES OF MODERN RIVER-SEDIMENT DISCHARGE

The first problem of applying modern stream-load data to long-term estimates of transport rates in geochemical cycles is in the accuracy of today's measurements. Not only does each individual measurement have to be accurate, but enough measurements have to be made to average out the marked temporal and spatial heterogeneities of riverine sediment transport. At its most operational level, an accurate measurement of sediment in rivers begins with overcoming the heterogeneous distributions of velocity and sediment concentration in river cross sections. Velocities are greatest near the water surface in midchannel, and they diminish toward the bed and banks of the river. Suspended sediment is more concentrated near the river bed than near the river surface; this is especially true of suspended sand (bed-material load) but in some rivers is also true of suspended silt (Colby, 1963, p. 25; Curtis et al.,

1979). Furthermore, because of incomplete mixing of tributary inputs or because of in-channel hydraulic effects, suspended sediments may not be distributed uniformly across a river channel (Jordan, 1965, p. 60-65; Meade, 1985, fig. 7; Meade et al., 1983, p. 1141). Accurate sampling of suspended sediment, or of any constituent adsorbed onto sediment particles, in large rivers requires composite depth-integrated samples or large numbers of individual point samples collected at multiple depths through a cross section (Guy and Norman, 1970; Nordin et al., 1983). The all too common practice of merely dipping a bottle into the river surface can entail sampling errors of several hundred percent.

3.1. Temporal Variability of Sediment Discharge

Even with accurate sampling techniques, however, the estimation of long-term suspended-sediment discharge in rivers is made more difficult by the variability of sediment concentration with time. In an ideal study, a river should be sampled for a period of decades to assess the long-term average sediment discharge. Samples should be collected frequently (more than once a day, ideally) when water discharge is changing rapidly, and less frequently (once or twice a week) when water discharge is low. In reality, however, this type of sampling program is expensive and, in many cases, impractical. The more usual course is to collect sediment data over as full a range of water discharges as possible in the time available for a study, and use these data to construct a relation—called a sediment rating curve—between sediment discharge and water discharge. The sediment rating curve is then combined with the available record of water discharge (which is almost always more detailed and lengthy than the measured sediment record) to synthesize a record of sediment discharge. This method and its inaccuracies have been discussed in detail by Walling (1977a, 1978) and Walling and Webb (1981), who show that it can easily entail errors of 50 percent or more in the estimation of long-term sediment discharges.

Suspended-sediment concentration and suspended-sediment discharge are approximate power functions of water discharge in many streams; that is, they tend to plot in a more or less linear pattern on log-log scales (Nordin, 1985). Because the sediment discharge is the product of water discharge and sediment concentration, it increases even more sharply than concentration with increasing streamflow. Consequently, most of the sediment carried by rivers of moderate size is transported during only a few weeks of the year. Over periods of 2 to 3 decades, for example, the larger rivers of the Atlantic drainage of the United States (small- to moderate-sized rivers on a world scale) transported half their sediment loads in 1 percent of the time and 85 to 90 percent of their sediment in 10 percent of the time (Meade, 1982). This means that much more attention should be given to the 10 percent of daily flows that carry 90 percent of the sediment than to the 90 percent of the daily flows that carry only 10 percent of the sediment. There exists the additional difficulty of sampling the very infrequent flow events that carry especially large sediment loads—events such as the flood of Christmas 1964 during which the Eel River of California carried as much sediment in 10 days as it had carried in the previous 10 years (Brown and Ritter, 1971; Meade and Parker, 1985, p. 58-59).

In most rivers, large and small, the relation between sediment concentration or sediment discharge and water discharge is not a mathematically perfect power function. In many rivers, there are marked seasonal differences in the relations between sediment concentration and water discharge: in summer they are different from (but often parallel to) those in winter (Gregory and Walling, 1973, p. 215-219; Meade, 1982, p. 237; Meybeck, 1985, p. 111). In many rivers, the seasonal relation between sediment discharge and water discharge, when plotted on a log-log scale graph with sediment discharge as the ordinate, forms a clockwise loop. That is, sediment concentrations on rising water discharges are greater than at equal discharges when the river stage is falling. This relation usually is explained as the "depletion" or "exhaustion" effect: fine-grained sediment, which is stored on channel beds and along river banks during low-water periods, is in plentiful supply as the river rises, but the stored material is soon resuspended, and it eventually becomes depleted before the river reaches its maximum discharge. Such clockwise looped relations are typical of many large rivers, among them the Amazon, Fraser, Mekong, Mississippi, Nile, Orinoco, and Yellow Rivers (Meade, 1985, fig. 6; Milliman, 1980, fig. 4; Whitfield and Schreier, 1981; Meybeck, 1985, fig. 8; Robbins, 1977, figs. 39-41; Shahin, 1985, p. 460; Meade et al., 1983, p. 1137-1138; Shi et al., 1985, p. 13), as well as many other rivers of moderate and smaller size (Nordin and Beverage, 1965; Temple and Sundborg, 1972; Walling, 1977b; Wood, 1977). One major exception to this pattern is the Yangtze River of China in which the sediment rating loop is counterclockwise because of the seasonal delay of inputs from its more sediment-laden tributaries (Shi et al., 1985).

3.2. Areal Variability of Sediment Discharge

The areal variability of river-sediment discharge is related most directly to the physical properties of the rocks and soils from which the sediments are being eroded, and it is evident at many scales—river basin, continental, and global. The reader is referred to the comprehensive collection of maps of erosion and sediment yield compiled by Jansson (1982). A few examples, most of which are not included in Jansson's compilation, will suffice here.

Spatial variability of sediment discharge is strongly evident in two river basins of moderate size (about 20,000 km^2) on the flanks of the Great Plains of the north central United States. In both river basins, most of the runoff is derived from mountainous areas underlain by crystalline rocks, whereas most of the suspended sediment is derived from areas of lower relief underlain by shales and siltstones. Underlying rock type is a more important factor than relief in controlling sediment yield from these basins. In the Wind River basin, more than half the water but less than 1 percent of the sediment is derived from the 14 percent of the area that is underlain by crystalline Precambrian rocks, whereas 63 percent of the sediment but less than 3 percent of the water is derived from the 16 percent of the drainage area that is underlain by poorly consolidated Quaternary deposits (Table I). Likewise, in the Powder River basin, most of the water but very little sediment is derived from the mountainous areas underlain by Precambrian crystalline rocks and Paleozoic limestone and dolomite, whereas most of the sediment but very little water is derived from areas of less relief underlain by Mesozoic and Cenozoic siltstones and shales (Hembree et al., 1952, p. 47-48).

Similar spatial variability of sediment yield is evident in large river basins. Gibbs (1967, p. 1218) estimated that 82 percent of the total suspended solids discharged by the Amazon River is supplied from the 12 percent of the total drainage basin that is underlain by the Andes. More recent data (Meade, 1985; Meade et al., 1985, fig. 2) suggest that more than 90 percent of the suspended sediment in the Amazon system is derived from the Andes—even though most of the water in the river is derived from the low-lying areas of the basin. Likewise, in the Orinoco River basin, the water is derived in roughly equal amounts from the areas underlain by the Andes and the Guayana Shield, whereas at least 90 percent of the suspended sediment is derived from the Andes (Meade et al., 1983). In the Yellow River basin of China, more than

Table I. Yields of water and sediment from five geologic areas in percentage of total water and sediment discharges, 1946-52, Wind (Bighorn) River at Thermopolis, Wyoming (Colby et al., 1956, p. 85).

Types of Underlying Rocks	Areas (%)	Total Yield	
		Water (%)	Sediment (%)
Precambrian	14	59	0.5
Paleozoic	10	9	0.5
Mesozoic	12	2.5	8
Tertiary	48	27	28
Quaternary	16	2.5	63
Total	100	100	100

90 percent of the sediment in the river comes from the loess region that occupies no more than 40 percent of the total basin area (Long and Xiong, 1981, fig. 4).

On a global scale, a recent compilation by Walling and Webb (1983, p. 77) shows that "maximum sediment yields are associated with the loess areas of China and the Cenozoic mountain areas around the Pacific margin. High values are also to be found in other mountain areas and in regions with Mediterranean and semiarid climates, and in the seasonally humid tropics." Especially low sediment yields are found in Australia, Siberian Asia, and Saharan Africa. This global pattern is corroborated in the compilation by Milliman and Meade (1983), who also point out the large sediment yields from areas (such as the mountain ranges of south-central Alaska) that are covered by active glaciers. With the notable exception of the loess region of China, the greatest sediment yields on a global scale generally are associated with areas of most intense tectonic activity (see also Stallard, this volume).

4. HUMAN INFLUENCES ON RIVER-SEDIMENT LOADS

Another factor that has to be taken into account in using modern stream-load data to estimate long-term rates of denudation is the influence of man. Much has been written on this subject, particularly on those human activities that increase river-sediment loads (see Douglas, 1967; Jansson, 1982; p. 38-41; Gregory and Walling, 1973, p. 342-358; Judson, 1968; Meade, 1969, 1982). On a global scale, the most massive anthropogenic increases in river sediment loads have come about as a result of crop farming, especially in areas where forests have been cut down to make way for croplands. The conversion of forests to croplands that accompanied the European settlement of the eastern United States between the years 1700 and 1900 caused a tenfold increase in the sediment loads carried by many of the rivers draining the region (Meade, 1969; see also Thomas, this volume). What have been the consequences of thousands of years of intensive cultivation of the heavily populated river basins of Asia? Although we can only guess at pre-cultivation sediment yields in these river basins, and although the evidence of such factors as climatic change are lost in antiquity and obscured by later effects, we can surmise that the anthropogenic additions made over the millennia to the natural sediment loads of rivers such as the Yellow, Ganges, and Indus must have been large.

Another human activity whose influence on sediment loads has been large on a river-basin scale is surface mining. Mining wastes have clogged the channels and inundated the flood plains of rivers of at least moderate size. Perhaps the best known example is that of the hydraulic gold-mining wastes in the Sacramento River valley of California, described in the classic paper by Gilbert (1917). Hydraulic-mining activities during a 60-year period (1849-1909) contributed 1.2×10^9 m^3 of sediment to the Sacramento River system, but natural sources during the same period contributed only 0.1×10^9 m^3. A more recent example is that of the copper-mining wastes in the Kawerong and Jaba Rivers of Papua New Guinea, described by Pickup et al. (1979, 1983). Mining activities during an 8-year period (1968-1976) contributed 0.2×10^9 tons of sediment to a small river system whose natural sediment load was negligible by comparison. One third of this vast quantity of sediment was deposited on flood plains and in the delta of the Jaba River.

The principal human activity that decreases the loads of sediments carried by rivers is the construction of dams and reservoirs. Dams are built to impound water for various purposes, and the reservoirs they form interrupt the downriver flow of sediment. Although the river water that enters a reservoir is released eventually (through a powerplant, into a diversion canal, or over a spillway), much of the sediment is trapped permanently in the reservoir. A reservoir that is only large enough to hold one hundredth of the water that flows into it each year can trap half the sediment that flows into its upstream end. A reservoir that can retain a tenth of the annual water inflow can trap 80 to 90 percent of the inflowing sediment (Brune, 1953). The most striking examples among the world's major sediment-transporting rivers are the reductions of the sediment discharges of the Colorado and Nile Rivers from about 125×10^9 kg/yr each to less than 3×10^9 kg/yr each by the construction of Hoover Dam and the high dam at Aswan (Meade and Parker, 1985, p. 54-55; Shahin, 1985, p. 460). The construction of large dams since the late 1940's in the basins of the Mississippi and Indus Rivers has reduced the discharges of sediment to the deltas of these great rivers by half or more (Meade and Parker, 1985, p. 51-52; Milliman et al., 1984).

5. STORAGE OF SEDIMENT IN RIVER SYSTEMS

Perhaps the most confounding factor in the use of modern sediment-load data to infer long-term denudation rates of the continents is the large-scale storage of sediment in river systems. Because of this storage, very little direct linkage is visible between the erosion of uplands and the discharges of sediment at the mouths of large rivers. In the conterminous United States, for example, upland erosion rates exceed the rates of river-sediment discharge into the coastal zones by a factor of ten (Curtis et al., 1973; Holeman, 1981). Ninety percent of the sediment presently being eroded off the land surface of the United States is being stored somewhere in the river systems between the uplands and the sea. Furthermore, the time lags between erosion and sediment transport are such that the sediments being carried by large rivers today may represent episodes of erosion that occurred decades, centuries, or even millennia ago.

Excellent summaries of the problems involved in understanding the linkage (or lack of it) between soil erosion and river-sediment transport have been prepared by Schumm (1977) and Walling (1983). Further examples are given below to illustrate patterns and problems of sediment storage at different time scales. As Schumm (1977,

p. 99) puts it, "Confusion can only be avoided if discussion is restricted to a consideration of one time span at a time."

5.1. Storage on Seasonal Time Scales

On a seasonal time scale, rivers generally tend to deposit sediments on their beds and along their banks at low or falling discharges, and to resuspend these sediments at high or rising discharges. But there are some exceptions to these general tendencies, as we shall see below in examples taken from the three biggest rivers of North and South America. Whether the storage and resuspension of sediments on seasonal time scales has any residual effect on the longer-term storage of river sediment is not clear.

In the lowermost 500 km of the Mississippi River of the United States, suspended sediment is stored and remobilized in response to seasonal changes in water discharge and river slope. At less than average water discharges and water-surface slopes, suspended sediment is dropped by the river and stored on the river bed. At average water discharges, sediment is neither stored nor resuspended, on a net basis, in this reach. At greater than average water discharges and water-surface slopes, at least part of the previously stored sediment is resuspended from the river bed. The pattern of storage and resuspension in this reach exaggerates the direct correlation between water discharges and sediment loads: at lower discharges, the already small sediment loads being transported into the upper end of the reach are further reduced by deposition and storage on the river bed; at higher discharges, the already large sediment loads are increased further by resuspension from the river bed (Meade and Parker, 1985).

In a 750-km reach of the lower Amazon River of Brazil, the seasonal pattern of storage and remobilization of suspended sediment is more closely correlated with river slope than with water discharge. Because of the timing of inputs from several large tributaries, the slope of the river surface in this reach is, contrary to the usual pattern, smaller on rising stages than on falling stages. Consequently, sediment is stored during rising river stages and resuspended during falling river stages, in quantities on the order of 10^6 tons per day. This pattern of sediment storage and resuspension, in contrast to the pattern observed in the lower Mississippi, regulates the discharge of sediment to the sea. During rising river stages, when sediment loads from the upper Amazon are large, the lower Amazon stores part of the sediment before it reaches the sea. During falling stages, when sediment loads from the upper river are smaller, the lower river augments the loads being transported seaward by resuspending sediment from the river bed (Meade et al., 1985).

A 200-km reach of the middle Orinoco River of Venezuela shows still another pattern of seasonal storage and resuspension. When the river stage is near its annual peak, this reach of the Orinoco forms a large shallow lake that includes the lower reaches of a number of tributaries as well as large tracts of intervening low-lying flood plains. During peak stage, nearly half the inflowing sediment is stored in this reach. During subsequent falling stages and during the early part of the next rising stage, this reach of the Orinoco yields sediment from storage to be transported farther downriver. Sediment loads in the lower Orinoco, therefore, are smaller at peak water discharge than on rising or early falling stages (Meade et al., 1983).

5.2. Storage on Decade to Century Time Scales

The classic case study of the movement and storage of sediment in a river system on time scales of 10 to 100 years is that of the hydraulic-mining debris in the

Sacramento River valley of California (Gilbert, 1917; Kelley, 1959). Between 1855 and 1885, enormous quantities of coarse sediment were washed into some of the tributaries of the Sacramento River during hydraulic mining for gold. The resulting problems that developed downstream (flooding, filling of navigation channels, destruction of flood-plain farms) became so serious that hydraulic mining was curtailed by a court decision in 1884. By that time, however, the large mass of sediment, characterized as a "wave" by Gilbert (1917), was already into the stream channels and was moving slowly down the tributaries and in the Sacramento River. As the mass of sediment advanced, it raised the levels of the channel beds, much as an ocean swell raises the level of the sea as it passes through. Bed levels rose 5 m in the tributary Yuba River at Marysville and nearly 3 m in the Sacramento River at Sacramento. The riverbeds at these towns reached their highest elevations 10 to 20 years after the mining was stopped, and then they declined steadily during the next 30 to 40 years to their previous elevations. All in all, the great wave of hydraulic-mining debris took nearly a century to pass through the channels of the Sacramento River system and finally to reach San Francisco Bay (Meade, 1982, p. 243-244).

5.3. Storage on Century to Millennium Time Scales

The pattern of storage and remobilization described in the preceding paragraph, however, applies only to the sediment in and near the river channels. It does not apply to the debris that overflowed onto the flood plains. The hydraulic-mining debris that was carried out of the river channel during floods and deposited on the flood plains was sufficient in many places to cover entire houses and orchards (Kelley, 1959, p. 134-135, 203-204). Most of that debris still remains where it was deposited a century ago. The time required to remove sediment from storage on the flood plain is much longer than the century that was required to remove the debris from the main river channels. Flood-plain deposits are removed mainly by erosion of channel banks as streams slowly migrate laterally, a process that proceeds at a substantially slower pace than the vertical removal of material stored in the bottom of the river channel. The complete remobilization of flood-plain deposits may require time periods of an order of a millennium or more (Leopold et al., 1964, p. 328). However, as the studies cited by Schumm (1977, p. 131-132) indicate, there is considerable variation in the length and time scales at which flood plains are remobilized.

Many of the problems associated with the prediction of storage and remobilization of sediment on time scales of 100 to 1000 years were demonstrated in a study carried out on Coon Creek, a small tributary of the Mississippi River that drains 360 km^2 of southwestern Wisconsin. Originally covered by forests, Coon Creek basin was settled by European immigrants and cleared for farming about 1850. As the forests were cleared and the land was plowed, a cycle of erosion and sedimentation began, the consequences of which are still strongly in effect today. Sediments were eroded at a greatly accelerated rate from upland and tributary areas, and were transferred to the lower hillslopes and valleys of the creek basin. Much less than 10 percent (5×10^6 tons) of the sediment eroded off the uplands during the years since 1850 was exported out of the basin by the creek. More than 90 percent (80×10^6 tons) of the sediment was deposited along the way, on hillslopes and flood plains, where most of it still remains in storage. Upland erosion rates, therefore, are not reflected in the sediment yields at the mouth of Coon Creek (Trimble, 1983).

In tectonically active regions of the world, cycles of erosion, storage, and remobilization of sediment in river valleys may be greatly accelerated. Grant (1985) presents evidence for 8 major episodes of erosion and alluvial sedimentation in New

Zealand during the last 1,800 years. He believes that the major episodes of erosion and sedimentation were related to fluctuating magnitudes of major rainstorms and floods. During periods of more intense rainfall, sediment was eroded from steep hillslopes and deposited as alluvium in the stream valleys. During intervening periods of less intense rainfall, the alluvial deposits were degraded, leaving behind terraced valleys. These valleys were then refilled with sediments during the next episode of heavier rainfall. In tectonically less active regions of the world, such episodes of erosion and sedimentation proceed more slowly (Leopold and Miller, 1954; Judson, 1963).

5.4. Storage on Multimillennial Time Scales

On time scales of 1,000 to 10,000 years, the most significant factor in the storage of sediment in large river systems is the eustatic rise of sea level that has accompanied the melting of the Late Pleistocene ice sheets. Some 15,000 years ago, at the end of the last major ice advance (the Wisconsin or Wurm stage), the world's big rivers were graded to a sea level 100 to 150 m lower than today's, and their lower valleys were 100 to 150 m deeper than they are today. During the last 15,000 years, as the great glaciers melted and sea level rose (the Flandrian transgression), the lower valleys of the large rivers have accumulated and stored vast quantities of sediment. The sequence of storage of alluvial sediment in the lower Mississippi valley, for example, has been reconstructed by Fisk (1944, 1947) from an extensive collection of data from bore holes drilled into the flood plain. (Fisk's reports are no longer in print, but a succinct summary, including block diagrams, is given by Schumm, 1977, p. 164-167). A similar sequence of sediment storage in the Amazon valley during the Flandrian transgression has been described by Sioli (1957, 1984) and Irion (1976) on the basis of surficial evidence (few subsurface data are available from the Amazon flood plain). Even today, large rivers like the Brahmaputra, Yellow, and Indus continue to deposit large proportions (20 to 70 percent) of their sediment loads onto the alluvial plains that separate their source areas from the sea (Goswami, 1985; Long and Xiong, 1981, fig. 4; Milliman et al., 1984). In the Yellow and Indus River basins, the Holocene sedimentation sequence takes the form of a large alluvial fan rather than a confined valley fill (Ye et al., 1983).

Although this discussion is not intended to extend to time spans much longer than 10,000 years, we should remember that we live in a time of marine transgression, and that the filling of the big river valleys discussed in the preceding paragraph represents only the transgressive half of a single regression-transgression cycle. The great ice sheets advanced and receded in at least four major pulses during the Pleistocene, and the sea level must have fallen and risen an equal number of times by 100 m or more. We can assume that the large alluvial fills that occupied the major river valleys were eroded and deposited an equal number of times. The well developed sequences of terraces, such as those described as "fossil flood plains" in the Amazon valley by Klammer (1984), attest to a repetitive series of major episodes of storage and remobilization of alluvial sediments in major river valleys during the last several hundred thousand years.

5.5. Perceptions of Sediment Storage Cycles

At least two factors bias our perceptions of storage and remobilization of sediment in river systems. The first bias is the natural emphasis on the time scale of 10 to 100 years which is, after all, the secular scale, or the scale of a human life span. Any direct sensory perceptions we may have of changes in rivers are confined to processes that operate over periods of 100 years or less. The second factor that biases our perceptions is that we live in a transgressive period, a time of rising sea level. Based on what we are able to observe today, the major river valleys are being aggraded, and large quantities of sediment are being stored between upland sources and the sea. But conditions were not always as they are today. During periods when sea level was falling, the large river valleys were being incised rather than filled, and the rivers must have discharged more sediment at their mouths than was being eroded from the uplands of their drainage basins.

To conclude the discussion of sediment storage, and to place the foregoing considerations into perspective, we borrow once more from Schumm, who pointed out in a recent paper (1985) that the perceived importance of an event depends strongly on the time span in which the event is being considered. An event of overwhelming importance on a seasonal time scale becomes secondary on an annual time scale, is even less important on a 10-year time scale, and is virtually insignificant on a 100-year time scale. Table II places the storage events discussed above into a framework of time scales ranging from 10^{-1} to 10^5 years. The table can be extended to either the right or left to place other geochemical cycles into a comprehensive temporal perspective.

6. CONCLUDING REMARKS

Our knowledge of the sediment loads of the world's major rivers is based almost entirely on data collected within the last 50 to 100 years. If we are to use a mere century's worth of data to estimate long-term rates of global sediment fluxes, we need more than the reassurance of the uniformitarianist dictum that "the present is the key to the past." We need to understand the reasons why today's observable conditions produce today's observable sediment yields before we can use today's rates to infer the rates in the geologic past.

A question remains whether there is any validity at all in the extrapolation of modern denudation rates to long periods of the geologic past. The tectonic recycling of the island of Taiwan, mentioned in the chapter by Stallard (this volume), is a case in point. If the material of which Taiwan consists actually is passing through a complete cycle of erosion, deposition, subduction, uplift, and re-erosion every 10^6 to 10^7 years, then the present-day sediment yields from Taiwan, which are among the largest in the modern world, have meaning only at time scales significantly shorter than 10^6 years. In the context of even longer time scales of global recycling, the scales of 10^8 to 10^9 years discussed in the chapters by Garrels and Veizer (this volume), one could argue that today's sediment-transport rates are mostly irrelevant. Whatever its purpose or outcome may be, any discussion of mechanical denudation in geochemical cycles should proceed only within the context of a clearly defined time scale.

7. REFERENCES CITED

Brown, W. M. III and Ritter J. R., 1971, Sediment transport and turbidity in the Eel River basin, California: *U. S. Geological Survey Water-Supply Paper* **1986**, 70 pp.

Brune, G. M., 1953, Trap efficiency of reservoirs: *American Geophysical Union Transactions,* **34,** 407-418.

Colby, B. R., 1963, Fluvial sediments—a summary of source. transportation, deposition, and measurement of sediment discharge: U. S. Geological Survey Bulletin **1181-A**, 47 pp.

Colby, B. R., Hembree, C. H., and Rainwater, F. H., 1956, Sedimentation and chemical quality of the surface waters in the Wind River basin, Wyoming: *U. S. Geological Survey Water-Supply Paper* **1373**, 336 pp.

Curtis, W. F., Culbertson. J. K., and Chase. E. B., 1973, Fluvial-sediment discharge to the oceans from the conterminous United States: *U. S. Geological Survey Circular* **670**, 17 pp.

Curtis, W. F., Meade, R. H., Nordin, C. F., Jr., Price, N. B., and Sholkovitz, E. R., 1979, Non-uniform vertical distribution of fine sediment in the Amazon River: *Nature,* **280,** 381-383.

Douglas, Ian, 1967, Man, vegetation, and the sediment yields of rivers: *Nature,* **215,** 925-928.

Emmett, W. W., 1981, Measurement of bed load in rivers, *in* Erosion and sediment transport measurement: *International Association of Hydrological Sciences Publication* **133,** 3-15.

Fisk, H. N., 1944, *Geological investigation of the alluvial valley of the lower Mississippi River*: Vicksburg, U. S. Corps of Engineers Mississippi River Commission, 78 pp. plus 80 figs., 33 pl.

Fisk, H. N., 1947, Fine-grained alluvial deposits and their effects on Mississippi River activity: Vicksburg, U. S. Corps of Engineers Mississippi River Commission Waterways Experiment Station, **1**, 82 pp.; **2**, 74 pl.

Gibbs, R. J., 1967, The geochemistry of the Amazon River system: Part I. The factors that control the salinity and the composition and concentration of the suspended solids: *Geological Society of America Bulletin,* **78,** 1203-1232.

Gilbert, G. K., 1917, Hydraulic-mining debris in the Sierra Nevada: *U.S. Geological Survey Professional Paper* **105**, 154 pp.

Goswami, D. C., 1985, Brahmaputra River, Assam, India: Physiography, basin denudation, and channel aggradation: *Water Resources Research,* **21,** no. 7, 959-978.

Grant, P. J., 1985, Major periods of erosion and alluvial sedimentation in New Zealand during the Late Holocene: *Royal Society of New Zealand Journal,* **15,** no. 1, 67-121.

Gregory, K. J., and Walling, D. E., 1973, *Drainage basin form and process*: New York, John Wiley & Sons, 456 pp.

Guy, H. P., and Norman, V. W., 1970, Field methods for measurement of fluvial sediment: *U. S. Geological Survey Techniques of Water-Resources Investigations,* Book **3**, Chapter C2, 59 pp.

Hadley, R. F., Lal, R., Onstad, C. A., Walling, D. E., and Yair, A., 1985, *Recent developments in erosion and sediment yield studies*: United Nations Educational, Scientific and Cultural Organization, International Hydrological Programme, Technical Documents in Hydrology, 127 pp.

Hembree, C. H., Colby, B. R., Swenson, H. A. and Davis, J. R., 1952, Sedimentation and chemical quality of water in the Powder River drainage basin, Wyoming and Montana: *U. S. Geological Survey Circular* **170**, 92 pp.

Holeman, J. N., 1981, The national erosion inventory of the Soil Conservation Service, U. S. Department of Agriculture, 1977-79, *in* Erosion and sediment transport measurement: *International Association of Hydrological Sciences Publication* **133,** 315-319.

Irion, Georg, 1976, Die Entwicklung des zentral- und oberamazonischen Tieflands im Spat-Pleistozan und im Holozan: *Amazoniana,* **6,** no. 1, 67-79.

Jansson, M. B., 1982, Land erosion by water in different climates: *Uppsala Universitet Naturgeografiska Institutionen Rapport* **57**, 151 pp.

Jordan, P. R., 1965, Fluvial sediment of the Mississippi River at St. Louis, Missouri: *U. S. Geological Survey Water-Supply Paper* **1802**, 89 pp.

Judson, Sheldon, 1963, Erosion and deposition of Italian stream valleys during historic time: *Science,* **140,** 898-899.

Judson, Sheldon, 1968, Erosion of the land, or what's happening to our continents?: *American Scientist,* **56,** no. 4, 356-374.

Kelley, R. L., 1959, *Gold vs. Grain—The hydraulic mining controversy in California's Sacramento Valley*: Glendale, Calif., Arthur H. Clark, 327 pp.

Klammer, Gerhard, 1984, The relief of the extra-Andean Amazon basin, *in* Sioli, Harald, editor, *The Amazon. Limnology and landscape ecology of a mighty tropical river and its basin*: Dordrecht, Netherlands, W. Junk Publishers, pp. 47-83.

Leopold, L. B., and Miller, J. P., 1954, A postglacial chronology for some alluvial valleys in Wyoming: *U. S. Geological Survey Water-Supply Paper* **1261**, 90 pp.

Leopold, L. B., Wolman, M. G., and Miller. J. P., 1964, *Fluvial processes in geomorphology*: San Francisco, W. H. Freeman & Co., 522 pp.

Long, Y-q., and Xiong, G.-s., 1981, Sediment measurement in the Yellow River, *in* Erosion and sediment transport measurement: *International Association of Hydrological Sciences Publication* **133,** 275-285.

Meade, R. H., 1969, Errors in using modern stream-load data to estimate natural rates of denudation: *Geological Society of America Bulletin,* **80,** no. 7, 1265-1274.

Meade, R. H., 1982, Sources, sinks, and storage of river sediment in the Atlantic drainage of the United States: *Journal of Geology,* **90,** no. 3, 235-252.

Meade, R. H., 1985, Suspended sediment in the Amazon River and its tributaries in Brazil during 1982-84: *U.S. Geological Survey Open-File Report* **85-492**, 39 pp.

Meade, R. H., Dunne, Thomas, Richey, J. E., Santos, U. de M., and Salati, Eneas, 1985, Storage and remobilization of suspended sediment in the lower Amazon River of Brazil: *Science,* **228,** 488-490.

Meade, R. H., Nordin, C. F., Jr., Perez-Hernandez, David, Mejia-B., Abel, and Perez-Godoy, J. M., 1983, Sediment and water discharge in Rio Orinoco, Venezuela and Colombia, *in* Proceedings of the Second International Symposium on River Sedimentation, 11-16 October 1983, Nanjing, China: Beijing, Water Resources and Electric Power Press, p. 1134-1144.

Meade, R. H. and Parker, R. S., 1985, Sediment in rivers of the United States, *in* National water summary 1984: *U. S. Geological Survey Water-Supply Paper* **2275,** 49-60.

Mertes, L. A. K., 1985, Floodplain development and sediment transport in the Solimoes-Amazon River, Brazil: University of Washington (Seattle), M.S. thesis, 108 pp.

Meybeck, Michel, 1985, Variabilite dans le temps de la composition chimique des rivieres et de leurs transports en solution et en suspension: *Revue Francaise des Sciences de l'Eau,* **4,** 93-121.

Milliman, J. D., 1980, Sedimentation in the Fraser River and its estuary, southwestern British Columbia (Canada): *Estuarine and Coastal Marine Science,* **10,** 609-633.

Milliman, J. D., and Meade, R. H., 1983, World-wide delivery of river sediment to the oceans: *Journal of Geology,* **91,** no. 1, 1-21.

Milliman, J. D., Quraishee, G. S., and Beg, M. A. A., 1984, Sediment discharge from the Indus River to the ocean: past, present and future, *in* Haq, B. U., and Milliman, J. D., editors, *Marine geology and oceanography of Arabian Sea and coastal Pakistan*: New York, Van Nostrand Reinhold Co., pp. 65-70.

Nordin, C. F., Jr., 1981, The sediment discharge of rivers—a review, *in* Erosion and sediment transport measurement: *International Association of Hydrological Sciences Publication* **133, vol. 2,** 3-47.

Nordin, C. F., Jr., 1985, The sediment loads of rivers, *in* Rodda, J. C., editor, *Facets of Hydrology II*: New York, John Wiley & Sons., pp. 184-203.

Nordin, C. F., Jr., and Beverage, J. P., 1965, Sediment transport in the Rio Grande New Mexico: *U. S. Geological Survey Professional Paper* **462-F**, 35 pp.

Nordin, C. F., Jr., Cranston, C. C., and Mejia-B., Abel, 1983, New technology for measuring water and suspended-sediment discharge of large rivers, *in* Proceedings of the Second International Symposium on River Sedimentation, 11-16 October, 1983, Nanjing, China: Beijing, Water Resources and Electric Power Press, p. 1145-1158.

Pickup, G., and Higgins, R. J., 1979, Estimating sediment transport in a braided gravel channel—the Kawerong River, Bougainville, Papua New Guinea: *Journal of Hydrology (Amsterdam),* **40,** 283-297.

Pickup, G., Higgins, R. J., and Grant, I., 1983, Modelling sediment transport as a moving wave—the transfer and deposition of mining waste: *Journal of Hydrology (Amsterdam),* **60,** 281-301.

Richards, Keith, 1982, *Rivers—form and process in alluvial channels*: London, Methuen, 358 pp.

Robbins, L. G., 1977, *Suspended sediment and bed material studies on the lower Mississippi*

River: U. S. Army Engineer District, Vicksburg, Potamology Investigations Report 300-1, 29 pp. plus 17 tables, 6 plates, 69 figures, 12 photos.

Schumm, S. A., 1977, *The fluvial system*: New York, John Wiley & Sons, 338 pp.

Schumm, S. A., 1985, Explanation and extrapolation in geomorphology: seven reasons for geologic uncertainty: *Japanese Geomorphological Union Transactions,* **6,** no. 1, 1-18.

Shahin, Mamdouh, 1985, *Hydrology of the Nile basin*: Amsterdam, Elsevier, 575 p.

Shi, Y-l., Yang, W., and Ren, M.-e., 1985, Hydrological characteristics of the Changjiang and its relation to sediment transport to the sea, *in* Milliman, J. D., and Jin, Q.-m., editors, Sediment dynamics of the Changjiang estuary and the adjacent East China Sea: *Continental Shelf Research,* **4,** no. 1/2, 5-15.

Sioli, Harald, 1957, Sedimentation im Amazonasgebiet: *Geologische Rundschau,* **45,** 608-633.

Sioli, Harald, 1984, The Amazon and its main affluents: hydrography, morphology of the river courses, and river types, *in* Sioli, Harald, editor, *The Amazon. Limnology and landscape ecology of a mighty tropical river and its basin*: Dordrecht, The Netherlands, W. Junk Publishers, pp. 127-165.

Temple, P. H., and Sundborg, Ake, 1972, The Rufiji River, Tanzania, hydrology and sediment transport: *Geografiska Annaler,* **54**, ser. A, no. 3/4, 345-368.

Tricart, Jean, 1962, Les discontinuites dans les phenomenes d'erosion: *International Association of Scientific Hydrology Bulletin,* **59,** 33-43. (English translation published 1982 in Laronne, J. B., and Mosley, M. P., editors, Erosion and sediment yield (Benchmark Papers in Geology, **63**): Stroudsburg, Pennsylvania. Hutchinson Ross Publishing Company, 34-43.)

Trimble, S. W., 1975, Denudation studies: Can we assume stream steady state?: *Science,* **188,** 1207-1208.

Trimble S. W., 1977, The fallacy of stream equilibrium in contemporary denudation studies: *American Journal of Science,* **277,** 876-887.

Trimble, S. W., 1983, A sediment budget for Coon Creek basin in the Driftless area, Wisconsin, 1853-1977: *American Journal of Science,* **283,** 454-474.

Vanoni, V. A., editor, 1975, Sedimentation engineering: *American Society of Civil Engineers Manual and Report on Engineering Practice* **54**, 745 pp.

Walling, D. E., 1977a, Limitations of the rating curve technique for estimating suspended sediment loads, with particular reference to British rivers, *in* Erosion and solid matter transport in inland waters: *International Association of Hydrological Sciences Publication* **122,** 34-48.

Walling, D. E., 1977b, Suspended sediment and solute response characteristics of the River Exe, Devon, England, *in* Research in fluvial geomorphology—Proceedings of the 5th Guelph Symposium on Geomorphology, 1977: Norwich, England, Geo Abstracts Ltd., p. 169-197.

Walling, D. E., 1978, Reliability considerations in the evaluation and analysis of river loads: *Zeitschrift fur Geomorphologie*, suppl. vol. **29,** 29-42.

Walling, D. E., 1983, The sediment delivery problem, *in* Rodriguez-Iturbe, Ignacio, and Gupta, V. K., editors, Scale problems in hydrology: *Journal of Hydrology (Amsterdam),* **65,** 209-237.

Walling, D. E., and Webb, B. W., 1981, The reliability of suspended sediment load data, *in* Erosion and sediment transport measurement: *International Association of Hydrological Sciences Publication* **133**, 177-194.

Walling, D. E., and Webb, B. W., 1983, Patterns of sediment yield, *in* Gregory, K. J., editor, *Background to palaeohydrology*: New York, John Wiley & Sons, pp. 69-100.

Whitfield, P. H., and Schreier, Hanspeter, 1981, Hysteresis in relationships between discharge and water chemistry in the Fraser River basin, British Columbia: *Limnology and Oceanography,* **26**, no. 6, 1179-1182.

Wood, P. A., 1977, Controls of variation in suspended sediment concentration in the River Rother, West Sussex, England: *Sedimentology,* **24,** no. 3, 437-445.

Ye, Q.-c., Jing, K., Yang, Y.-f., Chen, Y.-z., and Zhang, Y.-f., 1983, Changes of the river course of the lower Yellow River with respect to erosion of loess plateau (in Chinese), *in* Proceedings of the Second International Symposium on River Sedimentation, 11-16

October, 1983, Nanjing, China: Beijing, Water Resources and Electric Power Press, p. 597-607.

Table II. Perceived importance of sediment-storage events, as a function of time scale. Modified after Schumm (1985).

Relative Magnitude of Event	Time scale, in years					
	10^{-1} to 10^{0}	10^{0} to 10^{1}	10^{1} to 10^{2}	10^{2} to 10^{3}	10^{3} to 10^{4}	10^{4} to 10^{5}
Megaevent	Storage and remobilization in response to channel hydraulics	Sustained aggradation or degradation of river channel	Full cycle of channel aggradation and degradation; anthropogenic sediment wave	Remobilization of flood plain; multiple cycles of channel aggradation and degradation	Aggradation of valley in response to rising sea level	Multiple cycles of aggradation and degradation of valley with changing sea level
Mesoevent	Channel scour; migration of large sand waves	Storage and remobilization in response to channel hydraulics	Sustained aggradation or degradation of river channel	Full cycle of channel aggradation and degradation; anthropogenic sediment wave	Remobilization of flood plain; multiple cycles of channel aggradation and degradation	Aggradation of valley in response to rising sea level
Microevent	Movement of small sand dune on river bed	Channel scour; migration of large sand wave	Storage and remobilization in response to channel hydraulics	Sustained aggradation or degradation of river channel	Full cycle of channel aggradation and degradation	Remobilization of flood plain; multiple cycles of channel aggradation and degradation
Nonevent	Movement of sand ripple	Movement of small sand dune on river bed	Channel scour; migration of large sand wave	Storage and remobilization in response to channel hydraulics	Sustained aggradation or degradation of river channel	Full cycle of channel aggradation and degradation

INFLUENCE OF ACID RAIN ON WEATHERING RATES

Richard F. Wright
Norwegian Institute for Water Research
Box 333, Blindern
0314 Oslo
Norway

ABSTRACT

Acid rain has resulted in the acidification of lakes and streams over large areas of Europe and North America, and has been implicated in the recent widespread forest dieback in Europe. Chemical weathering releases base cations to the soil cation-exchange complex and thus represents the ultimate defense against acid rain. Determination of weathering rate for whole catchments is required to predict future acidification of soils and surface waters. Weathering rate can be estimated by several independent methods which range from laboratory experiments with pure minerals to input-output budgets at whole catchments. Inasmuch as congruent dissolution of many rock-forming minerals involves the consumption of hydrogen ions, acidification of soils by acid rain can be expected to result in an increase in weathering rate. Evidence from field and laboratory experiments, however, is so far insufficient to judge whether the rate of chemical weathering does indeed change due to the deposition of acid from the atmosphere. Recent models of soil and water acidification suggest that acid deposition causes only small changes in the pH of soil solution, and thus changes in weathering rate due to changes in soil chemistry may be minor. Research on the influence of acid rain on weathering rates has been given high priority in both Europe and North America.

1. INTRODUCTION

Lakes and streams in large regions of Europe and eastern North America have been acidified and fish populations lost due to the deposition of strong acids from the atmosphere (acid rain) during the past decades (Likens et al., 1979; Overrein et al., 1980). Recently extensive areas of central Europe have been affected by a new type of forest dieback that also appears due to air pollution and acid deposition (Ulrich, 1984; Ashmore et al., 1985).

Freshwater acidification occurs principally in areas of "sensitive" geologic terrain. Common to these sensitive areas is that natural rates of chemical weathering are low, and thus freshwaters are of low ionic strength. Sensitive environments are dominated by carbonate-free highly siliceous bedrock, overburden, and soils. Such materials in turn can be found under three sets of circumstances: (1) glaciated areas on granitic or other highly siliceous bedrock with overburden and soils derived from material of similar lithology, (2) areas with thick overburden of siliceous sands

A. Lerman and M. Meybeck (eds.), Physical and Chemical Weathering in Geochemical Cycles, 181–196.

regardless of bedrock lithology, and (3) areas with relatively old, deeply weathered and leached soils, again regardless of bedrock lithology (Wright, 1983a).

In both Europe and North America fresh waters are found in all three types of sensitive terrain (Figures 1 and 2; Wright, 1983a; Galloway and Cowling, 1978). Most of the thousands of lakes dotting the glaciated Fenno-Scandian and Canadian shields are of the first type. Lakes and streams on the siliceous sand plains of Denmark and Holland in Europe, and southern New Jersey and northern Wisconsin (U.S.A.) are of the second type. Streams in several areas of West Germany and the Appalachian Mountains of the southeastern U.S.A. are of the third type.

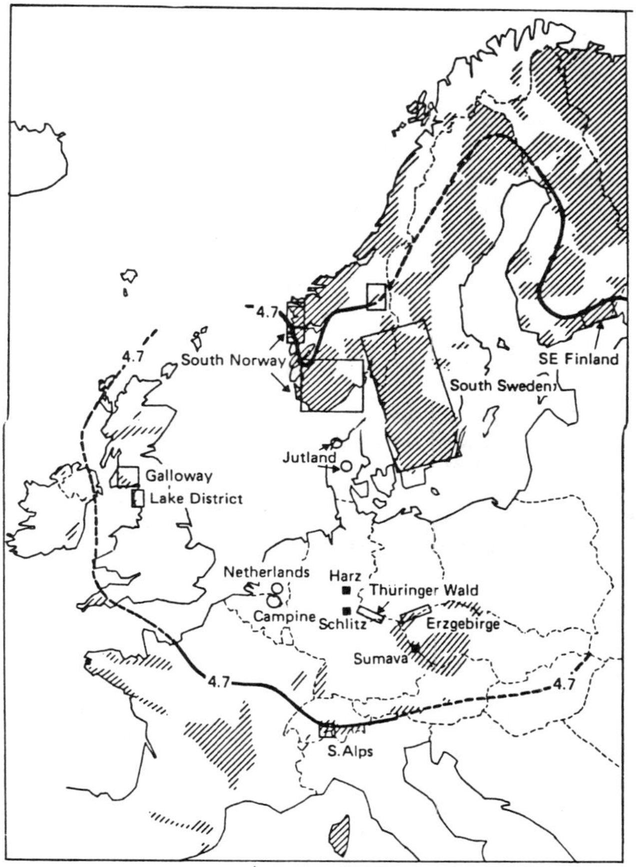

Figure 1. Map of Europe showing areas sensitive to freshwater acidification (from bedrock geology). Areas within the pH isoline receive precipitation of acidity exceeding the "threshold" for ecological effects in the most sensitive waters. Areas from which acid freshwaters (pH <5) have been reported lie on sensitive bedrock (squares) or on siliceous sands (circles) (from Wright, 1983a). Since this map was drawn acidified waters have been reported from additional areas.

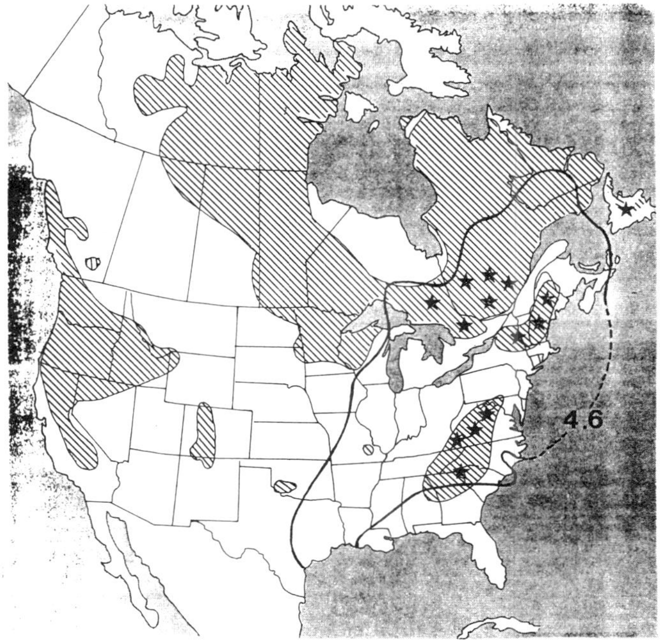

Figure 2. Map of eastern North America showing location of sensitive areas and pH=4.6 isoline for precipitation acidity. Acidified waters have been reported from areas marked with stars (J.N. Galloway, pers. comm.).

Acidification of freshwaters entails a change from the Ca^{2+}-Mg^{2+}-HCO_3^- chemical composition typical of waters in sensitive areas not receiving acid precipitation to H^+–Al^{3+}–Ca^{2+}–Mg^{2+}–SO_4^{2-} waters (Wright and Gjessing, 1976; Wright, 1983b). Inasmuch as most incoming precipitation percolates through soil before emerging in streams and lakes, the change in chemical composition of freshwaters indicates change in soil chemistry. In particular the base status of the soil is of critical importance in determining the chemical composition of soil solution and thus runoff. The base status of soil is a function of the fraction of the cation-exchange sites occupied by the four base cations Ca, Mg, Na, and K (usually expressed as percent base saturation). Although the processes affecting the base status of soils are not fully understood, the flux of base cations into and out of the soil is of central importance. Besides influx of cations from the atmosphere, release of base cations in chemical weathering represents the major source of cations to soils (Figure 3). Deposition of strong acids usually results in an increase in ionic strength of soil solution, and thus initially an increase in loss of base cations to runoff (Reuss and Johnson, 1985). If the rate of chemical weathering cannot keep pace with the loss of base cations to runoff and to growing biomass, a gradual decrease in the base status of the soil occurs, which in turn affects the chemical composition of the soil solution. Chemical weathering thus provides a catchment's ultimate defense against the inputs of acidifying compounds from the atmosphere.

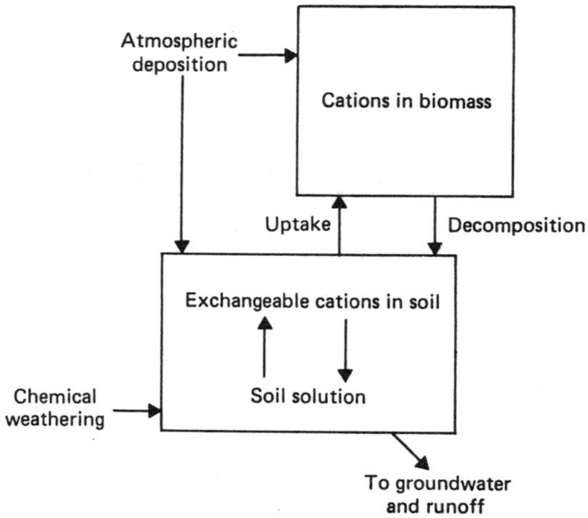

Figure 3. Schematic view of the fluxes of base cations into and out of the soil.

2. CHEMICAL WEATHERING

Chemical weathering can be defined as the release of base cations (Ca, Mg, Na, K) from primary minerals to soil solution and the cation-exchange complex. This definition is useful in the context of soil and water acidification and makes the distinction between release of cations from the cation-exchange complex (a reversible process) and the release of cations from mineral lattices (an irreversible process). In practice, of course, there is a continuum between cations held on particle surfaces to cations bound in particle interiors.

In the past, chemical weathering rates (CWR) for entire catchments have been estimated from cation denudation rates. Geologists first simply used the dissolved flux of base cations in river runoff (R) to calculate cation denudation rate which was then assumed equal to chemical weathering rate (Livingstone, 1963).

$$\text{chemical weathering rate (CWR)} = \text{flux in runoff (R)} \tag{1}$$

In the late 1960's, Meade (1969) pointed out that the flux of chemical components from the atmosphere and discharge of pollutants within the catchment must be subtracted from the runoff output to obtain cation denudation rate. Direct discharge of pollutants is usually minor for catchments in which freshwater acidification occurs. In sensitive areas the atmospheric flux, however, can comprise a significant fraction of the flux out in runoff.

$$\text{CWR} = \text{R} - \text{atmospheric input (I)} \tag{2}$$

Uptake and release of base cations in the biomass must also be taken into account. Studies at forested catchments such as the H.J. Andrews Experimental Forest in Oregon, U.S.A. (Sollins et al., 1980) and the Hubbard Brook Experimental Forest in New Hampshire, U.S.A. (Likens et al., 1977), show that a substantial fraction of the base cations released by weathering may be taken up in the growing biomass and then released following forest disturbance. Thus,

$$\text{CWR} = \text{R} - \text{I} + \text{change in biomass storage } (\Delta\text{B}) \tag{3}$$

It is now clear that the change in the chemical climate due to acid rain may also change the cation denudation rate, either by changing the rate of chemical weathering or by altering the amount of cations stored within the soil. Acid deposition alters the cation-exchange equilibria in soils (Reuss and Johnson, 1985), and thus changes the chemical composition of soil water and hence runoff. The increased flux of base cations from catchments receiving acid deposition may, therefore, reflect a net decrease in the pool of base cations stored on the cation-exchange complex in the soil.

$$\mathrm{CWR} = \mathrm{R} - \mathrm{I} + \Delta\mathrm{B} + \text{change in soil pool } (\Delta\mathrm{S}) \tag{4}$$

As each of these fluxes must be determined accurately over at least several years, measuring the rate of chemical weathering for an entire catchment becomes a major task. The chemical weathering rate is obtained indirectly as the net sum of several large numbers, each of which varies seasonally and from year to year. It is even more difficult to determine the change in weathering rate due to acid deposition. Yet precisely this information is required to predict acidification of soils and waters in response to future changes in acid deposition.

3. DETERMINING WEATHERING RATES

There exist several independent methods by which weathering rates can be determined. These range from laboratory experiments with pure minerals and acids to whole-catchment investigations. Each method has its advantages and shortcomings.

3.1 Present-day weathering rate from input-output budgets

3.1.1. Watersheds. If long-term input-output data and soil-chemistry data are available, the approach using equation (4) above provides perhaps the closest estimate. This approach has been used to estimate weathering rates at small calibrated catchments, *in situ* lysimeters and soil columns under controlled conditions (Fölster, 1985). The study of Sollins et al. (1980) at a small forested pristine catchment on andesite bedrock in western Oregon, U.S.A., provides a good example. Here all the fluxes of base cations to and from the biomass, and into and out of the catchment were measured. The ΔS term was assumed to be 0, a reasonable assumption for this area which has not experienced a recent change in chemical climate (that is, no acid rain). This method is extremely data intensive, as is illustrated by the number of different fluxes measured (Figure 4). The weathering rate for this site derived from equation (4) is 8.2 keq/ha/yr (Sollins et al., 1980). This value is considerably higher than the weathering rates estimated for catchments on granitic bedrock, presumably because of the relatively easily weathered andesitic bedrock.

A similar analysis has been conducted at the calibrated catchments at the Hubbard Brook Experimental Forest, New Hampshire, U.S.A. (Likens et al., 1977). These catchments are on siliceous gneiss with deciduous forests. This area receives considerable acid rain with average pH about 4.0. Here the ΔS term probably cannot be neglected because the recent change in chemical climate has resulted in increases in ionic strength of soil solution and hence increased leaching of base cation, some of which may be due to increased rate of chemical weathering but some of which may instead reflect a net decrease in the size of pool of base cations stored on the ion exchange complex. Unfortunately the ΔS term cannot be assessed independently because no measurements over time are available. The CWR reported by Likens et al. (1977) for the four base cations of 1.8 keq/ha/yr includes the loss of cations from the

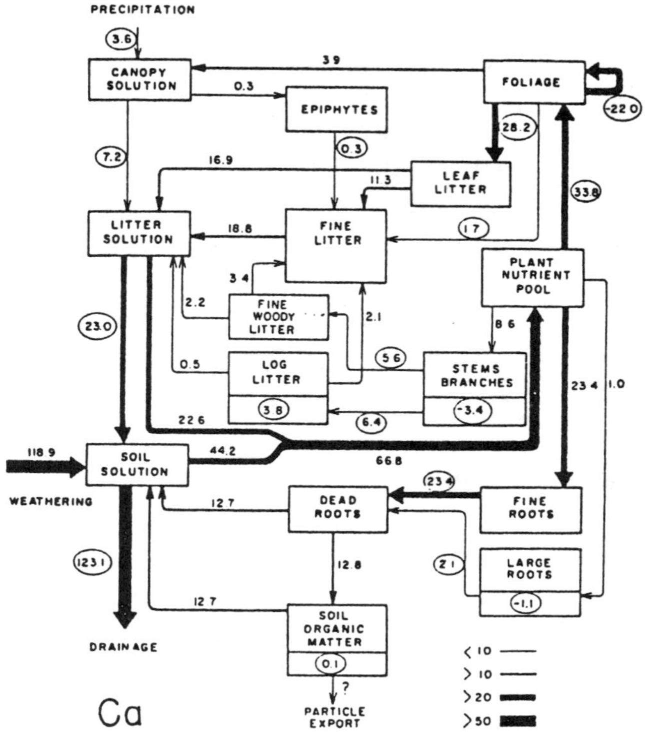

Figure 4. Flux of Ca (kg/ha/yr) to and from and within an old-growth coniferous forest in Oregon, U.S.A. (Sollins et al., 1980). The circles indicate measured fluxes; the other fluxes are estimated by difference.

soil and is thus an upper limit.

3.1.2. Soil columns. This same approach has been used to estimate weathering rate and change of weathering rate due to acid deposition for soil columns (Bjor and Teigen, 1980; Cronan, 1985). Bjor and Teigen (1980) conducted a 31 month study with soil columns in a greenhouse. They applied "rain" at four pH levels, 2.5, 3.0, 4.0, and 5.3. They measured the flux of base cations into and out of the lysimeters during the experiment and the amount of exchangeable base cations stored in the soil at the beginning and end of the experiment. The cations stored in the vegetation were measured after the experiment, but not before the experiment, so that the term ΔB can only be estimated, and the uncertainty in this estimate is reflected in rather large uncertainties in the calculated weathering rates (Table I). The experiment shows clearly that application of acid can reduce the base saturation of the soil (Figure 5). The weathering rates calculated from the results of Bjor and Teigen's experiment point to roughly similar rates for the soil columns receiving "rain" with pH in the range 3.0 to 5.3 (Table I). Only the pH=2.5 treatment gave a significantly higher weathering rate, but even this result is difficult to interpret because the vegetation was damaged by the high acidity in this treatment and thus the biological cycling of cations was severely disrupted.

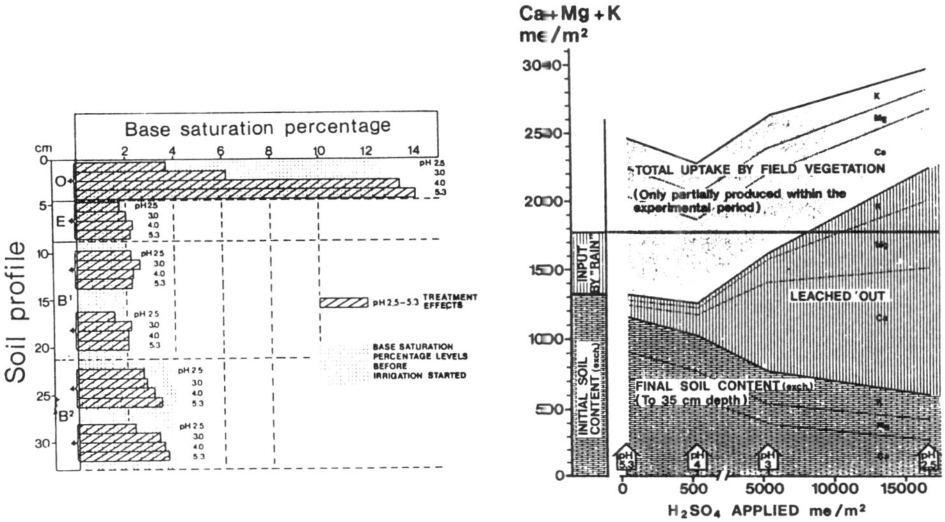

Figure 5. Experimental acidification of soil columns (Bjor and Teigen, 1980). (left) Base saturation measured in soil columns after 31 months of treatment with artificial acid rain of pH 2.5, 3.0, 4.0, and 5.3. Base saturation is significantly lower in the 0 horizons of the pH 2.5 and 3.0 treated columns. (right) Mass balance for the soil columns after treatment.

Table I. Chemical weathering rates (CWR) obtained from soil column experiments of Bjor and Teigen (1980) using the equation CWR = R–I+ΔS+ΔB, where R is the flux out in leachate, I is the flux in in precipitation, ΔS is the change in pool of exchangeable base cations in the soil, and ΔB the change in base cations stored in biomass. ΔB was apparently not measured prior to treatment; the range in CWR reflects estimates of ΔB from 0 to 100%. Units in eq/ha yr.

Treatment	R	I	ΔS	ΔB		CWR	
				min.	max.	min.	max.
pH 5.3	40	100	-40	0	290	-100	190
pH 4.0	55	100	-75	-35	260	-155	340
pH 3.0	210	100	-130	-40	250	-65	430
pH 2.5	400	100	-180	-100	190	+20	310

Cronan (1985) conducted an experiment with soil columns of similar design to that of Bjor and Teigen (1980) (Table II). He then compared weathering rates obtained by this method with those obtained for the same soils by several other methods. The range of values obtained varied greatly, due to the uncertainties in each of the terms measured and used in equation (4). With respect to the question of change in weathering rate with change in acid deposition Cronan's experiments do not give a clear picture. For two of the soils the calculated weathering rates were highest at the lowest pH of applied rain (pH 3.5), whereas for the soil with the lowest weathering rate initially (and thus representing the most sensitive condition with respect to water acidification) there was no apparent relationship between weathering rate and pH of treatment. The major source of uncertainty in Cronan's experiment is the base saturation of the soils before and after treatment; the wide range in the measurements meant that no significant difference was obtained, and thus he assumed that ΔS was zero for the experiments. With this assumption the increased flux of cations in the leachate necessitates a higher calculated weathering rate. If, on the other hand, the weathering rate was assumed constant then the change in base saturation required to explain the results falls well within the range of uncertainty in the measured base saturation.

Table II. Comparison of weathering rates for three soil types based upon several different estimation techniques. Weathering is taken to be the sum of Ca + Mg + K + Na released from primary minerals in eq/ha/yr (from Cronan, 1985).

Method/Soil Type	Adams	Becket	Unadilla
Mass Balance (Maximum)[A]	1,160	1,510	7,876
Mass Balance (Minimum)[A]	54	80	2,800
Sodium Flux Ratio[B]	320	1,030	2,590
Silica Flux Ratio[C]	60	250	330
Mineral Depletion[D]	--	620	--
Watershed Budget			
ILWAS[E]	--	1,480	--
Hubbard Brook[F]	--	2,000	--

[A] From the soil column mass balance experiments.

[B] Because sodium is considered to be relatively conservative, the solution Na flux and bulk mineral chemistry were used to calculate the release of Ca + Mg + K necessary to account for the observed Na flux.

[C] This technique makes the tenuous assumption that the solution silica flux is directly related to the rate of release of Si + Ca + Mg + K + Na from primary minerals.

[D] This estimate from April and Newton (1984) is based on a long-term weathering rate determined from mineral depletion in the Becket soil profile of Woods watershed.

[E] This estimate is based on data from the Integrated Lake Watershed Acidification Study (ILWAS (Chen et al., 1983) watersheds where the Becket Spodosol was obtained. The estimate combines the annual stream export of cations from Woods Lake watershed (April and Newton, 1984) with an estimate of the annual biomass increment of cations (Cronan, unpub. data).

[F] The Hubbard Brook Experimental Forest in New Hampshire (Likens et al., 1977) contains the same Becket Spodosol as the Adirondack soil used in this study.

3.1.3. In situ lysimeters. The input-output budget approach has also been used to estimate weathering rates from *in situ* lysimeters in forests. Fölster (1985) estimated the present-day weathering rate in a podzolic soil at Solling, near Göettingen, F.R.G., for the 15-year period 1968-83. Here the pools of cations stored in the soil and biomass were measured and the fluxes into and out of the system followed. Weathering rate for the four base cations was less than 1 keq/ha/yr at Solling, and this rate was less than the ΔS for the 15-year period of 1 keq/ha/yr under spruce and 1.6 keq/ha/yr under beech (Fölster, 1985). Folster concludes from this study that the rate of weathering of silicate minerals at Solling cannot keep up with the rate of deposition of acid from the atmosphere.

3.2. Historical weathering rate from soil profiles

The mean long-term weathering rate can be estimated from the mineralogy of soil profiles of known age (Fölster, 1985). This method compares the composition of the weathered soil profile with the composition of the unweathered parent material. For several West German soils developed on loess deposited during the last glaciation the long-term weathering rate determined by this method is 0.4-1 keq/ha/yr for the four base cations (Fölster, 1985). This method requires an inert mineral phase or element that provides the reference between the weathered and unweathered material. For these loess soils quartz was assumed inert. Other references may be provided by zircon or titanium dioxide.

Because of the differential susceptibility of different minerals to chemical weathering (Johnson, 1984), much of the base cation release in soils may be due to the weathering of a specific mineral that may be present in minor or trace amounts in the bulk soil. In soils derived from granitic material, the dark minerals such as amphiboles, pyroxenes and micas are generally more readily weathered than are the feldspars and quartz. The long-term weathering rate can then be estimated from the relative abundance of key dark minerals in the soil profile. April and Newton (1984) used the abundance of hornblende to estimate the historical weathering rate for the Becket soil in the catchments of Woods and Panther Lakes in the Adirondack Mountains of New York, U.S.A. (Figure 6). The weathering rate yielded by this study was about 0.6 keq/ha/yr. This soil is one of the same for which Cronan (1985) carried out weathering experiments with soil lysimeters in the laboratory, and in which the rate obtained was in the range 0.08-1.5 keq/ha/yr (Table II).

One difficulty with interpreting such long-term rates is that the weathering rate probably was not constant over the entire period, and may be quite different today than the mean over several thousand years. In particular the cation denudation rate of freshly deglaciated materials is much higher than that in well-developed soils several thousand years of age. This is well demonstrated by soil chronosequences such as those studied by Crocker and Major (1955) on a series of moraines 0-200 years in age at Glacier Bay, Alaska, U.S.A.. The soil pH, organic matter content and base saturation all change rapidly within the first hundred years. Johnson et al. (1972) point out that cation denudation rates in recently deglaciated terrain in the Cascade Mountains of Oregon, U.S.A., are at 9.3 keq/ha/yr, much higher than in areas of similar mineralogy in New England, U.S.A. (1-2 keq/ha/yr), presumably because of the high reactivity of the fresh mineral surfaces.

3.3. Use of Sr isotopes to estimate present-day weathering rate

The ratio of the stable isotopes of strontium, $^{87}Sr/^{86}Sr$, can be used to estimate

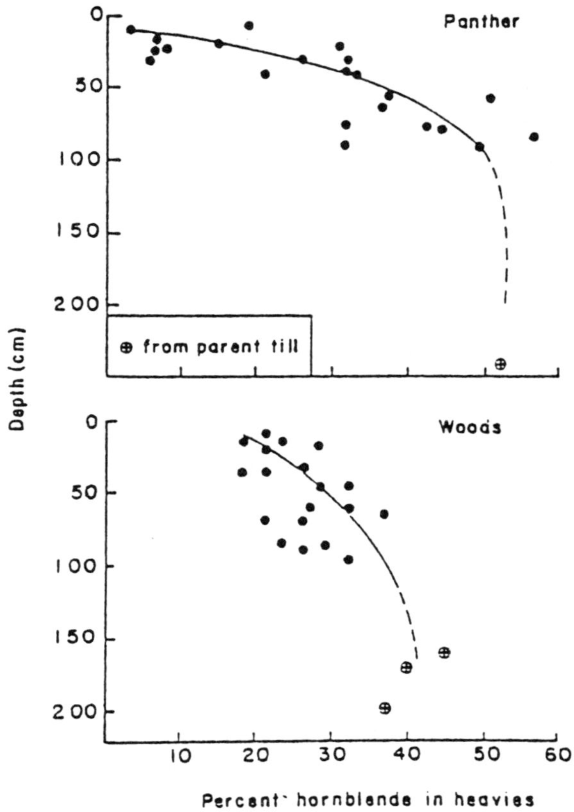

Figure 6. Abundance of hornblende in soils at Woods and Panther Lakes, Adirondack Mountains, New York, U.S.A.. The depletion at the top of the soil indicates loss due to chemical weathering (from April and Newton, 1983).

present-day weathering rate for Sr (Dasch, 1969). When combined with information on the Sr/Ca ratio, and under the assumption that Sr behaves similarly to Ca during weathering, this method yields an estimate for the rate of release of Ca during weathering. ^{87}Sr is formed by the radioactive decay of ^{87}Rb, whose half-life is 47 billion years. ^{87}Rb is chemically similar to potassium and thus is relatively abundant in minerals such as potassium feldspar. As ^{87}Rb decays to ^{87}Sr the ratio of ^{87}Sr/^{86}Sr increases and thus the ratio ^{87}Sr/^{86}Sr is highest for crystalline rocks of Precambrian age. Jacks (pers. comm.) has used this method to estimate the rate of weathering at a small catchment on Precambrian rocks in central Sweden (Figure 7). His preliminary data show differences in the ^{87}Sr/^{86}Sr ratio in atmospheric precipitation, soils, bedrock and streamwater, and from this information he estimated that Ca is released by weathering at this catchment at the rate of 0.1-0.25 keq/ha/yr.

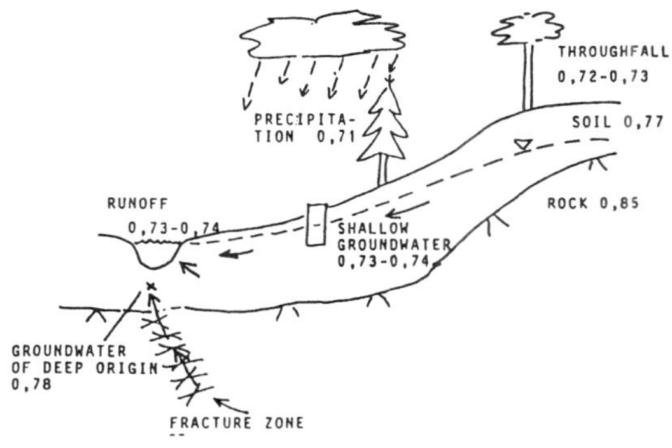

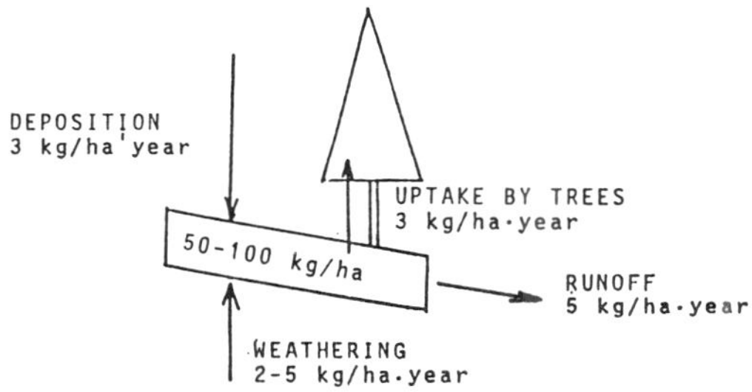

Figure 7. (top) Ranges of Sr-87/Sr-86 in various media at the Buskbäcken catchment, Kloten, central Sweden. (bottom) A tentative Ca budget for the Buskbäcken catchment obtained by the Sr isotope method (G. Jacks, pers. comm.).

3.4. Weathering rates obtained in the laboratory

Weathering rates and weathering mechanisms have been studied extensively in the laboratory in experiments with pure minerals under controlled conditions. An example of this type of work is the recent investigation by Wollast and Chou (1985 and this volume) on the weathering of albite, a sodium feldspar mineral (Figure 8). Although such studies give useful information on weathering mechanisms, the absolute rates obtained are difficult to extrapolate to field conditions. In the field not only are the mineral surfaces not fresh, but the chemical and physical microenvironment under which the actual weathering occurs is difficult to define. Weathering rates obtained from experiments with fresh minerals in the laboratory may be several orders of magnitude higher than rates in nature. Laboratory experiments are useful, however, in exploring the dependence of weathering rate on such factors as pH, temperature, and concentration of chelating ligands. This information bears on the question of change in weathering rate due to acid deposition.

3.5. Weathering rates using models for soil and water acidification

Several process-oriented models for soil and water acidification include weathering rate as a central process. These models can be used to set limits on the weathering

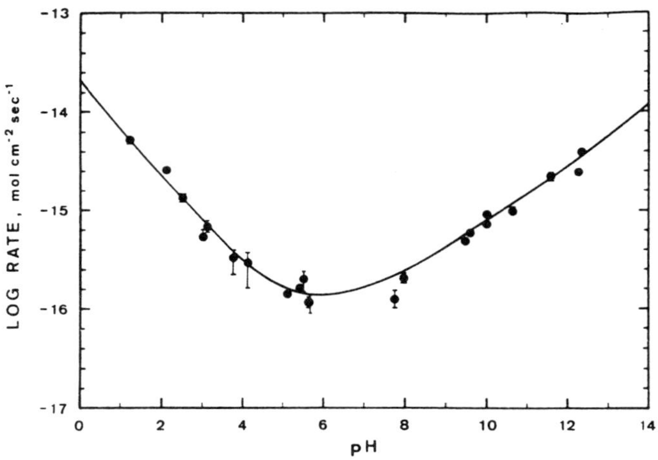

Figure 8. Influence of pH on the rate of congruent dissolution of albite during steady-state from experiments conducted in a fluidized-bed reactor in the laboratory (from Wollast and Chou, 1985).

rate, and on the importance of the pH-dependence of weathering rate. MAGIC (Model for Acidification of Groundwater In Catchments) is such a model (Cosby et al., 1985). MAGIC consists of (1) a soil-soil solution chemical equilibria section in which the concentrations of major ions are assumed governed by simultaneous reactions involving sulfate adsorption, cation exchange, dissolution and precipitation of aluminum, and dissolution of inorganic carbon, and (2) a mass-balance section in which the flux of major ions to and from the soil is assumed governed by atmospheric inputs, chemical weathering inputs, net uptake in biomass and loss to runoff (Figure 3). MAGIC was originally developed with data from White Oak Run, a small forested catchment in the Shenandoah National Park, Virginia, U.S.A. (Cosby et al., 1985), and has been since applied to several other catchments in Europe and North America (Wright et al., 1986a).

In MAGIC the weathering rates for the four base cations can either be computed or specified as independent variables and used to calculate other soil properties, such as the cation exchange selectivity coefficients. MAGIC has been used to delimit the weathering rate and change in weathering rate at several catchments that have been acidified by acid precipitation (Wright et al., 1986a). As Reuss and Johnson (1985) point out and MAGIC demonstrates quantitatively for soils in sensitive areas, although changes in the pH of surface waters induced by acid deposition may be large, changes in the pH of soil solution are small. Therefore changes in weathering rate due to changes in soil chemistry may be unimportant in predicting catchment responses to acid deposition. For example, for two sites, Lake Hovvatn in Norway, and Lake Gardsjön in Sweden, both of which have been acidified and lost their fish populations, a 140-year historical reconstruction using the MAGIC and a constant weathering rate throughout gives almost the same water pH over time as MAGIC with a pH-dependent weathering rate (Wright et al., 1986a) (Figures 9 and 10). Both reconstructions are consistent with independent historical data such as the diatom stratigraphy in lake sediments, historical pH measurements, and fisheries records.

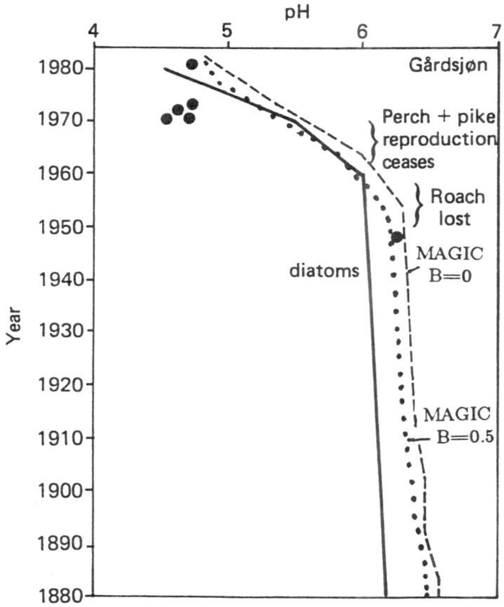

Figure 9. Reconstructed pH history for Lake Gardsjön, west coast of Sweden, from diatom stratigraphy and MAGIC. Also shown are historical pH measurements and fisheries information. With a pH independent weathering rate MAGIC gives the curve labelled 0. With a rate dependent on hydrogen-ion concentration in soil solution raised to the 0.5 power MAGIC gives the curve labelled 0.5 (from Wright et al., 1986a).

Other such models also contain the weathering rate as a central element, and perhaps also a weak link. The ILWAS model (Integrated Lake Watershed Acidification Study) (Chen et al., 1983) assumed that present-day weathering rates at 3 sites in the Adirondack Mountains, New York, U.S.A., are the same as the historical weathering rates obtained from mineral depletion curves in the soils (April and Newton, 1984). Furthermore the weathering rate was assumed to be pH dependent; rate $= k[H^+]^B$, where B is set to 0.5. The only evidence offered in support of this value is from laboratory studies. The "trickle-down" model of Schnoor et al. (1982) also chooses the value of 0.5 for B, again on the basis of laboratory studies.

4. CONCLUSIONS

Because hydrogen-ion is consumed in the dissolution of many of the major minerals in soils, an increase in the concentration of hydrogen ion in soil solution should *a priori* be expected to increase the rate of dissolution, but only if hydrogen ion is involved in the rate-limiting step in the weathering mechanism. Johnson (1984) suggests that perhaps hydrolysis, a pH-independent process, is the rate-limiting step. Laboratory studies, such as that of Wollast and Chou (1985), indicate a pH-dependent rate, but the dependence is very weak in the pH range of soil solution in areas sensitive to

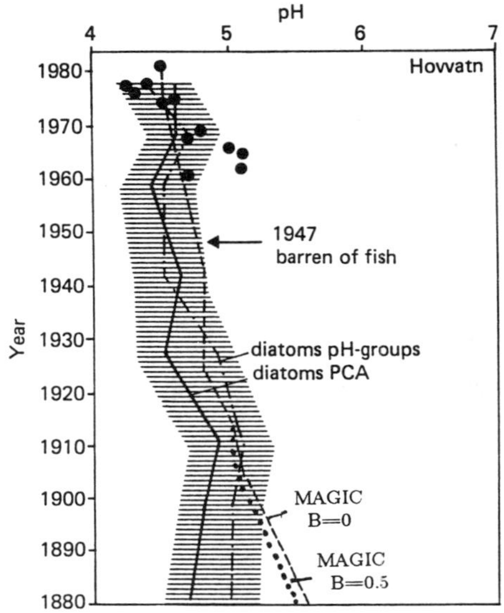

Figure 10. Reconstructed pH history for Lake Hovvatn, southernmost Norway, from diatom stratigraphy and from MAGIC (see Figure 9 for details) (from Wright et al., 1986a).

acidification (pH 3.5-6.0, Figure 8). The experiments with soil columns carried out by Bjor and Teigen (1980) and Cronan (1985) do not show significant differences in weathering rate except perhaps at pH levels in applied precipitation below pH 3.5. Even these results are difficult to interpret because the most acid treatment apparently caused direct damage to the vegetation and hence disrupted the normal biological cycling of elements.

Input-output budgets at catchments or *in situ* lysimeters do not provide information as to change in weathering rate, because before and after data are not available. Much of the present-day cation denudation at catchments receiving acid deposition may be due to the stripping of base cations from the soil cation-exchange complex rather than release from chemical weathering of soil minerals. It is thus difficult to use geographic comparisons of similar catchments across gradients of acid deposition to infer changes in weathering rate over time. Possibly such information will be available from large-scale experimental manipulations, such as the ongoing RAIN project (Reversing Acidification In Norway), in which the acid deposition to whole catchments is changed and the effects of soil and water chemistry followed (Wright et al., 1986b).

In summary, it appears that it is simply not yet known if acid deposition changes the rate of chemical weathering. Determination of weathering rate and its pH dependence is the single most important "unknown" hindering quantitative understanding of the acidification of soils and waters. Prediction of future acidification requires this information. Research on this subject has been given high priority in both North America and Europe.

5. REFERENCES

April, R. and Newton, R.M., 1984, The geology and geochemistry of the ILWAS lake-watersheds, *in* R.A. Goldstein (ed.) *The Integrated Lake-Watershed Acidification Study: Summary of Major Results*, vol. 4, Electric Power Research Institute, Palo Alto, California.

Ashmore, M., Bell, N., and Rutter, J., 1985, The role of ozone in forest damage in West Germany, *Ambio,* **14**, 81-87.

Bjor, K. and Teigen, O., 1980, Effects of acid precipitation on soil and forest: 6. Lysimeter experiment in greenhouse, *in* D. Drablos and A. Tollan (eds.) *Ecological Impact of Acid Precipitation,* pp. 200-201, SNSF Project, P.O. Box 61, 1432 As-NLH, Norway.

Chen, C.W., Gherini, S.A. Hudson, H.J.M., and Dean, J.D., 1983, *The integrated lake-watershed acidification study: Model principles and application procedures,* vol. 1, Electric Power Research Inst., Palo Alto, Calif.

Cosby, B.J., Hornberger, G.M., Galloway, J.N. and Wright, R.F., 1985, Modeling the effects of acid deposition: Assessment of a lumped parameter model of soilwater and streamwater chemistry, *Water Res. Res.,* **21**, 51-63.

Cosby, B.J., R.F. Wright, G.M. Hornberger, and J.N. Galloway, 1985, Modelling the effects of acid deposition: Estimation of long-term water quality responses in a small forested catchment, *Water Res. Res.,* **21,** 1591-1601.

Crocker, H.L. and J. Major, 1955, Soil development in relation to vegetation and surface age at Glacier Bay, Alaska, *Jour. Ecol.,* **43**, 427-448.

Cronan, C.S., 1985, Chemical weathering and solution chemistry in acid forest soils: Differential influence of soil type, biotic processes, and H^+ deposition, *in* J.I. Drever (ed.) *The Chemistry of Weathering,* pp. 175-196, NATO ASI Series Mathematical and Physical Sci., vol. 149, D. Reidel, Dordrecht, 324 pp.

Dasch, E. J., 1969, Strontium isotopes in weathering profiles, deep-sea sediments and sedimentary rocks, *Geochim. Cosmochim. Acta,* **33**, 1521-1552.

Fölster, H., 1985, Proton consumption rates in Holocene and present-day weathering of acid forest soils, *in* J.I. Drever (ed.) *The Chemistry of Weathering,* p. 197-210, NATO ASI Series Mathematical and Physical Sci., vol. 149, D. Reidel, Dordrecht, 324 pp.

Galloway, J.N. and Cowling, E.B., 1978, The effects of precipitation on aquatic and terrestrial ecosystems: A proposed precipitation chemistry network, *J. Air. Pollut. Control Assoc.,* **28**, 229-235.

Johnson, N.M., 1984, Acid rain neutralization by geologic materials, *in* O.P. Bricker (ed.) *Geological Aspects of Acid Deposition,* p. 37-53, Butterworths, Boston, 143 pp.

Johnson, N.M., R.C. Reynolds, and G.E. Likens, 1972, Atmospheric sulfur: Its effect on the chemical weathering of New England, *Science,* **177**, 514-516.

Likens, G.E., R.F. Wright, J.N. Galloway, and T.J. Butler, 1979, Acid rain, *Sci. Amer.,* **241**, 43-51.

Likens, G.E., F.H. Bormann, R.S. Fierce, J.S. Eaton, and N.M. Johnson, 1977, *Biogeochemistry of a Forested Ecosystem,* Springer-Verlag. N.Y., 146 pp.

Livingstone, D.A., 1963, Chemical composition of rivers and lakes, *in* Data of Geochemistry, 6th ed., *U.S. Geol. Surv. Prof. Paper* **440-G**, G1-G64.

Meade, H.R., 1969, Errors in using modern stream-load data to estimate natural rates of denudation, *Geol. Soc. Amer. Bull.,* **80**, 1265-1274.

Overrein, L.N., H.M. Seip, and A. Tollan, 1980, Acid precipitation—effects on forest and fish, *Final Report SNSF Project 1972-1980,* **FR 19/80**, 175 pp.

Reuss, J.O., and D.W. Johnson, 1985, Effect of soil processes on the acidification of water by acid deposition. *J. Environ. Qual.,* **14**, 26-31.

Reuss, J.O. and D.W. Johnson, in press, *Acid Deposition, Soils, and Water: An Analysis,*

Springer Verlag, N.Y.

Schnoor, J.L., G.R. Carmichael, and F.A. Von Schepen, 1982, An integrated approach to acid rainfall assessments, *in* L.H. Keith (ed.) *Energy and Environmental Chemistry: Acid Rain*, vol. 2, pp. 225-243, Ann Arbor Science, Boston.

Sollins, P., C.C. Grier, F.M. McCorison, K. Cromack, and R. Fogel, 1980, The internal element cycles of an old-growth Douglas-fir ecosystem in western Oregon, *Ecol. Monog.*, **50**, 265-280.

Ulrich, B., 1984, Effects of air pollution on forest ecosystems and waters—the principles demonstrated at a case study in central Europe, *Atmos. Environ.*, **18**, 621-628.

Wollast, R., and L. Chou, 1985, Kinetic study of the dissolution of albite with a continuous flow-through fluidized bed reactor, *in* J.I. Drever (ed.) *The Chemistry of Weathering*, pp. 75-96, D. Reidel, Dordrecht, 324 pp.

Wright, W.F., 1983a, Acidification of freshwaters in Europe, *Water Qual. Bull.*, **8**, 137-142.

Wright, R.F., 1983b, Predicting acidification of North American lakes, *Acid Rain Res. Rept.* **4/1983** Norwegian Institute for Water Research, Oslo, 165 pp.

Wright, R.F., and Gjessing, E.T., 1976, Acid precipitation: Changes in the chemical composition of lakes, *Ambio*, **5**, 219-223.

Wright, R.F., B.J. Cosby, G.M. Hornberger, and J.N. Galloway, 1986a, Comparison of paleolimnological with MAGIC model reconstructions of water acidification, *Water Air Soil Pollut.*, **30**, 367-380.

Wright, R.F., Gjessing, E., Christophersen, N., Lotse, E., Seip, H.M., Semb, A. and Sletaune, B., 1986b, Project RAIN: Changing acid deposition to whole catchments—the first year of treatment, *Water Air Soil Pollut.*, **30**, 47-63.

FRESHWATER CARBON AND THE WEATHERING CYCLE

Stephan Kempe
SCOPE/UNEP International Carbon Center
Geologisch-Paläontologisches Institut
Universität Hamburg
2000 Hamburg 13
Federal Republic of Germany

ABSTRACT

Carbonic acid is the most important chemical weathering agent. The CO_2-pressure of an aqueous solution governs the reactions with carbonate minerals and silicates. In addition to the inorganic reactions, life processes extract CO_2 from water photosynthetically or add it when organic matter is consumed by respiration. Both labile and stable organic molecules enter the water, the labile compounds being responsible for fueling respiration, whereas the stable species are transported unaltered. Six case studies follow the fate of carbon from small headwaters through lakes and rivers into estuaries. (1) The first study introduces the Plitvice Lakes, which are ponded behind actively growing travertine dams. Lakes are fed by small springs from carbonate terrains. The rates of release of CO_2 and of calcite precipitation are given. (2) The second case study introduces a mountainous bog releasing large quantities of organic matter. (3) The Amazon is discussed as an example of interactions between a large tropical river and its floodplains. (4) The fourth example deals with the Nile and how the closure of the Aswan Dam altered its hydrology and carbonate chemistry. (5) The Rhine serves as an example of a heavily polluted river. Here, in the limnic environment of Lake Constance, the P_{CO_2} is depressed by eutrophication while in the fluvial environment P_{CO_2} is increased due to internal respiration of labile organic matter. (6) Finally the estuary of the Elbe serves to show that high P_{CO_2} of a polluted river is buffered by dissolution of suspended carbonates.

1. INTRODUCTION

Carbon is a highly mobile element cycling through the atmosphere, biosphere, hydrosphere and lithosphere (Bolin et al., 1979; Degens et al., 1984). Its ability to change sides between the inorganic and the organic world and its many oxidation states (—4 to +4) enables carbon to form gaseous, dissolved, crystalline and organic phases. This paper follows the pilgrimage of carbon (Fig. 1) from groundwater sources, down rivers, through lakes and into the estuaries in a series of case studies. The impact of technological "civilization" on the geochemistry of fresh waters is highlighted. The data derive mainly from the SCOPE/UNEP Project "Transport of Carbon and Minerals in Major World Rivers" that started in 1980. So far three volumes of data on river and lake carbon chemistry have been published (Degens, 1982; Degens et al., 1983; Degens, et al., 1984).

A. Lerman and M. Meybeck (eds.), Physical and Chemical Weathering in Geochemical Cycles, 197–223.

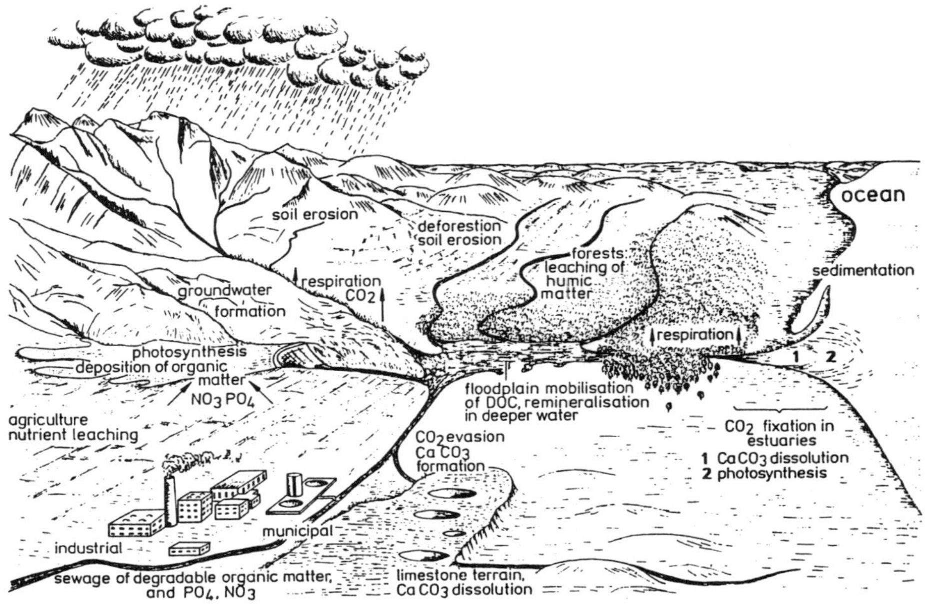

Figure 1. Scheme of a river system from source to sea. Natural and anthropogenic sources and processes affecting carbon and nutrient fluxes are indicated (Kempe 1982a).

2. CARBON GEOCHEMISTRY OF FRESH WATERS

Dissolved CO_2 is the most important substance for weathering processes. It can be supplied to waters by the following processes:

(1) Dissolution of atmospheric CO_2 in rain or other surface waters
(2) Dissolution of CO_2 liberated by root and microbial respiration in infiltrating water or other fresh water bodies
(3) Volcanic CO_2 dissolved in hydrothermal water
(4) Liberation of CO_2 during oxidation of certain minerals (e.g., $FeCO_3$)
(5) Liberation of CO_2 during mixing of waters due to thermodynamic causes.

Water charged with free CO_2 forms carbonic acid (H_2CO_3) which can react with carbonate and silicate minerals to release ions. The reaction involves a series of equilibria described by a set of mass balance equations (Garrels and Christ, 1965; Stumm and Morgan, 1970, 1981; Drever, 1982; Pytkowicz, 1983; Wigley, 1971, 1973a, 1973b; Wigley and Plummer, 1976). A review of the reactions and equilibrium constants needed for calculation of the various carbonate system parameters was given by Degens et al. (1984). In practice, it has been found useful to calculate the P_{CO_2} and the saturation index with respect to calcite for discussion of the carbonate chemistry of fresh waters. The P_{CO_2} gives the partial CO_2 pressure of a solution at equilibrium with the atmosphere. The P_{CO_2} is either given in ppmv (part per million per volume) or as its negative log (pP_{CO_2}). The calcite saturation index is given by the logarithm of the ratio of the ion activity product of the Ca^{2+} and the CO_3^{2-} ions and the solubility product of calcite at the same temperature. The calculations of the saturation index based on a chemical speciation model (Wigley, 1971) are best done by a computer program, such as given by Kempe (1975).

The biochemical reduction of CO_2 (C-oxidation state +4) to hydrocarbons (-1) by photosynthesis can extract CO_2 from water while the biochemical oxidation of hydrocarbons by respiration can add CO_2 to water. In most fresh waters these reactions determine the P_{CO_2} and thus their capacity to weather minerals. If photosynthesis prevails, the P_{CO_2} is depressed and the water becomes more alkaline as carbonate ions form to keep the charge balance which is upset by the removed CO_2 and bicarbonate ion. On the other hand, respiration liberates CO_2, increases the P_{CO_2}, consumes the carbonate ions and makes the water more acid.

Various sources supply organic matter to fresh water:

(1) Fresh terrigenous plant matter, such as wood, leaves and grass, enter the rivers as particulate carbon;
(2) Degraded soil material enters the rivers in particulate and dissolved form;
(3) Authigenic production in the water forms organic matter and excretes certain dissolved substances;
(4) Man discharges a variety of organic substances, dissolved and particulate, to the fresh waters.

Processes (1) and (4) will tend to increase P_{CO_2} greatly. Process (3) will decrease P_{CO_2}, but later respiration of products will release CO_2 again. Process (2) will cause only a slight increase of P_{CO_2} because most of this matter is highly degraded already and cannot serve as an energy source for bacteria.

The preceeding sources deliver organic material of a highly variable composition. In the early sixties, the first reliable techniques for dissolved (DOC) and particulate (POC) organic carbon were developed (Menzel and Vacaro, 1964; van Hall et al., 1963). Later, techniques for routine analysis of certain functional groups (e.g., carbohydrates, amino acids, hydrocarbons, lipids, phenols) were also developed. Methods to fractionate all of the DOC and to determine bulk characteristics became available only recently (Leenheer and Hufman, 1979; Leenheer, 1985). The analysis of organic fractions is based on the bulk element determination (C, H, 0, N, P, S), infrared spectrometry, and solid state ^{13}C nuclear magnetic resonance (NMR) spectrometry. Such methods cannot yet be applied routinely to large numbers of samples, but they can formulate hypotheses as to the structure of bulk DOC. Comparing seven samples from the Rio Negro System, Brazil, with ten samples of the White River system, Utah, Leenheer (1985) shows that the hydrophobic fraction is higher in the tropical black water rivers (59%) than in the temperate river system (43%). Among the hydrophilic fraction the neutral compounds representing the carbohydrates change from 4% in the tropics to 14% in the White River, suggesting that Rio Negro material is more highly degraded (more humic in nature). These data allow for the first time to make a model of strong hydrophobic acid fraction (Fig. 2) accounting for all analytical results. The unit on the lower right suggests a repeating structure; the branched methyl structure may indicate the importance of isoprenoid hydrocarbons as source materials for the hydrophobic-acid component of DOC.

Data on the composition of POC are also becoming available. Particulate amino acids and carbohydrates in the major rivers of the world account for 5-22% and 2-11%, respectively, of the POC. The St. Lawrence, Orinoco and Parana rivers show the highest concentrations of these two classes (Degens and Ittekkot, 1985). Amino acids and carbohydrates can be viewed as labile compounds. The St. Lawrence is characterized by a high labile fraction, possibly because most of its organic particulates are derived from plankton living in the epilimnion of Lake Erie, the source of

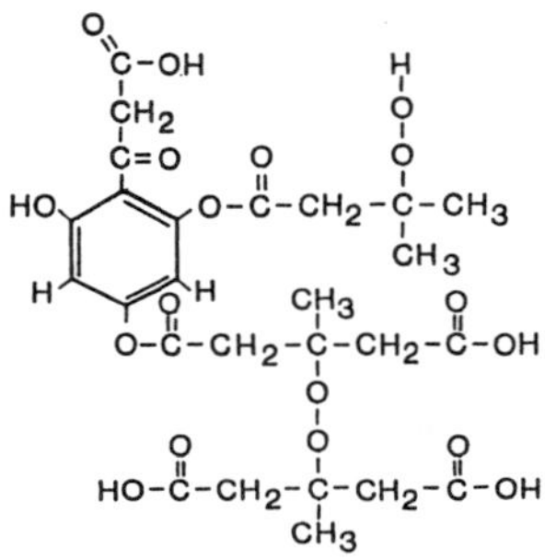

Figure 2. Average structural model of strong hydrophobic acid fraction of samples from the White River system, Utah (Leenheer, 1985).

the St. Lawrence. High proportions of labile substances in the particulate fraction of the Parana and Orinoco is somewhat more of a surprise, but could be connected to rapid interchange of matter between the main stems of the rivers and their large floodplains.

A similar conclusion is drawn from the organic data collected during the CAMEX cruises (Carbon in the Amazon River Experiment). As in the Amazon, there also occurs a substantial exchange of organic matter between the main channel and floodplain (várzea) (Richey et al., 1985). This is evidenced by high respiration rates measured in the river (0.5-4 μM O_2/hr/l). This material must have residence times of only a few days and it causes the CO_2-pressure of the Amazon to be very high, about 15 times that of atmosphere. Analysis of POC for lignin oxidation products shows that the coarse particulate matter (fraction greater than 63 μ) may be a mixture of 75-80 wt% of leaf and 20-25% of wood residues and a few percent of várzea grass (Hedges et al., 1985), resulting in phenol-carbon proportions of around 1%. Thus this material is fairly recent, a conclusion substantiated by its high $\Delta^{14}C$ (220 ppt) content. The fine particulate matter, however, which comprises most of the transported particulates, contains only 2% of phenolic carbon. In this fraction vascular plant matter seems to be highly diluted by degraded N-rich soil material amounting to at least 50%. Also $\Delta^{14}C$ concentration is much lower (40 ppt), suggesting greater age of this fraction (Hedges et al., 1986). Relatively low C/N values (between 15 and 10) are characteristic for particulate organic matter in most rivers (SCOPE/UNEP data base; Meybeck, 1982) and are believed to derive primarily from soils, the C/N ratio of which are similar (Stevenson, 1972; Bowen, 1979).

3. CASE STUDIES

This section illustrates how the carbonate system and organic carbon react to various climatic, hydrographic, geologic and pollutional settings. The discussion deals in sequence with a karst area of high mineralization, a mountain bog, a large tropical river (the Amazon), and a river altered by a dam (the Nile). Finally, the Rhine will illustrate what pollution does to the water chemistry of a river and how the carbonate system responds to the mixing of river and sea water, using the example of the Elbe Estuary.

3.1. Karst Systems: Plitvice and Postojna (Yugoslavia)

The hydrochemistry of the two most famous karst areas in Yugoslavia—the National park of Plitvice and the cave area of Postojna—was investigated in April and October 1983 (Kempe and Emeis, 1985). Figure 3 shows the conductivity (as a measure of total dissolved carbonates) and temperature of the two hydrologic systems for April. In case of the Plitvice System (solid lines) the karst springs of the Bijela Rijeka, the Crna Rijeka, the Rjecica and the Potol Plitvica deliver water of high conductivity and low temperature. As the water flows downstream, it warms up and looses $CaCO_3$ as evident from the decreasing conductivity. The combined discharge forms the Korana River, which leaves the park. The $CaCO_3$ deposits form tuffa dams behind which up to 47 m deep lakes have been ponded (Fig. 4). Growth of barriers continues today. This dammed lake complex occupies a former canyon cut into limestone and dolomite.

In case of the Postojna System (stippled lines, Fig. 3) surface streams collect water relatively low in conductivity and much warmer than in Plitvice. As the water sinks into the Postojna Cave Systems, it cools and increases in conductivity. Within the Planinska Jama both the Pivka arm and the Rak arm meet and mix. They exit from underground at about the same temperature and conductivity as the Korana water at Plitvica. Thus, at Plitvice, $CaCO_3$ is deposited while at Postojna it is being dissolved. Dissolution of the Postojna system is rather small, as the water volume of the underground rivers is relatively large and the time of underground travel tends to be short. Also, the surface waters already carry a large fraction of dissolved ions before they enter the caves. At Plitvice the change of water chemistry is much more dramatic. Calculated P_{CO_2} values show that CO_2 is quickly degassed in the rapidly flowing creeks, causing high supersaturation in calcite. This supersaturation is maintained throughout the lake passage even though more and more $CaCO_3$ is extracted. Supersaturation seems to be a prerequisite for any subaerial calcite precipitation to proceed at appreciable rates. Further CO_2 is removed from water by photosynthesis in the lakes and by physical degassing at the waterfalls. At one 78 m high water fall a P_{CO_2} drop from 580 to 440 ppmv and from 1120 to 740 ppmv was measured in April and October, respectively. Between the karst springs and the Korana, the P_{CO_2} drops from over 5000 to 470 ppmv.

The total amount of calcite precipitated from the system cannot, however, be calculated by taking the differences in concentrations between springs and the outflow of the system, as it must take into account the water volumes. From gauging data, made available by the National Park Administration, a model of the water balance was constructed (Fig. 4). Using this information and averages of carbonate concentrations, a calcite precipitation balance was calculated (Table I). Summing up inputs and outputs, a total precipitation rate of 333 g $CaCO_3$/s or 10,500 t $CaCO_3$/yr was estimated. Assuming a total area of lakes and dams of 2 km^2, such rates would amount to a layer of solid limestone 20 m deep in 10,000 years. The rate of CO_2-removal (degassing plus photosynthesis) needed to match the calcite deposition amounts to 50 mol m^{-2} yr^{-1}.

Calcite settles in two forms: as fine-grained Seekreide (lacustrine chalk) in the lakes and as solid tuffa at the dams. Holocene Seekreide is up to 8 m thick; thus—considering our model growth rates—dams could have formed during the Postglacial. Despite the high supersaturation of lake waters, calcite does not seem to form inorganically but by biological activity. In the lakes, the epilimnion becomes depleted with calcite even to the point that the dolomite saturation index becomes larger than the calcite index. Scanning electron microscope (SEM) investigation shows that grains

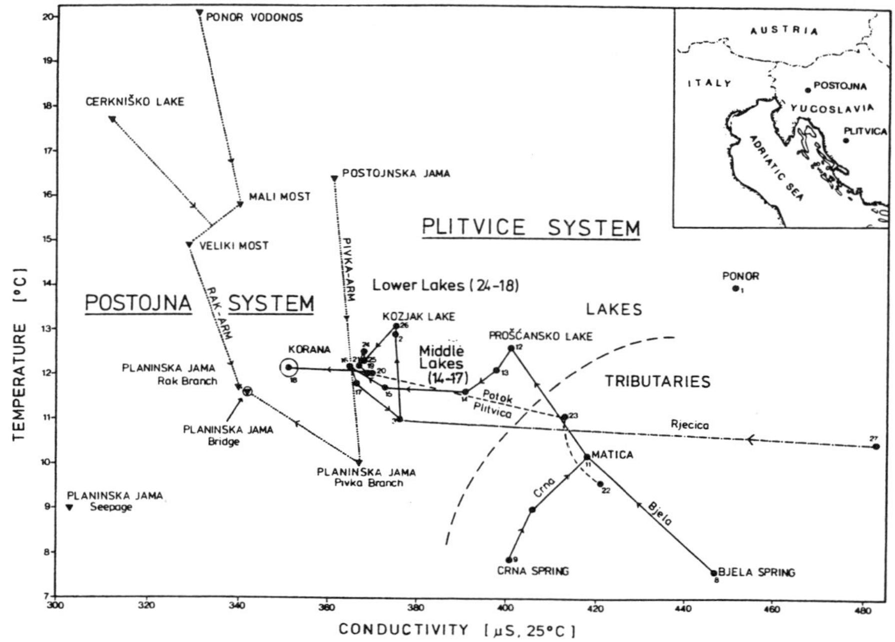

Figure 3. Comparison of Plitvice and Postojna system in terms of conductivity and temperature, April 1983.

are about 2 μ in size and euhedral to hypidiomorphic (Plate 1). At the dams, *Bryum* and *Cratoneuron* mosses grow. Under the SEM the young green leaves reveal that they are densely covered by pennate diatoms and filamentous blue-green algae (Plate 2). Strands of mucopolysaccharide glue them into place (Plate 2). Further down the leaf the mucopolysaccharides have caught calcite grains (Plate 3), which later grow into a solid, sparitic layer of calcite (Plate 4). A simple experiment shows that agglutination in fact is the first and essential step in calcite growth on surfaces. Plastic, iron and copper sponges were exposed to lake waters; plastic and iron sponges were soon covered by calcite, but not the copper sponge (Plate 5, 6). Copper is toxic for the epiphytic diatoms. Without diatoms, no mucopolysaccharides can form and no initial grains can be caught. Without seeds, no further growth occurs.

It is known that at high Ca concentrations cells excrete large amounts of their metabolic products to bind the Ca. Under nitrate stress, mucopolysaccharides are excreted instead of amino acids. At Plitvice, extremely low DOC values of 0.4 to 1.0 mg/l were measured, and these values are even lower than those often recorded in the ocean. Thus the special situation of the Plitvice system (high Ca-concentration and low nutrient supply; visibility in the lakes is several meters and phosphate measurements show low concentrations upstream and values near zero in the Korana) may favor diatom excretion of mucopolysaccharides rather than of other more soluble molecules. This glue is possibly essential to the rapid encrustation of the mosses and the building of the dams of Plitvice.

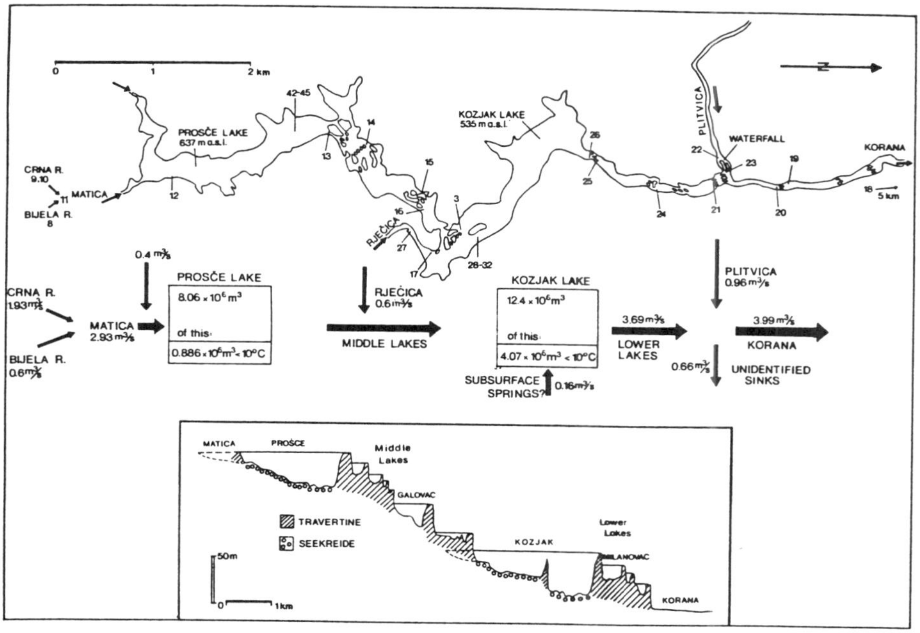

Figure 4. Plan and profile of lakes and tuffa dams at Plitvice. Model figures of discharge are given.

Table I. Mass balance model of Plitvice Lakes, based on total carbonate (Sum of Ca, Mg, HCO_3) concentrations (Kempe and Emeis, 1985).

River	Measured Concentrations mg/l	Average mg/l	1980-1981 Discharge m^3/sec	Transport g/sec
Input:				
Matica	366;369;365	367	2.93	1074
Rjecica	401;423;407	410	0.60	246
Plitvica	362;352; -	357	0.96	343
Kozjak source	(estimated)	370	0.16	59
			total	1722
Output:				
Korana	315;295;286	299	3.99	1192
sinks in stream bed		299	0.66	197
			total	1389

Loss in Plitvice system (input–output): 333 g total carbonates/sec = 10,500 t/yr.

3.2. Central Harz Sphagnum Bog

In the central Harz mountains of Germany, an 18.4 km^2 large raised bog feeds some of the headwater tributaries of the River Weser. The bog, up to 4 m deep, sits on granite. The long-term average annual precipitation in the area is 1510 mm, of which 1070 mm runs off.

Hartmann (1983) monitored the carbon output of this bog for the discharge year 1980-1981 by monthly sampling the Rehberg Graben, representing a drainage area of 10.8 km^2. The waters are very low in solutes, with conductivity values ranging from 63 to 90 μS. Even though in March 1981 a flood five times the long-term March average occurred, conductivities did not decrease substantially during this month. This illustrates that the water does not run off from the surface of the bog but rather percolates through the bog. Similar to conductivity, pH values do not show correlation with discharge. They are always low, between 4.9 and 5.9. In this pH range organic acids contribute to the hydrogen ion concentration. Sulfate is the main anion. The dissolved inorganic carbon (DIC) ranged from 3.2 mg/l in February to 1.0 mg/l in May, also showing no dilution during the month of flood. DIC values are high during the cold seasons and low during the warm period. The concentrations of dissolved organic carbon (DOC) are higher during the warm period, but low during the cold season. These opposing trends of DIC and DOC suggest that in summer CO_2 is consumed by photosynthesis, decreasing the inorganic carbon, but increasing the organic carbon output due to excretion from cells. In winter respiration prevails, producing inorganic carbon but reducing the organic output.

The export of total organic carbon (TOC) from the bog is 6.9 t km^{-2} yr^{-1}, 80% of which was in the dissolved phase (DOC 5.7 + POC 1.2). Only 5% of the DOC are carbohydrates, suggesting that most of the DOC are complex humic substances. The total net primary productivity (NPP) was estimated to be 450 t C km^{-2} yr^{-1} (Ajtay et al., 1979); thus, only 1.6% of NPP were exported as organic carbon. Nevertheless, the specific TOC discharge of the Harz bog is much higher than the specific TOC discharge of temperate rivers (Columbia 1 t C km^{-2} yr^{-1}, Mississippi 0.8; Kempe, 1982a) and even larger than the highly polluted Central European rivers: Elbe 2.7 t C km^{-2} yr^{-1}; Rhine 4 (Kempe, 1982a); Tiber 1.7 (Pettine et al., 1983). The specific discharge of the bog is best compared to the TOC discharges of tropical floodplain rivers such as the Amazon, 6.8 t C km^{-2} yr^{-1} (Richey et al., 1985); the Orinoco, 7.3 (Paolini et al., 1983), and the Paraná, 6.4 (Depetris and Cascante, 1985). Thus bog areas serve as important sources of dissolved humic substances to temperate river systems.

3.3. The Amazon River

By all standards the Amazon River is the largest river system on Earth. It drains 7.18×10^6 km^2 (6.2% of peripheral continental area). Total discharge of the river is close to 6,000 km^3/yr (Baumgartner and Reichel, 1975), or 15% of world discharge. Last station above tidal influences at which to measure discharge is Obidos (representing 4.69×10^6 km^2). Much confusion exists in the literature as to the rating curve of this station (Kempe, 1982a). Recent measurements for 1983 suggest a discharge of 4540 km^3/yr (data of Richey et al., in press). However, 1983 was a year of below-average discharge (Fig. 5), suggesting that long-term average discharge at Obidos is somewhat higher. Oltman (1968) calculated a long-term average for the 1928-1947 stage record of 4950 km^3/yr, which agrees with the more recent measurement. The shape of the discharge curve is a relatively regular sinusoidal curve and upstream and downstream stations peak at about the same time. This unusual phenomenon is due to floodwaves in the Rios Madeira, Tapajós and Xingu, which reach the lower Amazon earlier than the Rio Solimoes (synonym for upper Amazon) floodwave of Andean waters (Meade et al., 1985; see also Meade, this volume). The floodwave is also dampened by the large floodplain—the várzea—that borders the river from Iquitos downstream.

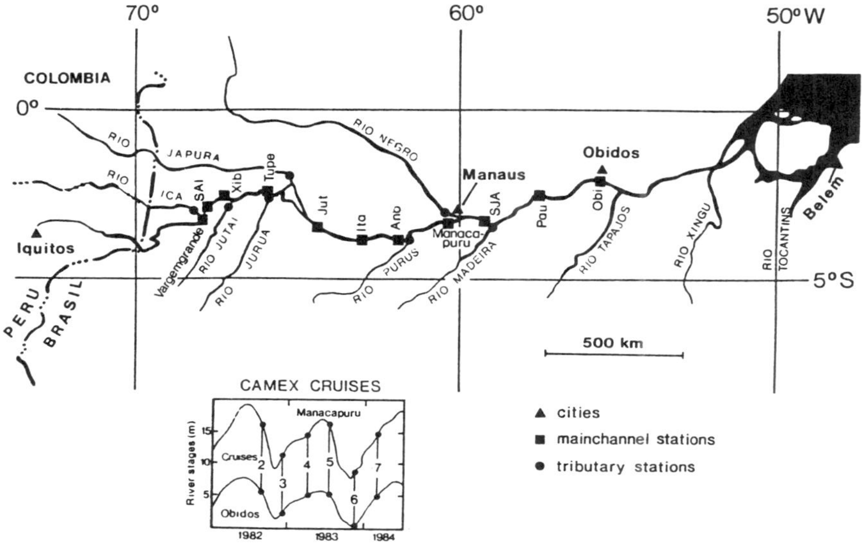

Figure 5. Sketch of Amazon River system and location of stations mentioned in text (after Richey et al., 1985). Amazon stage curves at Manacapuru (upstream station, top) and Obidos (downstream station, bottom). Distance between stations 750 km. CAMEX Cruises 1-8 are indicated (after Meade et al., 1985).

The objectives of a study of the behaviour of carbon in the Amazon system are: (1) to obtain carbon transport numbers for the largest river on Earth, (2) to understand transformations of carbon in a very long drainage channel under tropical conditions, and (3) to study the interaction of the 100,000 km^2 of virtually pristine floodplains with the chemistry of the main channel. To pursue the above aims a joint American and Brazilian "Carbon in the Amazon River Experiment" (CAMEX) was launched (Richey, 1982). A total of eight cruises (see Fig. 5) were conducted at different seasons. At 11 sections between Vargem Grande and Obidos and in six tributaries discharge, sediment load, and hydrochemistry were monitored. This unique set of data allows for the first time to picture the envelopes of concentrations of certain species in time and space. Figure 6 depicts the discharge diagram along with diagrams of concentrations of suspended matter and various carbon species (Richey et al., 1985). These data also allow calculation of a first crude carbon balance of the Amazon mainstem, given in Table II (Richey et al., 1985). Concentrations of fine particulate organic carbon (FPOC) closely follow the concentration of fine suspended solids (less than 63 μm), both decreasing downstream. This suggests that low-land rivers dilute particulate carbon of upstream origin, as already suggested on the basis of C/N measurements. This material is probably rather inert soil matter and an integral fraction

Figure 6. Variation of discharge, fine suspended matter (FSS <63 μm)) concentration and concentration of various carbon species with time in the Amazon and downstream. Result of CAMEX cruises 1-7 (Richey et al., 1985). FPOC = fine particulate organic carbon, DOC = dissolved organic carbon, DIC = dissolved inorganic carbon.

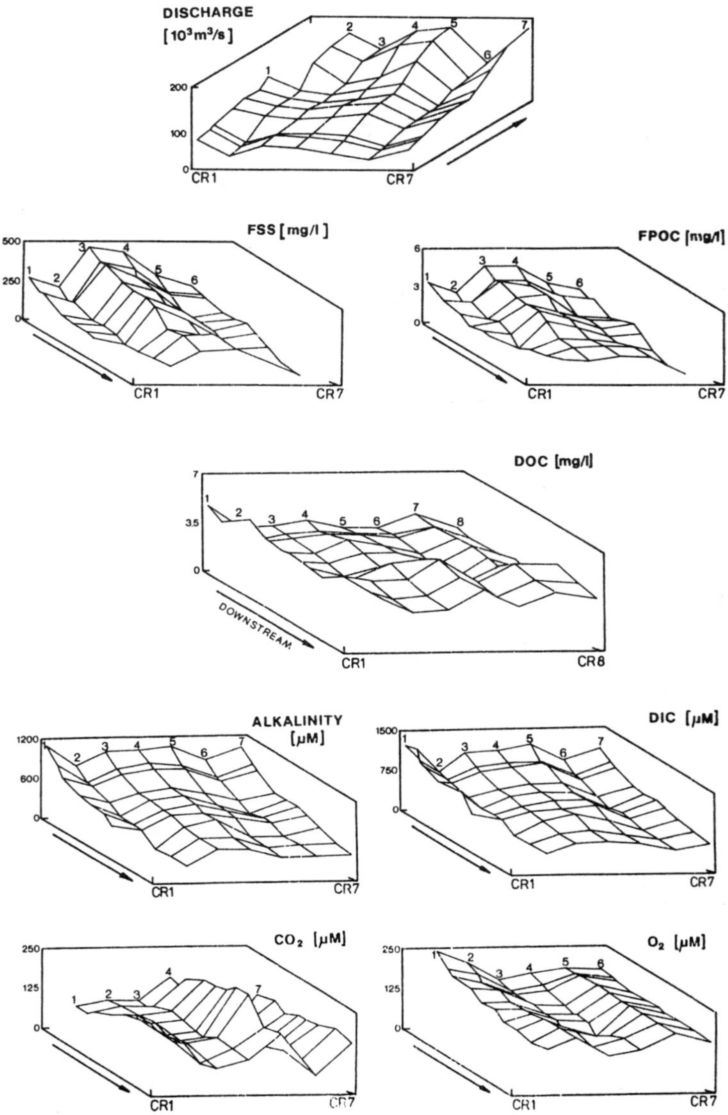

of suspended matter. Highest concentrations are encountered at early rising water (cruise 3) and lowest at early falling water. Balance of FPOC and FSS shows (Meade et al., 1985; Richey et al., 1985) that during rising level both FPOC and FSS are stored in the floodplain and that they are removed from the vа́rzea at near-peak and falling stage. In addition to the release of temporarily stored POC, the vа́rzea acts as a net source for POC as shown in Table II.

Table II. Amazon River: Carbon budget for CAMEX cruises 1982/83 (Richey et al., 1985). All figures in 10^3 mol C/sec.

Cruise	2	3	4	5	average
Dissolved Inorganic Carbon					
Vargem Grande	25.2	61.3	56.9	47.2	47.7
Tributaries	33.9	20.6	41.7	45.1	35.3
Obidos	77.6	79.1	111.2	108.9	94.2
Dissolved Organic Carbon					
Vargem Grande	8.3	15.3	15.5	9.8	12.2
Tributaries	35.7	20.5	33.0	51.9	35.3
Obidos	55.1	31.7	44.3	65.7	49.2
Particulate Organic Carbon					
Vargem Grande	9.0	26.0	24.4	12.7	18.0
Tributaries	3.8	11.8	29.5	10.7	14.0
Obiedos	31.7	25.5	41.6	41.1	35.0
Total Carbon					
Vargem Grande	9.0	102.6	96.8	69.7	78.0
Tributaries	73.4	52.9	104.2	107.7	84.6
Obidos	164.4	136.3	197.1	215.7	178.4
Excess DIC	18.5	-2.8	12.6	16.6	11.2
Excess DOC	7.5	-4.1	-4.2	4.0	0.8
Excess POC	18.9	-12.3	-12.3	17.7	3.0
Excess TC	44.9	-19.2	-3.9	38.0	14.9
CO_2 Evasion	24.3	15.1	19.7	41.6	25.2
Respiration	44.0	165.5(?)	99.6	67.1	94.6
Resp*	42.8	12.3	32.3	58.2	

Preliminary carbon budget calculated as concentration times discharge, except respiration (rate times volume) and atmospheric exchange (calculated using a stagnant boundary layer thickness of 35 μ. Excess values represent the difference between the output at Obidos and the sum of the Vargem Grande plus tributary inputs. Resp* is respiration calculated as excess DIC at Obidos plus CO_2 evasion.

Vargem Grande and Obidos denote total flux through cross section of Amazon at these stations, Tributaries denote total added flux of all major tributaries between Vargem Grande and Obidos.

(?) denotes high values possibly due to errors in method.

DOC concentrations are fairly uniform throughout the season and downstream showing that low-land tributaries play a much more important role as sources for DOC than for FPOC. Again in the balance throughout the year there is a net source of DOC from the várzea, especially at falling water.

Dissolved inorganic carbon (DIC) is the most important carbon species transported. Its concentration varies with discharge. Except for the very early rising water stage (cruise 3) the várzea functions as a DIC source, surpassing the POC source by far. As no carbonates are available for dissolution in the soils of the Amazon floodplain, it must be assumed that this inorganic carbon is formed by the total remineralization of plant matter decaying in the várzea lakes. This conclusion is substantiated by tank experiments and by the observation that DIC accumulates in lakes cut off from water exchange with the main channel (Furch and Junk, 1985). It is not clear what cation provides charge balance for this increase in DIC that correlates wit an increase in alkalinity, but potassium is a likely candidate, also liberated by decaying plant matter.

The concentration of free CO_2 in the water, and hence the P_{CO_2}—shows a sharp peak during early falling water (cruise 5). This could be the result of the deeper, more highly mineralized and cooler water from the várzea lakes which are seemingly drained first to the mainstem. P_{CO_2} seems to increase downstream, marking the influence of the várzea on the carbon chemistry of the Amazon. If one calculates the amount of CO_2 degassed from the Amazon (using a mean boundary layer thickness of 35 μ) then (Table II) it is found that almost as much CO_2 is lost as POC is exported downstream of Obidos. This amount of carbon also does not originate upstream, but must be derived from respiration of organic matter entering the mainstem from the floodplains. Thus the sum of the excess DIC and the CO_2-evasion gives a measure of the amount of labile carbon derived from the floodplains. In Table II this is marked by Resp*. It is further interesting to compare this derived measure of respired organic carbon with direct measurements of respiration, also given in Table II. These measurements prove to be even higher than the derived remineralization rate, suggesting that the CO_2 loss may be even higher than calculated with the assumed boundary layer thickness (if one accepts the respiration measurements as being experimentally correct). These figures suggest that the várzea delivers highly labile organic carbon which is very quickly remineralized, having residence times of a few days only. Similar to carbon, the várzea also serves as a net-source for dissolved phosphate and nitrate, an unexpected result of the CAMEX investigation (Richey et al., 1985).

Using the data of Table II one can calculate the carbon export of the Amazon below Obidos (in 10^6t C/yr): 35.6 (52%) for DIC, 18.6 (28%) for DOC and 13.2 (20%) for POC, yielding a total of 67.5×10 t C/yr (100%). Total CO_2-evasion amounts to 9.5 10^6t C/yr.

One can also assess the total export from the várzea (including minor tributaries) by subtracting the transport through the section at Vargem Grande plus the major tributaries from the export at Obidos. The following differences (in 10^6t C/yr) are found: 4.3 for DIC, 1.3 for POC+DOC, totalling 5.6×10^6 t C/yr or 8.3% of total export at Obidos. To this we may add the CO_2-evasion of 9.5 and compute a total excess derived from the floodplains as 15.1×10^6t C/yr. 13.8×10^6 t C/yr (9.5 CO_2-evasion + 4.5 DIC) have already been remineralized and only 1.4×10^6 t C/yr várzea carbon reaches Obidos in organic form. However, if one looks at the measured respiration rates, the CO_2 loss may be as large as 35.8×10^6 t C/yr. Junk (1985) provides an independent estimate of the carbon amount available for export from the Amazonian floodplains. He assumes the várzea (100,000 km^2) yields a NPP of 154×10^6 t C/yr. Of this a maximum of 47×10^6 t C/yr may be available for export. However, most will be remineralized in the floodplain so that probably only 17×10^6 t C/yr reach the Amazon for further downstream transport. This figure is strikingly close to the CAMEX estimate of total excess plus CO_2-evasion (15×10^6 t C/yr).

3.4. The Dammed Nile

In 1964 the High Dam at Aswan was closed. From 1965 to 1975 the Nile filled Lake Nasser (5500 km^2), one of the largest man-made reservoirs in the world. Average discharge of the Nile at Aswan was 83.2 km^3/yr for the period 1912-1947. The dam serves well its prime aim, to store a full annual flood. Before the dam was closed the Nile flooded its banks in Egypt from late July to November. This flood originates in the Ethiopian mountains and delivers 78% of total discharge. During this time the Blue Nile and the Atbara swell dramatically and superimpose a flood wave on the base flow provided by the White Nile that has its lowest flow in July. The White Nile originates in Lake Victoria. In this lake, dissolved salts are already somewhat concentrated by evaporation. Later, as the White Nile crosses the Sudd region, more water is lost and the total ion concentration increases further. Thus the White Nile water carries the higher salt content of the two Niles.

If one compares the average seasonal hydrographs at Asyut (400 km below Aswan) for the period before the dam was closed (1960-1964) with the period after the dam was closed (1969-1982) (Kempe, 1983), it becomes apparent that the old floodwave does not arrive at Asyut at present. This has probably three reasons: 1) additional loss of water from the surface of the reservoir, 2) increased consumption of water for agriculture above Assyut, and 3) a decline in water input due to Sahelian drought (1972) and the Ethiopian drought in later years. The loss of water due to evaporation from the reservoir is probably not higher than 10 km^3/yr, thus additional consumption and reduced input are needed to explain the non-appearance of the old floodwave. As it stands, Egypt might have suffered similar famines as Ethiopia and Sudan in recent years, were it not for the dam that stores enough water to sustain a minimum flow during lean years.

Table III. Nile River: Comparison of discharge weighted mean concentrations before 1959-60 and after 1969-70, the closure of the High Dam (Kempe, 1983).

	pH	Alk	SO_4^{2-}	Cl^- (meq/l)	Ca^{2+}	Mg^{2+}	Na^+
1959-1960	7.74	2.02	0.22	0.25	1.14	0.68	0.69
1969-1970	8.07	2.58	0.58	0.90	1.44	0.93	1.48
Factor of increase	(1.04)	1.28	2.64	3.60	1.26	1.37	2.14

	K^+	TDS* (mg/l)	P_{CO_2} (ppmv)	Calcite (Saturation Index)	Dolomite (Saturation Index)	Gypsum (Saturation Index)
1959-1960	0.09	193	2070	-0.33	-0.38	-2.88
1969-1970	0.17	299	1265	0.16	0.16	-2.45
Factor of increase	1.89	1.55	0.61	δ: +0.49	+0.49	+0.43

* TDS = total dissolved solids.

Chemical data as measured at Giza (after Hafez, 1962; Fathi and Soliman, 1977); discharge curves are those of the Asyut record.

However, the consequences of the dam building are, as a whole, considerable. Of a former suspended load of 125×10^6 t/yr now only $1\text{-}2\times10^6$ t/yr pass the dam (Schamp, 1983), most of that probably of planktonic origin (Emeis, 1985). Lacking suspension, the river degradates its bed (by $2\text{-}3\times10^6$ t/yr), thereby endangering the barrages, a series of weirs, at which channels divert onto the floodplain. Costs to replace these barrages would probably be higher than those building the Aswan dam. Also irrigation today has to be regulated by pumping, at the expense of much energy. However two to three crops can be gathered where formerly only one to two crops were harvested.

Important alterations can also be observed with the hydrochemistry of the Nile. Figure 7a compares the measurements of total dissolved solids at Giseh (Cairo) in three different periods, 1933-1936, 1959-1960 and 1969-1970 (Kempe, 1983). Before the closure of the dam the salinity peaked in summer when the White Nile water predominated. Today, the salinity is much higher and shows an inverted seasonality. Table III gives the discharge weighted annual means for the two most recent sets of measurements: total dissolved salts have risen by factor of 1.55. This is much higher than what could be expected by the loss of water through evaporation in Lake Nasser alone (a factor of 1.2-1.4 should be expected). Thus additional concentration of salts must occur due to the new irrigation techniques. Looking at the individual ions, one notices that the geochemically most conservative ion, Cl^-, shows an increase of factor 3.6 , while HCO_3^-, Ca^{2+} and Mg^{2+} increased by only a factor of 1.3. These relations suggest that—relative to chloride—Ca and Mg carbonates are lost from the river water and that without this loss, the increase of salinity in the Nile should have been by about a factor of 3.6. Loss of Ca and Mg carbonates is in fact possible, as suggested by the saturation of the Nile water with respect to calcite and dolomite, calculated from the two chemical records (Fig. 7b). Before the dam was built, the Nile water reached supersaturation only in July. Today, the river appears to be supersaturated with respect to both minerals for most of the year. One can roughly calculate how much carbonate is lost from Nile water (Kempe, 1983). Results amount to 60% of the older carbonate transport, roughly 0.6 t/yr for every hectar of upper Egyptian farmland. By using Nile water already high in Ca^{2+}, Mg^{2+} and HCO_3^- due to evaporative concentration upstream for irrigation, the Nile water becomes supersaturated with respect to the carbonate minerals. From the irrigated fields, surplus water drains back to the Nile, enriching the Nile further with Cl^-, Na^+ and K^+, while Ca^{2+}, Mg^{2+} and HCO_3^- precipitate on fields and stay behind. In accordance with this model explaining the differences in enrichment factors for the individual ions, the P_{CO_2} of the Nile appears to have dropped from 2070 to 1260 ppmv (Table III). Today, labile organic carbon cannot enter the Nile to any great extent, neither from upstream sources because of interception by Lake Nasser, nor from lateral sources, as no erosion of top soil occurs on the former floodplain. The Nile serves as an example of how man can unintentionally alter the hydrochemistry of a river by building dams.

3.5. The Polluted Rhine

The Rhine is one of the most important rivers of Europe. Its tributary area (159,680 km^2) collects a discharge volume of 72 km^3/yr (average of 1936-1968, data of gauging station Rees; see Figure 8 for location). Forty-three percent of this water derives from the area above Basel (near station Kembs in Figure 8); thus the alpine areas are of the same importance for the Rhine discharge as the lowland tributaries. The seasonal discharge curve is very irregular due to the low retention capacity of the channelized rivers and due to highly variable weather conditions in Central Europe. Highest

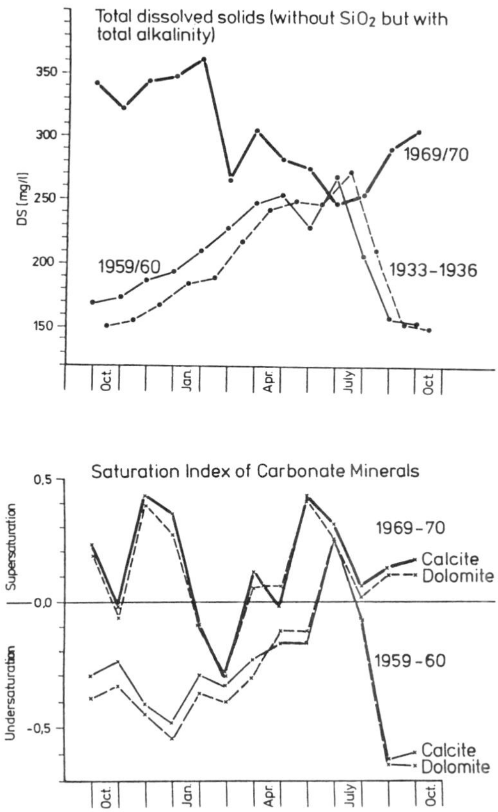

Figure 7. (a) Total dissolved solids of Nile water at Giza calculated by summing mg/l concentrations of Ca^{++}, Mg^{++}, Na^{+}, K^{+}, HCO_3^-, Cl^- and SO_4^-. Data 1933-1936 after W. T. H. Williamson, published in Hurst (1952), data 1959-1960 after Hafez (1962), data 1969-1970 after Fathi and Soliman (1977).
(b) Calcite and dolomite saturation index calculated for the chemical records of Nile water at Giza 1959-1960 and 1969-1970 (Kempe, 1983).

discharge occurs in January (lowland snow melt) and lowest in October. A secondary peak in June-July results from alpine snowmelt. The Rhine serves 41 million people as a sewage channel and some of the busiest metropolitan and industrial centers in the world are situated in its basin. By discharge it ranks about fortieth, but by load of pollutants and nutrients it would rank among the ten largest rivers on Earth.

Since 1963, multiparameter monitoring results are regularly published by the International Commission for Protection of the Rhine against Pollution for several stations (see Fig. 8). These data allow insight into the behaviour of carbon with respect to downstream, seasonal and interannual changes (Kempe, 1982a).

Figure 9 displays the downstream profile of long-term average P_{CO_2} of the Rhine. When the Rhine leaves Lake Constance, it has a P_{CO_2} of about ambient air. Then the P_{CO_2} increases gradually for 600 km until it reaches, at station Braubach, a 16-year average P_{CO_2} of 6300 ppmv, 20 times the pressure of ambient air. Below Braubach the

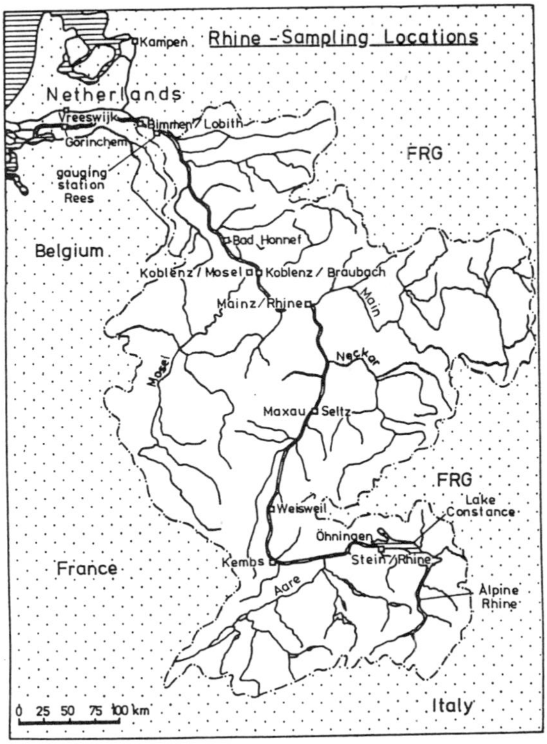

Figure 8. Sketch map of Rhine River basin and location of monitoring stations.

P_{CO_2} declines somewhat to the three delta stations of Kampen, Gorinchem, and Vreezwijk in the Netherlands. In 1979 the increase in P_{CO_2} extended to the Dutch-German border, while upriver stations showed decreasing P_{CO_2} trends.

This P_{CO_2} pattern is caused by the interplay of photosynthesis and respiration in the limnic and fluvial environment and their reaction to pollution with nutrients and organic matter (Fig. 10). The limnic environment is represented by Lake Constance (Station Stein, see Fig. 8). In Lake Constance, pollution with nutrients causes a stimulation of phytoplankton growth consuming dissolved CO_2 and increasing the pH of the surface waters. In Figure 11 it is shown that vigorous photosynthesis consumes CO_2 in summer so rapidly that P_{CO_2} drops below the atmospheric P_{CO_2} of 340 ppmv. During the warm season the lake becomes supersaturated with respect to carbonate minerals. The statistically significant regression of the P_{CO_2} with time in Figure 11 suggests that the nutrient pollution in Lake Constance increased during the time of the record, from 1974 to 1977. In the fluvial environment, on the other hand, respiration is the dominating process. Due to the high turbulence in a flowing medium and the high turbidity the primary productivity is lower than in lakes. Organic matter swept out of the epilimnion of the upstream alpine lakes (Lake Constance and Aare lakes) and organic substances added by man will be respired causing the observed increase in P_{CO_2} and acidifying the water

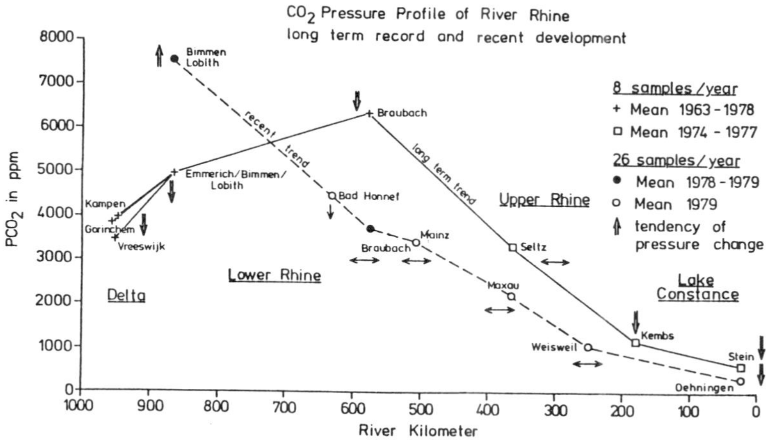

Figure 9. Variation of long-term average P_{CO_2} (in ppmv) along the Rhine River (solid line) and the P_{CO_2} in 1979 (broken line). (Kempe, 1982a).

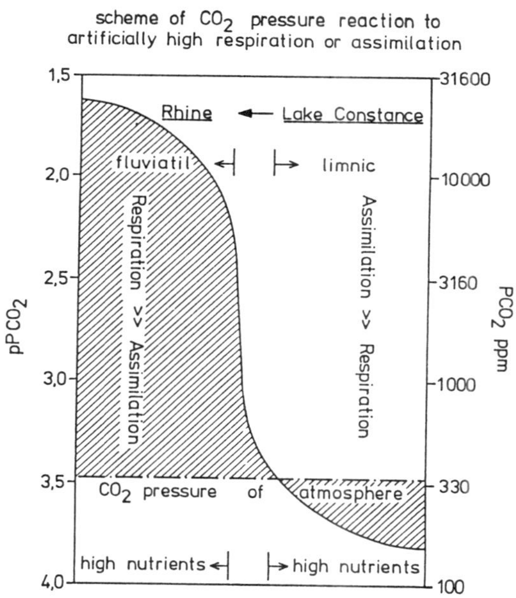

Figure 10. Scheme of P_{CO_2} reaction to pollution in the lacustrine and the fluvial environment.

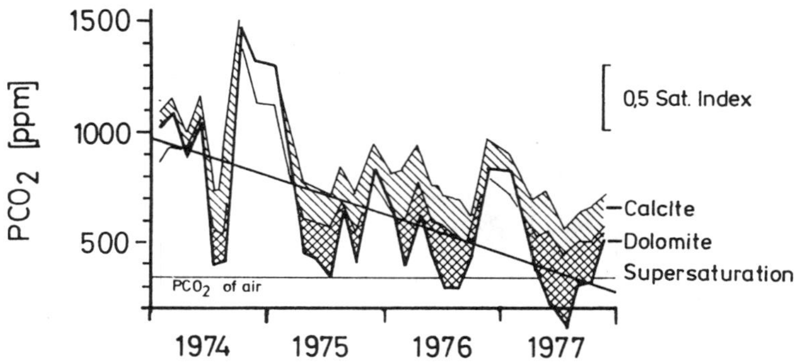

Figure 11. Trend of P_{CO_2} variation of Lake Constance outflow to Rhine. Note the extremely low P_{CO_2} in summer due to photosynthesis. Increasing pollution with nutrients seem to depress P_{CO_2} further.

Figure 12 compares the average seasonal P_{CO_2} of station Stein and station Braubach. In the outflow of Lake Constance the lowest P_{CO_2} values occur in July-August and the highest in December-January, typical for photosynthetical governed limnic systems in the northern hemisphere. At Braubach the phase is shifted and lowest values occur in April-May and highest in October when discharge is lowest. The same seasonal pattern is also evident for station Lobith. This shift in phase between the limnic and the fluvial P_{CO_2} seasonal pattern occurs between stations Kembs and Seltz, where the Rhine looses its "memory" of its limnic source about 200 km downstream of Lake Constance.

The Rhine water also shows significant long-term trends in its P_{CO_2}. Until 1972 the P_{CO_2} was steadily climbing, indicative of an increasing pollution with labile organic carbon; Figure 13 shows P_{CO_2} for two stations and a similar pattern is also characteristic of other sites upstream. Since 1972 the P_{CO_2} has been declining, suggesting that the enforced building of sewage plants by the countries of the Rhine basin has finally reversed the trend. Similar breaks in trend can also be shown for parameters like the chemical and biological oxygen demand, and oxygen (note high positive correlation coefficient of COD, BOD, TOC with P_{CO_2} in Fig. 14) (Kempe, 1982a).

The extremely high P_{CO_2} of Rhine waters is somewhat of an enigma. It cannot be explained by simple oxidation of organics due to consumption of oxygen. Using the oxygen and temperature records, the apparent oxygen deficit and the P_{CO_2} that should result from such an oxygen deficit can be computed. For most of the Rhine samples, a much higher P_{CO_2} is observed than could be expected from the apparent oxygen deficit. It is suggested that nitrate serves as a second oxygen source for oxidation of organic matter. Nitrate concentrations in the Rhine are still increasing and have reached levels of about 20 mg/l. Correlation and factor analysis show that nitrate does in fact show a significant positive correlation with oxygen and a significant negative correlation with P_{CO_2} (Fig. 14) (Kempe, 1982a; Kempe and Lammerz, 1983). It appears that denitrification delivers a large part of the oxygen needed for respiration. This process seems to proceed at low concentrations of dissolved oxygen, Putting it cynically, one can say that the Rhine possibly would have turned anaerobic, were it not for its high pollution with nitrate.

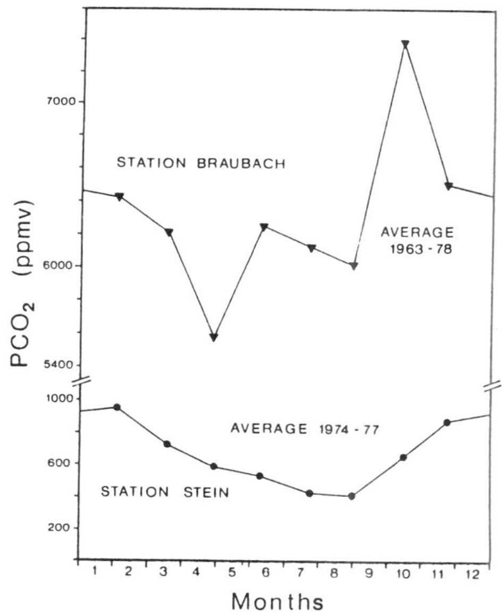

Figure 12. Comparison of P_{CO_2} seasonal variation for two Rhine stations. Station Stein monitors the outflow of Lake Constance, the limnic environment, while station Braubach represents the fluvial environment governed by respiration.

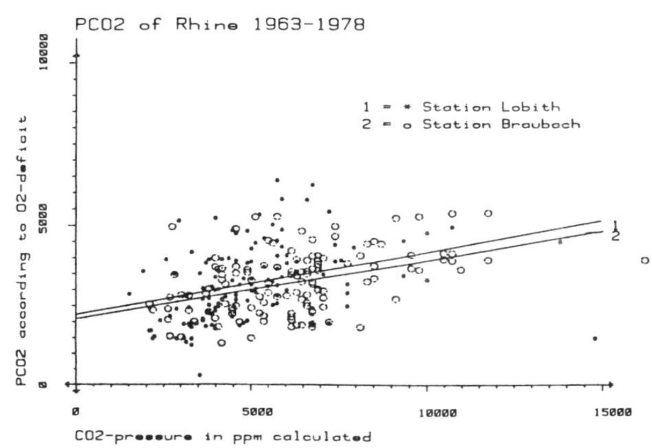

Figure 13. CO_2 in Rhine River water formed by oxygen consumption from different sources. See text for details.

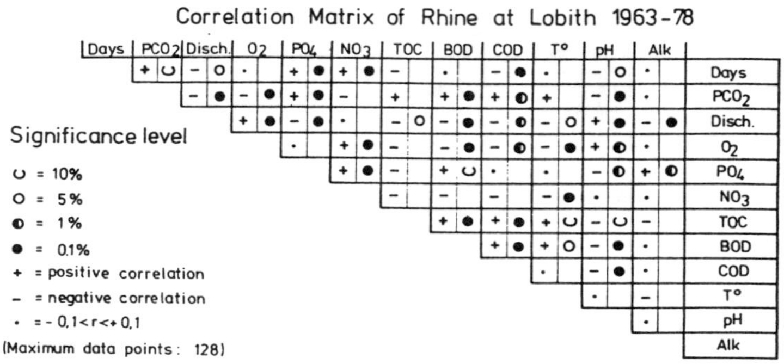

Figure 14. Correlation matrix for some hydrochemical parameters of station Lobith for data record 1963-1973 (Kempe, 1982a)

3.6. Carbonate Dissolution in the Elbe Estuary

Similar to the lower Rhine, the highly polluted River Elbe has a high P_{CO_2} of around 4000 ppmv (Kempe, 1982a). Values are again lowest in early spring and highest in late summer, as is indicative of a fluvial respiration-controlled environment. Due to this high P_{CO_2}, the river water is undersaturated with respect to calcite. Upon mixing such a water with surface seawater—which is always supersaturated with respect to calcite—the undersaturation is lost.

The undersaturation with respect to calcite in the estuary derives from two sources: (1) the primary undersaturation due to the river water of high P_{CO_2} and (2) undersaturation caused by the release of additional CO_2 if two waters of different P_{CO_2} are mixed (mixing corrosion). In Figure 15 an Elbe water is mixed in a numerical experiment with water of the North Sea (Kempe, 1982b). This experiment is done using the WATMIX program of Wigley and Plummer (1976). The calculation assumes that no calcite dissolution or precipitation nor CO_2-degassing is occurring, and mixing is done by conserving mass. Due to the mixing corrosion effect the undersaturation of the Elbe water should persist until it is diluted to more than 50% by North Sea water. The program also allows calculation of the amount of calcite needed to compensate the undersaturation and make the water saturated. Results indicate that at the maximum of the corrosion potential, about 100 μmol/l of calcite could dissolve, increasing the alkalinity of the water by 0.2 meq/l.

In October 1981 the alkalinity and pH distribution in the Elbe Estuary was monitored on a cruise of the R/V Valdivia. Alkalinity was determined by potentiometric Gran-titration using the programs of Bradshaw et al. (1981) of the GEOSECS project to evaluate the titration curve (Kempe, 1982b).

The results of the alkalinity titration are plotted in Figure 16 versus salinity. It is clear that alkalinity shows a steep increase at very low salinities. The corrosion potential is consumed very early in the mixing process of Elbe with North Sea water. The step height is of the predicted size, about 0.2 meq/l. The alkalinity finding is corroborated by the pH measurements (Fig. 16). In addition, pH increases in a steplike mode. Thus the high P_{CO_2} brought into the estuary by the river is buffered by the dissolution of calcite in the inner estuary. Calcite is most probably of marine origin, mixed into the estuary with suspended matter. Thus high CO_2 concentration of a polluted river is not mixed conservatively into seawater nor is it lost by degassing to air,

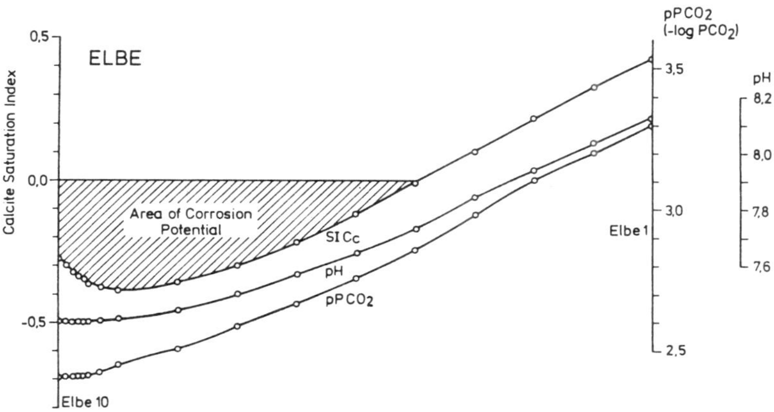

Figure 15. Chemical equilibrium model of mixing Elbe fresh water with seawater. Samples were obtained during the RV Valdivia Cruise, October 1981 (Kempe, 1982b). SICc = calcite saturation index.

but is fixed permanently into additional marine alkalinity by way of calcite dissolution.

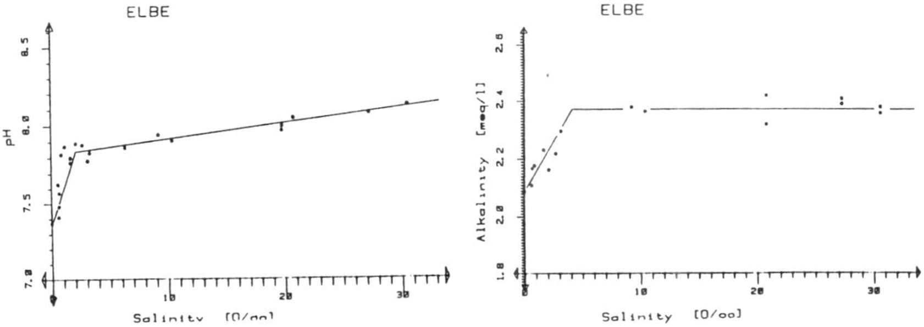

Figure 16. Alkalinity and pH profile of Elbe estuary as measured during RV Valdivia Cruise, October 1981 (Kempe, 1982b).

4. GLOBAL CONSIDERATIONS

When comparing major world rivers in terms of their average P_{CO_2} and calcite saturation index (Fig. 17) a certain pattern becomes evident (Kempe, 1982a, 1984). Rivers deriving their water from watersheds with a low percentage of sedimentary rock, such as the South American and many African rivers, have a low calcite saturation index. The P_{CO_2} in these rivers appears to be highly variable, depending on the presence of floodplains and their importance for the mainchannel chemistry.

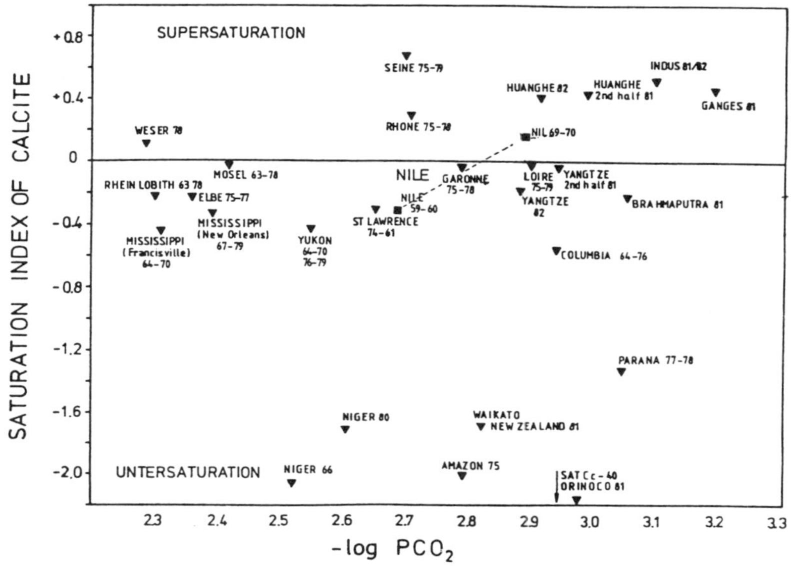

Figure 17. Plot of annual or long-term discharge weighted averages of P_{CO_2} and calcite saturation indices for various major world rivers (Kempe, 1982a, 1984).

Next, several rivers group at high P_{CO_2} and at about saturation with calcite to the left in Figure 17. These rivers (Mississippi, Rhine, Mosel, Elbe, Weser) are highly polluted with labile organic substances, thereby attaining high P_{CO_2}.

The rest of the rivers group at relatively low P_{CO_2} levels; some of them show very high calcite supersaturation levels; others are slightly undersaturated. These rivers mostly derive their waters from drainage basins rich in carbonate rocks. Some of these rivers carry calcite in suspension which effectively buffers them against P_{CO_2} buildup by internal respiration of organic matter. The Seine, for example, is heavily polluted but does not show high P_{CO_2} due to internal buffering with suspended calcite. The Huanghe River (Yellow River) is another river with high calcite supersaturation values. It also is the river with the highest alkalinity of the large rivers (3.2 meq/l, Gan Wei-Bin et al., 1983). Possibly irrigation of rice paddies introduces much labile organic matter to the river causing calcite dissolution of suspended loess-calcite and thus increases alkalinity. On the other hand, the large supersaturation may cause precipitation of calcite on river banks, thus cementing them. It could be this geochemical process which contributes to the aggradation of the Huanghe channel which causes the river to break its levees from time to time, causing catastrophic inundation of surrounding lowlands.

The Nile is shown with two points, one representing its composition before the Aswan Dam was closed and the other representing its chemistry after the dam was closed. The change towards supersaturation is evident.

Total transport of inorganic carbon by rivers gives a minimum estimate of continental weathering induced by carbonic acid. Including continental basins this transport amounts to 0.5×10^{15} g C/yr of which about 1/4 or 0.11×10^{15} derives from silicate weathering, (Garrels and MacKenzie, 1971; Kempe, 1979). This amount plus half of the inorganic carbon derived from carbonate minerals, a total of 0.3×10^{15} g C/yr, was originally present as carbonic acid.

The CO_2 for this carbonic acid must derive largely from biotic sources and cannot be sequestered directly from the atmosphere. Rain in equilibrium with ambient air (pP_{CO_2}=3.5) is capable of dissolving 0.05 mmol CO_2/kg (Stumm and Morgan, 1981). Total continental precipitation is 1.081×10^{17} kg H_2O (Baumgartner and Reichel, 1975). Thus some 0.065×10^{15} g CO_2-C/yr are available for weathering. If weathering occurs on limestone as an open system then about 0.4 mmol/l $CaCO_3$ could be dissolved, of which 0.2 mmol/l would be atmospheric CO_2. Thus, 0.26×10^{15} g CO_2-C could be consumed as an absolute upper limit per year. However, most rain does not fall on open rock surfaces and most rain reevaporates shortly after falling (70% of it is evaporated in total). Even if it falls on rock, this rock is most likely non-carbonate (only 4% of continents are karst areas).

If in fact half of the dissolved inorganic carbon in rivers were of biotic origin, then its isotopic composition should lie between those of limestones ($\delta^{13}C$ = +1 o/oo) and average vegetation ($\delta^{13}C$ = -25 o/oo), at about $\delta^{13}C$ = -12 o/oo. Spiker (1981) gives a few $\delta^{13}C$ measurements of DIC in rivers which show values from -12 to -7.5 o/oo. These findings suggest that exchange of riverine DIC with atmospheric CO_2 (-7 o/oo) is relatively effective.

Case studies show that biogenic CO_2 enters the freshwater system a) in the soil zone (see Plitvice as an example of a headwater system) and b) by internal respiration in rivers (examples Amazon and Rhine). Man has strongly altered the second source. It is estimated that the total change in carbon in the freshwater system may amount to 1×10^{15} g C/yr (Kempe, 1984). The major part of this alteration, however, involves deposition of eroded particulate carbon on land—at slope feet, in alluvial fans, floodplains, inland deltas, reservoirs and lakes (Mulholland, 1981)—and deposition of organic matter produced due to fertilization of lakes, reservoirs and coastal seas. The effects of this change in the P_{CO_2} of rivers and lakes was illustrated by the case studies of the Rhine and Elbe rivers. Fertilization in lakes can cause $CaCO_3$ supersaturation and precipitation; pollution in rivers can induce $CaCO_3$ dissolution either when calcite is available in suspension or when marine carbonates mix into estuaries. The net effect on the rate of weathering of continents, however, is probably small compared to the natural weathering induced by high P_{CO_2} in soil air.

5. ACKNOWLEDGEMENTS

Grateful acknowledgement is given to the Bundesministerium für Forschung und Technologie (BMFT), the Scientific Committee on Problems of the Environment (SCOPE), and to the United Nations Environment Program (UNEP) for continuous financial support.

6. REFERENCES

Ajtay, G.L., P. Ketner and P. Duvigneaud, 1979, Terrestrial primary production and phytomass, *in* B. Bolin et al. (eds.), The Global Carbon Cycle, *SCOPE Rep.* **13**, 129-181.

Baumgartner, A. and E. Reichel, 1975, *Die Weltwasserbilanz.* Munchen, Wien, R. Oldenburg, 179 pp.

Bolin, B., et al. (eds.), 1979, The Global Carbon Cycle. *SCOPE Rep.* **13**, 491 pp.

Bowen, H.J.M., 1979, *Environmental Chemistry of the Elements*, New York, Academic Press.

Bradshaw, A.L., P.G. Brewer, and D.K. Shafer et al., 1981, Measurements of total carbon dioxide and alkalinity by potentiometric titration in the GEOSECS program. *Earth Planet. Sci. Lett.*, **55**, 99-115.

Degens, E.T. (ed.), 1982, Transport of Carbon and Minerals in Major World Rivers, Pt. 1, Mitt. Geol.-Palaeont. Inst. Univ. Hamburg, *SCOPE/UNEP Sonderbd.* **52**, 766 pp.

Degens, E.T. and V. Ittekkot, 1985, Particulate organic carbon—an overview, *in* E.T. Degens, S. Kempe and R. Herrera (eds.), Transport of Carbon and Minerals in Major World Rivers, Pt. 3, Mitt. Geol.-Palaeont. Inst. Univ. Hamburg, *SCOPE/UNEP Sonderbd.* **58**, 7-28.

Degens, E.T., S. Kempe and R. Herrera (eds.), 1985, Transport of Carbon and Minerals in Major World Rivers, Pt. 3, Mitt. Geol.-Palaeont. Inst. Univ. Hamburg, *SCOPE/UNEP Sonderbd.* **58**, 645 pp.

Degens, E.T., S. Kempe and H. Soliman (eds.), 1983, Transport of Carbon and Minerals in Major World Rivers, Pt. 2, Mitt. Geol.-Palaeont. Inst. Univ. Hamburg, *SCOPE/UNEP Sonderbd.* **55**, 535 pp.

Degens, E.T., S. Kempe and A. Spitzy, 1984, Carbon dioxide, a biogeochemical portrait. *in* O. Hutzinger (ed.) *The Handbook of Environmental Chemistry, Vol. 1/C*, New York, Springer, 127-215.

Depetris, P.J. and E. Cascante, 1985, Carbon transport in the Parana River. *in* E.T. Degens, S. Kempe and R. Herrera (eds.), Transport of Carbon and Minerals in Major World Rivers, Pt. 3, Mitt. Geol.-Palaeont. Inst. Univ. Hamburg, *SCOPE/UNEP Sonderbd.* **58**, 299-304.

Drever, F.I., 1982, *The Geochemistry of Natural Waters*, Englewood Cliffs, N.J., Prentice Hall, 388 pp.

Emeis, K., 1985, Particulate suspended matter in major world rivers, II. Results on the rivers Indus, Waikato, Nile, St. Lawrence, Yangtze Parana Orinoco, Caroni, and Mackenzie, *in* E.T. Degens, S. Kempe and R. Herrera (eds.), Transport of Carbon and Minerals in Major World Rivers, Pt. 3, Mitt. Geol.-Palaeont. Inst, Univ. Hamburg, *SCOPE/UNEP Sonderbd.* **58**, 593-617.

Fathi, H. and F.H. Soliman, 1977, Variations in soluble and insoluble matter after the High Dam. *Proc. Sci. Symp. on Effects of the High Dam on Soil Fertility and Water Quality in Egypt, April 16-18, 1977,* **1**, 53-75.

Furch, K. and W.J. Junk, 1985, Dissolved carbon in a floodplain lake of the Amazon and in the river channel, *in* E.T. Degens, S. Kempe and R. Herrera (eds.), Transport of Carbon and Minerals in Major World Rivers, Pt. 3, Mitt. Geol.-Palaeont. Inst. Univ. Hamburg, *SCOPE/UNEP Sonderbd.* **58**, 285-298.

Gan Wei-Bin, Chen Hui-Min and Han Yun-Fang, 1983, Carbon transport by the Yangtze (at Nanjing) and Huanghe (at Jinan) Rivers, People's Republic of China, *in* E.T. Degens, S. Kempe and H. Soliman (eds.), Transport of Carbon and Minerals in Major World Rivers, Pt. 2, Mitt. Geol.-Palaeont. Inst. Univ. Hamburg, *SCOPE/UNEP Sonderbd.* **55**, 459-470.

Garrels, R.M. and C.L. Christ, 1965, *Solutions, Minerals, and Equilibria*, San Francisco, Freeman, 450 pp.

Garrels, R.M. and F.T. MacKenzie, 1971, *Evolution of Sedimentary Rocks*, W.W. Norton Co.,

397 pp.

Hafez, A.A., 1962, Seasonal variations in both irrigation and drainage water in Giza District, M.Sc. Thesis, Fac. of Agric., Cairo Univ. A.R.E.

Hartmann, R., 1983, Carbon and dissolved matter output from a central Harz sphagnum bog, *in* E.T. Degens, S. Kempe and H. Soliman (eds.), Transport of Carbon and Minerals in Major World Rivers, Pt. 2, Mitt. Geol.-Palaeont. Inst. Univ. Hamburg, *SCOPE/UNEP Sonderbd.* **55**, 353-363.

Hedges, J.I., W.C. Clark, P.D. Quay et al., 1985, Compositions and fluxes of particulate organic material in the Amazon River, *Limnol. Oceanogr.*, **31**, 717-738.

Hurst, H.E., 1952, *The Nile—a General Account of the River and the Utilizations of its Waters*, London, Constable, 326 pp.

Junk, W.J., 1985, The Amazon floodplain—a sink or source for organic carbon? *in* E.T. Degens, S. Kempe and R. Herrera (eds.), Transport of Carbon and Minerals in Major World Rivers, Pt. 3, Mitt. Geol.-Palaeont. Inst. Univ. Hamburg, *SCOPE/UNEP Sonderbd.* **58**, 267-284.

Kempe, S., 1975, A computer program for hydrochemical problems in karstic water. *Ann. Speleol.*, **30** (4), 699-702.

Kempe, S. 1979, Carbon in the rock cycle, *in* B. Bolin (ed.) The Global Carbon Cycle, *SCOPE Rep.* **13**, 343-377.

Kempe, S. 1982a, Long-term records of CO_2 pressure fluctuations in fresh waters, *in* E.T. Degens (ed.) Transport of Carbon and Minerals in Major World Rivers, Pt. 1, Mitt. Geol.-Palaeont. Inst. Univ. Hamburg, *SCOPE/UNEP Sonderbd.* **52**, 91-332.

Kempe, S., 1982b, Valdivia cruise, October 1981, carbonate equilibria in the estuaries of Elbe, Weser, Ems and in the Southern German Bight, *in* E.T. Degens (ed.) Transport of Carbon and Minerals in Major World divers, Pt. 1, Mitt. Geol.-Palaeont. Inst. Univ. Hamburg, *SCOPE/UNEP Sonderbd.* **52**, 719-742.

Kempe, S., 1983, Impact of Aswan High Dam on water chemistry of the Nile, *in* E.T. Degens, S. Kempe and H. Soliman (eds.), Transport of Carbon and Minerals in Major World Rivers, Pt. 2, Mitt. Geol.-Palaeont. Inst. Univ. Hamburg, *SCOPE/UNEP Sonderbd.* **55**, 401-423.

Kempe, S., 1984, Sinks of the anthropogenically enhanced carbon cycle in surface fresh waters, *J. Geophys. Res.* **89**, 4657-4676.

Kempe, S. and K. Emeis, 1985, Carbonate chemistry and the formation of Plitvice Lakes, *in* E. T. Degens, S. Kempe and R. Herrera (eds.), Transport of Carbon and Minerals in Major World Rivers, Pt. 3, Mitt, Geol.-Palaeont. Inst. Univ. Hamburg, *SCOPE/UNEP Sonderbd.* **58**, 351-384.

Kempe, S. and U. Lammerz, 1983, Statistical interpretation of hydrochemical data from major world rivers, *in* E.T. Degens, S. Kempe and H. Soliman (eds.), Transport of Carbon and Minerals in Major World Rivers, Pt. 2, Mitt. Geol.-Palaeont. Inst. Univ. Hamburg, *SCOPE/UNEP Sonderbd.* **55**, 39-54.

Leenheer, J.A., 1985, Assessment of organic-solute distribution in water *in* E.T. Degens, S. Kempe and R. Herrera (eds.), Transport of Carbon and Minerals in Major World Rivers, Pt. 3, Mitt. Geol.-Palaeont. Inst. Univ. Hamburg, *SCOPE/UNEP Sonderbd.* **58**, 179-190.

Leenheer, J.A. and E.W.D. Hufman Jr., 1979, Analytical method for dissolved organic carbon fractionation, *U.S. Geol. Surv., Water Res. Invest.* **79-4**, 16 pp.

Meade, R.H., T. Dunne, J.E. Richey et al., 1985, Storage and remobilization of suspended sediment in the lower Amazon River of Brazil, *Science,* **228**, 488-490.

Menzel, D.W. and F.F. Vacaro, 1964, The measurement of dissolved organic and particulate carbon in seawater, *Limnol. Oceanogr.*, **9**, 138-142.

Meybeck, M., 1982, Carbon, nitrogen, and phosphorus transport by world rivers, *Am. J. Sci.*, **282**, 401-450.

Mulholland, P.J., 1981, Deposition of river-borne organic carbon in floodplains, wetlands and deltas, *in* Flux of Organic Carbon by Rivers to the Oceans, U.S. Department of Energy, Washington D.C., 142-172.

Oltman, R.E., 1968, Reconnaissance investigations of the discharge and water quality of the Amazon River, *U.S. Geol. Surv. Circ.* **552**, 1-16.

Paolini, R., R. Herrera and A. Nemeth, 1983, Hydrochemistry of the Orinoco and Caroni Rivers, *in* E.T. Degens, S. Kempe and H. Soliman (eds.), Transport of Carbon and Minerals in Major World Rivers, Pt. 2, Mitt. Geol.-Palaeont. Inst. Univ. Hamburg, *SCOPE/UNEP Sonderbd.* **55**, 223-236.

Pettine, M., T. la Noce, R. Puddu et al., 1983, Tiber River: a picture of its water quality, *in* E.T. Degens, S. Kempe and H. Soliman (eds.), Transport of Carbon and Minerals in Major World Rivers, Pt. 2, Mitt. Geol.-Palaeont. Inst. Univ. Hamburg, *SCOPE/UNEP Sonderbd.* **55**, 281-293.

Pytkowicz, R.M., 1983, *Equilibria, Nonequilibria, and Natural Waters, Vols. I* and II, New York, Wiley, 351 pp. and 353 pp.

Richey, J.E., 1982, The Amazon River system, a biogeochemical model, *in* E.T. Degens (ed.) Transport of Carbon and Minerals in Major World Rivers, Pt. 1, Mitt. Geol.-Palaeont. Inst. Univ. Hamburg, *SCOPE/UNEP Sonderbd.* **52**, 365-378.

Richey, J.E., R.M. Meade, E. Salati et al., in press, Water discharge and suspended sediment in the Amazon River: a sampling procedure and distributions, *Water Res. Res.*

Richey, J.E., E. Salati and U. Dos Santos, 1985, Biogeochemistry of the Amazon River: an update, *in* E. T. Degens, S. Kempe and R. Herrera (eds.), Transport of Carbon and Minerals in Major World Rivers, Pt. 3, Mitt. Geol.-Palaeont. Inst. Univ. Hamburg, *SCOPE/UNEP Sonderbd.* **58**, 245-258.

Schamp, H., 1983, Sadd el-Ali, der Hochdamm von Assuan I, II: Fakten, Ziele, Konsequenzen. *Geowiss. in unserer Zeit* **1/2**, 51-59 and **1/3**, 73-85.

Spiker, E.C., 1981, Carbon isotopes as indicators of the source and fate of carbon in rivers and estuaries, *in* CO_2, Flux of Organic Carbon by Rivers to the Oceans, Washington, D.C., U.S. Dept. of Energy, 75-108.

Stevenson, F.J., 1972, Nitrogen: element and geochemistry, *in* R.W. Fairbridge (ed.) The Encyclopedia of Geochemistry and Environmental Sciences, New York, Van Nostrand Reinhold, Vol. IV, 795-801.

Stumm, W. and J.J. Morgan, 1970, 1981, *Aquatic Chemistry,* New York, Wiley, 1st and 2nd ed., 583 pp. and 780 pp.

van Hall, C.E., J. Safranka and V.A. Stenger, 1963, Rapid combustion method for the determination of organic substances in aqueous solution, *Anal. Chem.*, **35**, 315-319.

Wigley, T.M.L., 1971, Ion pairing and water quality measurements, *Can. J. Earth Sci.*, **8**, 468-476.

Wigley, T.M.L., 1973a, Chemical solution of the system calcite-gypsum-water, *Can. J. Earth Sci.*, **10**, 306-315.

Wigley, T.M.L., 1973b, The incongruent solution of dolomite, *Geochim. Cosmochim. Acta,* **37**, 1397-1404.

Wigley, T.M.L., and L.N. Plummer, 1976, Mixing of carbonate waters, *Geochim. Cosmochim. Acta,* **40**, 989-995.

PLATES

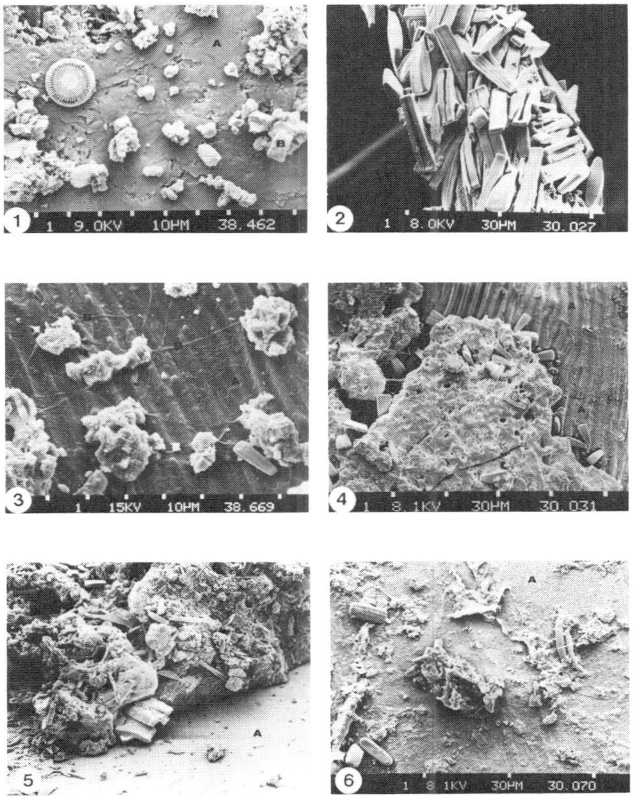

Plates 1-6: Scanning electron-micrographs of Plitvice sediments, mosses, and artificial substrates. Distance between marks on lower bar is indicated in microns. Electron-micrographs prepared by Emeis (Kempe and Emeis, 1985).

1: *Seekreide* particles and cyclic diatom cell from a sediment core (0.2 m subbottom). Calcite particles of hypidiomorphic habitus form aggregates and clusters. A = surface of sample stub (carbon)

2: A sprout of *Bryum*-moss densely settled by epiphytic, non-motive pennate diatoms (*Achnanthes, Gomphonema, Navicula*).

3: Surface of moss (A). Note filaments of mucopolysaccharide excretions agglutinating $CaCO_3$ particles on surface (B).

4: *Bryum* surface (A) and advancing calcite cover in different stages of solidification.

5: Surface of a stainless-steel wire (A) exposed to the water for 2 weeks displays similar incrustations as those on *Bryum* moss.

6: Settling of diatoms and bacteria on the surface of a copper wire (A) exposed next to the steel wire is inhibited due to the toxidity of Cu to organisms. On the surface amorphous CuO, clay minerals, and minor diatoms are visible, but no $CaCO_3$.

WEATHERING AND EROSION IN THE HUMID TROPICS

Robert F. Stallard
Department of Geological and Geophysical Sciences
Princeton University
Princeton, New Jersey 08544
U.S.A.*

ABSTRACT

In the humid tropics, the primary factor which distinguishes different weathering regimes is tectonic setting, especially the contrast between tectonically active areas and stable cratonic regions. In tectonically active environments, most of the material that is exposed to weathering has undergone rapid uplift involving brittle deformation, and volcanism is sometimes important. Uplift rates are on the order of 10^3-10^4 mm/ka, while erosion rates are about the same to an order of magnitude less. As a result, easily weathered lithologies are exposed on steep slopes and weathering rates are lithology dependent. Where carbonates and evaporites are present, their weathering products dominate the river solution chemistry. Similarly, weakly cemented clastic rocks dominate solid load inputs. Aluminosilicate rocks do not weather completely to phases depleted in the major cations (Na, K, Mg, Ca); instead, some unweathered primary minerals remain, and various secondary cation-rich clays frequently form. On the cratons, sea level changes exert a major control on long term weathering rates and landscape development. When sea level is high, sediment is deposited on the passive margins, continental platforms, and intracratonic basins. At the same time near-sea-level planation surfaces develop on shield areas. As sea level drops erosion is not regionally uniform, but is localized on scarps which advance into the older erosion surfaces. For South American shields average uplift rates approach 10-20 mm/ka, and it seems reasonable that their long-term average erosion rates are close to this value. Weathering on the craton is much more complete, and the solid weathering products are cation depleted: quartz, kaolinite, and sesquioxides. After cyclic salt corrections, major cations in rivers are in bedrock proportions.

1. INTRODUCTION

The humid tropics make a particularly important contribution to global erosion. According to Meybeck [47], the humid tropical zone presently occupies about 25% of the earth's land surface, and supplies about 65% of the dissolved silica and 38% of the ionic load delivered by rivers to the ocean. Data compiled by Milliman and Meade [48] indicate that the same region contributes roughly 50% of the total river solid load to the ocean. Both studies show that the bulk of this material is derived

* Present address: U.S. Geological Survey, MS 413, Denver, Colorado 80225, U.S.A.

A. Lerman and M. Meybeck (eds.), Physical and Chemical Weathering in Geochemical Cycles, 225–246.

from active orogenic belts and island arcs. The present paper examines climatic and tectonic controls on erosion in the tropics and the implications of these observations regarding the composition of erosion products.

Many erosion-related phenomena are often episodic, infrequent or so exceedingly slow that the time scale of human existence is inadequate for direct observation. Consequently, their rates and effects are very hard to gauge directly. The key to understanding weathering and erosion in the tropics is to decipher the relationship between landforms, the processes that produce them, and the chemistry and discharge of river-borne materials. To a first approximation, landscape forming processes can be viewed as being controlled by climatic and tectonic factors. Climate delimits where tropical weathering can occur, while tectonics controls how the effects of weathering are expressed. The relationship between tectonic setting and tropical weathering is emphasized in the following discussion because tectonic history is often critical in controlling erosion rates and the composition of erosion products.

Examination of erosion on hill slopes provides a means of linking the composition of erosion products to processes that control landscape development. To a first approximation, mass movement processes on slopes occupy a continuum between weathering-limited and transport-limited [9,72,74]. For the former, the capacity of the transport process exceeds the rate at which weathering generates material, while for the latter, the rate of supply of material by weathering exceeds the capacity of transport processes to remove the material. Weathering-limited situations are associated with thin soils and steep straight slopes which often undergo parallel retreat at a threshold angle. Since erosion rates are controlled by susceptiblity to weathering, and since fresh rock is always near the surface, rapid removal of partially weathered material is typical. In contrast, thick soils and slight convexo-concave slopes are characteristic of regions where transport-limited conditions dominate. With time, slopes tend towards increasing flatness. Weathering rates are ultimately limited by the formation of soils which are sufficiently thick or impermeable to restrict free access by water to unweathered material. Erosion rates are low, and soils and solid weathering products are cation-poor.

These two styles of weathering present an interesting parallel to weathering reaction rate controls on mineral surfaces. Again a continuum can be defined, here between surface reaction control and transport (diffusion) control [4]. In the former case, reaction rates are controlled by local surface energy on the crystals. Under the microscope, surfaces appear rough and pitted often on zones of obvious crystallographic defects. In the latter case, reactions are limited by diffusion of solution products away from the crystal surfaces, which appear smooth with rounded edges.

The effects of vegetation are complex [72]. Vegetation can greatly accelerate erosion on steep slopes by anchoring soil. The anchored soil traps moisture and bioacids, thereby facilitating further weathering and soil accumulation. This in turn leads to instability, landslides, and a new cycle of accumulation. On slight slopes, over very extended periods of time, vegetation may reduce weathering rates by allowing very thick soils to accumulate.

2. CLIMATE AND VEGETATION IN THE TROPICS

The high temperatures, moist conditions, and associated luxuriant vegetation of the tropics appear ideal to promote rapid chemical weathering. Accordingly, climate history of the moist tropics can be best described using vegetational patterns. Walter [86] states that rain forests occur where moisture supply is continuous, either on the meteorological equator or in areas of orographic precipitation. Moving away from the equator, annual precipitation decreases and the duration of the dry season increases. Correspondingly, one first crosses a zone of deciduous forests, then dry savannas and thornbush, and finally desert. If the soil is too well-drained or overly nutrient-depleted, grasslands may occur in areas that otherwise would be forested. Tropical forests and moist savannas are generally found where precipitation exceeds 1000 to 2000 mm/a and mean annual temperatures are greater than about 18 °C.

The regional extent of tropical vegetation zones has varied considerably through time in response to climate change. According to Wolfe [87], since the Cretaceous there has been an irregular long-term cooling trend which ended with the major short-term (20-100 ka) temperature oscillations associated with the Pleistocene glaciations. In the mid-Cretaceous, tropical vegetation grew in what are today temperate regions. During the early Cenozoic, cooling was gradual, except that the Miocene appears to have been a brief warmer episode. The most prominent cooling occurred in the late Eocene accompanied by an increased seasonality of temperature. Climate style changed dramatically in the Pleistocene (about 3 Ma BP) with the onset of the major continental glaciations.

The distribution of tropical vegetation from the last glacial epoch until the present is reasonably well understood through palynology, geomorphology, and the examination of the distribution of relict flora and fauna [12,20,42,50,61,79, 82,85,86,89]. During the peak glaciation (18 ka BP), deserts were much more extensive and tree lines were much lower than they are now. At the same time, extensive areas of rain forest in Amazonia and other tropical areas appear to have been replaced by savanna. Bracketing 6 ka BP, in what is referred to as the hypsithermal or climate optimum, tropical forests and savannas expanded considerably into what are now drier vegetation zones, and tree lines were higher than their present elevations. These changes may have been associated with a stronger monsoonal type of circulation than at present. Areas of moist vegetation subsequently retreated to their current limits. It is reasonable to expect that roughly the above pattern has been repeated for each of the Pleistocene glacial cycles.

3. TECTONISM, SEA LEVEL, AND EROSION

Tectonic regimes related to tropical erosion cannot be as easily outlined as climatic regimes. In the following sections, tectonically active and tectonically quiet areas are discussed separately to highlight differences in the nature of erosion between these two types of settings. An important aspect of this difference is the degree to which eustatic sea level fluctuations influence erosion.

The ocean surface represents the master base level for continental erosion and sedimentation. Given a sufficient period of time in the absence of tectonic processes, continents would presumably be planed off roughly to sea level. Eustatic sea level, however, is known to fluctuate significantly [76]. If these fluctuations are superimposed on crustal uplift or subsidence, the resultant base level changes often, producing characteristic erosional and depositional features. In the case of sustained uplift, an apparent sea level still stand occurs whenever the rate of eustatic sea level rise matches the rate of uplift. With the base level so fixed, either depositional or erosional terraces may form. Presumably the length of the still stand is related to the degree of terrace development. After sea level falls and rises again, the terrace may have risen sufficiently to be preserved upslope. Episodic uplift produces similar terraces, but these should not be synchronous with terraces developed on other widely separated terrains, unless episodes of uplift were being controlled by some deep-seated process.

Figure 1 displays uplift curves, representing the elevation of various sea level indicators versus time, for both tectonically active (A, B, C) and quiet (D, E, F) regions. Note that uplift rates for Taiwan, a fold and thrust belt, exceed estimates for shield areas in South America by almost three orders of magnitude. If we assume that the hypsographic curves for any of these regions have remained fixed through time (that is, the region was in a steady state), then uplift rates translate into erosion rates. Because the climatic and the tectonic processes that govern landform development do not operate constantly and continuously, actual contemporary erosion rates may differ significantly from these values.

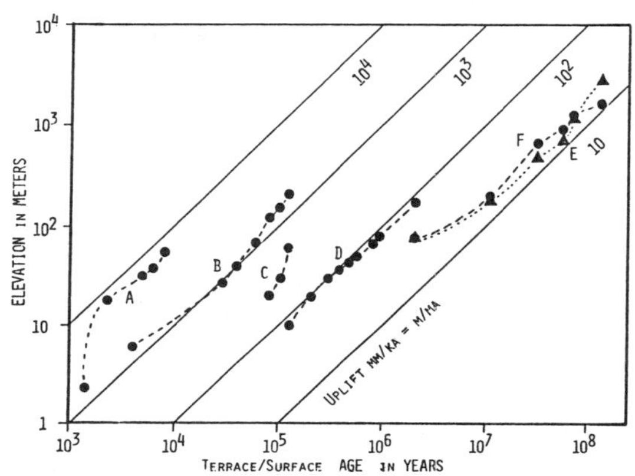

Figure 1. Graph of the elevation of terraces and other sea level indicators versus their age for several locations. To derive actual uplift rates from this type of data, one needs to know the age of the terrace and the eustatic sea level at that time. A: Uplift data for Taiwan constructed using data from Peng et al. [53] for the Hengchun Peninsula, Taiwan, and Eastern Coast Range, corrected for glacio-eustatic effects and grouped by roughly 1500 year intervals. Uplift appears discontinuous, but averages to about 5000 mm/ka. B: Calculated composite terrace elevations for the Huon Peninsula, New Guinea using data from Bloom et al. [6]. Uplift is continuous and ranges from 940-2560 mm/ka; the calculation used 1620 mm/ka. C: Data for Barbados (Clermont traverse) from Matthews [44]. The rate of uplift is approximately 400 mm/ka. D: Data for depositional terraces in the Amazon Trough from Klammer [37]. For this figure, ages are estimated from those of the most pronounced interglacials on the standard oxygen isotope curve. The calculated uplift rate is about 80 mm/ka. Klammer argues this is not uplift but instead reflects a sea level drop since the Pliocene. E and F: Erosion surfaces from the Guyana and Brazilian shields respectively (King [35], McConnell [45], and Aleva [2]). Uplift rates are 10-20 mm/ka. Locations for C, D, E, and F are given in Figure 2.

Because of sediment storage as alluvium or colluvium, solution chemistry of rivers provides the best gauge of current weathering for a particular terrain. Total dissolved solid concentrations (TDS) can be used to estimate ranges of erosion rates for particular regions. Histograms of dissolved solid concentrations for different morphotectonic regions of the Amazon and Orinoco basins (Figure 2) are presented in Figure 3. The very approximate erosion scale is calculated as the product of dissolved load concentrations, mean annual runoff (1000 mm/yr), and a correction factor to account for high suspended to dissolved load ratios in Andean rivers and for the higher than average annual precipitation in the lowland shields. The factor was treated as a linear function of dissolved solids and ranged from two for the most dilute rivers to four for the Andean Rivers. Bedrock density was assumed to be 2.65 gr/cm^3. These denudation rates are in general agreement with more rigorous calculations [21,22,71]. Comparison with Figure 1 shows a reasonable match between erosion rates and uplift rates for a particular type of terrain. The most concentrated water samples and highest erosion rates are observed in river basins in tectonically active areas.

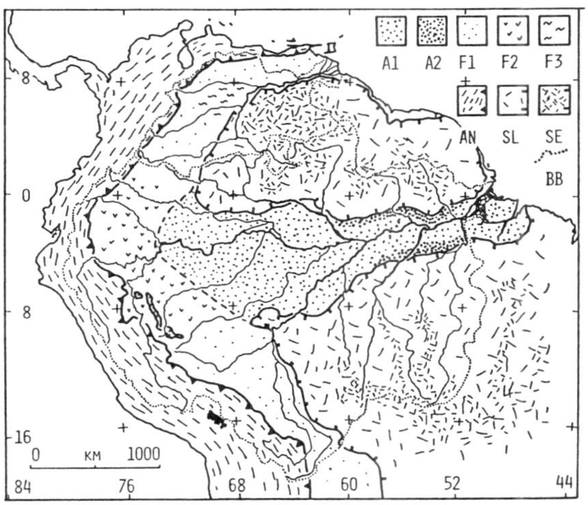

Figure 2. Morphotectonic map of northern South America [72]. In the east is the South American craton consisting of the Guyana and Brazilian shields (SL = low-lying; SE = elevated erosional remnants), between which is located the Amazon Trough intracratonic basin (A1). The northeast coast is a typical passive margin. On the eastern end the trough has exposed Paleozoic and Mesozoic sediments including carbonates and evaporites (A2). To the west is the Andean foreland basin (F1; F2 = sediments with volcanic ash; F3 = dunes) and the Andean Cordillera (A1). BB indicates a drainage basin boundary. In addition, x = location of Pleistocene terraces described by Klammer [37]; y = Merida Andes described by Giegengack; z = Barbados described by Matthews [44].

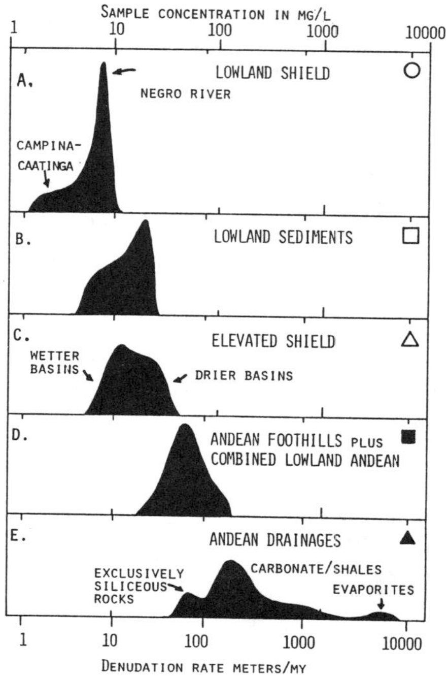

Figure 3. Histogram of sample dissolved solid concentrations with corresponding denudation rate scales [72]. Symbols used in subsequent figures are indicated in the upper right of each panel.

4. EROSION IN TECTONICALLY ACTIVE REGIONS

To a first approximation, fast tectonic processes occur at plate boundaries, while slow ones occur in plate interiors. The primary exceptions to this are incipient rifts and the sites of the so-called hot spots or mantle plumes, both of which are associated with uplift and volcanic activity in plate interiors.

Suppe [77], Davis et al. [16], and Dahlen et al. [15] have modeled the effects of brittle deformation in accretionary fold and thrust mountain belts such as Taiwan and the Andes. The basis of their model is that rock deformation is governed by pressure-dependent, time-dependent brittle fracture or frictional sliding—"Coulomb behavior". These investigators argue that the overall topographic profile of such belts evolves into a stable form such that erosional outputs balance accretional inputs [77,16]. Consequently, there should be a continuous supply of easily eroded material so long as accretion continues. Suppe [pers. com.] believes that mountain belts, for example the Andes, might tend to be wider where erosion rates are reduced such as in a region of dry climates.

In many active mountain belts, both long term rapid uplift and erosion are sustained by tectonic recycling of sediments. For example, the island of Taiwan (Figure 1, A), which is a fold and thrust belt, has been formed over the last several million years during an ongoing collision between the Luzon island arc and the Eurasian continental margin [11,13,77]. During this orogeny, metamorphosed and diagenetically altered sediments have been uplifted, eroded, and deposited, only to be reincorporated into the cycle once more [13]. Some of the highest erosion rates (13,000 t/km^2/a solid

+ 650 t/km^2/a dissolved, equivalent to 5150 mm/ka) for river catchments anywhere have been reported for the central range of Taiwan [40]. This probably reflects the poorly lithified highly tectonized nature of the sedimentary rocks which compose the island.

Episodes of very rapid uplift following prolonged erosion seem to be characteristic of some cordilleran mountain systems. Giegengack [24], for example, has identified a paleo-oxisol high in the Venezuelan Andes that may have been part of a late Neogene (about 2 Ma) near-sea level erosion surface formed after previous episodes of deformation [45]. Remnants are found at 3000 m near Pico Bolivar and project over its top at 5000 m. If it is assumed that the paleosol was raised 5000 m over the last 1-2 Ma, the uplift rate would be 2500-5000 mm/ka. Valleys are now deeply incised into the surface defined by the paleosol, suggesting an erosion rate perhaps an order of magnitude less than the uplift rate.

River chemistry is consistent with weathering-limited erosion. In the Andes, sedimentary rocks constitute the principal basement lithology, and the river chemistry agrees with catchment geology [71,72,74]. For example, black shales have particularly high Mg:Ca ratios, and Bolivian rivers which drain black shales are exceptionally magnesium-rich. Rivers draining evaporites have the highest total cation (TZ+) concentration, followed by those draining carbonates, and finally by those draining only siliceous rocks. This is illustrated by the trimodal nature of the histogram of Andean TDS in Figure 3-E and by the ternary diagram in Figure 4. Here, the parameters chloride + sulphate, alkalinity, and silica are used as input markers for the respective lithologies, bearing in mind that alkalinity is also produced by silicate weathering and sulfate by pyrite weathering. Total cations increase systematically from silica to alkalinity to chloride + sulfate. Many Andean rivers are slightly supersaturated with respect to calcite and to smectite [71,75], an indication that saturation with respect to these phases ultimately limits the supply of alkalinity and dissolved silica to the rivers. Similar limits do not apply to Na, K, Mg, and Ca because of additional inputs from silicates and evaporites. Erosional contributions by carbonates and especially by evaporites greatly exceed those expected if inputs are assumed to be proportional to the fraction of the catchment area covered by these lithologies. Much of the evaporite input is sustained by actively extruding salt diapirs [3,71].

The presence of unstable and cation-rich minerals in the suspended and bed loads of rivers which drain the Andes [19,21,22,27,29,59] also suggests that extraordinarily rapid erosion is occurring. This is evidenced by the presence of carbonates in the bed load of the lower Amazon [19,59], the only source of which could be the Andes. Inputs of clays and other minerals correlate with geology. Irion [27,29,30] shows that much of the smectite in foreland basin rivers is derived from the weathering of volcanic glass. Andean rivers often acquire an intense red color only after crossing regions with red beds. Furthermore, the suspended load of rivers in the Peruvian Andes which have not yet crossed the red beds is rich in vermiculite and mica, while in the lower courses of the same rivers it is rich in smectite and kaolinite. Finally, the marine shales which are abundant in the Bolivian, Colombian, and Venezuelan Andes are very micaceous [71], and rivers draining these regions are particularly illite-rich.

Data from New Guinea show similarly high erosion rates dominated by inputs from susceptible lithologies. The Puari River catchment in Papua is described in detail in an excellent book edited by Petr [54]. Pain [51] notes that "large stratovolcanoes, bold limestone scarps and deeply incised gorges are marked features of the landscape. Slopes are predominantly steep ($>35°$) and lead down to rivers that almost always flow as swift muddy streams." Average annual precipitation is 3500

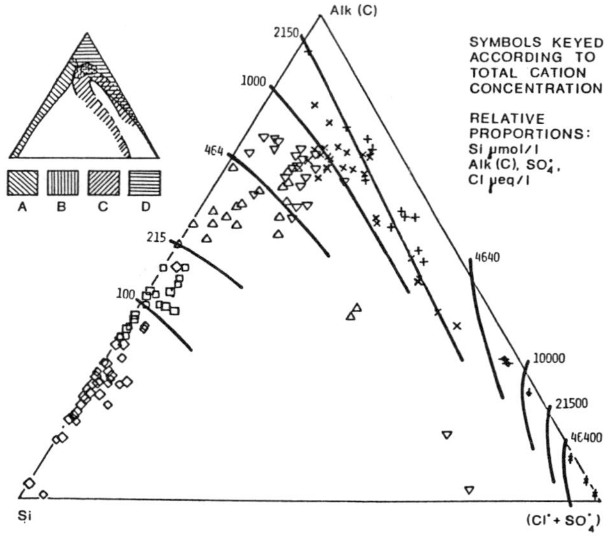

Figure 4. Ternary diagram relating silica, carbonate alkalinity, and chloride + sulfate in the Amazon and Orinoco systems. Analyses are cyclic salt corrected. Curves are numbered in total cation concentration (μeq/l). The predominant symbol within each interval corresponds to samples whose concentrations fall in that interval. A - cratonic; B - Amazon mainstem; C - Andean, black shales; D - Andean, carbonate and evaporite rich sediment.

mm and runoff is 2700 mm [18]. River-water chemistry is dominated by carbonate weathering [55], while solids contain abundant cation-rich clays (smectites, illite, chlorite) [31,56]. Erosion rate by solution is about 200 tons/km^2/a (excluding 50% bicarbonate load as atmospheric) [55] and solid erosion rate is about 3000 tons/km^2/a [57]. This converts to a roughly 1200 mm/ka denudation rate. Pain [51] notes that poorly lithified and highly tectonized shales erode more rapidly than any other rock type and that the volcanic ash cover in parts of the basin may actually reduce solid erosion rates owing to its high permeability. Ruxton and McDougall [62] used erosional remnants on interfluves to reconstruct the original contours of the Hydrographers volcano in Papua-New Guinea. Then, using K/Ar dating and present topography, they calculated denudation rates averaged over the last 650 ka. These ranged from 80 mm/ka at 60 m of original relief to 750 mm/ka at 760 m of relief.

5. EROSION OF THE CRATONS

The term craton is generally used to refer to tectonically quiet or stable continental areas. Major subdivisions include shields where long-term erosion has exposed extensive areas of old crystalline basement, platforms where shields have a flat-lying sedimentary veneer, intracratonic basins where slow long term subsidence has led to thicker deposits on the craton, and passive continental margins where continental crust has rifted, separated, cooled, and subsided. The boundary between cratons and mountain belts often involves a foreland basin where basement has slowly subsided as a result of sedimentary loading and tectonic downwarping. Intracratonic basins, passive margins, and foreland basins, represent loci of long-term sediment accumulation and storage [60,65,66,68,80,81]. Intracratonic basins subside through especially long stretches of geologic time. The Amazon Trough, for example, has been active for all of the Phanerozoic [68]. Several investigators have argued that there is a major tectonic component to the subsidence histories of passive margins and intracratonic basins [58,66,67,76]. Sloss and Speed [67] and Sloss [66] noted that subsidence episodes appear to be globally synchronous and coincident with high sea levels. They argue that simple sedimentary loading caused by deposition at times of high sea levels is not an adequate explanation and suggests that some common deep driving mechanism controls both sea level and subsidence.

5.1. Terraces and Erosion Surfaces

Two principal models have been put forward to describe the development of extremely eroded topography following a drop in base level in a region that had once been planed flat [32]. One is the classic Davis cycle [17], the other is a model developed by King after Penck [33,36]. These differ in their prediction of how hill slopes erode into the raised topography.

According to Davis, rivers first cut steep valleys into the landscape, then their valleys broaden, and regional slopes are flattened until the last remnants of the original surface are worn away at interfluves. At this point the landscape is said to be mature. Slopes continue to flatten until formerly elevated interfluves become low swells. This is a peneplain. In King's model, initial slopes are formed along the edges of the uplift and the sides of large penetrating river valleys. Slopes then evolve into steepened equilibrium forms which separate the old erosion surface from an incipient new surface. As time proceeds, these slopes undergo parallel retreat into the older surface which is consumed while the younger surface is extended. The end product is a pediplain. The steep slopes are the locus of most of the erosion and the remaining terrain would be almost flat, being drained by rivers having little erosive capacity. In the Davis model, no remnants of previous surfaces can exist. In King's model, however, successive erosional levels can remain stacked and can even actively expand through parallel retreat into the older surfaces long after the initiating change in base level has been superseded.

A search for stacked erosion surfaces became the centerpiece of work by King [36], who compiled spectacular examples on a continental scale in South America, Africa, India, and Australia. In a sense, these surfaces are super-terraces, each falling into a narrow range of altitudes for a particular region. Surfaces nearest sea level are well defined and undissected, while the most elevated are usually remnants of small extent or are simply delineated by a large number of hills having peaks which fall into a narrow altitude range and which are topped by deeply weathered soils. In South America (Figures 1 and 2) and Africa at least five levels seem to be identifiable. King showed that to a first approximation the surfaces are globally synchronous (within about 10 Ma) and that the oldest ones are of great age, perhaps predating the rifting of the South Atlantic (late Jurassic to early Cretaceous).

Subsequent geomorphic studies and work related to bauxite exploration has led to better dating and descriptions of relationships between erosion surfaces in South America [1,2,45,46,88]. The highest grade bauxites are found on the Tertiary (55 Ma) surfaces. This is because older surfaces are of small extent and very dissected, while on younger surfaces the bauxite becomes progressively more ferruginous. Where the substrate is quartzose, quartz persists through the entire weathering profile, which may be as much as 50 m thick on the 55 Ma surface [46].

These surfaces are dated by classical stratigraphic techniques [1,2,34,35,36,45]. Deposits of fluvial, lacustrine, or eolian sediments containing dateable pollen or vertebrates are occasionally found on the surface. As this is not common, dating is usually done by following a surface into an area of subsidence, such as a passive margin or an intracratonic basin. Occasionally, the surfaces themselves become sufficiently downwarped to be buried under marine sediments. More often, such as off of the northeastern South American coast, erosion surfaces coincide with hiatuses in the sedimentary section, these are frequently overlain by deposits of quartz and bauxite gravels and quartz sands. This is thought to be indicative of the early dissection of deep and highly weathered soils following a drop in base level. Such hiatuses are not unlike those associated with sea level high stands seen in coastal onlap sequences. Accordingly, Aleva [2] notes that the major surfaces appear to coincide with major high stands on the late Mesozoic-Cenozoic sea level curve.

The most recent and youngest surfaces are 10 or so glacio-eustatic terraces from the Amazon valley described by Klammer [37] (Figures 1 and 2). The uppermost terrace is equivalent to the youngest of the major erosion surfaces. Klammer notes that in the Amazon Trough this surface is purely depositional and asserts that this also must have been the case for the erosion surface on the shields. As no extensive Late Neogene sedimentary deposits are found on the shields, this scenario seems unlikely. It would be hard to see why sediment deposited on crystalline basement should be entirely removed by erosion while that deposited on preexisting sediment is moderately intact. Instead, on the shield, crystalline rocks of different lithologies are completely beveled off indicating very effective subaerial erosion controlled by a fixed base level.

5.2. Erosion and the Slow Uplift of the Shields

It appears that vast areas of cratons are undergoing a slow but sustained rise relative to sea level. As they rise, they erode, leaving behind remnants of erosion surfaces as evidence of their former thickness. It would be interesting to speculate that this tendency to rise has held true for many shield areas through much of geologic time, explaining the high metamorphic grade of rocks exposed on the surface. In this sense, shields would be dynamical opposites to intracratonic basins. Continental platforms are then areas that neither rise nor sink dramatically. Tectonic modes change with time. For example, the thick quartzites that top many of the highest peaks of the Guyana shield are evidence that this area once behaved as either a platform or a basin.

Part of the rising and sinking of cratons may be due to isostatic adjustments in response to erosion and to sediment accumulation. Part, however, may be due to deep-seated processes. Burke and Wilson [7] have argued that the exceptionally elevated nature of the African craton is due to its slow motion movement over the mantle relative to other continents. This presumably reduces heat loss via mantle convection and leads to heating and thermal expansion of the African crust. Crough [14] proposed that uplift and erosion of shields is caused by the randomly repeated passage of continental crust over mantle plumes. As a segment of crust passes over a mantle plume, it becomes hot and more buoyant. This produces a fairly rapid uplift followed by a slow reduction of elevation as a result of combined erosion and cooling.

If erosion is sufficiently intense, the segment of crust becomes a depositional basin once it has completely cooled. Morgan [49] used this idea to examine the relationship between hot spots and regional geology in the continents bracketing the Atlantic. He located a hot plume under the Guyana shield to explain its presently elevated nature.

The appearance of globally synchronous erosion surfaces and slow uplift suggests that models based on localized heating of the crust and mantle are not adequate to explain uplift of shields. To produce near simultaneous surfaces on separate continents, King [36] invoked global epeirogeny. Once scarps are initiated, however, they may become self sustaining to some degree, for as they retreat, isostatic adjustment would raise the topography [34], thereby stabilizing the effects of the initial base level change. The isostatic uplift would be cumulative, thus explaining the stacked topography. Global sea level change seems to be a more reasonable initiating event, but it is still a matter of speculation as to whether the uplift is continuous or episodic and self sustaining or tectonically driven. There is, however, an interesting parallel between King's view of erosion surfaces, and Sloss and Speed [67] and Sloss's [66] idea that there is a tectonic component to the subsidence of sedimentary basins at times of high sea level.

5.3. Weathering-Limited Erosion on The Elevated Shield

The best exposures of elevated crystalline basement in the humid tropics are the Brazilian and Guyana shields (Figure 2). The topography can be spectacular with steep slopes, either of bare rock or thinly vegetated, often topped by high plateaus, some of which are quite expansive. Inselbergs are common in some areas. Great talus piles do not accumulate below cliffs and at the base of slopes. Much of this material probably just dissolves. A karst-like topography developed on quartzite, gneiss, and granite is found on some of the higher and presumably most ancient areas [5,78]. The karst on quartzite is established to such a degree that stalactites consisting of opal, cristobalite, tridymite, and quartz have formed locally [10]. Aerial and radar photographs show that all but the youngest surfaces are dissected to various degrees, and that this dissection is geologically controlled. Many rivers appear to follow fault systems and dyke swarms. This would indicate that the older surfaces are no longer expanding in the manner described by King, since the dissected surfaces can no longer serve as fixed base levels for overlying scarps. Especially resistant lithologies such as quartzites and some intrusives persist as high elevations. Thus, even though weathering rates are very slow (Figure 3, C), the observation that lithology plays an important role in landscape morphology suggests that weathering limited erosion is occurring.

Rivers from the shields typically carry very little dissolved or solid load, and the solids tend to be more resistant mineral phases [21,22,27,29,71,72]. River sands include more potassic feldspars as might be expected if weathering-limited erosion is occurring on the steep slopes. Erosion rates in these terrains are much lower than in active mountain belts like the Andes, even though elevations in the Guyana Shield reach 3000 m, and 2000 m in the Brazilian Shield. Both lithology and basement structure contribute to the low erosion rates in elevated crystalline basement terrains [72]. Rocks are frequently massive and made of minerals, mostly quartz, potassium and sodium feldspars, and micas, that are somewhat resistant to weathering. Uplift apparently involved ductile (plastic) deformation and was followed by a long history of weathering which may have removed all less stable mineral assemblages. The basement of active orogenic belts is not so massive owing to faulting and brittle deformation. Furthermore, active uplift provides a continuous supply of fresh material for weathering.

5.4. Transport-Limited Erosion on the Lowlands

Vast tracts of the South American (Figure 2) and West African lowlands can be classified as having dominantly transport-limited denudation regimes [71,72,74]. These regions represent the flattest and youngest erosional and depositional surfaces of the Late Neogene. On the shield, the substrate is crystalline basement, whereas in the intracratonic basins, the Andean foreland basin, and the coastal plains much of the sedimentary substrate consists of strongly weathered fluvio-lacustrine sediment [70]. Soils and solid loads of rivers are rich in quartz, kaolinite, and iron sesquioxides, all of which are cation-depleted phases. River sands are rich in quartz [19,59], and the suspended load is largely kaolinite plus quartz with or without iron sesquioxide coatings [21,22,23,27,29]. Aluminium sesquioxides (gibbsite) are much less common. Most of the cation load is in solution. Dissolved solid concentrations, and by inference denudation rates, are exceedingly low (Figures 3-A,B and 4). This agrees with the diminution of weathering rates in conjunction with the development of thick soils. Dissolved loads are lower on the shields than on the fluvio-lacustrine sediments, even though rocks of the shield are composed of less stable minerals. However, the low permeabilities of the rocks and many of the soils of the shields compared to those of the sediments apparently counterbalances this.

Where easily weathered rocks, such as carbonates and evaporites are near the surface, as in the lower Amazon valley (Figure 2), their contribution to the rivers appears minor, probably because thick residual soil covers have developed. This indicates that susceptability to weathering is indeed less important in controlling lowland erosion rates. Carbonate platforms, such as southern Florida and the Yucatan in North America, are exceptions because of deep water circulation through the carbonate karst.

In the lowlands of South America, the relationship between soil, geology, vegetation, and climate is complicated. Where the protolith is either crystalline rock or fluvio-lacustrine sediments of the Amazon Trough and parts of the Andean foreland basin, soils consist of kaolinite, quartz, and iron and aluminum sesquioxides [51,70,83]. The proportions of these components may in some way be related to zones of vegetation and climate [71,72,74]. Areas with the driest climate and the longest dry season have the most aluminous soils; the wettest areas are the most siliceous. In the savanna, rates of physical denudation are greater [63,89], and soils are definitely more aluminous than forest soils [8,71]. Abundant gibbsite is reported in some savanna soils; although, classic laterites and bauxites are rare [70,83]. Except in areas of unusual substrate, forest soils consist mainly of kaolinite and quartz, and they tend to be richer in organic matter than savanna soils [43,64,69]. Soils consisting of almost pure quartz sand (tropical podzols, spodosols) are found straddling the meteorological equator, where there is little dry season. These soils form on quartz-containing bedrock in very flat areas and are associated with so-called campina and caatinga vegetation. Rivers draining these regions are exceptionally brown, acidic, cation-poor, and sediment-free [38].

The positive correlation of soil silica content to wetness is opposite to what is conventionally assumed to be the case. This relationship is quite rough, as both the savannas and the more aluminous soils are on the older erosion surfaces, which are coincidently drier. Moreover, drier areas were far more extensive in the tropics during glaciations. To an approximation, the amount of water moving through soils in wet as opposed to dry regions probably varies by a factor of about four. If it is assumed that the rate of quartz dissolution varies to the same degree, then the high quartz content of wet lowland soils must reflect more effective loss of iron and aluminum.

A primary factor contributing to increasing loss of aluminum and iron from soils in progressively wetter regions is probably organic acids and complexing agents [71]. Laboratory studies have shown that at temperatures and pH values characteristic of the lowlands, recrystallization of iron [39] and aluminum [84] is inhibited by organics in solution. Moreover, organics may assist in the formation of crystalline kaolinite, perhaps by facilitating the formation of Al-O-Si bonds [26,41], and if organics are not present, amorphous aluminum hydroxide or gibbsite often forms. The prevalence of kaolinite in forest soils over savanna soils may be related to higher levels of organics. The role of organic matter is clearly seen in the case of iron, where Lopez and Bisque [43] note that in surface horizons of savanna and forest soils, amorphous iron oxides recrystallize into goethite only where the carbon content is less than 0.5%. The role of organic matter is most pronounced in the tropical podzols where Al and Fe have been completely removed.

In many areas along the Andean foreland basin (Figure 2), sedimentary substrates contain carbonates and unstable aluminosilicates, and both soils and sediments sometimes have a volcanic ash component. Consequently, soils are much richer in cations, and streams draining these areas have higher TDS (Figure 3-D) as well as smectites and illites in their solid load [28,30,31,70,71,74].

Table I. Mineral Stability in Tropical Soils

MOST STABLE
Quartz* >>
K-Feldspar*, Micas* >>
Na-Feldspar >
Ca-Feldspar, Amphiboles >
Pyroxenes, Chlorite >
Dolomite >
Calcite >
Gypsum, Anhydrite+ >>
Halite+
LEAST STABLE

* Quartz, micas, and K-feldspars are stable in hard saprolites developed on acid to intermediate igneous and metamorphic rocks. Quartz frequently persists through entire soil profiles.

\+ Gypsum and anhydrite are observed in soils developed on salt domes in the Huallaga Basin; halite, however, is not [72].

6. ELEMENTAL PARTITIONING

There are two principal ways to partition elements between the dissolved and solid loads: by the formation of secondary phases that are enriched or depleted in certain elements, relative to bedrock, and by the selective weathering of particular phases [72]. A stability sequence for common primary minerals based on data compiled by Pasquali et al. [52], Lopez and Bisque [43], and Stallard [71] is given in Table I; it is very similar to the sequence defined by Goldich [25]. Minerals may weather congruently to produce only dissolved products or incongruently to produce both dissolved cations (Na, K, Mg, Ca) and solid cation-depleted products. Common minerals that always weather congruently are halite, gypsum, aragonite, calcite, dolomite, and quartz. Halite, anhydrite, and gypsum are so highly soluble that they are almost never found in the solid load of rivers. Aragonite, calcite, and dolomite are sometimes found in alkaline rivers, supersaturated with respect to these minerals. Quartz is the most persistent of all common primary minerals and is found in most river sediments. Minerals which contain iron and aluminum usually weather incongruently to produce clays, and iron and aluminium sesquioxides. Magnesium, potassium, and to a lesser extent calcium and sodium are retained in cation-rich clays such as the smectites, illites, and vermiculites.

Aggradation or degradation of biomass or soil reservoirs may also lead to what appears to be fractionation. This is owing to the elemental ratios in vegetation or soil reservoirs that can be very different from those of bedrock. Sufficiently large and rapid changes in these reservoirs would be evident in river chemistry.

Different styles of erosion are associated with different degrees of partitioning of elements between dissolved and solid load. As rocks weather chemically, they lose their structural integrity. On crystalline rocks and indurated sediments, solifluction, soil avalanching, and sheet runoff erosion in moist vegetated areas involve poorly cohesive material (solum and soft saprolite), leaving a cohesive hard saprolite behind [71,72,74]. A few minerals such as zircon, quartz, potassium feldspars, and micas remain in the soft saprolite; thus, K and Mg would be enriched in solid erosion products where mechanical erosion is sufficiently intense. Na and Ca would be enriched in solution. Mg is incorporated into the lattice of many secondary clays, further accentuating its retention in the solid phase. When only kaolinite or gibbsite or other cation poor phases are forming, cation ratios in solution should match those in bedrock.

When K and Mg are preferentially retained in solids during weathering, the relative mobility trend for Na, K, Mg, and Ca is quite unlike the composition trend for common igneous rocks [71,72,74]. In the latter, as one goes from basic to acidic rocks, the Mg/(Mg+Ca) ratio decreases and the K/(K+Na) ratio increases. Consequently, the effect of partitioning is clearly seen when K*/(K*+Na*-Cl*) is plotted against Mg*/(Mg*+Ca*) for Amazon and Orinoco samples (Figure 5). The "*" refers to cyclic salt corrected values [73], and Cl* is subtracted to correct for halite, so the K*/(K*+Na*-Cl*) ratio represents inputs solely from silicate weathering. The Mg*/(Mg*+Ca*) ratio, however, also includes contributions from carbonates and evaporites. Two features can be seen. First, samples from particular regions plot as distinct groupings. Second, samples from rivers in cratonic areas plot in a general field encompassing common igneous and sedimentary rock compositions (no fractionation), while Andean samples are sodium and calcium enriched compared to typical rock compositions. In the Andean samples, calcium enrichment is largely due to contributions from limestones and evaporites; however, the (Na*-Cl*) enrichment can only be due to the partitioning of potassium into solid phases of soils and river suspended load. This is particularly pronounced in rivers draining lower Paleozoic shales where the cation ratios in water are quite low compared to "average" shale and black shale ratios, reflecting the stability of potassium-bearing micas. Because of evaporite and carbonate contributions, stability trends for calcium and magnesium need to be confirmed using river-borne solids.

Like K and Mg, Si release into solution may also be limited by the formation of cation-rich clays in the Andes [71,72]. When feldspars weather to kaolinite, two moles of Si enter solution for every mole of (Na+K). If the products are cation-rich clays instead of kaolinite, this ratio is less than 2:1, while if gibbsite is forming, the ratio is greater than 2:1. When dissolved Si is plotted against (K*+Na*-Cl*) (Figure

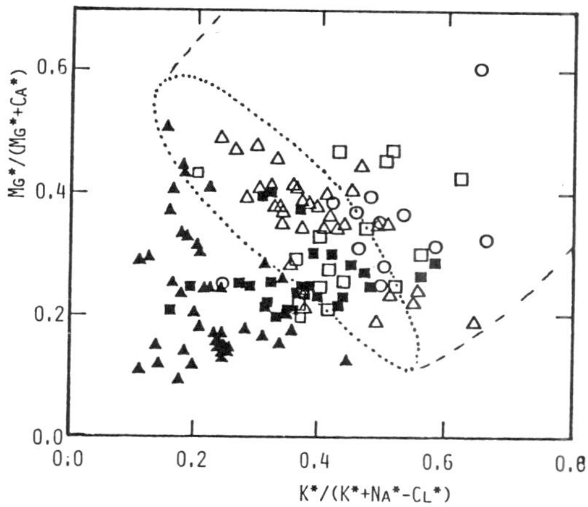

Figure 5. Comparison of Mg*/(Mg*+Ca*) versus K*/(K*+Na*-Cl*) in lithologies representative of the Amazon and Orinoco bedrock with dissolved material in surface waters. Symbols for dissolved phases are given in Figure 3. The dotted oval represents the range of analyses for common igneous rocks; the dashed lines that extend off of the igneous field represent the composition field for common sedimentary rocks. Open symbols are for rivers draining cratonic areas; solid symbols are for samples from the Andean Cordillera.

6), the data fall roughly parallel to a 2:1 trend. This is suggestive of a system dominated by the weathering of feldspars to kaolinite combined with background silica coming from quartz dissolution, kaolinite dissolution, or gibbsite formation. Waters from most regions are silicon depleted when compared to igneous rocks, river sediments are silicon enriched. Some of the lowland rivers fall within the rock field, suggesting that in these regions silica quantitatively dissolves. Almost all of the data from Andean rivers fall below the 2:1 trend (Figure 6, insert), suggesting that cation-rich clays must be forming.

The degree to which basement rock in the wettest lowland areas is dissolving is best displayed by aluminum and iron data from the Amazon system [71,72]. High concentrations of dissolved iron and aluminum are found in rivers in the wettest regions [71]. Fe and Al concentrations greatly exceed those predicted by thermodynamic calculations for inorganic systems. Furthermore, Al and to a lesser degree Fe are stable for several years in untreated samples, suggesting that they exist as either organic complexes or organically stabilized colloids. Iron concentrations correlate most strongly with color (as a measure of a class of dissolved humic materials) and aluminum with hydrogen ion concentration. When (Al+Fe)/(Si+Al+Fe) is plotted against K*/(K*+Na*-Cl*) very few rivers plot near the rock field (Figure 7). Those that do have the lowest dissolved loads; solid loads are negligible. Presumably, the lack of abundant cations in the soil environment allows for high levels of organic acids and facilitates the mobilization of iron and aluminum.

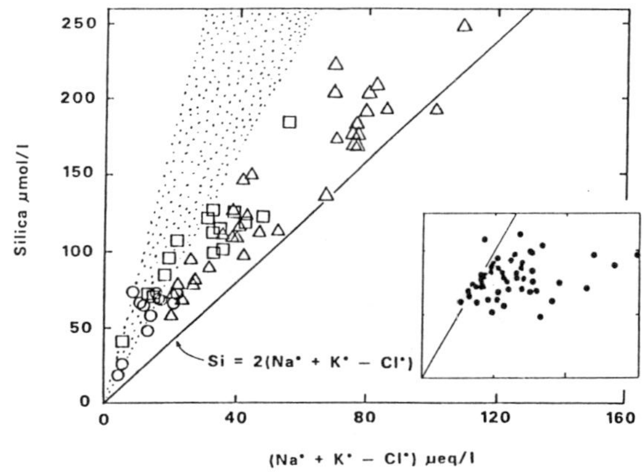

Figure 6. Si versus (Na*+K*-Cl*) for samples from the South American craton. Data are cyclic salt corrected. Symbols in Figure 3. Stippled area represents zone of complete dissolution of common igneous rocks.

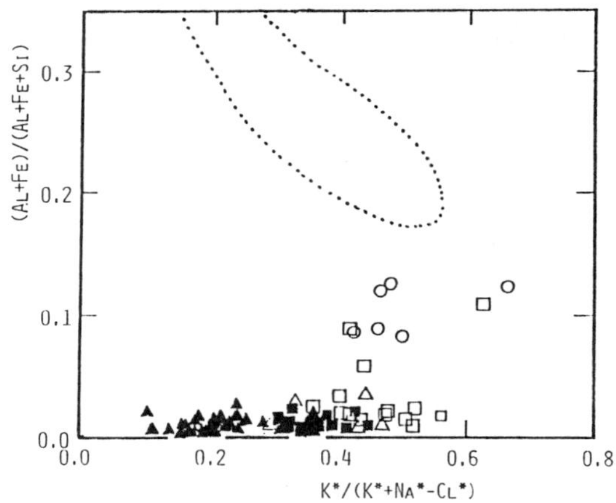

Figure 7. Comparison of (Al+Fe)/(Al+Fe+Si) versus K*/(K*+Na*-Cl*) in lithologies representative of the Amazon and Orinoco bedrock with the same quotients in dissolved material in surface waters. Symbols are given in Figures 3 and 5.

7. SUMMARY

The humid tropical zone is characterized by considerable moisture, luxuriant vegetation, and active chemical erosion. This zone is now more confined that it has been for much of the Cenozoic owing to a long-term climatic cooling trend, but it is much more expansive than it was during peak glaciations when the present forest was savanna and desert. In tectonically active areas, erosional rates are lithology dependent and show no strong relationship to sea level, which only becomes important after uplift and volcanism cease. In areas of sluggish tectonics, such as the stable cratons, the effects of sea level change on a 10 Ma time scale appear to be especially important in affecting landscape morphology.

In tectonically active environments, most of the material exposed to weathering has undergone rapid uplift involving brittle deformation; volcanism is sometimes important. Owing to feedbacks in the deformation processes, erosion may actually exert a major control on the deformation history of mountain ranges. Uplift rates are on the order of 10^3-10^4 mm/ka, while erosion rates vary from about the same values to an order of magnitude lower. As a result, easily weathered lithologies are exposed on steep slopes and weathering rates are lithology dependent. Where carbonates and evaporites are present their weathering products dominate the river water chemistry. Similarly, weakly cemented clastic rocks dominate the solid load inputs. Aluminosilicate rocks do not weather completely to phases depleted in the major cations (Na, K, Mg, Ca); instead, some unweathered primary minerals remain and various secondary cation-rich clays (e.g., smectites and illites) form in soils. Solid load is enriched in K over Na and in Mg over Ca. The ratio of Si to (Na+K) in the dissolved load is low because Si is also retained in the cation-rich clays.

On the cratons, sea level changes exert a major control on weathering rates and landscape development. When sea level is high, sediment is deposited on the passive margins, continental platforms, and intracratonic basins. At the same time near-sea-level planation surfaces develop in elevated shield areas. When developed completely, the landscape would be planed off and erosion on land and deposition in coastal waters would be minimal. This situation must resemble that of some lowland areas of the Amazon and Orinoco basins of today (Figure 3, A and B). Here erosion on exposed crystalline basement proceeds slowly, at about 1-10 mm/ka, and apparently involves substantial dissolution of bedrock and its secondary weathering products. During major sea level high stands, such as during the Cretaceous, the erosional contribution from the cratons to the oceans must therefore have been very low.

Erosion surfaces take a long time to form following a drop in base level. Aleva [2] estimates that roughly 25 million years is required. As base level drops, it seems reasonable to envision greatly accelerated erosion. Soils, recently deposited cratonic sediments, and vulnerable lithologies in the shields would erode first, after which erosion would stabilize at rates typical of elevated shield areas today. In these areas today (Figure 3, C), rates are about 10-40 mm/ka. For South American shields, average uplift rates approach 10-20 mm/ka (Figure 1), and it seems reasonable that average long-term erosion rates are close to this value. The erosion is not regional but is localized on scarps. Because these scarps form a perimeter which steadily advances into the core of the craton, mean denudation rates for the shield as a whole should decrease roughly linearly with time. This process apparently occurs repeatedly, generating a series of stacked erosion surfaces that appear to correlate with highs on the sea level curve. A potentially interesting area of research would involve establishing whether the uplift tectonics is entirely driven by erosion and simultaneous isostatic adjustment.

Three factors contribute to low erosion rates on the cratons. First, sluggish tectonics limits the supply of weatherable materials. Second, transport-limited conditions reduce erosion rates in flatter regions. Third, in uplifted areas basement rocks are more massive than those in mountain belts because of a long history of metamorphism and uplift largely involving plastic deformation.

Weathering on the craton is much more complete. The solid weathering products are cation depleted: quartz, kaolinite, and sesquioxides. Cation ratios in river waters are close to bedrock values and when Si is plotted against Na+K, data fall parallel to a 2:1 trend indicating the importance of kaolinite formation from primary minerals.

In summary, simple weathering reactions can be used to describe the composition of the dissolved and solid load in tropical rivers, after correction for cyclic salt inputs. In the most dilute rivers of the humid lowlands, elemental ratios in the dissolved load resemble those of bedrock, suggesting that all minerals are weathering by congruent solution, albeit by a number of steps. In drier regions Si, Fe, and Al are strongly retained in the soil, while Na, K, Mg, Ca, and some Si enter solution. Preferential retention of K and Mg over Na and K in solid phases is seen in rivers from mountain belts. Even though there have been substantial changes in climate, vegetation, and the role of humans from the Pleistocene to the present, it appears that changes in the cation reservoirs of vegetation and soil have not been sufficient to have had a major impact on the chemistry of these rivers. Thus, to a first approximation, both biomass and cation-rich components of soils can be assumed to be in steady state.

8. ACKNOWLEDGMENTS:

This paper benefits from almost a decade of working with John Edmond (MIT), Robert H. Meade (USGS), and Carl Nordin, Jr. (CSU), first on the Amazon and now on the Orinoco. Conversations with graduate students and faculty at Princeton also contributed much to its content. Water analyses of the Amazon are published [73,74]; those of the Orinoco are courtesy of John Edmond. Funding for this work comes from NSF grant EAR-8407651, from the Venezuelan Ministerio del Ambiente y Recursos Naturales Renobales: Proyecto Orinoco-Apure (Abel, Mejia, dir.), and from the author's department including a Dusenbury Preceptorship.

9. REFERENCES

1. Aleva, G. J. J., 1979, Bauxite and other duricrusts in Surinam: A review, *Geol. Mijnb.*, **58**, 321-336.
2. Aleva, G. J. J., 1984, Laterization, bauxitization and cyclic landscape development in the Guiana Shield, *in*, Jacob, L., Jr., ed., *Bauxite, Proceedings of the 1984 Bauxite Symposium, Feb. 27-Mar. 1, 1984*, Amer. Inst. Mining Metall. Petrol. Engineers, New York, pp. 297-318.
3. Benavides, V., 1968, Saline deposits of South America, *Geol. Soc. Amer. Spec. Paper* **88**, 249-290.
4. Berner, R. A., 1978, Rate control of mineral dissolution under earth surface conditions, *Amer. J. Sci.*, **278**, 1235-1252.
5. Blancaneaux, P., and Pouyllau M., 1977, Formes d'alteration pseudokarstiques en relation avec la geomorphologie des granites precambriens du type Rapakivi dans le territoire Federal de l'Amazone, Venezuela, *Cah. O.R.S.T.O.M. Ser. Pedol.*, **15**, 131-142.
6. Bloom, A. L., W. S. Broecker, J. M. A. Chappell, R. K. Matthews, and K. J. Mesolella, 1974, Quaternary sea level fluctuations on a tectonic coast: New $^{230}Th/^{234}U$ dates from the Huon Peninsula, New Guinea, *Quarternary Res.*, **4**, 185-205.
7. Burke, K. and J. T. Wilson, 1972, Is the African Plate stationary? *Nature*, **239**, 387-390.
8. Camargo, M. N., and I. C. Falesi, 1975, Soils of the Central Plateau and Transamazonic Highway, *in*, E. Bornemisza and A. Alvarado eds., *Soil Management in Tropical America*, University Consortium on Soils of the Tropics, Raleigh, North Carolina, pp. 25-45.

9. Carson, M. A., and M. J. Kirkby, 1972, *Hillslope, Form and Process,* Cambridge University Press, Cambridge, 475 pp.
10. Chalcraft, D. and K. Pye, 1984, Humid tropical weathering of quartzite in southeastern Venezuela, *Z. Geomorph. N. F.,* **28**, 321-332.
11. Chi, W. R., J. Namson, and J. Suppe, 1981, Stratigraphic record of plate interactions in the coastal range of eastern Taiwan, *Mem. Geol. Soc. China,* **4**, 491-530.
12. Colinvaux, P., 1979, The Ice-age Amazon, *Nature,* **278**, 399-400.
13. Covey, M. C., 1984, *Sedimentary and tectonic evolution of the western Taiwan foredeep,* Ph.D. thesis, Princeton University, 152 pp.
14. Crough, S. T., 1979, Hotspot epeirogeny, *Tectonophysics,* **61**, 321-333.
15. Dahlen, F. A., J. Suppe, and D. Davis, 1984, Mechanics of fold-and-thrust belts and accretionary wedges: Cohesive Coulomb theory, *Jour. Geophys. Res.,* **89**, 10087-10101.
16. Davis, D., J. Suppe, and F. A. Dahlen, 1983, Mechanics of fold-and-thrust belts and accretionary wedges, *J. Geophys. Res.,* **88**, 1153-1172.
17. Davis, W. M., 1932, Piedmont Benchlands and Primarrumpfe, *Geol. Soc. Amer. Bull.,* **43**, 399-440.
18. Evesson, D. T., 1983, The climate of the Puari River basin, *The Purari-Tropical Environment of a High Rainfall River Basin,* Petr, T., ed. Dr. W. Junk Publishers, The Hague, The Netherlands, pp. 9-26.
19. Franzinelli, E., and P. E. Potter, 1983, Petrology, chemistry, and texture of modern river sands, Amazon River system, *Jour. Geology,* **91**, 23-39.
20. Garner, H. F., 1959, Stratigraphic-sedimentary significance of contemporary climate and relief in four regions of the Andes Mountains, *Geol. Soc. Am. Bull.,* **70**, 1327-1368.
21. Gibbs, R. J., 1965, *The geochemistry of the Amazon River Basin,* Ph. D. Thesis, Geology, University of California, San Diego, 96 pp.
22. Gibbs, R. J., 1967, The geochemistry of the Amazon River system, Part I, The factors that control the salinity and composition and concentration of suspended solids, *Geol. Soc. Amer. Bull.,* **78**, 1203-1232.
23. Gibbs, R. J., 1977, Transport phases of transition metals in the Amazon and Yukon rivers, *Geol. Soc. Amer. Bull.,* **88**, 829-834.
24. Giegengack, R., 1984, Late Cenozoic tectonic environments of the Central Venezuelan Andes, *in*, Bonini, W.E., R. B. Hargraves, and R. Shagam, eds., The Caribbean - South American Plate Boundary and Regional Tectonics, *Geol. Soc. Amer.* **Mem.162**, Geological Society of America, Boulder, Colorado, pp. 343-364.
25. Goldich, S. S., 1938, A study in rock weathering, *J. Geol.,* **46**, 17-58.
26. Hem, J. D., and C. J. Lind, 1974, Kaolinite Synthesis at 25 °C, *Science,* **184**, 1171-1173.
27. Irion, G., 1976, Mineralogisch-geochemische Unterschungen an der pelitischen Fraktion amazonischer Oberboden und Sedimente, *Biogeographica,* **7**, 7-25.
28. Irion, G., 1978, Soil infertility in the Amazonian rain forest, *Naturwissenschraften,* **65**, 515-519.
29. Irion, G., 1984, Sedimentation and sediments of Amazonian rivers and evolution of the Amazonian landscape since Pliocene times, *in* Sioli, H., ed., *The Amazon, Limnology and Landscape Ecology of a Mighty Tropical River and its Basin,* Dr. W. Junk Publishers, The Hague, The Netherlands, pp. 201-214.
30. Irion, G., 1984, Clay minerals of Amazonian soils, *in* Sioli, H., ed., *The Amazon, Limnology and Landscape Ecology of a Mighty Tropical River and its Basin,* Dr. W. Junk Publishers, The Hague, The Netherlands, pp. 537-579.
31. Irion, G. and T. Petr, 1983, Clay mineralogy of selected soils and sediments of the Puari River basin, *in* Petr, T., ed., *The Purari-Tropical Environment of a High Rainfall River Basin,* Dr. W. Junk Publishers, The Hague, The Netherlands, pp. 87-107.
32. James, P. E., 1959, The geomorphology of eastern Brazil as interpreted by Lester C. King, *Geogr. Rev.,* **49**, 240-246.
33. King, L. C., 1953, Canons of landscape evolution, *Geol. Soc. Amer. Bull.,* **64**, 721-751.
34. King, L. C., 1956, A geomorfologia do Brasil oriental, *Rev. Brasileira de Geogr.,* **18**, 147-265.

35. King, L. C., 1957, A geomorphological comparison between eastern Brazil and Africa (central and southern), *Geol. Soc. London Quart. Jour.*, **112**, 445-474.
36. King, L. C., 1967, *The Morphology of the Earth,* 2nd. ed., Oliver and Boyd, Edinburgh, 762 pp.
37. Klammer, G., 1984, The relief of the extra-Andean Amazon basin, *in* Sioli, H., ed., *The Amazon, Limnology and Landscape Ecology of a Mighty Tropical River and its Basin,* Dr. W. Junk Publishers, The Hague, The Netherlands, pp. 47-83.
38. Klinge, H., 1967, Podzol soils: A source of black water rivers in Amazonia, *Atas do Simposio sobre a Biota Amazonica,* **3**, pp. 117-125.
39. Kodama, H., and M. Schnitzer, 1977, Effect of fulvic acids on the crystallization of Fe(III), *Geoderma,* **19**, 279-291.
40. Li, Y.-H., 1976, Denudation of Taiwan Island since the Pliocene Epoch, *Geology,* **4**, 105-107.
41. Linares, J., and F. Huertas, 1971, Kaolinite: Synthesis at room temperature, *Science,* **171**, 896-897.
42. Livingstone, D. A. and T. Van der Hammen, 1978, Palaeogeography and palaeoclimatology, in Tropical Forest Ecosystems, *UNESCO Natural Res. Res.* **XIV**, UNESCO/UNEP/FAO, Paris, pp. 61-90.
43. Lopez Eyzaguirre, C., and R. E. Bisque, 1975, Study of the weathering of basic, intermediate and acidic rocks under tropical humid conditions, *Quart. Colo. Sch. Mines,* **70**, pp. 1-59.
44. Matthews, R. K., 1973, Relative elevation of late Pleistocene high sea level stands: Barbados uplift rates and their implications, *Quaternary Res.*, **3**, 147-153.
45. McConnell, R. B., 1968, Planation surfaces in Guyana, *Geograph. Jour.*, **134**, 506-520.
46. Menendez, A and A. Sarmentero, 1984, Geology of the Los Pijiguaos bauxite deposits, Venezuela, *in* Jacob, L. Jr., ed., *Bauxite, Proceedings of the 1984 Bauxite Symposium, Feb. 27 - March 1, 1984,* Amer. Inst. Mining Metall. and Petrol. Engineers, New York, pp. 387-406.
47. Meybeck, M., 1979b, Concentrations des eaux fluviales en éléments majeurs et apports en solution aux océans, *Rev. Géogr. Dynam. Géol. Phys.*, **21**, 215-246.
48. Milliman, J. D., and R. H. Meade, 1983, World-wide delivery of river sediment to the oceans, *J. Geol.*, **91**, 1-21.
49. Morgan, W. J., 1983, Hotspot tracks and the early rifting of the Atlantic, *Tectonophysics,* **94**, 123-139.
50. Mousinho de Meis, M. R., 1971, Upper Quarternary process changes of the Middle Amazon area, *Geol. Soc. Amer. Bull.*, **82**, 1073-1078.
51. Pain, C. F., 1983, Geology and geomorphology of the Purari River catchment, *in* Petr, T., ed., *The Purari-Tropical Environment of a High Rainfall River Basin,* Dr. W. Junk Publishers, The Hague, The Netherlands, pp. 27-46.
52. Pasquali, Z. J., C. Lopez E., and H. Meinhard, 1972, Meteorización de rocas del escudo de Guyana en ambiente tropical, Congreso Geológico Venezolano, Memoria IV, Tomo IV, *Boletín de Geología, Publicación Especial* **5**, 2245-2302.
53. Peng, T.-H., Y.-H. Li, and F. T. Wu, 1977, Tectonic uplift rates of the Taiwan Island since the early Holocene, *Geol. Soc. China Mem.*, **2**, 57-69.
54. Petr, T., ed., 1983, *The Purari-Tropical Environment of a High Rainfall River Basin,* Dr. W. Junk Publishers, The Hague, The Netherlands, 624 pp.
55. Petr, T., 1983, Limnology of the Purari basin Part 1. The catchment above the delta, *in* Petr, T., ed., *The Purari-Tropical Environment of a High Rainfall River Basin,* Dr. W. Junk Publishers, The Hague, The Netherlands, pp. 141-177.
56. Petr T. and G. Irion, 1983, Geochemistry of soils and sediments of the Puari River basin, *in* Petr, T., ed., *The Purari-Tropical Environment of a High Rainfall River Basin,* Dr. W. Junk Publishers, The Hague, The Netherlands, pp. 109-122.
57. Pickup, G., 1983, Sedimentation processes in the Purari River upstream of the delta, *in* Petr, T., ed., *The Purari-Tropical Environment of a High Rainfall River Basin,* Dr. W. Junk Publishers, The Hague, The Netherlands, pp. 141-177.

58. Pitman W. III, 1978, Relationship between eustasy and stratigraphic sequences of passive margins, *Geol. Soc. Amer. Bull.*, **89**, 1389-1403.

59. Potter, P. E., 1978, Petrology and chemistry of modern big river sands, *J. Geol.*, **86**, 423-449.

60. Ronov, A. B., A. A. Migdisov, and N. V. Barskaya, 1969, Tectonic cycles and regularities in the development of sedimentary rocks and paleogeographic environments of sedimentation of the Russian Platform (an approach to a quantitative study), *Sedimentology,* **13**, 179-212.

61. Rossignol-Strick, M., 1983, African monsoons, an immediate climate response to orbital insolation, *Nature,* **303**, 46-49.

62. Ruxton, B. P. and I. McDougall, 1967, Denudation rates in northwest Papua from potassium-argon dating of lavas, *Amer. J. Sci.,* **265**, 545-561.

63. Sarnthein, M., 1978, Sand deserts during glacial maximum and climatic optimum, *Nature,* **272**, 43-46.

64. Scott, G. A. J., 1975, Soil profile changes resulting from the conversion of forest to grassland in the montana of Peru, *Great Plains-Rocky Mountain Geogr. J.,* **4**, 124-130.

65. Sloss, L. L., 1963, Sequences on the cratonic interior of North America, *Geol. Soc. Amer. Bull.,* **74**, 93-114.

66. Sloss, L. L., 1979, Global sea level changes: A view from the craton, in *Geological and Geophysical Investigations of Continental Margins,* Watkins, J. S., L. Montardert, and P. W. Dickerson, eds., *Amer. Assoc. Petrol. Geol. Mem.* **29**, Amer. Assoc. Petrol. Geol., Tulsa, Oklahoma, pp. 461-467.

67. Sloss, L. L. and R. C. Speed, 1974, Relationships of cratonic and continental-margin tectonic episodes, in *Tectonics and sedimentation,* Dickenson, W. R., ed., *SEPM Spec. Publ.* **22**, 98-119.

68. Soares, P. C., P. M. B. Landim, and V. J. Fulfaro, 1978, Tectonic cycles and sedimentary sequences in the Brazilian intracratonic basins, *Geol. Soc. Amer. Bull.,* **89**, 181-191.

69. Sombroek, W. G., 1966, *Amazon Soils,* Centre for Agricultural Publications and Documentation, Wageningen, 292 pp.

70. Sombroek, W. G., 1984, Soils of the Amazon region, in *The Amazon, Limnology and Landscape Ecology of a Mighty Tropical River and its Basin,* Sioli, H., ed. Dr. W. Junk Publishers, The Hague, The Netherlands, pp. 521-535.

71. Stallard, R. F., 1980, *Major element geochemistry of the Amazon River system,* Ph.D. Thesis, Jan. 1980, MIT-WHOI Joint Program in Oceanography, **WHOI-80-29**, 366 pp.

72. Stallard, R. F., 1985, River chemistry, geology, geomorphology, and soils in the Amazon and Orinoco basins, *The Chemistry of Weathering, Proceedings of a NATO Advanced Research Workshop, July 2-6, 1984, Rodez, France,* J. I. Drever, ed., D. Reidel Pub. Co., Dordrecht, Holland.

73. Stallard, R. F., and J. M. Edmond, 1981, Geochemistry of the Amazon I: Precipitation chemistry and the marine contribution to the dissolved load at the time of peak discharge, *J. Geophys. Res.,* **86**, 9844-9858.

74. Stallard, R. F., and J. M. Edmond, 1983, Geochemistry of the Amazon 2. The influence of the geology and weathering environment on the dissolved load, *J. Geophys. Res.,* **88**, 9671-9688.

75. Stallard, R. F., and J. M. Edmond, 1987, Geochemistry of the Amazon 3. Weathering Chemistry and Limits to Dissolved Inputs *J. Geophys. Res.*, **92**, 8293-8302.

76. Steckler, M., 1984, Changes in Sea Level, *Patterns of Change in Earth Evolution,* Holland, H. D. and A. F. Trendall, eds., Springer-Verlag, Berlin, pp. 103-121.

77. Suppe, J., 1981, Mechanics of mountain building in Taiwan, *Mem. Geol. Soc. China,* **4**, 67-89.

78. Szczerban, E., 1976, Cavernas y simas en areniscas precambricas del Territorio Federal Amazonas y Estado Bolivar, *Venez. Dir. Geol. Bol. Geol. Pub. Esp.,* **7,** 1055-1072.

79. Tricart, J., 1974, Existence de periodes seches au quaternaire en amazonie et dans les regions voisines, *Rev. Geomorphol. Dynam.,* **4**, 145-158.

80. Vail, P. R., R. M. Mitchum, R. G. Todd, J. M. Widmier, S. Thompson III, J. B. Sangree, J. N. Bubb, and W. G., 1977, Seismic stratigraphy and global changes of sea-level, *in*

Seismic-stratigraphy Applications to Hydrocarbon Exploration, C. E. Payton, ed., *Amer. Assoc. Petrol. Geol. Memoir* **26**, Amer. Assoc. Petrol. Geol., Tulsa, Oklahoma, pp. 49-212.

81. Vail, P. R. and J. Herdenbol, 1979, Sea-level changes during the Tertiary, *Oceanus*, **22**, 71-79.
82. Van der Hammen, T., 1974, The Pleistocene changes of vegetation and climate in tropical South America, *J. Biogeography*, **1**, 3-26.
83. Van Wambeke, A., 1978, Properties and potentials of soils in the Amazon Basin, *Interciencia*, **3**, 233-242.
84. Violante, A, and P. Violante, 1980, Influence of pH, concentration, and chelating power of organic anions, *Clays Clay Miner.*, **28**, 425-434.
85. Vuilleumier, B. S., 1971, Pleistocene changes in the fauna and flora of South America, *Science*, **173**, 771-780.
86. Walter, H., 1979, *Vegetation of the Earth and Ecological Systems of the Geo-biosphere*, Springer-Verlag, New York, 274 pp.
87. Wolfe, J. A., 1978, A paleobotanical interpretation of Tertiary climates in the northern hemisphere, *Amer. Sci.*, **66**, 694-703.
88. Zonneveld, J. I. S., 1969, Preliminary remarks on summit levels and the evolution of the relief in Surinam (S. America), *Verhandelingen Kon. Ned. Geol. Mijnbouwk. Gen.*, **27**, 53-60.
89. Zonneveld, J. I. S., 1975, Some problems of tropical geomorphology, *Z. Geomorph. N. F.*, **19**, 377-392.

HOW TO ESTABLISH AND USE WORLD BUDGETS OF RIVERINE MATERIALS

Michel Meybeck
Institut de Biogéochimie Marine
Ecole Normale Supérieure
92120 Montrouge
France

1. INTRODUCTION

Average river water composition is often taken as a reference of surface water chemistry (Stumm and Morgan, 1981), and global river inputs have been used to set up the elemental mass balance of oceans (Burton and Liss, 1973; Berner and Raiswell, 1983; Martin, Burton and Eisma, 1981; Wollast and Mackenzie, 1983). The sodium budget has been used to estimate the age of the ocean (Clarke, 1924; Conway, 1942; Livingstone, 1963b). The rivers are a major pathway in the global geochemical cycles of elements, and they are probably one of the better known parts of the external cycle, since extensive work has been done on the composition and origins of river loads (Garrels and Mackenzie, 1971; Berner, 1971; Holland, 1978; Ivanov, 1981; Kempe, 1979). A recent trend in geochemistry is the modelling of past cycles based on today's data, an approach that is also of great interest in geology and sedimentology (Berner, Lasaga and Garrels, 1983; Garrels and Lerman, 1984). However, this extrapolation can only be achieved through careful consideration of present river loads which are now very much affected by human activity. The coupling between elemental cycles, as discussed in SCOPE publications (Likens, 1981), makes clear the role of the atmosphere and biosphere in the global cycles and, hence, in the river fluxes. Finally, the river loads have long been considered as the essential information basis for land erosion rates (Meade, 1969, and this volume)

There has been a constant need for reliable river data (water discharges, and water and suspended matter composition) since the pioneer work of Clarke (1924), whose results have been extensively used for forty years (Conway, 1942; Erikson, 1960). The newer work of Livingstone (1963a) on river water composition has become a basic reference for most western scientists, while the data of Alekin and Brazhnikova (1960, 1968) are used by East European geochemists. As most major rivers were studied during 1960's and 1970's, it became necessary to reconsider the budgets and numerous papers have been presented in the last ten years on major dissolved elements (Meybeck 1979; Arauda 1980; Meybeck, 1983), nutrients and organic carbon (van Bennekom and Salomons, 1981; Schlesinger and Melack, 1981; Meybeck, 1982; Kempe, 1985), particulate load (Holeman, 1968; Jansen and Painter, 1974; Alekin, 1978; Milliman and Meade, 1983), composition of particulate matter (Gordeev and Lisitzin, 1978; Martin and Meybeck, 1979; Gordeev, 1983), dissolved trace elements (Figuères, Martin, Thomas, 1982; Martin and Salvadori, 1983). The first published analyses of the Chinese rivers, Huang Ho (Yellow River) and Chiang Jiang (Yang Tse

A. Lerman and M. Meybeck (eds.), Physical and Chemical Weathering in Geochemical Cycles, 247–272.

Kiang), have also become available (Li et al., 1984; Hu Ming Lui et al., 1982; Milliman et al., 1984) and can be used in world budgets.

In-depth analysis of previous river budgets reveals great discrepancies concerning basic data (water discharge, areas of endoreic or inland drainage), influence of atmospheric inputs, human activities (pollution, river daming, enhanced land erosion), and also the quantity and quality of analyses both for dissolved and particulate matter. The purpose of this paper is to review some historical budgets, to propose different methods for setting up improved world budgets, and last but not least, to discuss critically the use of world budgets and global averages.

2. SOME BASES OF GLOBAL RIVER BUDGETS

2.1. Some early budgets (Table I)

Clarke (1924, p. 114) gives the first data set on river water chemistry that mainly represents the USA, the Amazon and Uruguay rivers for South America, the Nile, and some minor rivers in Great Britain, France and Sweden. There are no data on Asian nor Canadian rivers. The total documented area and discharges are respectively 20% and 15% of the part of the continents drained to the ocean, mainly represented by the Amazon. Clarke calculates the average specific dissolved transport Td, the amount of total dissolved ions plus silica carried by rivers from a unit area usually given in metric tons/km^2/year, then extrapolates to the 80% of remaining area; this step in the budget is due to the lack of global water budget on earth while the total area drained to the ocean (exoreic area) is known (102.4×10^6 km^2). For each continent, Clarke gives an average proportion of dissolved elements, the world average is then computed by weighting concentrations with continent areas. Denudation is derived from this figure after correction for atmospheric CO_2 and rain inputs. A first attempt to determine the ocean age was based on sodium budget (p. 151) for which a correction for pollution was applied. The four basic features already exist in this budget: i) difference between endoreism and exoreism, ii) extrapolation of specific rates, iii) atmospheric corrections, iv) pollution correction.

Livingstone's data (1963a) are seemingly not more representative (13% of water discharge, 22% of area) than Clarke's. However, the analyses used are much more comprehensive than Clarke's, and they include the Asian, Canadian and African rivers. In order to avoid pollution effects as much as possible, some older data, already used by Clarke, were considered. Both continental and world averages given by Livingstone (tables 75 to 81) have been obtained by water-discharge weighting. The total water input to oceans was estimated to be 32,350 km^3/yr for an area of 101.6×10^6 km^2. There seems to be a minor error concerning this budget: Livingstone also calculates specific transport loads for the total area of each continent, but considers only the exoreic transport. This leads to an underestimation of the average denudation rates of each continent, a figure that has been widely used since then (Garrels and Mackenzie, 1971).

The budget of Alekin and Brazhnikova (1960, 1968) is mostly based on rivers of the Soviet Union from which a typology of river transport is derived and then extrapolated to the rest of the world. Dissolved carbonates are expressed as CO_2 (as in Clarke), sodium is combined with potassium, silica is not given. In order to compare this budget with the previous ones, we have taken the average ionic proportions for Soviet Union plus the USA (from Clarke's data) as was suggested by these authors. The water budget used here is 36,400 km^3/yr.

Table I. Average contents (mg/l) of dissolved major elements in rivers discharging to the oceans.

		SiO_2	Ca^{++}	Mg^{++}	Na^{+}	K^{+}	Cl^{-}	SO_4^{--}	HCO_3^{-}	$\Sigma_{i(1)}$
Clarke (1924)	(2) Ct	-	20.4	3.4	5.8	2.1	5.7	12.1	71.5	121
Conway (1942)	Ct	-	29.8	5.0	8.4	3.1	8.3	17.7	104	176.3
Livingstone (1963a)	(2) Ct	13.1	15	4.1	6.3	2.3	7.8	11.2	58.4	120
Alekin and Brazhnikova (1962-1968)	(3) Ct	-	18.5	4.9	9.0		8.4	16.4	31.1	88
Alekin (1978)	(2) Ct	-	-	-	-	-	-	-	-	78
Aurada (1980)	(4) Ct	-	16.3	3.3	7.4		5.0	10.9	60.1	103
Meybeck (1976)	(5) Ct	11.6	14.6	3.8	5.1	1.35	5.3	8.5	57.7	96.4
Meybeck (1979)	(2) Ce	10.4	13.4	3.15	3.65	1.25	3.05	7.85	52	84.3
	Cp	0	1.3	0.3	2.05	0.1	2.5	3.25	1	10.5
	Ca	0	0.05	0.2	1.5	0.07	2.7	0.4	0	4.9
	Ct	10.4	14.7	3.65	7.2	1.4	8.25	11.5	53	99.7
Meybeck (1983)	(2) Ca	0	0.25	0.38	2.75	0.13	4.95	1.2	0	11.8
	Ct	10.4	14.95	3.85	8.45	1.50	10.5	12.3	53	106.6

Ce: river water concentration that can be attributed to natural chemical erosion, including atmospheric CO_2 used in weathering, Cp: concentration attributed to overall human activity (estimated here in 1970), Ca: additional contents related to ocean-derived atmospheric precipitation, Ct: average river composition including all sources of dissolved elements (Ce+Ca+Cp).

(1) Sum of major cations and anions in mg/l, (2) see text for details, (3) the ionic proportions are derived from Soviet Union and USA rivers as suggested by the authors who only give $\sum_i$, (4) mostly based on Alekin and Brazhnikova (1960) and Livingstone (1963), (5) based on 50% of river waters, simple extrapolation, no correction for human activity or atmospheric input.

Meybeck's (1979) budget of dissolved major elements mostly takes into account river analyses that were made after Livingstone's review: Mackenzie, Orinoco, Parana, Amazon, Chari, Zaire, Niger, Zambezi, Danube, Ganges, Indus, Brahmaputra, Mekong, and numerous regional averages in Southern India, Indonesia, Papua New Guinea, plus averages for all major U.S.S.R. regional seas (Alekin and Brazhnikova, 1960, 1968). The sample of rivers for which analyses are provided is up to 63% for both drainage area and water discharged to the ocean. The basic water budget used is the one of Baumgartner and Reichel (1976) which is detailed in Table II and preferred to others (Soviet IHD, 1978) because it gives, among others, the water balance by ocean watershed areas and by latitudinal zones. The non-documented area is then extrapolated on the basis of a global typology of dissolved transport rates for twelve

Table II. Global water budget (from Baumgartner and Reichel, 1976).

	EXOREIC DRAINAGE		ENDOREIC DRAINAGE	
	A	Q	A	Q
EUROPE	8 273	2 564	1 752	
ASIA	31 453	12 467	12 680	
AFRICA	17 565	3 409	12 220	
OCEANIA	4 698	2 394	4 197	
NORTH AMERICA (1)	21 500	5 528	901	
SOUTH AMERICA	16 424	11 039	1 461	
TOTAL NON-GLACIATED AREA	99 913	37 400	33 200(2)	2 000
ANTARCTICA	14 062	1 987		
GROENDLAND	1 730	313		
ALL CONTINENTS	115 705	39 700	33 200	

A=area in 10^6 km^2; Q=runoff in km^3/yr.

(1) excluding Greenland, (2) 65% of this area is areic; 20% is arid (runoff <1 l s^{-1} km^{-2}) or semi-arid (from 1 to 4 l s^{-1} km^{-2}).

major morphoclimatic environments (based on runoff and temperature). This budget gives separate numbers for four climatic zones (cold, temperate, arid and tropical), and for each ocean, as well as a breakdown of dissolved river load according to various origins (chemical weathering, atmospheric CO_2, oceanic aerosols contribution, anthropogenic inputs).

As concerns the amount of suspended matter carried by rivers, about a dozen different budgets have been proposed, ranging from 13.5×10^{15} g/yr (Milliman and Meade, 1983) up to 58.1×10^{15} g/yr (Fournier, 1960). There is a great confusion in most reviews since many authors have extrapolated average transport rates (Ts in metric ton km^{-2} yr^{-1} to the entire continental area, including the glaciated part of the continents (about 16×10^6 km^2) and the endoreic drainage (at least 30×10^6 km^2). This is the case of the budgets of Fournier (1960), and Jansen and Painter (1974) that are based on multiregression analysis of Ts with monthly and yearly rainfall intensity, altitude and drainage area (Fournier, 1960), and with rainfall, temperature, rock hardness, altitude, relief ratio, and vegetation cover (Jansen and Painter, 1974). As an example, the global sediment transport of the latter authors, estimated as 26.7×10^{15} g/yr, is related to an area of 123.6×10^6 km^2, i.e. with most of the endoreic area included.

Holeman's budget (1968) is based on about eighty rivers, representing about one third of the exoreic drainage. For each continent, an average specific transport load (Ts) was computed, then simply extrapolated to the non-documented area. The budget for Asia is probably overestimated (average Ts $\simeq$ 550 t $km^{-2}yr^{-1}$) due to under-representation of the lowland Siberian rivers that have much lower Ts values than south East Asia rivers (20 and > 600 t $km^{-2}yr^{-1}$). Latest estimates for Asia are much lower: 340 t $km^{-2}yr^{-1}$ (Soviet IHD, 1978, p. 529) and 380 t km^{-2} yr^{-1} (Milliman and Meade, 1983).

The sediment budget of Milliman and Meade (1983) represents the most up-to-date attempt. Their sample of known rivers is about 50% of the exoreic area, and it includes recent data for the Chinese rivers, the Magdalena and the Amazon. For the remaining 50×10^6 km^2, the average loads have been estimated for forty-four different regions on the basis of nearby-studied river basins. These authors did not extrapolate continental averages as in many previous studies. However, this budget must be considered with great care since they took into account the present day levels of suspended load which suffers from two main human influences: i) increased erosion due to agriculture, mining, road building, and similar activities, and ii) storage in flood plains and reservoirs. The combined effect of these two processes is still not clearly understood on a global scale (see Meade, this volume).

2.2. Common difficulties in establishing budgets

The first difficulty is to obtain for each river reliable averages of suspended and dissolved concentrations. In most cases, these data are obtained with monthly surveys which are usually not sufficient to set up a reliable average with less than 20% error. The number of samples needed in such surveys is inversely proportional to the size of the river, and increased sampling is needed during floods, particularly for the solid load (see Meade, this volume, and examples of seasonal variations and budgets in Meade et al., 1979, and in Carbonnel and Meybeck, 1975). One should note that the water discharge of the Amazon, estimated at about 100,000 m^3/s when Livingstone established his budget, is now estimated at 175,000 m^3/s, a figure that has been taken into account in the latest global water balances.

Extrapolation of river data cannot be avoided in any global budget: if 50% of the exoreic area is drained by 35 rivers, 200 rivers are needed to raise the coverage to 70%. For water discharge, the figure is even worse: 200 rivers for 60% and up to 1000 for 70% (Figure 1). According to what we know or guess for the first 100 rivers, classified according to their dissolved load, they would only contribute 60% of the total load going to the ocean. The great majority of authors have extrapolated on a continent basis, i.e. the average values determined for some major rivers were taken as representative of the whole continent. This approach is not valid since the great variety of environments (climate, vegetation, lithology, relief) in any single continent is responsible for variations of transport rates from two orders of magnitude (dissolved loads) to three orders of magnitude (suspended loads) (Meybeck, 1976, 1979; Milliman and Meade, 1983; Stallard, 1980). The variation of dissolved contents are even much greater than the specific loads: up to three orders of magnitude, even for extended area (Meybeck, 1981, 1984b; Stallard and Edmond, 1983). Typologies of concentrations or specific transport rates must be set up on a regional basis (Milliman and Meade, 1983), on a morphoclimatic basis (Meybeck, 1979), or on the basis of watershed lithology (Meybeck, 1986).

Two major natural processes may greatly influence our knowledge of river loads and transports to the oceans: the atmospheric precipitation input and eventual

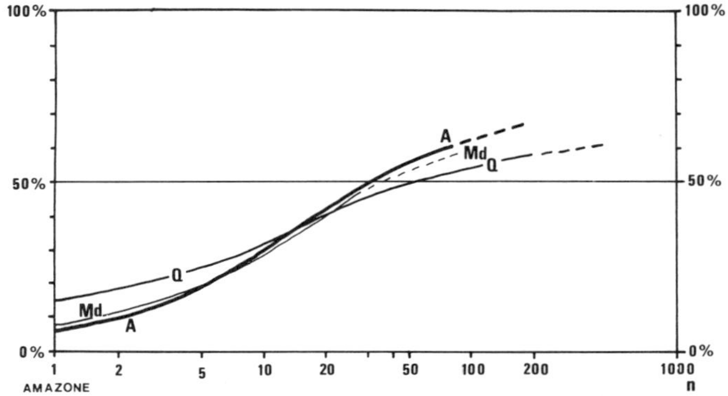

Figure 1. Cumulative water discharge (Q), drainage area (A) and dissolved load (Md), ranked in increasing order, of rivers flowing into the ocean in percent of total amount. n=number of rivers considered. The Amazon is the first ranked river in any case.

trapping or release at the continent-ocean boundary in the estuarine system. The atmospheric contribution is very difficult to take into account, mainly for lack of a global network for rain chemistry, particularly in the Southern Hemisphere and in the coastal zone at a distance less than 100 km from the sea. Whether this coastal stretch is considered or not, the global budget of dissolved elements carried to the sea is much influenced by such elements as Cl , Na and Mg . Estuaries and coastal zones trap the majority of suspended matter (see Eisma, this volume) and, in some cases, can also alter the load and chemical speciation of some dissolved elements (Martin, Burton, Eisma, 1981; Burton, this volume). Therefore, all budgets that have been previously presented are only valid up-stream from the estuary.

Human influence is now present everywhere, even in remote places through atmospheric transport, and any world budget has to answer two questions: what were the background levels and rates, and what are the ones at the present day?

3. HOW TO BUILD GLOBAL RIVER BUDGETS

Several different types of river budgets can be proposed (Meybeck, 1982), depending on i) the nature of the material considered (dissolved or particulate), ii) the influence of man's activities, and iii) the amount of available data. The first step is to check the representative values of the available river data by considering their average water runoff (Q) that should be as close as possible to the world value 11.8 l s^{-1} km^{-2} (i.e., 37,400 km^3/yr for 99.9×10^6 km^2). All climatic zones should be represented, from the arctic to equatorial regions, and from arid regions ($q<1$ l s^{-1} km^{-2}) to the wet tropics ($q>30$ l s^{-1} km^{-2}). A more difficult check concerns the average lithological composition which should ideally be 66% of sedimentary rocks, of which 23% are carbonate rocks. This is seldom realized in a relatively small area, and even a big continent can present a lithological bias. For example, in Europe, 87% of surficial rocks are of sedimentary origin, mostly carbonates (Blatt and Jones, 1975). Another screening of data is to check the presence of both mountain ranges and lowlands, major sedimentary platforms or shields. Finally, as much as possible, the heavily contaminated rivers in the USA and Europe must be avoided. One must note that most budgets now include the Amazon river. When this river is not counterbalanced by a corresponding water input (175,000 m^3/s) and drainage area (about 6×10^6 km^2) from temperate and cold regions, requiring a dozen other major rivers, there is a systematic bias in the global budget in favour of the wet tropical waters, as the Amazon alone represents about 15% of the water discharged to the ocean (Figure 1).

Reliable river water discharges are now available through the International Hydrological Decade, followed by the International Hydrological Program (Unesco, 1969, 1971, 1979) and suspended sediment data through the International Association of Scientific Hydrology (Fournier, 1969; Unesco, 1974), although these data should be updated. Specific papers (Milliman and Meade, 1983; Meybeck, 1976, 1979, 1982; Kempe, 1982, 1985) give extended list of references and data on both suspended and dissolved contents. Most probably, the most comprehensive review of river input and estuarine geochemistry is so far the one by Gordeev (1983).

3.1. Simple extrapolation budget

If the river data are representative of a major part of the drainage area and water discharge, about 50%, a simple extrapolation can be made globally. Two major hypotheses have to be considered at this stage: either a constant concentration hypothesis, the water discharge then being used for extrapolation (as realized by Livingstone, 1963a), or a constant specific load hypothesis using the drainage area for extrapolation, as done by Alekin and Brazhnikova (1968), Holeman (1968), and Meybeck (1979). The second hypothesis should be preferred because specific loads are less variable than concentrations (Meybeck, 1976), and the drainage areas are more accurately known than global water budgets that keep changing. As an example, Alekin used a global river discharge to the ocean of 36,450 km^3/yr in his earliest papers, then 40,300 km^3/yr in 1979 (Soviet IHD, 1978, p. S22) while this number is 37,400 km^3/yr in Baumgartner and Reichel (1976). Some budgets which used the simple extrapolation method are probably in error for lack of sufficient data (Corbel, 1964; Meybeck, 1977), or have used the constant concentration hypothesis (Kempe, 1985).

3.2. Extrapolation based on typology

This approach may be used when the rivers studied are not representative of the world, such as the data coming from one political or environmental unit (USSR or USA rivers; Amazon plus some smaller rivers). The documented rivers should provide information to set up a typology of specific transport for all known environments (preferable to the concentrations typology). Depending on the available data, the number of representative averages considered is highly variable: 12 for major elements (Meybeck, 1979), 6 for nutrients (Meybeck, 1982), and about 30 for the suspended matter (Milliman and Meade, 1983). Early attempts of solid load typology date back to Parde (1953). Schlesinger and Melack (1981) considered 10 environments to set up the organic carbon budget.

3.3. Use of support parameters

When the collected data are too scanty and not representative, global budget can be based on the relationship between the poorly studied element and a well-known environmental variable (such as watershed temperature or river runoff) or another water quality variable (conductivity, suspended solids, calcium). Such a method has been used in some global budgets: particulate nitrogen with support of particulate organic carbon (POC; Meybeck, 1982), POC with support of suspended matter

(Meybeck, 1982), dissolved uranium with support of total dissolved solids (Figuères et al., 1982), and total organic carbon with support of terrestrial primary production (Mantoura and Woodward, 1983; Meybeck, 1982). Usually, the accuracy of this type of budget is rather poor. But this approach should be used with success in models that reconstruct the evolution of river transports during geological times.

3.4. Multivariable analysis and statistics

Some authors have tried to relate directly the concentrations or the specific transport rates to a set of environmental parameters. This method has been considered for the sediment transport by Fournier (1960) and by Jansen and Painter (1974) with four to seven variables. Statistics on suspended sediment content, established on a set of major rivers, have also been used to determine the global sediment budget (Meybeck, 1982; see section 7).

3.5. Pure guesses

In some budgets, particularly in the early ones on trace elements (Turekian, 1971), the average levels or loads were mostly authors' own estimates, based on very scanty data which provided no more than correct orders of magnitude. A good example is the dissolved Rare Earth Elements (REE) supply to the ocean, by Martin et al. (1976), that was only known for the Gironde estuary, or about 1% of the world water discharge, while the average REE content in suspended matter was known in the Amazon, Zaire, Ganges, and Mekong rivers.

3.6. Anthropogenic influence

This is an essential point that must be considered in global budgets. Anthropogenic effects must first be assessed by consideration of rivers in identical climatic and geological conditions but exposed to different human influences. A measure of this influence is a simple index: population size multiplied by the average per capita energy consumption, called the Demophoric Index (Vallentyne, 1978). Excess nutrient loads have been related to this index (Meybeck, 1982), and an average excess load per unit of population can be determined for the major polluted rivers.

Other evaluation methods of human influence consider direct measurements of urban or industrial wastes (Meybeck, 1979), but they lack the integrating character of the previous one. Finally, some authors have used the mining statistics to estimate the global human influence on rivers. This should be avoided because i) some wastes can directly reach the ocean, and ii) there may be an important retention of some elements on the continents, as is the case of phosphorus (Meybeck, 1982). An example of this method is the global budget of excess fluoride determined by Martin and Salvadori (1983) on the basis of the world production of phosphogypsum, the major source of anthropogenic fluoride, and the excess of fluoride in the Seine River related to P-fertilizer plants in the watershed.

4. MAJOR ELEMENTS

A selection of budgets from Clarke (1924) to Aurada (1980) is presented in Table I. Major discrepancies in data presented in Table I are due to differences in basic river quality data, construction of these budgets and, most of all, a lack of clarity about

Table III. Types of dissolved transport based on morphoclimatic criteria: specific runoff, air temperature and relief.

q → / t(°C) ↓	mm/yr: 30 $l\ s^{-1}\ km^{-2}$: 1	mm/yr: 30–120 $l\ s^{-1}\ km^{-2}$: 1–4	mm/yr: 120–280 $l\ s^{-1}\ km^{-2}$: 4–9	mm/yr: 280–630 $l\ s^{-1}\ km^{-2}$: 9–20	mm/yr: 630 $l\ s^{-1}\ km^{-2}$: 20
4			TAIGA AND TUNDRA Bering, Kara, Laptev, Yukon, Nelson, Mackenzie. (A 20 ; Q 10.7)	HUMID TAIGA Fraser, Barentz, Finland, Quebec. (A 3.15 ; Q 3.4)	WET TAIGA British Columbia, Norway, Iceland. (A 0.2 ; Q 0.6)
4–15		SEMI-ARID TEMPERATE Aral, Caspian, Black Sea, Colorado. (A 3.35 ; Q 1.0)	TEMPERATE Poland, Iberia, Mississippi, Danube, Baltic, Shatt el Arab. (A 6.7 ; Q 3.35)	HUMID TEMPERATE Alps, Columbia, Sweden, St. Lawrence. (A 7.45 ; Q 7.75)	WET TEMPERATE (HIGHLANDS) Brahmaputra, Japan, New Zealand. (A 4.5 ; Q 15.3)
15–25	ARID Orange, Texas, White Nile, Murray. (A 17.2 ; Q 0.65)				
20		SAVANNA Parana, Zambezi, Niger, Carpentaria Gulf, Chari, Senegal, Volta (A 13.25 ; Q 5.85)		HUMID TROPICS Chao Phrya, Sri Lanka, Burma, Zaire. (A 9.2 ; Q 8.85)	WET TROPICS (LOWLANDS) Guyana, Orinoco. (A 6.9 ; Q 18.45) WET TROPICS (HIGHLANDS) Mekong, Magdalena, Papua N.G. (A. 7.95 ; Q 24.05)

Each type is characterized by a set of watersheds for which average air temperature (t in °C) and runoff q have been determined. The wet tropics type is split into two classes according to relief intensity. Lithological factors are not considered here. The contributions (in % of total area A and total runoff to ocean Q) of each type are given in parentheses. Listed rivers are those used in the definition of specific transport rates on which the extrapolation of major elements budget is based. Some of them may belong to endoreic basins (Meybeck, 1979).

Figure 2. Example of geographic distribution of morphoclimatic types used in the budget of major dissolved elements. See Table III for definition of each type. 1: glaciated area, 2: tundra, 3: taiga, 4: humid taiga, 5: wet taiga, 6: wet temperate, 7: humid temperate, 8: temperate, 9: semi-arid temperate, 10: desert, 11: arid, 12: savanna, 13: humid tropics, 14: wet tropics (lowlands), 15: wet tropics (highlands), 16: limits of continents (for Asia-Oceania), 17: endoreic basins. (Meybeck, 1979)

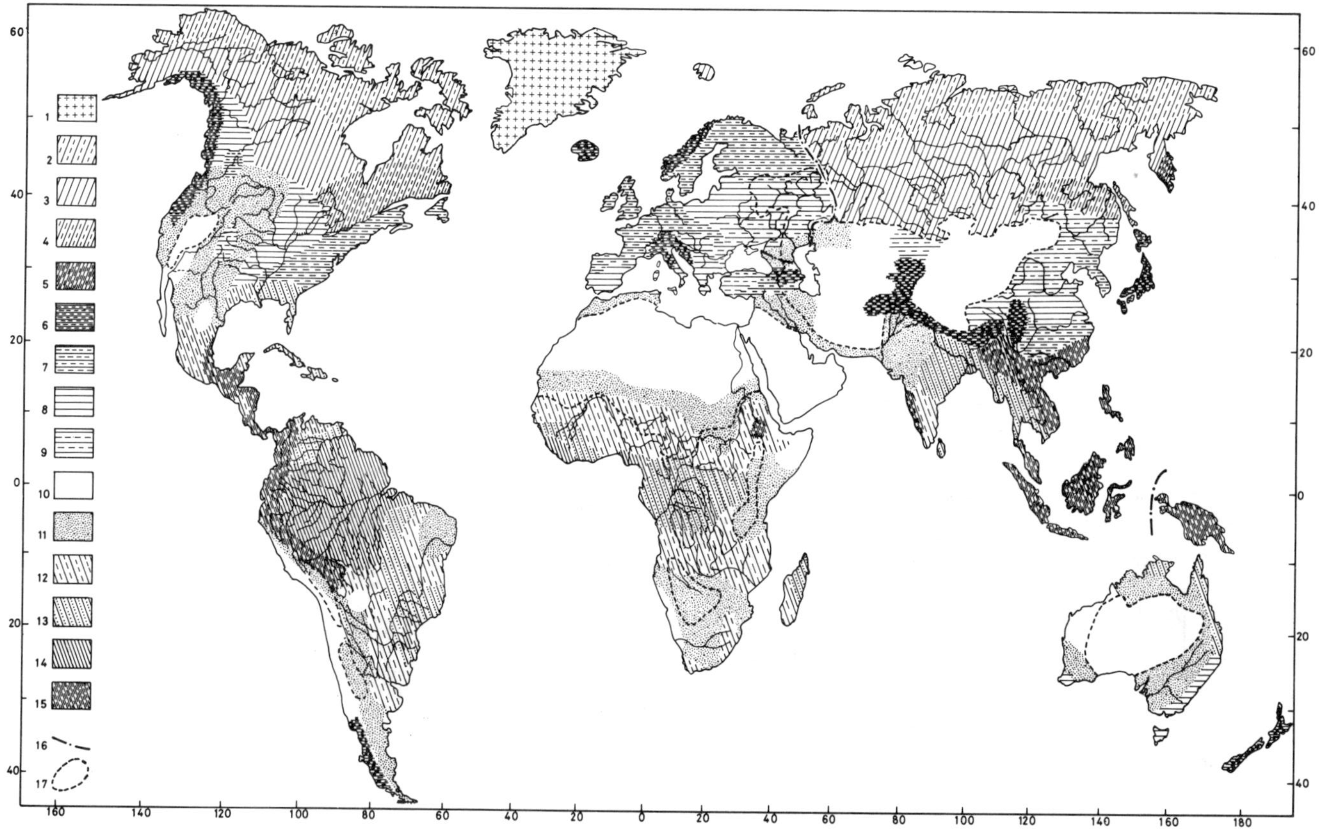

the major sources of dissolved elements: the chemical weathering (Ce), the atmospheric contribution from oceanic salts (Ca) and the overall pollution (Cp). As has been discussed, we do not recommend computation and use of continental averages. However, one must note that the differences between budgets of Table I are even greater at this scale, not only for the poorly known continents such as Africa, for which the average ionic content values are Ct = 121 (Livingstone, 1963a), 72 (Lvovic, 1972), 88 (Aurada 1980), 45.8 (Meybeck, 1979), 151 (Alekin, 1978) mg/l, but also for the better studies area of North America where Ct = 142, 69, 208, 126, 90. This example indicates the difficulties of establishing and using global budgets.

Meybeck's (1979) budget based on specific dissolved transport and mean temperatures of the watershed is shown in Table III and Figure 2. Two major criticisms can be made of this budget: i) extrapolation does not take into account the watershed lithology, and ii) the influence of oceanic salts is not considered. However, it is likely that the budget is still valid for two reasons: i) river specific runoff, q, is the second factor, after watershed lithology, that regulates ionic export and probably the first factor in silica export (Meybeck, 1976, 1981, 1986); ii) by coincidence in most cases, there is some lithologic zonality, and the data from the "unknown" rivers of the temperate zone, such as the Chinese rivers, are very similar to the extrapolated values used in 1979. Therefore, the inclusion of Hu Ming Hui et al. (1982) data on these rivers in the 1979 budget does not change the budget significantly (on the order of 1%).

In the corrected budget presented in Table II, the atmospheric contribution (Ca) has been revised and a distinction has been made between coastal inputs (100 km stretch from the ocean) and inland inputs (details in Meybeck, 1983). The 1979 budget has then been corrected for Ca and Ct; the weathering source (Ce) and the pollution inputs (Cp) have remained unchanged.

The result is a doubling of Na and Cl fallout on land. It is mainly caused by the assumption that average coastal fallout of Cl was 6 t $km^{-2}yr^{-1}$ over an area of 25×10^6 km^2. This is extremely difficult to check and eventual studies may find the atmospheric contribution (Ca) to be intermediate between the 1979 and 1983 estimates, based on the Cl transport rates in coastal unpolluted regions and on islands, such as Iceland (4.8 t $km^{-2}yr^{-1}$), Guyana (3.9), New Zealand (8.6), Norway (4.7), Papua New Guinea (8.5), and Indonesia (5.5). Better consideration of this contribution would not change much our understanding of geochemical cycles but would greatly clarify our interpretation of river water quality at any scale.

Improvement of major elements budgets will next come from water quality surveys in some very humid parts of the world still poorly known, from more study of anthropogenic inputs to surface waters, which includes long term trends and pristine level determination, and from the development of a global rainfall quality network still very sparse, particularly in the Southern Hemisphere.

5. NUTRIENTS AND ORGANIC CARBON

The DOC, nitrate, and phosphate budgets are much influenced by anthropogenic activities, and it is useless to determine river transport at a given time if separate budgets of natural and anthropogenic inputs are not considered. The natural levels (Table IV) are derived from major tropical rivers still unpolluted and from small representative watersheds for the temperate regions (Meybeck, 1982). For anthropogenic N and P inputs the budget type is the one described in section 3.4.

Table IV. Average nutrient levels in rivers (in μg of N, P, C/liter)

	NO_3^- (1)	NH_4^+ (1)	NO_2^- (1)	DON (2)	PO_4^{3-} (1)	DOC (3)	POC (3)(4)	DOC/DON (3)	POC/PON (3)
NATURAL STATE	100	15	1	170	10	5 500	4 500	30	8.5
PRESENT STATE (1) Cp	←	190	→	?	10	?	?		

DON=dissolved organic nitrogen; DOC and POC=dissolved and particulate organic carbon. Cp=additional content in 1970.

(1) Meybeck (1982); (2) Ittekot et al. (1983); (3) update of Meybeck (1982), see text; (4) weighted average: 54% of average river waters have POC content <3 mg/l.

On the basis of the SCOPE Carbon Program data (Kempe, 1985), the DOC averages for climatic zones can be updated: tundra, 2 mg/l (same as in Meybeck, 1982); taiga, 7 mg/l (average of Mackenzie, Yukon and Alaskan rivers); temperate, 4 mg/l (average of St Lawrence, Columbia, USA West Coast rivers); wet tropics, 8 mg/l (average of Orinoco, Caroni, Zaire); dry tropics, 3 mg/l (Niger); semi-arid, 1 mg/l (Orange). Taking into account the relative amount of water discharged from these environments, respectively (see Table III): 3.0, 11.7, 27.5, 51.3, 5.8, 0.7%, the discharge-weighted DOC average is 5.5 mg/l compared to 5.75 mg/l, an earlier estimate. Degens and Ittekot (1985) estimate DOC between 3 and 6.7 mg/l. Similar updating of natural nitrate and phosphate budgets cannot be made with the SCOPE data; the influence of man is much greater than for DOC and this program does not provide much new information on natural levels.

As concerns TOC (Total Organic Carbon), the last budget of Kempe (1985) is based on 100 rivers representing 18,369 km^3/yr, or about 49% of river waters. The total TOC load was simply extrapolated to the rest of the world; the resulting TOC input to ocean is 280×10^6 t/yr . Schlesinger and Melack's (1981) first budget (370×10^6 t/yr) was based on an average TOC content determined by an autocorrelation between TOC loads and water discharge, while the Mantoura and Woodward (1983) estimate (780×10^6 t/yr) was simply derived from the world terrestrial primary production, about 100 times the TOC export rate (Meybeck, 1982), a relationship checked in the Severn watershed only. On the basis of these data, an updated (Table V) total input is lowered from 380 to 300×10^6 t/yr, a value closer to Kempe's estimate. However, one must note that these specific loads are mostly determined on lowland rivers. Consideration of highland rivers from South East Asia could increase this estimate.

Budgets of Particulate Organic Carbon (POC) should take into account a well-established inverse correlation between the POC content and suspended sediment content (SS) of rivers (Meybeck, 1982; Milliman et al., 1984; Degens and Ittekot, 1985). This relation is presented in Figure 3. The decrease of POC in highly turbid rivers is limited to values approaching the average content of shales, around 0.5%. Such lower values can be reached in some rivers at SS>500 mg/l (Degens and Ittekot, 1985); on the contrary, Milliman et al. (1984) found much higher limits (1 to 2%) for the highly turbid Yang Tse Kiang. New measurements of the Huang Ho, the

Table V. TOC budget to the ocean, an update.

Environment	TOC export rates ($t\ km^{-2}\ yr^{-1}$)		TOTAL AREA $10^6\ km^2$	TOC load ($10^6\ t\ yr^{-1}$)	
	Meybeck (1982)	This work		1982	This work
tundra	0.6		7.55	4.5	
	1.9	2.1(1)	23.4		49.1
taiga	2.5		15.85	39.6	
temperate	4.0	3.1(2)	22.0	88	68.2
wet tropics		7.15(3)	24.1		172.3
	6.5			241	
dry tropics		0.7(4)	13.1		9.2
semi-arid	0.3	0.2(5)	15.5	4.6	3.1
desert	0	0	1.7	0	0
TOTAL	3.8	3.0	99.9	378	302

(1) average for Siberian rivers, $16.7\times10^6\ km^2$ (Romankevich and Artemeev, 1985), (2) weighted average Yang Tse, Mississippi and Parana (Kempe, 1985), (3) weighted average Amazon, Zaire, Orinoco (Kempe, 1985), (4) average of Niger and Gambia (Kempe, 1985), (5) value for Orange River (Kempe, 1985).

world's highest SS levels of major rivers, confirm low levels (around 0.6%, G. Cauwet, personal communication). Therefore, it does not seem to be necessary to update the POC vs. SS variation as Milliman et al. (1984) did after their study (their estimate of global POC input to ocean is 280×10^6 t/yr). However, as the global suspended load is highly dependent on the Huang Ho figure, which keeps changing, any decrease of its load will decrease the estimated POC load; therefore, if a total Huang Ho load of 1080×10^6 t/yr is chosen (Milliman and Meade, 1983) for this river with an average POC content of 0.6%, the earlier estimate, based on a much higher sediment load (3180×10^6 t/yr) and a guess of 0.5% POC content, becomes 169 instead of 179 million tons per year.

The SS-weighted world averages for POC are 1.1% and 4.5 mg/l. However, due to the skewed distribution of SS, 70% of river waters are characterized by very different values: POC between 1.6 and 6% of SS, and from 1.5 mg/l to 5.0 mg/l.

The budget of Particulate Organic Nitrogen (PON) is strictly derived from the POC budget and a relationship between POC and PON in river waters; the C/N ratio (g/g) ranges from 7 to 10 in most rivers (Meybeck, 1982). Once again, the SCOPE data were used for updating (Degens, 1982; Degens et al., 1983; Degens et al.,

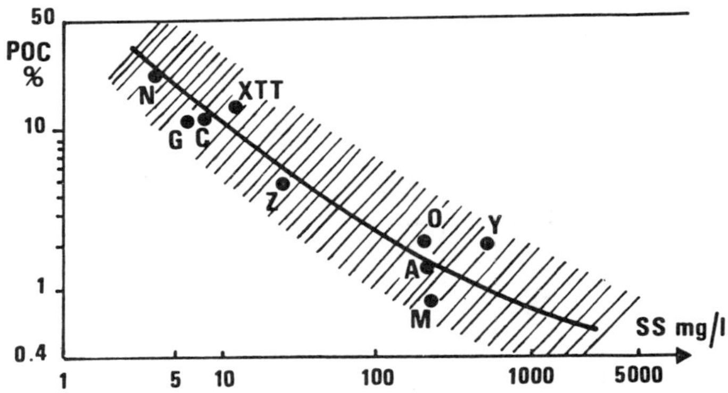

Figure 3. Evolution of particulate organic carbon (POC), in % of suspended matter (SS in mg/l), with SS. Shaded area: global variation given by Meybeck (1982). Update: N=Rio Negro, M=Madeira, XTT=Xingu Tapajos Tocantins, and A=Amazon (Gordeev et al., 1985); G=Gambia (Lesack et al., 1984); O=Orinoco and C=Caroni (Lewis and Weibezahn, 1981); Y=Yang Tse Kiang (Milliman et al., 1984), yearly average. Correction for Zaire incorrectly plotted in 1982 figure.

1985): Parana (C/N=11.1), St. Lawrence (C/N=8.5), Niger (C/N=13.3), Amazon (C/N=10.7), Gambia (C/N=8.6), Mackenzie (C/N=18.5), and Indus (C/N=16.7). The latter two values are somewhat higher, but other studies on the Yang Tse Kiang (C/N=8.9) by Milliman et al. (1984) and the Amazon (C/N=8.6) by Edmond et al. (1981) confirm our 8.5 average based on the Mississippi (C/N=8; Malcolm and Durum, 1976) and ten other rivers. The median value for 23 rivers is between 8 and 9. The resulting PON input to ocean is 20×10^6 t/yr. Ittekot et al. (1983) have estimated the PON budget on the basis of average content in waters sampled each month. Their sample of river waters (29% of exoreic inputs) is also biased by the weight of less turbid waters, the Amazon and Zaire; their world PON input (8.16×10^6 t/yr) is therefore much lower than our estimate. On the contrary, their dissolved organic nitrogen budget (6.3×10^6 t/yr or 0.17 mg N/l) is better documented than ours (0.26 mg N/l) and should be preferred as it is mostly based on slightly or not polluted rivers from various climatic types. A summary of C, N, and P updated contents in rivers is given in Table IV.

6. DISSOLVED TRACE ELEMENTS

For trace elements, the atmospheric inputs (Ca), weathering contribution (Ce), and pollution inputs (Cp) are difficult to decipher, and even natural levels (Ca+Ce) are still poorly known. In most cases the world averages are derived from few major rivers (Turekian, 1971). Recent reviews (Martin and Meybeck, 1979; Martin and Whitfield, 1983) revise some of the old values mostly established in the sixties. However, some major rivers, such as the Amazon, Orinoco, Zaire, Huang Ho, and Yang Tse Kiang, are presently under study and new estimates will have to be set up.

Three examples of budgets for strontium, fluoride and uranium are presented in this section.

A major objective in the strontium budget is to set up the contribution of each rock type, each characterized by different strontium isotope ratio (Wadleigh et al., 1985). Therefore, the budget developed in Table VI is based on the method of section

Table VI. The strontium budget based on Sr/Ca ratio in waters.

Rock type	Sr/Ca in waters (1) 10^{-3} mole/mole	Relative Ca^{++} contribution to river transport (2) %	Relative Sr^{++} contribution to river transport %
plutonic	5	0.6	1.2
gneiss + mica schist	5	1.1	2.2
other metamorphics (3)	6	2.8	6.6
volcanic	6	1.8	4.3
shale + sandstone	5	19.7	39.3
carbonate (4)	0.7	67	18.6
evaporites (gypsum)	10	7.0	27.8

(1) Meybeck (1986); (2) Meybeck (1984a and in prep.), global Ca^{++} input from weathering is 503×10^6 t/yr, see Table I (Ce); (3) serpentinite, amphibolite, marble excluded; (4) including marble and carbonate sandstone.

3.3 and the Sr/Ca ratio in homogeneous watersheds draining single rock-type areas studied in France (Meybeck, 1986). While Ca originates mainly from carbonate rocks which represent a minor part (16%) of surficial rocks, Sr comes mainly from non-carbonate sedimentary rocks (about 49% of earth surface), and from gypsum deposits (0.75% of earth surface). As a consequence, Sr isotopes input to ocean, of great interest for geochemists, will not depend much on crystalline rocks but on sedimentary rocks. The average river Sr content is 75 μg/l. It is difficult to build a Sr budget by classical methods (sections 3.1 and 3.2) for lack of data and due to the great variation of Sr contents: from 3.6 μg/l in the Rio Negro (Furch, 1984) to more than 500 μg/l in gypsum-rich watersheds, such as the Rhone river. A very rough estimate can be made with six rivers representative of major climatic environments and lithologies, as suggested in section 9: using the Indus (Sr = 370 μg/l), Ob (110), Orange (140), Mackenzie (210), Solimoes (22) and Negro (3.6), a discharge-weighted average is 106 μg/l. Average values previously given (Durum et al., 1960; Turekian, 1971; Martin and Meybeck, 1979) are somewhat lower (60 μg/l).

The fluoride levels in rivers can be dramatically influenced by direct pollution (Martin and Salvadori, 1982). When considering, as for strontium, the natural levels in 200 uncontaminated homogeneous watersheds in France, a range between 40 μg/l and 500 μg/l was observed (Meybeck, 1986). Due to the occurrence of higher contents mainly linked to evaporites, the discharge-weighted average (90 μg/l) in these streams is higher than median value (60 μg/l). The first figure is probably valid as a world average, as already proposed by Carpenter (1969) and confirmed by the levels in some unpolluted rivers: 95 μg/l in the Amazon, 20 μg/l in the Negro (Williams, 1968), and 100 μg/l in the Mackenzie (Reeder et al., 1972). Fluoride concentrations in waters from industrialized countries are sometimes 20 times higher and the actual input to the ocean may have already increased by 35% from the natural value (Martin and Salvadori, 1982). Fluoride contribution from oceanic atmospheric inputs is probably of the order of the weathering-derived fluoride (Carpenter, 1969).

Dissolved uranium has been analysed in rivers more than many other trace elements for obvious reasons. As for Sr and F, uranium is highly variable, from 0.08 μg/l in the Zaire to more than 4 μg/l (Figuères et al., 1982), a range similar to those of Na and Ca (Meybeck, 1981, 1984b). Figuères and co-workers demonstrated a direct relationship between uranium and total dissolved solids (TDS), which probably implies that U is mostly derived from the most soluble sedimentary rocks. These authors have divided the U budget into i) a set of documented major rivers, Amazon, Zaire, Mississippi and Yenissei, representing 21% of water discharge but definitively biased towards poorly mineralized waters of the humid tropics in which the average content is 0.16 μg/l and ii) an extrapolated average of 0.26 μg/l based on the U-TDS variation, the range of TDS in world rivers, and their respective contributions to the world budget. This type of budget is a combination of those described in sections 3.2 and 3.3. The final weighted average is 0.24 μg/1.

It is outside the scope of this paper to review the world average of about forty trace elements as given by Martin and Meybeck (1979) and Martin and Witfield (1983). These estimates are continuously changing, but on the basis of our present-day knowledge on more than ten rivers, including some major ones, the following ranges of world averages can be attributed to seventeen trace elements (contents in μg/l): Al 40±20, Ag 0.4±0.2, As 1±0.5 (Seyler, 1985), B 30±20, Co 0.1±0.05, Cr 0.8±0.3, Cs 0.05±0.03, Cu 2±1, F 100±20, Fe 50±30, Mn 10±5, Mo 0.8±0.4, Ni 0.4±0.3, Rb 1.5±0.5, Sr 80±20, U 0.26 (Figuères et al., 1982), Zn 10±5.

These levels are comparable to the latest estimates of Martin and Whitfield (1983) with three exceptions (Co and Zn are lower and Sr is higher). Other trace element averages are still what we have called "pure guesses," i.e., based on less than 20% of water discharge and/or no typology or supporting parameters.

7. SUSPENDED MATTER

The most complete review of sediment rivers load is the one by Milliman and Meade (1983) based on a documented area of 50×10^6 km^2 and extrapolated with regional typology (see section 3.2). Their estimate is 13.5×10^9 t/yr. However, this value should be handled with care as some river loads are affected by river damming (Columbia, Parana, Volta, Zambezi, Nile, Indus, Niger, Orange). Taking into account the loads before damming (Meybeck, 1976), a budget based on transport rate typology is given in Table VII. When compared to Milliman and Meade's estimate, the figure is higher (17.8×10^9 t/yr), but considering the amount of sediment annually stored in reservoirs (on the order of 1.5×10^9 t/yr), their value, if corrected, would be within 20% of the estimate in Table VII.

Another sediment budget was calculated for POC (Meybeck, 1982) by distributing the amount of particulate matter carried to ocean into nine classes of suspended matter (SS) from 46 major rivers draining 55×10^6km^2 (data in Appendix). For each class of SS, a partial budget was determined, giving the following breakdown of river inputs (in 10^6 t/yr):

Table VII. River transport of suspended load to the oceans. A budget based on zonal averages for 13 morphoclimatic regions.

MORPHOCLIMATIC REGIONS (1)	DOCUMENTED BASINS		ALL BASINS	
	RIVERS USED	average Ts (3) t km^{-2} $year^{-1}$	Total area 10^6 km^2	Total load 10^6 t $year^{-1}$
dry tundra and taiga	Mackenzie, Yukon ; Kara, Laptev, Bering seas watersheds.	16	20	320
humid taiga	Barentz watershed ; Finland	25	3.15	79
wet taiga	Author's estimate	350	0.2	70
semi-arid temperate	Black, Aral, Caspian watersheds (2) ;	360	2.6	940
	Huang He only (5)	1 500	0.75	1 080
dry temperate	Baltic watershed, Shatt el Arab, Indus, Mississippi, Danube	190	6.85	1 300
humid temperate	Alps watersheds ; Columbia, Yang Tse	365	7.7	2 820
wet temperate (highlands)	Ganges, Brahmaputra	850	4.5	3 820
wet tropics (lowlands)	Orinoco, middle and lower Amazon watersheds	56	7.05	395
wet tropics (highlands)	Mekong, Magdalena, Ucayali, Maranon	440	7.8	3 430
wet tropics (mixed)	Zaire, Chao Phrya, Burma rivers	115	8.95	1 030
savanna	Niger, Chari (2), Parana, Blue Nile, Zambezi	90	13.1	1 180
arid	Murray, Orange, Nile, Rio Grande	75	15.5	1 160
desert		0	1.7	0
TOTAL				
cold regions			23.4 % (4)	2.7 % (4)
temperate			22.4 %	56.5 %
tropic			37.0 %	34.2 %
arid			17.2 %	6.6 %
exoreic drainage			99.9	17 600

For the Indus, Niger, Nile and Rio Grande, the predamming data have been considered.

(1) Meybeck (1979); see Table III, (2) endoreic watersheds used to set up the average Ts, (3) Ts specific transport in t km^{-2} yr^{-1}, (4) in % of total, (5) Milliman and Meade (1983), most other data are in Meybeck (1976). See Appendix.

SS range (mg/l)	5-15	15-50	50-150	150-500	500-1500	1500-5000	5000-15000	15000 (1)
Total load to oceans	11	204	960	2260	7000	2010	208	2760
% of total SS	0.07	1.3	6.2	14.7	45.5	13.0	1.3	17.9
% of water budget	3.8	18.2	32	20.2	23.4	1.8	0.07	0.45

(1) including Huang Ho new load, 1080×10^6 t/yr (Milliman and Meade, 1983).

The total load by this method is 15.4×10^6 t/yr. One must note that only the pre-man natural river load to the ocean is of interest, and it is difficult to estimate from present-day river transport: human activities have enhanced mechanical erosion and most of its products are still stored on the continents in small reservoirs and on flood plains.

According to Gordeev (1983), the amount of terrigeneous material received by oceans is highly variable (in 10^6 t/yr): 7.15 for the Pacific, 3.19 for the Atlantic and Arctic, 8.19 for the Indian Ocean, which corresponds to a total of 18.5×10^9 t/yr, a figure very similar to those mentioned earlier.

8. SUSPENDED MATTER COMPOSITION

The contents of the main components (Al, Ca, Fe, K, Mg, Mn, Na, P, Si, Ti) of river particulate matter has been given by Gordeev and Lisitzin (1978) and Martin and Meybeck (1979). The least variable major element is the most abundant one, Si (from 215 mg/g in the Ob to 300 in the Mackenzie, Danube, Syr Darya, Orinoco; an exception is noted for the Colorado, 350 mg/g). The most variable is calcium, which is closely linked to watershed lithology, ranging from less than 5 mg/g (Orinoco, Niger) to more than 90 (Amu Darya and Syr Darya). Aluminium contents present two modes, at 75 (cold and temperate rivers) and at 120 (tropical rivers). These data indicate that three sets of rivers can be considered: i) mountainous watersheds in tectonically active regions of the temperate and cold climates (Rockies, Alps, Himalayas); ii) tropical rivers draining old shields or resistant sediments (quartz, sandstone) which have been exposed to weathering for a long time; iii) the Huang Ho river that carries mainly silt originating from extensive thick loess deposits (Catt, this volume), characterized by its very high calcium content (6% Ca; Li et al., 1984). The average contents of the main elements in suspended matter are given in Table VIII. These results are comparable to those given in Meybeck (1981).

For the trace elements, the estimates given here rely on 20 or fewer rivers. A weighted average of six representative rivers is discussed in the next section for the more extensively analyzed elements: Ba, Co, Cr, Cu, Ni and Zn. The resulting values are very close (Table IX) to the medians for these elements, based on 25 to 30 rivers, as reported by Martin and Meybeck (1979).

Table VIII. Average composition of the three major categories of river suspended matter (10^{-6} g/g)

	Al	Ca	Fe	K	Mg	Mn	Na	P	Si	Ti	Ms (1)
tropical and warm arid zones (2)	114 000	7 500	61 700	18 300	9 600	890	5 100	1 600	264 000	7 300	40 %
cold and temperate zones (3)	75 000	31 500	46 600	23 000	12 500	1 100	8 000	1 350	283 000	4 900	54 %
Huang He (4)	80 000	60 000	32 000	(21 000)	13 000	800	9 000	1 100	270 000	4 000	6 %
world average this method	91 000	23 600	51 800	20 900	11 400	1 000	6 900	1 400	274 000	5 800	100 %
Gordeev and Lisitzin (1978)	83 000	25 000	51 000	15 000	12 500	1 100	10 000	600	255 000	4 000	

(1) Relative contribution to the sediment budget in % (see Table VII).
(2) Arithmetic average of Amazon, Zaire, Marbada, Niger, Nile, Orenoco, Parana, Magdalena (Al content $\geq$ 90 000)
(3) Arithmetic average of Yang Tse Kiang (4), Colorado, Danube, Ganges, Mississippi, Mackenzie, St. Lawrence, Yukon, Ob, Pechora, and various European rivers (Al content $\leq$ 90 000). Most analyses are in Martin and Meybeck (1979).

9. WORLD BUDGETS FROM SIX SAMPLES

Many trace elements in dissolved and particulate forms have not, and will not be, studied in many rivers. Some authors have suggested that an average of the Yukon and Amazon concentrations would be close to the world mean. This approach is discussed here for six major unpolluted rivers that may well represent the world average lithology and climate. Used here are the Mackenzie (for the arctic and subarctic environments), the Huang Ho (for the semi-arid and dry temperate regions), the Yang Tse Kiang (for the humid temperate regions), the Brahmaputra (less affected than the Ganges by human activities; for the wet mountainous regions), the Zaire (for the lowland tropical regions), and the Solimoes, the Amazon upstream of the confluence with Rio Negro (for the wet tropical highland regions). The contributions of each environment to the water budget, for dissolved elements averages, and to the sediment budget, for particulate matter averages, are given in Table IXA. The data are derived from Tables III and VII. The weighted values obtained for major elements (Table IXB and C) are comparable to the averages based on dozens to 100 major rivers. In a way, this is a validation of the method, particularly for those elements sensitive to lithology. The particulate trace elements (Table IXD) are also in close agreement with other averages. As concerns the dissolved elements, there is probably a bias; the atmospheric inputs are not sufficiently taken into account since most stations are away from river mouths.

Table IX. A world budget with six samples.

A. Relative weights of each river (in % of total input to oceans) :

	Mackenzie	Huang Ho	Brahmaputra	Yang Tse Kiang	Zaire	Solimoes
water budget	14.7	1.05	15.3	12.15	33.1	24.1
sediment budget	2.6	14	21	22.8	20.8	18.8

B. Average of dissolved major elements (compare to Ce + Ca, Meybeck (1979) ; Table II) (in mg/l) :

	SiO_2	Ca^{++}	Mg^{++}	Na^+	K^+	Cl^-	SO_4^{--}	HCO_3^-
This method	7.9	14.5	3.4	3.9	1.45	4.7	8.5	53.8

C. Average of major elements in particulate matter (compare to Table VIII) (in 10^{-3} g/g) :

	Al	Ca	Fe	K	Mg	Mn	Na	P	Si	Ti
This method	97.4	26	50.7	19.6	11	1.08	6.6	1.18	267	6.0

D. Average of trace elements in particulate matter (10^{-6} g/g) :

	Ba	Co	Cr	Cu	Ni	Zn
This method	627	23	116	98	75	236
Median of 25 rivers	-	18	120	65	-	320
Martin & Whitfield (1983)	600	20	100	100	90	250
Gordeev & Lisitzin (1978)	380	18	130	80	84	310

10. CONCLUSIONS

All the data on global compositional averages and river inputs to oceans must be regarded with care for reasons of numerous uncertainties associated with them:

i) The natural variations of dissolved elements found in running waters are at least one order of magnitude, for major watersheds, and two orders of magnitude, for small watersheds and some species, such as sodium, sulfate and fluoride. Therefore, world averages should not be used as standard references of pristine levels. The same remark probably holds for trace elements in particles. Moreover, global means are sometimes affected by extreme values, such as suspended solids concentration, and medians are more appropriate reference central values.

ii) Any documented world budget of dissolved element should give the intensities of the three major sources: weathering, oceanic aerosols, pollution. Final figures may greatly depend on the way the last two sources are taken into account in budgets.

iii) Fluxes to the ocean cannot be simply derived from river transport without extensive corrections, even for such dissolved elements as calcium and silica that can be trapped in lake sediments.

iv) Rivers are the major but not the only pathway of elements from land to ocean, and global mass balance should also consider direct groundwater inputs, glaciers inputs, eolian transport to oceans, leaching of fresh volcanic material, and marine processes that may also affect the quantity and quality of riverborne material.

v) Present-day river inputs should be used with caution in the modelling of geochemical cycles through geological times. The postglacial Quaternary period is characterized by the development of temperate climate, mechanical erosion, and extended fluvioglacial and periglacial (loess) deposits. Past river inputs should be based on relationships between transport rates and climate (runoff, temperature), and lithology (masses of evaporitic and carbonate rocks). They should consider the past exoreic and endoreic area, and the position of continents, with respect to equator as well as the amount of glaciated land (Holland, 1978).

At any level (lithology, geography, climate), river concentrations and transport rates are found to be highly heterogeneous. Limestone and evaporitic rocks, which cover about 17% of the continents surface are responsible for 63% of the river dissolved load derived from weathering. As concerns the geographic origins, the following examples of contributions (in % of total inputs to oceans) can be proposed (from Meybeck, 1979, 1982; Milliman and Meade, 1983):

Region	Water (%)	Area (%)	Silica (%)	Ions (%)	TOC (%)	Suspended Matter (%)	Pollutants (%)
wet and humid tropics	51	24	65	38	57		
temperate regions	27.5	22	20	40			90
Southeast Asia (1)	20	12		27		35	

(1) from Huang Ho to Ganges, including islands.

In order to answer the basic questions on river transport of not only "how much" but "where from," "how," "controlled by which environmentai parameter," "what range," "links with other element," much more work is needed. Some directions can be proposed.

1) Some fair estimates of river inputs to oceans can be set up on the basis of very few selected major rivers.
2) Origins and links with lithology of river material have to be studied more. Small (100 km^2) homogeneous representative watersheds can be considered for this purpose.
3) Relationships with climate and vegetation should be studied more, in 10^4 to 10^5 km^2 watersheds.
4) The influence of atmospheric inputs, particularly near the coast is not yet understood on a global scale.
5) Changes in composition of dissolved and particulate matter resulting from man's activities are not known in sufficient detail. Further studies may be based on river trend stations, representative polluted watersheds, and direct analyses of wastewaters.
6) Natural ranges of many elements (with the exception of Si, Ca, Mg, Na, K, Cl, S, C, Al, Fe, Mn, P, Ti) in riverborne material are still poorly known, as well as the causes of their variation.
7) As a whole, the number of reliable analyses on suspended matter of major rivers is still two orders of magnitude less than those available on dissolved substances.

REFERENCES

Alekin, O.A., 1978. Water erosion of land surface, *in* World Water Balance and Water Resources of the Earth. *Studies and Reports in Hydrology* **25**, Unesco Press, 663 p.

Alekin, O.A., and Brazhnikova, L.V., 1960. Contribution to knowledge of dissolved matter runoff at the earth' s surface. *Gidrochim. Mat.*, **32**, 12-34 (in Russian).

Alekin, O.A., and Brazhnikova, L.V., 1968. Dissolved matter discharge and mechanical and chemical erosion. *Assoc. Int. Hydrol. Sci., Publ.* **78**, 35-41.

Aurada, K.D., 1980. Steuerungsmechanismen des Ionenabflusses und der chemischen Denudation. *Acta Hydrochim. Hydrobiol.*, **8**, 6, 525-559.

Baumgartner, A., and Reichel, E., 1976. *The World Water Balance.* Elsevier, 179 p.

van Bennekom, A.J., and Salomons, W., 1981. Pathways of organic nutrients and organic matter from land to oceans through rivers, *in* Martin J., Burton D., and Eisma D. (eds.), *River Input to the Ocean System*: Unesco-Unep-Scope Workshop, Rome, 1979, 33-51.

Berner, R.A., 1971. Worldwide sulfur pollution of rivers. *J. Geophys. Res.* **76**, 6597-6600.

Berner, R. A., Lasaga, A. C., and Garrels, R. M., 1983. The carbonate-silicate geochemical cycle and its effect on atmospheric carbon dioxide over the past 100 million years. *Am. J. Sci.*, **283**, 641-683.

Berner, R.A., and Raiswell, R., 1983. Burial of organic carbon and pyrite sulfur in sediments over Phanerozoic time: a new theory. *Geochim. Cosmochim. Acta* **47**, 855-862.

Blatt, H., and Jones, R.L., 1975. Proportions of exposed igneous, metamorphic and sedimentary rocks. *Geol. Soc. Am. Bull.*, **86**, 1085-1088.

Brass, G.W., 1978. Magnesium in continental runoff. *Geochim. Cosmochim. Acta* **42**, 1459-1462.

Burton, J.D., and Liss, P., 1973. Process of supply and removal of dissolved silicon in the oceans. *Geochim. Cosmochim. Acta* **37**, 1761-1773.

Carbonnel, J.P., and Meybeck, M., 1975. Quality variation of the Mekong river at Phnom Penh, Cambodia, and chemical transport in the Mekong basin. *J. Hydrol.*, **27**, 249-265.

Carpenter, R., 1969. Factors controlling the marine geochemistry of fluorine. *Geoch. Cosmochim. Acta* **33**, 1153-1167.

Clarke, F.W., 1924. Data of Geochemistry. 5th ed., *U.S. Geol. Surv. Bull.* **770**, 841 p.

Conway, E.J., 1942. The chemical evolution of the ocean. *Irish Acad. Sci. Proc.*, **4813**, 119-159 and 161-212.

Corbel, J. 1964. L'érosion terrestre étude quantitative. *Ann. Géographie,* **398**, 385-412.

Degens, E., 1982. Transport of carbon and minerals in major world rivers. *Mitt. Geol. Palaeont. Inst. Univ. Hamburg,* **52**, 766 p.

Degens, and Ittekot, V., 1985. Particulate organic carbon, an overview, *in* E. Degens, S. Kempe, and H. Herrera (eds.), Transport of Carbon and Minerals in Major World Rivers, Part III. *Mitt. Geol. Palaeont. Inst. Univ. Hamburg,* **58**, 7-28.

Degens, E., Kempe, S., and Soliman, H., 1983. Transport of carbon and minerals in major world rivers. Part 2. *Mitt. Geol. Palaeont. Inst. Univ. Hamburg,* **55**, 534 p.

Degens, E., Kempe, S., and Herrera, H. (eds.), 1985. Transport of carbon and minerals in major world rivers. Part 3. *Mitt. Geol. Palaeont. Inst. Univ. Hamburg,* **58**, 645 p.

Durum, W.H., Heidel, G., Tison, L.J., 1960. Worldwide runoff of dissolved solids. *Int. Ass. Sci. Hydrol., Publ.* **51**, 618-628.

Edmond, J.M., Boyle, E.A., Grani, B., and Stallard, R.F., 1981. The chemical mass balance in the Amazon plume. I. The nutrients. *Deep-Sea Res.* **28A,** 11, 1339-1374.

Erikson, E., 1960. The yearly circulation of chloride and sulfur in nature, meteorological, geochemical and pedological implications. *Tellus,* **12**, 63-109.

Figuères, G., Martin, J.M., and Thomas, A., 1982. Apport par les fleuves d'uranium dissous a l'océan; exemple du Zaire. *Oceanol. Acta* **5**, 141-147.

Fournier, F., 1960. *Climat et Erosion.* Presses Universitaires de France, Paris,

Fournier, F., 1969. Transports solides effectués par les cours d'eau. *Bull. Int. Ass. Sci. Hydrol.* **14**, 7-49.

Furch, K., 1984. Water chemistry of the Amazon. The distribution of chemical elements among freshwaters, *in* H. Sioli (ed.), *The Amazon. Limnology and Landscape Ecology of a Mighty Tropical River and Its Basin,* Junk and Pudoc, pp. 167-199.

Garrels, R.M., and Mackenzie, F.T., 1971. *Evolution of Sedimentary Rocks.* W.W. Norton, New York, 397 p.

Garrels, R.M., and Lerman, A., 1984. Coupling of the sedimentary sulfur and carbon cycles—an improved model. *Am. J. Sci.,* **284**, 989-1007.

Gordeev, V.V., and Lisitzin, A.P., 1978. Average chemical composition of suspended matter in world river and river inputs to the oceans (in Russian). *Dokl. Akad. Nauk. SSSR,* **238**, 1 225-228.

Gordeev, V.V., 1983. *River Inputs to the Oceans and Their Chemical Characteristics.* Nauka Publishers, Moscow, 160 pp. (in Russian).

Gordeev, V.V., Miklishansky, A.Z., Migdisov, A.A., Artemyev, V.E., 1985. Rare element distribution in the surface suspended material of the Amazon river, some of its tributaries and estuary, *in* Degens E., Kempe S., and Herrera H. (eds.), *Transport of Carbon and Minerals in Major World Rivers.* Mit. Geol. Palaeont. Inst. Univ. Hamburg, pp. 225-244.

Holeman, J.N., 1968. The sediment yield of the major rivers of the world. *Water Res. Res.* **4**, 737-747.

Holland, H.D., 1978. *The Chemistry of the Atmosphere and Oceans.* Wiley-Interscience, 351 p.

Hu, Ming Hui, Stallard, R.F., and Edmond, J.M., 1982. Major ion chemistry of some large Chinese rivers. *Nature,* **298**, 550-553.

Ittekot, V., Martins, O., and Seifert, R., 1983. Nitrogenous organic matter transported by the major world rivers. *Mitt. Geol. Palaeont. Inst. Univ. Hamburg,* **55**, 119-127.

Ivanov, M.V., 1981. The global biogeochemical sulfur cycle, *in* G.E. Likens (ed.), *Some Perspectives of the Major Biogeochemical Cycles.* SCOPE 17, John Wiley, 61-80.

Jansen, J.M.L., and Painter, R.B., 1974. Predicting sediment yield from climate and topography. *J. Hydrol.* **21**, 371-380.

Kempe, S., 1979. Carbon in the Rock Cycle, *in* B. Bolin, E.T. Degens, S. Kempe, and P. Ketner (eds.), *The Global Carbon Cycle*. SCOPE 13, John Wiley, 343-377.

Kempe, S., 1982. Long-term record of the CO_2 pressure fluctuations in fresh water, *in* E. T. Degens (ed.), Transport of Carbon and Minerals in Major World Rivers. Part I. *Mitt. Geol. Palaeont. Inst. Univ. Hamburg,* **52**, 91-332.

Kempe, S., 1985. Compilation of carbon and nutrients discharge from major world rivers, *in* E. T. Degens, S. Kempe, and H. Herrera (eds.), Transport of Carbon and Mineral in Major World Rivers, Part III. *Mitt. Geol. Palaeont. Inst. Univ. Hamburg,* **58**, 29-32.

Lesack, L.F.W., Hecky, R.E., and Melack, J.M., 1984. Transport of carbon, nitrogen, phosphorus, and major solutes in the Gambia river, West Africa. *Limnol. Oceanogr.*, 816-830.

Lewis, W.M., and Weibezahn, 1981. The chemistry and phytoplankton of the Orinoco and Caroni rivers. *Archiv. Hydrobiol.* **91**, 4, 521-528.

Li, Y.H., Teraoka, H., Yang, T.S., and Chen, J.S., 1984. The elemental composition of suspended particles from the Yellow and Yang Tse rivers. *Geochim. Cosmochim. Acta* **48**, 1561-1564.

Likens, G.E. (ed.), 1981. *Some Perspectives of the Major Biogeochemical Cycles*. SCOPE 17, John Wiley, 175 p.

Livingstone, D.A., 1963a. Chemical composition of rivers and lakes, *in* Data of Chemistry. *U.S. Geol. Survey Prof. Paper* **440** G, 1-64.

Livingstone, D.A., 1963b. The sodium cycle and the age of the ocean. *Geochim. Cosmochim. Acta* **27**, 1055-1069.

Lvovich, M.I., 1972. World water balance. *Int. Ass. Sci. Hydrol. Reading Symp.*, 401-415.

Malcolm, R.L., and Durum, W.H., 1976. Organic carbon and nitrogen concentration and annual organic load for six selected rivers of the USA *U.S. Geol. Survey Water Supply Paper* **1817**, 21 p.

Mantoura, R., and Woodward, E., 1983. Conservative behaviour of riverine dissolved organic carbon in the Severn estuary: chemical and geochemical implications. *Geochim. Cosmochim. Acta* **47**, 1293-1309.

Martin, J.M., Hogdahl, O., and Philippot, J.C., 1976. Rare earth elements supply to the ocean. *J. Geophys. Res.* **81**, 18, 3119-3124.

Martin, J.M., Burton, D., and Eisma, D. (eds.), 1981. *River Input to the Ocean System*. Unesco-Unep-Scope Workshop, Rome, 1979, 384 pp.

Martin, J.M., and Meybeck, M., 1979. Elemental mass-balance of material carried by world major rivers. *Marine Chem.*, **7**, 173-206.

Martin, J.M., and Salvadori, F., 1983. Fluoride pollution in French rivers and estuaries. *Est. Coast. Shelf. Sci.* **17**, 231-242.

Martin, J.M., and Whitfield, M., 1983. The significance of the river input of chemical elements to the ocean, *in* C.S. Wong (ed.), Trace Metals in Seawater. *Proc. NATO Adv. Res. Inst., March 1981,* Erice, Plenum Press, 265-296.

Meade, R.H., 1969. Errors in using modern stream load data to estimate natural rates of denudation. *Geol. Soc. Am. Bull.* **80**, 7, 1265-1274.

Meade, R.H., Nordin, C.F., Curtis, W.F., Costa Rodriguez, F.M., Do Vale, C.M., and Edmond, J.M., 1979. Sediment loads in the Amazon river. *Nature,* **278**, 161-163.

Meybeck, M., 1976. Total mineral transport by world major rivers. *Hydrol. Sci. Bull.,* **21**, 265-284.

Meybeck, M., 1977. Dissolved and suspended matter carried by rivers: composition, time and space variations and world balance, *in* H.L. Golterman (ed.), *Interaction between sediments and fresh waters*. Amsterdam, Junk and Pudoc, pp. 25-32.

Meybeck, M., 1979. Concentration des eaux fluviales en éléments majeurs et apports en solution aux océans. *Rev. Geol. Dyn. Geogr. Phys. 21,* **3**, 215-246.

Meybeck, M., 1981. Pathways of major elements from land to ocean through rivers, *in* J.M. Martin, J.D. Burton, and D. Eisma (eds.), *River Input to the Ocean System*. Unesco-Unep-Scope Workshop, Rome, 1979, 18-30.

Meybeck, M., 1982. Carbon, nitrogen and phosphorus transport by world rivers. *Amer. J. Sci.*, **282**, April, 401-450.

Meybeck, M., 1983. Atmospheric inputs and river transport of dissolved substances. in, Dissolved loads of rivers and surface water quantity/quality relationships. Proc. IASH Symp. Hamburg, August 1983, *Int. Ass. Hydrol. Sci. Publ.* **141**, 173-192.

Meybeck, M., 1984a. Les fleuves et le cycle geochimique des éléments. Thèse de Doctorat d'Etat, Univ. Paris VI, 500 pp.

Meybeck, M., 1984b. Variabilité de la composition chimique naturelle des eaux courantes. *Verh. Internat. Verein. Limnol.* **22**, 1766-1774.

Meybeck, M. 1986. Composition chimique des ruisseaux non pollués de France. *Sci. Géol. Bull. Strasbourg,* **39**, 1, 3-77.

Milliman, J.D., and Meade, R.H., 1983. World-wide delivery of river sediment to the oceans. *J. Geology,* **91**, 1-21.

Milliman, J.D., Quinchun, X., and Zuosheng, Y., 1984. Transfer of particulate organic carbon and nitrogen from the Yang Tse river to the ocean. *Am. J. Sci.*, 824-834.

Parde, M., 1953. La turbidité des rivières, ses facteurs géographiques. *Rev. Geogr. Alpine* **41**, 399-421.

Reeder, S.W., Hitchon, B., and Levinson, A.A., 1972. Hydrogeochemistry of the surface waters of the Mackenzie river drainage basin, Canada. I. Factors controlling inorganic composition. *Geochem. Cosmochim. Acta* **36**, 825-865.

Romankevich, E.A., and Artemyev, V.E., 1985. Import of organic carbon into seas and oceans bordering the territory of the Soviet Union. *Mitt. Geol. Palaeont. Inst. Univ. Hamburg,* **58**, 459-469.

Schlesinger, W.H., and Melack, J.M., 1981. Transport of organic carbon in the world's rivers. *Tellus,* **33**, 172-187.

Seyler, P., 1985. Formes chimiques et comportement de l' arsenic dans le milieu estuarien. Thèse de Doctorat, Univ. Paris VI, 193 p.

Soviet IHD, 1978. World water balance and water resources of the earth, by the U.S.S.R. Committee for the International Hydrological Decade. *Studies and Reports in Hydrology,* **25**, Unesco Press, 663 p.

Stallard, R. F., 1980. Major element geochemistry of the Amazon river system. Ph. D. Thesis, MIR-WHOI, *Woods Hole Inst. Oceanogr., WHOI-80* **29**, 362 p.

Stallard, R. F., and Edmond, J.M., 1981. Geochemistry of the Amazon. 1. Precipitation chemistry and the marine contribution to the dissolved load at the time of peak discharge. *J. Geophys. Res.,* **86**, 9844-9858.

Stallard, R.F., and Edmond, J.M., 1983. Geochemistry of the Amazon. 2. The influence of geology and weathering environment on the dissolved load. *J. Geophys. Res.,* **88**, 9671-9688.

Stumm, W., and Morgan, J.J., 1981. *Aquatic Chemistry,* 2nd Ed. Wiley Interscience, 780 p.

Turekian, K., 1971. Rivers, tributaries and estuaries, *in* D.W. Hood (ed.), *Impingement of Man on the Oceans*. Wiley, New York, 9-73.

UNESCO (1969 and 1971). Discharge of selected rivers of the world. *Studies and Reports in Hydrology,* Vol. **I, II, III.**

UNESCO, 1974. *Gross sediment transport into the oceans*. Unesco-IAHS, internal report, Division of Water Sciences.

UNESCO, 1979. *World register of rivers discharging to the oceans*. Division of Water Sciences, Technical Paper.

Vallentyne, J. R., 1978. Today is yesterday's tomorrow. *Verh. Intern. Ver. Limnol.,* **20**, 1-12.

Wadleigh, M., Veizer, J., and Brooks, C., 1985. Strontium and its isotopes in Canadian rivers: fluxes and global implications. *Geochim. Cosmochim. Acta* **49**, 1727-1736.

Williams, P.J.L., 1968. Organic and inorganic nutrients in the Amazon river. *Nature,* **218**, 937-938.

Wollast, R., and Mackenzie, F.T., 1983. The global cycle of silica, *in* Aston (ed.), *Silicon Geochemistry and Biogeochemistry.* Academic Press, 39-76.

Annex. General features of the world's biggest rivers (Meybeck, 1976, updated and corrected).

River	Discharge, Q [m³ s⁻¹]	Area, A [× 10³ km²]	Runoff, q [l. s⁻¹ km⁻²]	Total mineral content, S [mg l.⁻¹]	Dissolved transport, Td [t km⁻² year⁻¹]	Solid transport, Ts [t km⁻² year⁻¹]	$\frac{Ts}{Td}$	Annual mineral transport, Md [× 10⁶ t year⁻¹]	References‡
Amazon*	175 000	6 300	28	52.4	46.4	143	3.1	290	25; 25; k
Amur	11 000	1 850	6.0	57	10.9	13.6	1.2	20	61; 61; 61
Amu-Daria	1 450	450	3.2	600	60	209	3.5	27	61; 61; 61
Brahmaputra	19 300	580	33	124	130	1 370	10.5	75	71; 50; 71
Chari	1 320	600	2.2	64	4.4	6.5	1.5	2.65	56; 56; 56
Colorado	640	635	1.0	700	23	870	46	12	34; 34; 16
Columbia	7 960	670	11.9	139	52	43	0.8	34.8	72; 72; 34
Congo	39 200	4 000	9.8	38	11.7	13.2	1.1	47	21; a; 62
Danube	6 430	805	8.0	300	75	84	1.1	60	43; 43; 43
Dniepr	1 650	500	3.3	210	22	2.2	0.1	11	61; 61; 61
Don	870	420	2.1	495	33	13.6	0.4	13.8	61; 61; 61
Ganges	11 600	975	12.0	208	78	537	6.9	76	71; 50; 71
Huang Ho (Yellow)	1 480	745	3.4	450	45	1 080	24	–	30; f; 54; i; j
Indus	6 700	950	7.0	295	65	500	8	68	30; b; c
Irrawady	13 400	430	31.2	215	211	700	3.3	–	41; g; –
Kolyma	3 150	645	5.8	–	–	9.3	–	–	61; –; 61
Lena*	16 300	2 430	6.7	165	36	6.3	0.15	85	61; 61; 61
Mackenzie	9 600	1 800	5.3	230	39	65	1.7	70	9; 9; 9
Madeira*†	32 000	1 380	23	60	42	157	3.7	–	25; 25; 25
Magdalena	7 500	240	31	120	117	1 000	9	28	41; b; 52
Maranon†	11 000	407	27	105	90	250	2.8	–	25; 25; 25
Mekong	18 300	795	23	105	75	435	5.8	59	71; 50; 71
Mississipi*	18 400	3 267	5.65	223	40	94	2.3	131	16; 11; 11
Murray	740	1 070	0.68	380	8.2	30	3.7	8.8	–; 20; 44
Negro†	45 300	755	60	5	10	10.1	1.0	–	25; 25; 25
Nelson	3 500	1 150	3.1	280	27	–	–	31	45; 20; –
Niger	6 100	1 125	5.5	53	9.0	60	6.7	10	21; 29; 44
Nile*	2 830	3 000	0.88	208	5.8	37	6.4	17.4	21; e; 60
Ob*	12 350	2 500	4.9	130	20	6.3	0.3	50	61; 61; 61
Orange	300	800	0.39	182	1.8	150	80	12	18; 18; 44
Orinoco*	30 000	950	31.6	52	52	91	1.7	50	69; 45; 30
Parana*	18 000	2 800	6.4	100	20	40	2	56	17; 7; 17
Rio Grande	100	670	–	482	–	–	–	3.5	72; –; –
St. Lawrence	10 700	1 025	10.4	161	53	5.0	0.1	54	d; 20; 24
Sao Francisco	2 900	470	6.2	–	–	–	–	–	69; –; –
Si Kiang (Pearl)	7 800	1 350	22.3	140	98	–	–	–	71; i; 71
Shatt El Arab	1 750	410	4.3	320	43	250	5.8	18	69; b; 65
Tapajos†	7 200	500	14.4	8	3.6	1.2	0.3	–	25; 25; 25
Tocantins	11 000	900	12.2	–	–	–	–	–	41; –; –
Ucayali†	9 700	400	24.0	180	136	307	2.0	–	25; 25; 25
Uruguay	5 000	350	14.3	50	23	70	3.0	8	17; 45; 17
Volga	8 400	1 350	6.2	290	57	19	0.3	77	61; 61; 61
Xingu†	7 800	540	14.4	6.5	2.9	0.9	0.3	–	25; 25; 25
Yang Tse Kiang*	28 000	1 950	14.35	185	85	250	3.0	–	41; f; 54; i; j
Yenissei*	17 200	2 600	6.7	135	28	5.1	0.2	73	61; 61; 61
Yukon	6 200	770	8.0	174	44	103	2.3	34	69; 72; 9
Zambezi	7 100	1 340	5.3	70	11.5	75	6.5	15.4	41; 12; 44
Quebec	15 750	880	17.9	30	17	–	–	15	51; 51; –
Japan	17 400	370	47	128	58	–	–	47	36; 36; –
Finland	3 500	380	9.2	49	14	3.2	0.2	5.3	38; 38; 38

* 'Continental' rivers.
† Tributaries of the Amazon
‡ Respectively for A and Q, S, and Ts.

(b) Unpublished
(d) At Quebec City, recomputed from (51)
(e) Unpublished analysis plus (31)
(a) Unpublished analyses plus (66).
(c) Estimated

(f) SCOPE-carbon Degens ed. (1982)
(g) Kobayashi (1962)
(h) Milliman et Meade (1983)
(i) GEMS-WATER (1987)
(j) Hu Ming Hui *et al.* (1982)
(k) Meade *et al.* (1979)

TRANSPORT AND DEPOSITION OF SUSPENDED MATTER IN ESTUARIES AND THE NEARSHORE SEA

D. Eisma
Netherlands Institute for Sea Research
P. O. Box 59
1790AB Den Burg, Texel
Netherlands

1. INTRODUCTION

Rivers carry yearly about 13.5×10^9 tons of suspended matter towards the sea, together with 1 to 2×10^9 tons of bed load (sand and gravel) and about 4×10^9 tons of dissolved material (Meybeck, 1976; Milliman and Meade, 1983). The supplied sand and gravel go into the formation of banks, channel floors and beaches, sometimes with considerable longshore transport. At many river mouths, however, only very small amounts of bed load reach the sea, most of the supply being deposited in the lower reaches, of the river or in the estuary. There the sand and gravel beaches and near-shore sand and gravel deposits are primarily built up with reworked older material. Off some large rivers sand is moved towards the deep sea but does not reach it: at the Zaire the bed load moving into the canyon (medium to coarse sand) is stored somewhere in the canyon and very little of it is found on the submarine fan at 2000 to 5000 m water depth.

The material in suspension is partly retained in the river mouths, partly deposited on the shelf. Only a very small part (less than about 10%; Drake, 1976) is transported into the deep sea, which occurs mainly where a rivermouth is located near to, or at, the shelf edge (the Niger, the Mississippi) or where a canyon funnels suspended material downward (the Zaire, the Ganges-Brahmaputra). The supply of particulate organic matter from rivers is of the order of 0.4×10^9 tons per year, the average organic matter content of river-supplied suspended matter being between 1% and 5%, ranging from less than 0.5% to more than 20% (Meybeck 1981). This supply is small compared to the amount of organic matter yearly produced in the estuaries and the coastal sea, which is estimated to be of the order of 14×10^9 tons (De Vooys, 1979; Wollast and Billen, 1981).

Suspended sediment is also supplied to estuaries and the coastal sea from other sources. Considerable amounts of sediment can come from coastal runoff, from erosion of older deposits along the shore or in the near-shore sea, from the atmosphere (mostly dust), from glaciers, from waste dumping, and from organisms. In the estuaries of large rivers with a large suspended matter load (Amazon, Chang Jiang, Hoang He) these modes of supply are insignificant compared to the river supply, but in smaller estuaries they may be important. The runoff through small gullies and depressions comes only from a narrow strip of land near to the sea and its volume of water is negligible in quantity to the volume of water flowing seaward through rivers, but because suspended matter concentrations can be large, this supply can be of local

A. Lerman and M. Meybeck (eds.), Physical and Chemical Weathering in Geochemical Cycles, 273–298.

significance.

The importance of erosion and resuspension of older deposits depends very much on local or regional conditions, but in some areas can be appreciable, as along the US east coast or in the North Sea, where supply from erosion is of the same order or more than supply from the rivers (Meade, 1972; Eisma, 1981). Atmospheric supply can be important where arid or semi-arid areas occur near the coast, or where volcanoes, industries or mining activities produce much dust. In the North Sea the supply of atmospheric dust (chiefly from the densely populated and industrialized parts of northwestern Europe) is about 10% of the total amount of suspended sediment that is deposited in the North Sea. Glaciers reaching the coast can supply very large amounts of particulate material but these are usually deposited nearby. When icebergs are formed, however, and also when ice formed along the coast starts moving, frozen particulate material can be transported over large distances. Supply of particulate material from organisms—plankton or benthos—can be very large in the estuary as well as in the near-shore sea. Most of the organic matter and the opal (diatom frustules) produced by organisms, however, is consumed or mineralized so that only a very small amount, of the order of a few percent or less, is finally stored in bottom deposits. Carbonate-rich deposits (shell-beds, coral sands, carbonate muds) are formed where the sediment supply from other sources is small, or where carbonate material is sorted out. Finally, waste dumping can locally be an important source. In Long Island Sound, which receives waste from the New York-Boston area, waste discharges in 1960-1963 accounted for a suspended matter input of four times the natural river input, and of nearly nine times in 1964-1968 (Koppelman et al., 1976).

The recycling of sediment deposited in estuaries and the coastal sea can be almost immediate, when deposits are quickly eroded again, or it can take thousands, millions, or hundreds of millions of years. The minerals in the sediment (quartz, feldspars, clay minerals) are left over from weathering processes and are very resistant to further breakdown. The particles of biogenic origin (carbonate, opal, organic material) are usually less stable. The adsorbed or exchangeable substances in the suspended matter do not necessarily follow the bulk of the sediment. They can be mobilized again and follow much shorter cycles than the resistant sediment particles. Adsorbed substances are primarily present in the fine size fractions, which have a much larger specific surface and, because of the presence of clay minerals and organic matter, are far more surface-active than sand grains or gravel, although the latter can have coatings of organic matter, iron hydroxide or carbonate. The fine size fractions are transported in suspension and are dispersed more quickly and widely than the bed load; sand moving along the bottom is temporarily stored in ripples or banks of varying size, while gravel rolling over the bottom goes even slower. The transport velocity of suspended matter, which approximately follows the water transport, is therefore several orders of magnitude higher than the transport velocity of the bottom material. The fine size fractions also contain most of the particulate organic matter, which is a major food source for planktonic and benthic organisms. Through the inter action with the dissolved phase, either by exchange processes, dissolution or precipitation, and the interaction with the biological cycles, the suspended matter is a factor in other geochemical cycles as well, which is much less so for the sand- and gravel-sized bed load material. Because of this, and because the bedload supply reaching the coast is only 10 to 15% of the suspended matter supply, this paper is mainly concerned with the suspended material.

2. SUSPENDED MATERIAL AND SUSPENDED MATTER TRANSPORT

2.1. Particle size

The material in suspension in estuaries and near-shore seas is predominantly present in the form of flocs consisting of mineral particles strongly held together by organic matter, and as single mineral particles. Flocs have been observed in estuaries and coastal seas ranging from the tropics to temperate and arctic regions (Fig. 1). Usually also plant fibers and other plant debris are present and, when plankton is growing in the water, also particles consisting only of organic matter of planktonic origin. Floc-like particles without visible amounts of organic matter, but possibly with organic coatings, have been observed near to glaciers, where there is hardly any organic matter available (Syvitski and Murray, 1981). The size of the suspended flocs, when measured by a Coulter Counter, pipette analysis or under a microscope, is rarely larger than 125 μm. Mineral particles larger than 125 μm (sand, gravel) are only transported in suspension under extreme conditions with strong turbulence, as in the surf zone along beaches or on shallow sandbanks during a storm. Plant fragments commonly reach larger sizes. The lower size limit of suspended matter is conventionally put at 0.4 μm, but clay particles and particles of a cryptocrystalline or amorphous nature can be much smaller. Particles as small as 0.02 μm have been observed in the Rhine estuary.

Figure 1. Microfloc (Mahakam river estuary, Indonesia; SEM photograph)

Concentrations of suspended matter in estuaries and coastal seas range from less than 1 mg/l to more than 10 g/l. A change in the behavior of suspended matter occurs at concentrations of 5 to 10 g/l when the settling slows down to a very low rate even in quiet water. At such concentrations the particles are in each other's way, preventing settling at the normal rate for single particles: large loose flocs are formed and the suspension becomes a thick fluid mud. Fluid mud occurs locally in a number of estuaries and near-shore seas but normally suspensions of much lower concentrations are present.

Recent observations (with a plankton camera) of suspended particles *in situ* in the water in several northwest European and Canadian estuaries indicated that the *in situ* particle size is of the order of one magnitude larger than the particle size as determined after sampling (Fig. 2; Eisma et al., 1983; Kranck, 1984). The large particles, which have a maximum size of 3 to 4 mm, are fragile so that they come apart during the sampling process. Careful sampling with a special sampling device designed to collect the large particles intact on a filter, showed that they were large flocs formed by flocculation of the smaller flocs and single mineral particles normally found after sampling (Eisma et al., 1983). The large flocs (macroflocs), although easily broken up when disturbed mechanically (by sampling or size analysis), are stable in the water at current velocities of 1 to 1.5 m/s. The small flocs found after sampling (microflocs) are tightly packed and very stable with respect to physical disturbance. Only ultrasonic treatment or removal of the organic matter will take them apart. The plankton camera allows particle size measurements only of particles larger than 100 μm, so that the precise relation between the *in situ* particle size distribution of the macroflocs in the water and the particle size distribution after sampling is not known.

In order to flocculate, the suspended particles have to collide and to stick together. Several mechanisms may contribute to achieve this: neutralization of the negative particle charge by positive ions in the surrounding water so that the particles can collide and van der Waals forces can hold them together (salt flocculation); differences in particle settling rate; Brownian motion and turbulent motion; organic matter acting as glue and organisms actively clogging particles into flocs. The terms flocculation, aggregation, agglomeration, conglomeration and coagulation have been causing much confusion because different processes were indicated: "flocculation" is used here to designate all processes whereby particles are brought together into larger units called "flocs".

2.2. Flocculation

Flocculation of suspended material by adding salts or organic electrolytes can easily be demonstrated in the laboratory, and it has gained wide application in industry for removing particulate material from suspension. Field studies up to now have not shown real evidence for salt flocculation of particles larger than 1 μm in nature except perhaps in fjords, where the particulate material comes directly from a nearby glacier (Syvitski and Murray, 1981). Surfaces immersed in natural waters quickly acquire an organic coating which changes the surface characteristics of the particles (Hunter and Liss, 1982; Tipping and Cooke, 1982). Salt flocculation of particles smaller than 1 μm (colloid iron hydroxides, humates and associated substances) has been demonstrated by Sholkovitz and others (Sholkovitz, 1976, Sholkovitz et al., 1978), but quantitatively this amount is very small in comparison to particulate matter from other sources. This also applies to particles formed by inorganic precipitation.

Gravity affects all suspended material. In the range of 1 to 100 μm the fall velocity of grains of 100 μm is in the order of 10 larger than the fall velocity of particles of 1 μm, which makes differential settling of particles of different size a major mechanism for bringing particles together (McCave, 1984). Brownian motion affects only very small particles (8 μm at most). Turbulence limits the size of the macroflocs: the maximum macrofloc size is about the same as the size of the smallest whirls, which in estuaries are of the order of millimeters.

Macroflocs are formed under laminar flow conditions, the laminar shear varying in time in relation to turbulence, and by gravity and Brownian motion, with the

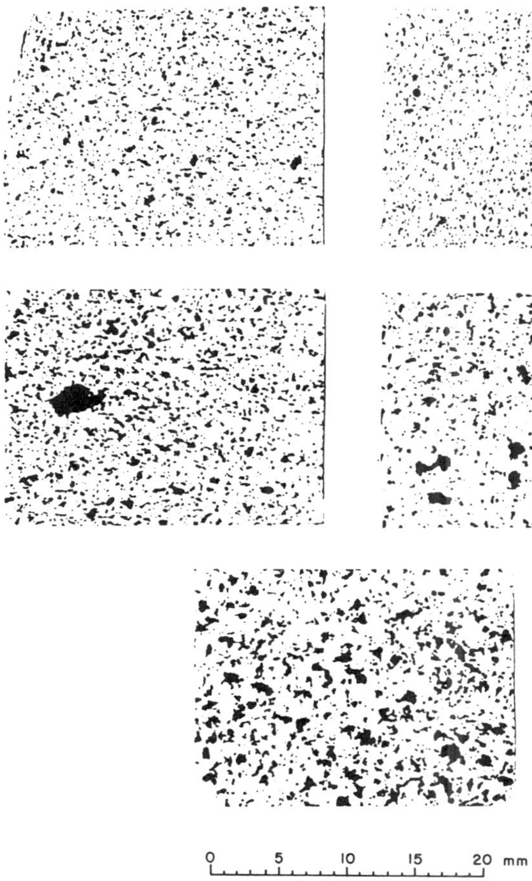

Figure 2. Different types of macroflocs, from planktoncamera photographs taken *in situ* in the Ems and Rhine Estuaries.

organic coatings producing weak bonds. When current velocities drop below 10 to 15 cm/sec the flocs start to settle out and much larger flocs are formed (depending on the suspended matter concentration). The microflocs that build up the macroflocs, show a strong binding between the mineral particles and the organic matter. Evidence from several West European estuaries indicates that mycopolysaccharides, produced by bacteria, algae and higher plants, play a large role in gluing particles together (Eisma et al., 1983). Organisms fixing suspended particles are probably a major factor in many estuaries. The importance of deposition of suspended matter by organisms was shown by Biggs and Howell (1984) for Delaware Bay, where zooplankton, oysters and mussels deposit about 200 times as much particulate matter as is supplied by the river. This indicates strong resuspension after deposition, and then redeposition, and so on. Although the microflocs, in contrast to the macroflocs, are very stable with regard to physical disturbance, the composition of the organic matter in the flocs changes from terrestrial material to estuarine or marine material (Eisma et al., 1984). This indicates that in estuaries microflocs are taken apart, presumably by organisms that consume part of the organic matter, and are built up again with new organic matter of local origin. The importance of organic matter for the flocculation process was pointed out by Schubel (1972), who found in Chesapeake Bay no evidence for flocculation in the absence of organic detritus, and by Kranck and Milligan (1980), who made artificial flocs of more than a few micron diameter only by mixing organic detritus with the mineral particles. The presence of organic matter is also important for the settling rate: experiments by Kranck and Milligan (1980) indicated that a 50%-50% mixture of organic matter and mineral particles settles much faster than the organic matter or the mineral particles alone (Fig. 3).

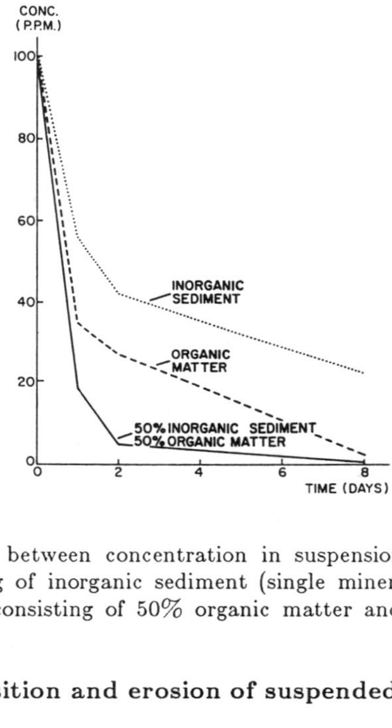

Figure 3. Relation between concentration in suspension and time for settling suspensions consisting of inorganic sediment (single mineral particles), or organic matter and of flocs consisting of 50% organic matter and 50% mineral particles (from Kranck, 1984).

2.3. Transport, deposition and erosion of suspended matter

The forces determining suspended matter transport are the flow rate, the concentration and settling rate of the particles in water, the bed shear stress, and other factors, such as the flow depth, the activity of organisms, and the composition of the suspended matter. At given flow conditions and at concentrations of 1 g/l or higher, a constant amount of suspended matter remains in suspension indefinitely and there is virtually no exchange of suspended material with bottom material. When the flow rate changes, a certain fraction of the material in suspension is deposited (at a change to a lower flow rate) or being picked up from the bottom (at a change to higher flow rate). This occurs at shear stresses well above the critical shear stress for deposition or erosion, but in the case of erosion the amount picked up also depends on the erodibility of the bottom sediment (Mehta and Partheniades, 1973). When the initial concentration is small, as is the case in most offshore waters, this probably is not very important, and when organisms are present that clog particles together, the whole system is changed, but at the high concentrations sometimes found in coastal waters and estuaries, this can be an important process.

2.4. Flow regimes and particle transport

The flow in estuaries, in contrast to the mostly unidirectional flow in rivers, is strongly influenced by the tides, by the wind, and by the relation between tidal flow and the density circulation. On the basis of the last relation a division is commonly made between stratified estuaries, partially mixed, fully mixed, and negative estuaries (Fig. 4; Postma, 1980). The landward limit of the estuary is taken either at the point where the river water meets the more saline water, or at the landward limit of the tidal influence. The seaward limit most often lies near the coastline but off larger rivers the estuarine area may include a large part of the adjacent sea. The density circulation is caused by low density river water flowing out in seawater of high density, or, in the so-called negative estuaries, by the formation (through evaporation) of water heavier than seawater inside the estuary or in the river (subtropical lagoons, the Salum river estuary in southern Senegal).

In a stratified (salt-wedge) estuary, where river outflow dominates over the tidal flow, most of the sediment, at least most of the fine grained material in suspension, is of river origin and carried seaward with the surface water. Part of the suspended material is dropped at the head of the estuary and forms a turbid bottom layer moving out into the coastal sea (Zaire, Rhone; Eisma et al., 1984b; Aloisi et al., 1979). Also the bottom load is dropped at the estuary head and gradually fills up the estuary (Meade, 1972). Inflow along the bottom from the sea is absent or low so that even when sediment is available in the coastal sea, it will not easily be moved into the estuary. In partially and fully mixed estuaries the inflow along the bottom is much stronger, compensating for the outflow of saline water along the surface. At the head of the estuary is an area of low velocities where the residual bottom currents—seaward in the river and landward in the estuary—converge. Here sediment is trapped resulting in a turbidity maximum (Postma, 1980). This maximum moves through the estuary with the tides and the seasons, pushed inward by the flood, outward by the ebb, sometimes being pushed entirely out of the estuary during a period of very high river discharge, or moving far inland when river discharge is very low. Local resuspension of temporarily deposited bottom sediment contributes considerably to the maximum.

A turbidity maximum can develop independently from any density circulation where tidal flow dominates and tidal asymmetry creates a sediment trap, as happens on tidal flats (Postma, 1980; Allen et al., 1980). Two kinds of tidal asymmetry can occur, or a combination of both (Fig. 5): one where the flood attains higher current velocities than the ebb but is of shorter duration, and one where the change from flood to ebb at high tide occurs more slowly, with a long period of low current velocities, than the change from ebb to flood at low tide. In the first case the flood tide can carry more suspended material inward than the ebb-current can carry outward, because the amount of transported suspended sediment increases, approximately according to a third power law with increasing current velocity. In the second case the period with low current velocities, during which suspended matter can settle, is much longer during high tide than during low tide, which favors mud deposition high on the tidal flats, as, for example, in the Waddensea (Postma, 1980). In estuaries where the effect of inward convergence on the incoming tidal wave is larger than the damping by friction, the tidal amplitude can attain a maximum within the estuary, which also contributes to the inward accumulation of sediment (Gironde; Allen et al., 1980). Where there is little or no tidal asymmetry the difference in water depth at high tide (on tidal flats) and at low tide (in the channels) plays a major role in the settling and accumulation of suspended matter around high tide level.

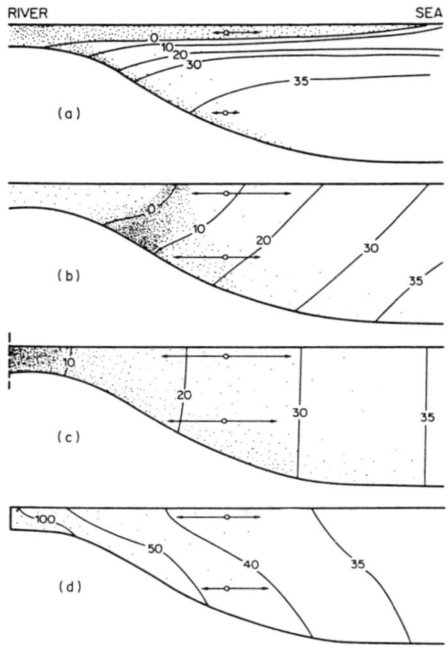

Figure 4. Schematic representation of types of estuaries. The dots indicate sediment concentration, the arrows net water movements over ebb and flood, and the numbers salinity. a) salt wedge estuary, b) partially mixed estuary, c) fully mixed estuary, d) negative estuary with excess evaporation and high salinities at the head (from Postma, 1980).

Secondary maxima can develop in long estuaries along the bottom (James river, Rappahannock, Thames; Officer and Nichols, 1980) or in the water column and not connected with the channel floor (St. Lawrence; d'Anglejean and Ingram, 1976). Their formation is related to secondary circulations within the estuary. Other marked gradients in the suspended matter concentration distribution are associated with salinity and temperature fronts. Such fronts, often marked by color changes and lines of foam, are common where major rivers discharge to the coastal sea or in estuaries where secondary circulations converge. The turbidity differences usually are incidental and do not contribute to the density gradients. Kirby and Parker (1982), however, describe a front in the Severn estuary where the density gradients are caused by the variations in suspended matter content and are not related to salinity and temperature differences. This indicates that although the water is well mixed, the suspended matter is not; the mechanism causing this is as yet not known. The overall result of all particle flow processes in an estuary will be a net-flux out of the estuary or

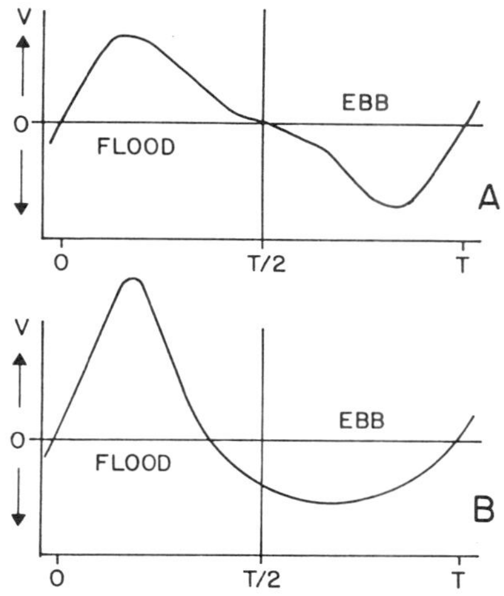

Figure 5. Two types of tidal asymmetry. A) a slow change from flood to ebb, a rapid change from ebb to flood; B) higher flood velocities but of shorter duration than the ebb velocities.

deposition on the bottom to balance the influx of sediment from the river. Variability can be high, however: in Chesapeake Bay 80% of the total river sediment supply between 1966 and 1976 was supplied in 10 days (Gross et al., 1978) so that long periods have to be considered to estimate net effects (see also Meade, this volume).

Further influence on the particle flux through estuaries is imposed by the activities of estuarine and marine organisms and by weather conditions. As mentioned above, organisms are a major factor in suspended matter deposition by clogging particles together, but they may also increase the erodibility of bottom deposits by increasing the water content or the bottom roughness (Nowell et al., 1981; Jumars et al., 1981). Biological stabilizing and destabilizing processes at the bottom may go on simultaneously and show seasonal changes: organisms on tidal flats (benthic diatoms forming mats, higher plants reducing water flow) probably have an over all favorable effect on the formation of muddy deposits, whereas in deeper water the over all effect probably is a higher erodibility.

The influence of variable wind conditions on the estuarine circulation can be very strong. In the Potomac estuary 2.5-day fluctuations in the circulation pattern have been attributed to the wind, which is able to completely reverse the expected normal flow pattern with seaward surface flow and landward bottom flow (Elliot 1978). Reversals occur about 20% of the time, the normal circulation 43 %, total in flow along the surface and the bottom about 20%, and in the remaining time there is total outflow or a more complex circulation pattern. Wind-induced surface waves are usually smaller in an estuary than in the adjacent sea and have a limited effect, but in large shallow tidal areas such as the Dutch Waddensea, wave action can be sufficient to prevent further building up of the tidal flats. In this way a tidal regime is maintained (Postma, 1980). Exceptionally severe storms and heavy river floods, occurring only rarely, may cause more changes in an estuary within a few days than happens otherwise during many years. This is well documented for the effects of hurricane-generated processes—strong currents, wave action and flooding resulting from heavy rain fall—in Chesapeake Bay, and other bays and inlets along the US east coast and the Gulf of Mexico (Hayes 1978).

A factor that is important in mud transport in estuaries and the coastal sea is the interaction between unidirectional (tidal) currents and surface waves. Temporary mud deposits have been observed during periods of low wave activity in shallow water (15 to 30 m deep) where tidal current velocities reach 30 to 100 cm/sec and the bottom is normally sandy (Eisma, 1981). This is related to two phenomena: 1) the difference between the velocity profiles formed in the bottom boundary layer by unidirectional currents and by surface waves (Teleki, 1972; Fig. 6), and 2) the general reduction in flow rate when surface waves are reduced. Near to the bottom, where the flow caused by unidirectional currents is rapidly decreasing with the height above the bottom, the flow caused by surface waves is comparatively strong (depending on the wave characteristics and the water depth). Reduction or absence of surface waves allows suspended matter to be accumulated in the near-bottom region of weak flow, which will be particularly marked in near-shore areas with high suspended matter concentrations.

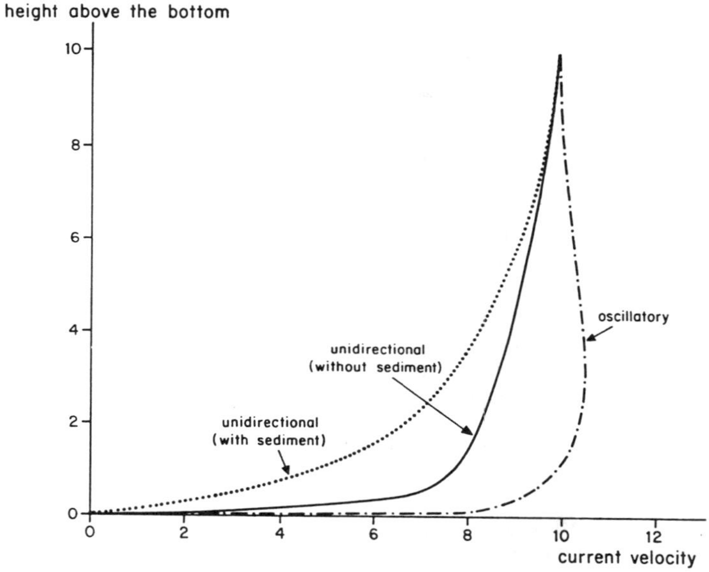

Figure 6. Generalized vertical velocity distribution in unidirectional flow with and without sediment, and at oscillatory flow. Units for height above the bottom and current velocity are arbitrary (after Teleki, 1972).

When a mud deposit is formed on the bottom, the near-bottom sublayer rises to the top of the mud layer so that a two-layer system develops with (fluid) mud below and the normal velocity profile above it. When this takes place on a normally sandy bottom this involves also a reduction of the bottom roughness and a reduction in the erodibility of the bottom. When wave activity increases again, the mud is picked up but this takes time and there is probably an intermediate period when the mud layer is not yet suspended but moving as a fluid muddy layer over the bottom. This fits in with observations at the Rotterdam harbor where deposition of fluid mud in the channels is strongest during stormy weather after a calm period.

2.5. Sedimentation and sedimentation rates

Mud deposits are formed in usually sheltered areas with low wave activity and (at least periodically) low current velocities. With time the deposited suspended matter consolidates, whereby water is squeezed out, and the concentration of particulate matter and the density of the deposit increase. The process goes quicker when there is sand admixed and when new deposits are formed on top, providing additional pressure. Erosion of deposited mud depends very much on their water content (or solid content) or on the consolidation time, and on their composition so that different muds have a different erodibility. This can be related to the presence of gas in the sediment, to the salt concentration, and the ratio of Na to Ca and Mg ions adsorbed onto the clay particles (Ca- and Mg-clays are more strongly bonded than Na-clays; Kandiah et al., 1975), the erodibility increasing with the presence of gas, with the salt content, and with the ratio of Na to Ca+Mg. In nature there is probably often a strong gradient in erodibility at the top of a sediment (Terwindt, 1977), which is difficult to measure, but when the mud contains benthic fauna (which is usually the case), the characteristics of the mud are completely altered because, as pointed out above, bottom organisms can both stabilize and destabilize bottom sediment.

The formation of mud deposits in coastal and near-shore areas is usually rather slow (of the order of 1 to 100 cm per 100 years), and is the net-result of repeated deposition and resuspension. Deposition rates are preferably estimated from series of soundings which allow calculation of the resultant deposition (or erosion) over a period of time. Where such series are not available, other indicators have been used, such as C-14 data of *in situ* peat layers or mollusk shells in life position, the occurrence of pieces of coal or charcoal, and more recently the concentration distribution of ^{210}Pb (halflife 22.3 years) and ^{238}Th (halflife 1.9 years)—both natural isotopes—and the concentration of isotopes coming from nuclear tests or from nuclear plants (^{137}Cs, ^{239}Pu, ^{240}Pu). From such data recent sedimentation rates covering the past 50 to 100 years and average sedimentation rates for longer periods can be determined, but this interpretation has to take into account that bottom sediment is mixed, mostly by burrowing organisms or by waves. The thickness of the mixed layer and the mixing rate within that layer influence the deposition rate estimates. Whether mixing occurs or not can easily be seen in the bottom sediment profile by a disturbance of the layering. When a sediment is homogeneous (which can be checked by taking an X-ray radiograph), the mixing rate is estimated from the distribution of short-lived isotopes, such as ^{234}Th (halflife 24 days) and ^{238}Th (halflife 1.9 years) in the top of the sediment, or from the fallout isotopes (Cochran, 1984). Isotope deposition rates are normally obtained from sediment cores: in contrast to soundings which follow transects, such data are only valid for one point and some method of extrapolation is required to obtain the deposition rate over a larger area.

3. PARTICLE FLUXES THROUGH ESTUARIES

As previously discussed, suspended sediment is supplied to estuaries from various sources and, depending on the local conditions, one or more of the sources will dominate the supply. Removal of particulate matter from the estuaries also takes place in different ways: downward (deposition) or sideward (towards the shore, a floodplain, tidal flats or salt marshes) and then again downward, outward into the coastal sea, or up-river by tidal dispersal. There can also be some loss of particles to the atmosphere but quantitatively this is negligible compared to the other forms of removal. In many estuaries large amounts of material are removed by dredging. In the Rhine mouth, where the annual river supply is about 3.5 million tons of sediment (fine sand and mud) and annual net inflow from the coastal sea is 13 million tons, about 10.5 million tons are dredged out yearly.

The various ways in which particulate matter is supplied to estuaries and is removed, are schematically summarized in Figure 7. In a steady state the total supply of sediment to the estuary, given in mass/time units, should be balanced by the total that is removed. The time during which the particles remain in the estuary, the residence time, can be estimated from the total quantity of particulate matter present in suspension or moving along the bottom, divided by either the supply rate or the removal rate. It can also be defined as the average time needed for particulate matter to move from the point of supply to the point of removal. Particulate matter in estuaries varies in size, density, composition, shape and surface characteristics, so that there are large variations in the way particulate matter moves through estuaries. Different types of material (sand, clay, organic matter) have different residence times as well as a different mass balance. Often, however, material from different sources cannot be clearly distinguished on the basis of size and composition so that only a gross average residence time or mass balance is estimated involving material of different origin and characteristics. Even the obvious distinction between the mass balances of particulate organic and inorganic matter is not always made. The data of Biggs (1970) for the upper and middle Chesapeake Bay show large differences between the budget for organic matter in suspension as compared with the budget for the inorganic suspended material (Table I).

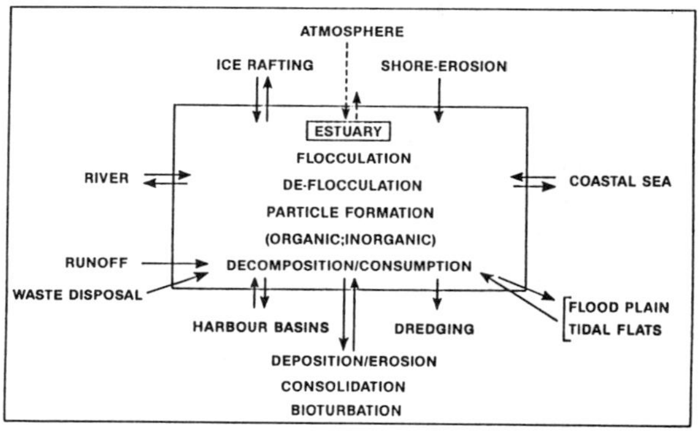

Figure 7. General scheme of supply and removal of sediment in estuaries.

3.1. Morphology of estuaries

The particle flux through estuaries is largely related to the type of estuary, but also depends on the general morphology of the area, the effects of the Holocene sea level rise, the tectonics of the area (uplift, subsidence) and, as was seen above, on the activity of estuarine organisms, on wind conditions, and other tidal characteristics. The main factors determining the geomorphology of an estuary are the river sediment input, the tides and wave action (Fairbridge, 1980). Deltas, which are built out into the sea, are predominantly formed by rivers with a high sediment load along coasts with a low tidal range and limited wave action (Mississippi, Orinoco, Niger, Nile,

Table I. Sources of the organic and the inorganic fraction of the suspended sediment in the upper and middle Chesapeake Bay, after Biggs (1970), in percent of the total amount of sediment in suspension.

	Organic %	Inorganic %
Upper:		
Skeletal	2	0
River supply	22	61
Shore erosion	0	13
Organic production	2	0
Middle:		
Skeletal	18	0
Shore erosion	0	52
Organic production	22	0
From other parts of the estuary	5	3

Rhone, Danube, Fraser, Yukon). On such coasts the estuaries are usually of the salt wedge type (at some rivers only during river floods: Mississippi, Orinoco) and the rivers with low sediment loads have open estuaries (Texas Bays). Where the tidal range is high, the V-shaped drowned-valley type of estuary usually is preserved because more sediment is swept out of the estuary towards the coastal sea. Such estuaries are often only partly filled up with sediment (Loire, Long Island Sound) and are of the partly or fully mixed type. All estuaries have been formed during the Holocene sea level rise and are up to 10,000 years old, having started their existence at a much lower sea level. Where sediment supply has kept pace with the rise of sea level the estuary is completely filled up, and further extension in the form of a delta became possible after the sea level rise levelled off about 6000 years ago. Fjords typically indicate conditions whereby the low sediment supply was not able to fill up the deep valleys, whereas the Amazon and the Chang Jiang, both rivers with a very large sediment supply, have filled up their river basin but did not form a delta because most of the sediment at the mouth is swept away over the shelf and along the coast (by the tides and by a longshore current). Most estuaries are in an intermediate stage of filling up an old river basin and in many estuaries regional subsidence, resulting in a relative rise of sea level also after the worldwide Holocene sea level rise had come to a halt, has retarded or prevented the filling up of the estuary. Thus the Thames estuary was already an estuary *ca.* 8900 years ago and has remained an estuary up to the present (Greensmith and Tucker, 1973).

Uplift enhances the filling-in because of reduced water depths and stronger erosion of the adjacent land. In tectonically active regions, local or regional movements can completely overshadow the effects of tides and waves. In the tectonically active area of Cook Inlet (Alaska), where earthquakes can alter large areas of the coastline in a few seconds, deltas are formed in spite of a tidal range of 4 to 6 m and currents of up to 7 m/sec (Hayes and Michel, 1982). The morphology of the river mouth and the types of sediment that dominate, can be expressed by the relative influence of the river supply versus the near-shore wave and tidal power. Where the river influence is large and wave and tidal power is low, a strongly protruding delta is formed (the classical delta) with a delta plain containing marshes, mangroves, bays, and localized sand ridges. Where the influence of river supply is small and wave and tidal power is strong, the coastline is straight (no delta is protruding) with barriers, beach ridges, dunes and thick sand sheets containing well-sorted clean sands (Wright, 1978). At intermediate conditions the river mouth protrudes somewhat and is smooth, arcuate, with small beach ridges and dunes alternating with marshes and/or mangroves, with moderately sorted sands, poorly sorted silts and clays. The mouths of some large rivers like the Amazon and the Ganges-Brahmaputra also belong to this intermediate type, forming a broad river entrance, large sedimentary plains with many channels and islands, which also are called "deltas."

3.2. Flux estimates

Particle fluxes through estuaries are estimated in several ways: from direct measurements of flow velocities and concentrations, from numerical models, from the distribution of natural and artificial tracers, and from a consideration of fluxes across the estuary borders resulting in a mass-balance, based on all available data. There is a large practical interest in estuarine sediment fluxes because they give an indication of the fate of polluted muds and waste discharges and because dredge spoils may constitute a large part of the total flux.

Direct measurements of flow and concentrations covering an estuary over a representative period of time are complicated because, particularly in the outer part of estuaries with channels and banks, large variations in the amounts and the direction of the residual transport occur over short distances. Near-bottom mud transport is hardly measurable, and measurements are often difficult to carry out in outer estuaries when there is shipping in the channels and during stormy weather. Residual transport and deposition rates in most cases cannot be reliably estimated from measurements made along transects through the estuary, because this involves calculations based on small differences in large numbers, and temporal and spatial variations in concentration are large. Thus in the Dutch Waddensea where on the average 15 to 20×10^6 tons of suspended matter (dry weight) is yearly moved in and out through the tidal inlets, only about 5% of this amount is yearly deposited (the deposition rate in this case was estimated from soundings). At the mouth of the Chao Phya river (Thailand), however, sediment transport is less variable and changes mainly with the wet and dry seasons. Here a good relation was found between the tidal level and the transport rates through several transects, and between the wind force and the deviations from the average transport rates (NEDECO, 1965). From these data it was estimated that during the wet season the resultant transport through the estuary, through the offshore channel and over the large shoals that lie off the river mouth, is outward (Fig. 8). During the dry season when river discharge is low and saltwater intrudes inward, sediment is chiefly moved in the opposite direction, into the river mouth and to the eastern parts of the shoals. From these data an overall yearly transport could be calculated.

3.3. Modelling and tracers

Numerical sediment transport models make use of general models of estuarine circulation and mixing, which are based on the fresh water or salt budget, on tidal mixing, on wind effects, and on a consideration of advection and diffusion in the estuary. An advection-diffusion equation in one form or another is the basis of numerical models. Where there is a strong density gradient, as is the case in many salt wedge estuaries, a two-layer model gives a more appropriate representation and in most estuaries, even

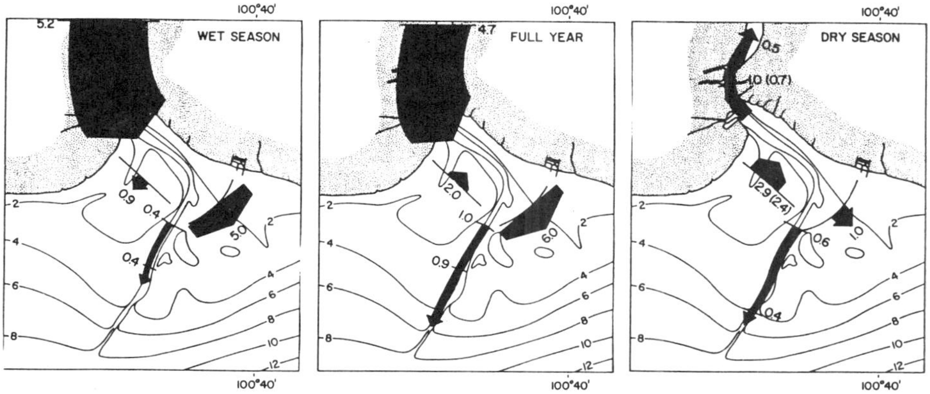

Figure 8. Sediment flow in the Chao Phya estuary, in 10^6 tons (after NEDECO, 1965).

in relatively simple channels, lateral flow across the estuary has to be taken into account. A full discussion of the circulation and mixing in estuaries was given by Officer (1976) and Bowden (1980).

Modelling of sediment transport in estuaries is more complicated because of the non-conservative behavior of particles, the different transport modes—along the bottom and in suspension—and the cohesiveness of the fine-grained material, the influence of organisms, and other factors. By introducing more or less empirical values for particle settling and erosion (resuspension) rates, and from the salinity distribution, the river flow and the estuarine geometry, models of estuaries have been made that give a reasonably correct indication of the areas of erosion and deposition, and of the concentration distribution of suspended matter, including a turbidity maximum (Fig. 9). However, sediment modelling remains largely descriptive; predicting sediment behavior involves large uncertainties because the required knowledge on transport, and deposition and erosion processes is lacking.

Natural or artificially introduced tracers give an indication of the percentage contribution of specific sediment sources to the estuarine sediment. From such data an idea can be obtained of sediment dispersal in the estuary and the coastal sea. Naturally occurring tracers are more useful than artificial ones: they cover an entire estuary whereas the use of introduced tracers is limited by the amount of tracer that can be added and usually gives only an indication of local conditions. Various sediment characteristics have been successfully used: particle mineralogy, elemental composition, organic composition, isotopic composition of certain sediment components (carbonate, organic matter, clay minerals, quartz), discharged radionuclides that are attached to suspended particles (^{137}Cs, ^{239}Pu, ^{240}Pu). Also combinations of methods have been useful: Sr-content and isotopic analysis of carbonates and clay minerals in the Rhine and Ems estuaries (Salomons, 1975; Salomons et al., 1975); C/N ratios, lignin concentrations and isotopic composition of particulate organic matter in the St. Lawrence estuary (Tan and Strain, 1979; Pocklington and Leonard, 1979); organic carbon isotopes, organic composition and elementary composition in the Ems and Gironde estuaries (Eisma et al., 1984a). Tracers are useful only when they are specific for a source and when they retain their characteristics in the estuary. Tracer elements

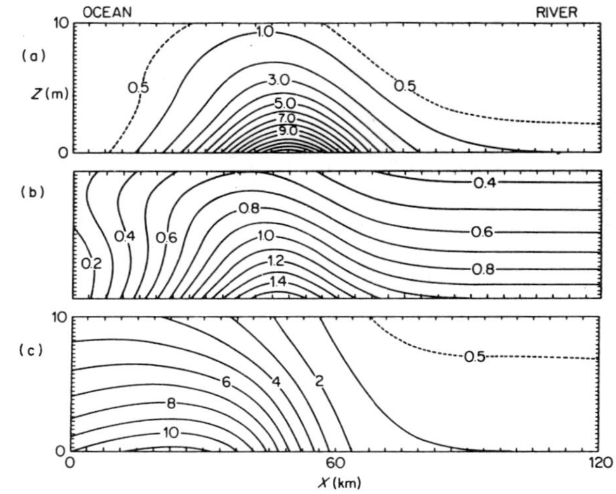

Figure 9. Steady state numerical model of the turbidity maximum in a partially mixed estuary: a) river to ocean source ratio 10, particle settling velocity 3×10^{-3} cm/sec, b) river to ocean source ratio 10, particle settling velocity 1×10^{-3} cm/sec, c) river to ocean source ratio 0.1, particle settling velocity 1×10^{-3} cm/sec (from Postma, 1980, after Festa and Hansen, 1978). Isolines of turbidity in relative units.

should not be mobilized from the sediment particles or be deposited from solution. Differential settling, sorting, and change in the percentage composition should be minimal. The tracers mentioned above to a large extent satisfy these requirements.

Clay minerals have been considered unsuitable as tracers because of flocculation effects and differential settling (Edzwald and O'Melia, 1975), or of diagenetic processes, as was postulated for the estuaries along the U.S. east coast. Clay minerals can change in estuaries because of uptake of potassium from the seawater or by the formation of interlayer hydroxide sheets of aluminum and iron, but both processes can readily be recognized. Flocculation effects and differential settling, which have been demonstrated in laboratory tests (Whitehouse et al., 1960), have not generally been found to influence clay mineral distributions in estuaries. In James river, Feuillet and Fleischer (1980) found that the estuarine circulation, causing inward transport of marine sediment, could account entirely for the observed clay mineral gradients. Gallenne (1974), however, found a strong increase in montmorillonite content in the turbidity maximum of the Loire estuary, probably because of a lower settling rate. This increase is reflected in the clay mineral composition of the mud banks in the estuary, so that there remains some caution as to the use of clay minerals as estuarine tracers.

Sediment balances are estimated from the fluxes across the estuary boundaries as well as through the estuary, using all available data and incorporating the results obtained from direct measurements, models and tracers. Such estuarine balances tend to be open ended. An example is the sediment budget for the Rhine mouth, which is based on data and estimates for the supply of suspended matter and (fine) sand from the river to the estuary, data on the composition and the quantities of sediment dredged from the harbor basins, and data on the outflow of suspended matter into the coastal sea (Fig. 10). Supply from the coastal sea to the estuary makes up the difference but is not based on measurements, because supply from the sea is largest during storms and probably takes place mainly by flow very near to the bottom. Because of the uncertainties related to the processes of sediment transport and erosion, it is for partially or fully mixed estuaries often difficult to decide whether the estuary is trapping sediment, or is a sediment source for the near-shore sea (salt wedge estuaries are always a source). In many river mouths the answer to this is complicated by the changes caused by dredging and by the construction of harbor basins and artificial channels. Even relatively small changes such as the removal or reshaping of bridge pylons may result in changes in tidal flow with a consequent change in sediment transport.

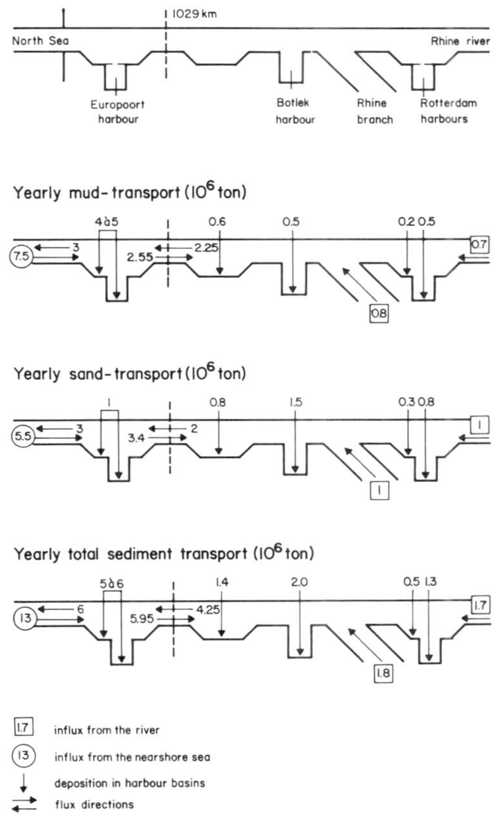

Figure 10. Sediment budget for the Rhine estuary (from Eisma et al., 1982).

4. DISPERSAL AND DEPOSITION OF SUSPENDED MATTER IN THE NEARSHORE SEA

Outside the estuaries the suspended matter concentration, except off large rivers, rapidly decreases with the distance from the coast, from more than 100 mg/l to less than 5 mg/l. The presence of Recent mud deposits likewise decreases. McCave (1972) distinguished muddy coasts, near-shore mud deposits, and mid- and outer-shelf mud deposits. Recent near-shore and coastal mud deposits are the most numerous. The mid- and outer-shelf are usually covered with relict or reworked sediments, often containing a more recent admixture of biogenic carbonate and of fine grained organic matter. The Recent mid- and outer-shelf mud deposits are mostly formed by sediment supply from a nearby large river (Zaire: Congo-Gabon shelf; Mississippi: NW Gulf of Mexico; Columbia river: Washington-Oregon shelf). Off Namibia (Southwest Africa), a near-shore to mid-shelf mud deposit is largely composed of diatom frustules produced in large quantities as a result of strong upwelling (Bremner, 1980). On the New England U.S.A. shelf, mid-shelf muds are formed in topographic depressions, the mud being supplied from the nearby banks where it is being winnowed out by storm waves. Similar winnowing out and deposition in deeper parts (in this case by the tides) occurs in the Celtic Sea. Mid-shelf deposits in the Gulf of Gascogne and off Southeastern Australia are most probably relict (subrecent) with possibly some recent deposition of river-supplied material (Rumeau and Varney, 1968; Davies, 1979). Mid-shelf deposits in the North Sea are almost all relict (Eisma, 1981).

The factors that keep suspended material nearshore are complex. Differences in density between the near-shore water and the water further offshore will often be present where there is fresh water supply from the land, if only by lateral dispersal from an estuary. Because of the greater buoyancy of the near-shore water a density stratification will develop with the near-shore water flowing outward along the surface and the offshore water flowing inward along the bottom. Suspended matter, flowing outward with the near-shore water will tend to settle and return with the bottom water. Such estuarine-like circulations are known from the southern North Sea and the eastern shelf of the USA (Drake, 1976). This circulation is constrained by the development of density fronts restricting the dispersal of surface water but inducing downwelling (Simpson et al., 1978), and by vertical mixing of the water by the tides and wind-generated waves. Onshore waves, however, drive water shoreward on the inner shelf and will tend to keep suspended matter near shore.

Since the coast is a barrier for any water flow, either wind-forced, tidal or Coriolis-forced flow, long shore currents will often develop with suspended matter transport dominating in the downcurrent direction parallel to the coast and little exchange with the water further offshore. So suspended matter will tend to stay near shore, once it is there, and deposition in near-shore areas (lagoons, river mouths, salt marshes, tidal flats, the near-shore sea) is favored. Since most sources of suspended matter are at the shore or near shore (where erosion by waves and tidal currents is strongest), high suspended matter concentrations and mud deposits will usually be found there. Where very large amounts of mud are being supplied, as along the north coast of South America from the Amazon river (Fig. 11) and along the Louisiana coast from the Mississippi, or where they are accumulated (west coast of Korea, southwest coast of India), soft fluid mud deposits develop because of the high mud concentrations, in spite of strong wave action and a large tidal range (Wells, 1983). The incoming waves are attenuated by the soft mud, but even small waves are able to resuspend the mud and large transport rates occur at relatively weak flow.

Additional concentration of mud takes place resulting in the formation of mud banks migrating along the coast (north coast of South America) or in more irregular patterns of accretion and erosion, consistent with seasonal cycles (Korea, Louisiana). Such concentration mechanisms are even more marked along coasts with a lower suspended matter content (for example, the North Sea, Fig. 12; Eisma and Kalf, 1987). This is partially due to deformation of the tidal wave by the local topography,

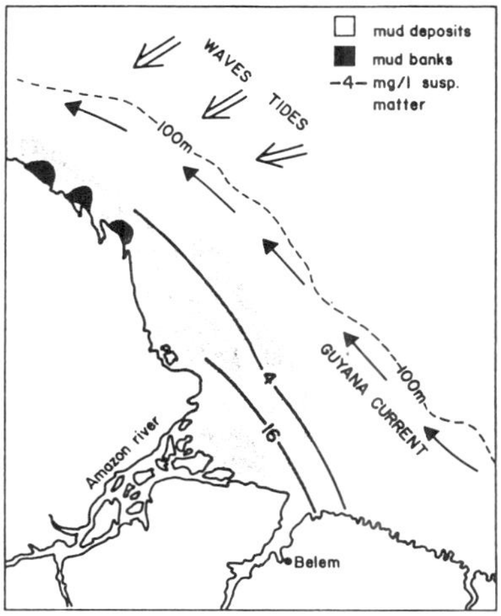

Figure 11. Dispersal and deposition of suspended matter from the Amazon River (after Delft Hydraulics Laboratory, 1962; Milliman et al., 1975).

particularly at shallow depths and in partly enclosed sea areas, which results chiefly in the two types of tidal asymmetry discussed above (Fig. 5). In the German Bight the tidal asymmetry results in concentration of suspended matter to more than 120 mg/l near to the coast. This mud patch, as well as other suspended matter concentrations and patchy mud deposits are associated with large tidally and topographically induced eddies (Bristol Channel, Ferentinos and Collins, 1979; North Sea, Fig. 12), as well as with areas where resultant currents converge (East Anglia, Belgian coast). The effect of the eddies is twofold: suspended matter, once it has reached such an area, tends to stay there, and organisms get the time to deposit the suspended matter. Once suspended matter is concentrated, with enough suspended matter to form a deposit, the amount of wave energy and not the tidal current velocities appears critical for the deposition of mud. In more exposed shallow areas where suspended matter concentrations are high, temporary mud deposits are formed during periods of low wave action, but are eroded again when wave activity increases. Only a small part of this mud remains behind as an admixture in the normally sandy bottom: on the beach and from there down to 10 to 18 m water depth along the Belgian-Dutch coast, or in between gravel off the East Anglia coast.

In most river mouths estuarine mixing is confined to a limited, predominantly inshore area, but at some very large rivers most or all of the mixing and the dispersal of river supplied sediment takes place in the coastal sea (Amazon, Zaire, Chang Jiang, Mississippi). The river water usually flows out as a plume of variable depth. Such a plume is maintained by the inertia of the outflowing river water and by its low density. Mixing occurs laterally and vertically by eddy diffusion and entrainment of salt water. The latter occurs also where the plume flows out over a salt wedge (Mississippi, Zaire). At the Amazon and the Mississippi a large bar is formed off the mouth because of the deceleration of the outflow, resulting in deposition of transported material (suspended matter as well as bed load). At the Zaire the river water flows out over a canyon filled with seawater that begins at the head of the estuary and is

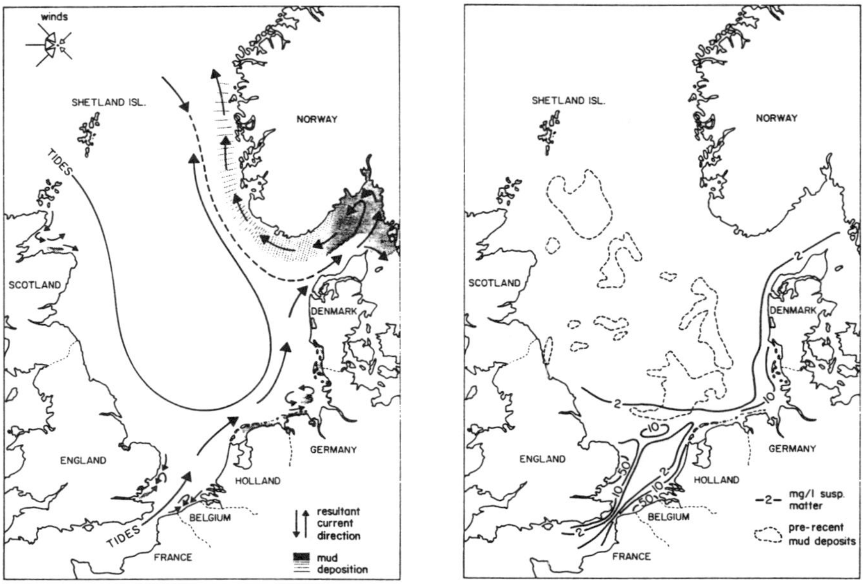

Figure 12. Dispersal and deposition of suspended matter in the North Sea (after Eisma, 1981; Eisma and Kalf, in press).

already 450 meters deep at the mouth (Fig. 13). There no bar can form, but within the river mouth already half of the suspended matter is lost from the outflowing river water and the bed load (large sandbars migrating seaward) is pushed into the canyon at its head. Almost all the suspended matter is deposited on the deep sea fan (70 to 80 % of the yearly supply; Eisma and Kalf, 1984) or on the shelf (20 to 30 %). Only a very small amount is moved in suspension into the ocean through the surface plume. The plume is not much deflected by the coastal current, probably because the saline, high density coastal current largely passes below the shallow low-density plume. The outflow of the Amazon, however, is much more deflected along the coast by the large Guyana current: low salinity water reaches as far as the Caribbean and the suspended matter is all deposited on the shelf (80 to 90%) off the river mouth, the remainder being transported further westward along the coast toward the Orinoco delta and the shelf north of Venezuela (Eisma et al., 1978; Milliman et al., 1982). At the Zaire and Mississippi plumes, and at many other river plumes, lateral mixing proceeds rather slowly, the outflowing water often being separated from the surrounding seawater by fronts, so that the more turbid low salinity water (which usually also has a different temperature and a different color) can be followed over long distances, particularly from the air or from satellites.

Continental shelf waters usually consist of a turbid surface layer and a near bottom turbid layer, which merge in the near-shore waters, but on the middle and outer shelf are separated by clear water. The surface turbid layer has usually as its lower boundary a thermocline or halocline where the settling of particles is delayed. Towards the outer shelf the surface turbid layer contains increasingly less mineral particles and more biogenic (mostly combustible) material. The near-bottom turbid

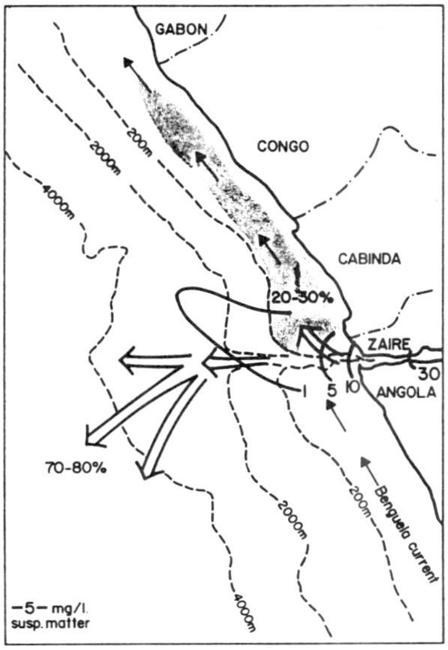

Figure 13. Dispersal and deposition of suspended matter from the Zaire River (after Eisma and Kalf, 1984). Shaded area: deposit of Zaire mud on the shelf.

layer contains chiefly mineral (non-combustible) particles, mainly resuspended from bottom sediment but in some areas also supplied from a nearby river. On the middle and outer shelf the mud in the bottom nepheloid layer tends to move outward toward the shelf edge (Drake, 1976). The concentrations in the bottom layer, however, are only a few mg/l so that when temporary deposition occurs, this will not quickly result in the formation of mud deposit of noticeable thickness. The outer shelf is not a favorable environment for mud deposition, because of an increased near-bottom flow related to the tides, to storm waves, internal waves and fronts, low suspended matter concentrations, and because most of the material in suspension (up to more than 90%) is organic matter and of low density. Most of the suspended material moving over the shelf bottom will move over the shelf edge to be deposited on the continental slope or on the adjacent ocean floor.

5. SUMMARY AND CONCLUSIONS

The sediment supply from rivers to estuaries is made of about 90% (by weight) of suspension and of 10% of bed load. The bed load (sand and gravel) goes into sand banks and beaches along the coast, the suspended matter goes more than 90% into estuarine, near-shore and some mid-shelf mud deposits. Only a small fraction of the suspended matter supply crosses the shelf and reaches the deep sea. Recycling of the deposited sediment may take from a few hours or a few days up to many million years, some of the sediment being taken up in short-term transport cycles, other sediment being taken up in long-term geological cycles.

Suspended matter is fine grained with a large specific surface area, is highly surface-active, and has a high organic matter content compared to coarser material. The specific surface and the surface reactivity of sand and gravel are very low. Suspended matter interacts strongly with the biological and chemical cycles in estuaries and the sea, whereas for sand and gravel this interaction is much less and chiefly by the formation (or dissolution) of coatings or through the building of sandy tubes and other structures by benthic organisms. Suspended matter is mostly transported in a flocculated state, flocs (macroflocs) reaching several millimeters in diameter. The macroflocs are fragile and built up from relatively stable microflocs, consisting of mineral grains strongly held together by organic matter, and single mineral grains, both with a maximum size of about 125 μm. Suspended matter is transported almost as fast as the water in which it is suspended but bedload is transported at a speed which is several orders of magnitude lower. Bed load or bottom transport is mainly in the form of ripples and banks, where the bed load is temporarily stored. Because of this the dispersal of sand and gravel is much more limited than the usually wide dispersal of suspended material.

Suspended matter transport is related to flow rate, the suspended matter concentration, the particle settling rate, the bed shear stress, the flow depth, the characteristics of the material in suspension, and the activity of organisms. In rivers the flow is largely undirectional, but in estuaries and coastal areas the tides, the density differences and wind-generated waves, variations in water level, and currents, determine suspended matter transport. In salt wedge estuaries a two-layer system develops, the suspended matter moving seaward through the surface plume and in a turbid layer near to the bottom. In partially and fully mixed estuaries a turbidity maximum is formed where the out flowing river water meets the inward flowing estuarine bottom water. A turbidity maximum can also be formed by tidal asymmetry and by the geometry of channels and tidal flats in areas without a river in-flux. The wind can temporarily reverse the normal flow in an estuary or the coastal sea. Waves have an important effect on water flow near to the bottom: the degree of wave action can decide whether a mud deposit is formed or not.

The settling rate of suspended matter is strongly influenced by the degree of flocculation, which is caused by several processes bringing particles together (laminar shear, gravity, Brownian motion) and by the presence of organic matter acting as glue. Organisms are an important factor in the deposition of sediment by clogging particles together (biodeposition), by inducing sediment deposition through reducing the water flow, or by inducing a greater erodibility of deposited mud by decreasing the cohesion and increasing the bottom roughness. On a larger scale the sediment transport through an estuary is very much influenced by the general morphology of the river mouth, the effects of the Holocene sea-level rise and by tectonic movements (uplift, subsidence).

Estuaries often receive sediment from other sources than the river: from the adjacent sea (derived from coastal or near-shore bottom erosion, or from another river by long-shore transport), from local runoff, from the atmosphere, from waste discharges, and from organic production. Sediment fluxes can be determined from direct measurements of currents and suspended matter concentrations, from numerical models, from the distribution of natural and artificial tracers, which together with other data can be combined into sediment balance estimates. Such estimates, based on all available data, give the most comprehensive and reliable results but are usually open ended where the influx from the sea is difficult to estimate (which is the case in many of the smaller tidal estuaries). Off large rivers with a large sediment supply (Amazon, Chang Jiang) influx from sources other than the river is negligible and the sediment flux pattern is shown by the dispersal of the river-supplied material.

In the coastal sea outside the estuary suspended matter tends to stay nearshore by the combined effect of a density circulation, waves and long-shore currents. Mud deposits are formed near shore where a) suspended matter is concentrated (by tidal asymmetry, in residual gyres, or where resultant currents converge), and b) also wave action is low (sheltered areas or at greater water depth). Mud can be deposited in spite of periodically high tidal current velocities. Deposition occurs predominantly on the inner shelf or inshore: on the middle and outer shelf Recent mud deposition is found near to some large rivers, in areas of large organic production, or where older deposits are winnowed out by waves and/or tidal currents. Most mud deposits seaward of the inner shelf are relict. Transport of recent river-supplied suspended matter to the deep ocean takes place mainly where a river mouth is located near, or at, the shelf edge or where a canyon funnels it downward. There is often a slow transport of relatively small amounts of suspended matter over the shelf through a low-concentration nepheloid layer near to the bottom, the suspended matter coming from the shore or being stirred up from the older shelf deposits. The suspended matter transported through the bottom nepheloid layer is chiefly deposited on the continental slope and in the adjacent ocean, the conditions on the outer shelf being unfavorable for mud deposition.

6. REFERENCES

Allen, G. P., J. C. Salomon, P. Bassoulet, Y. de Penhoat, and C. de Grandpre, 1980. Effects of tides on mixing and suspended sediment transport in macrotidal estuaries. *Sed. Geol.* **26**, 69-90.

Aloisi, J. C., A. Monaco, C. Millot, and H. Pauc, 1979. Dynamique des suspensions et mecanismes sedimentogenetiques sur le plateau continental du Golfe de Lion. *C. R. Acad. Sci. Paris,* **289**, ser. D., 879-882.

d'Anglejean, B., and R. O. Ingram, 1976. Time-depth variations in tidal flux of suspended matter in the St. Lawrence estuary. *Est. Coast. Mar. Sci.* **4**, 401-406.

Biggs, R. B., 1970. Sources and distribution of suspended sediment in northern Chesapeake Bay. *Mar. Geol.* **9**, 187-201.

Biggs, R. B., and B. A. Howell, 1984. The estuary as a sediment trap: alternate approaches to estimating its filtering efficiency, *in* V. S. Kennedy ed., *The Estuary as a Filter*, Acad. Press Inc., 107-129.

Bowden, K. F., 1980. Physical factors: salinity, temperature, circulation and mixing processes, *in.* E. Olausson, andI. Cato eds., *Chemistry and Biogeochemistry of Estuaries,* Wiley and Sons, 37-70.

Bremner, J. M., 1980. Physical parameters of the diatomaceous mud belt off South West Africa. *Mar. Geol.* **34**, M67-M76.

Cochran, J. K., 1984. The fates of uranium and thorium decay series nuclides in the estuarine environment, *in* V. S. Kennedy ed., *The Estuary as a Filter*, Acad. Press, 179-220.

Davies, P. J., 1979. Marine geology of the continental shelf off southeast Australia. *Bull. Bur. Miner. Resour. Geol. Geophys. Aust.* **195**.

Delft Hydraulics Laboratory, 1962. *Demerara Coastal Investigation.* Delft, 240 p.

Drake, D. E., 1976. Suspended sediment transport and mud deposition on continental shelves, *in* D. J. Stanley and D. J. P. Swift, eds., *Marine Sediment Transport and Environmental Management.* Wiley and Sons, Ch. 9, 127-158.

Edzwald, J. K., and Ch. R. O'Melia, 1975. Clay distributions in recent estuarine sediments. *Clays and Clay Min.* **23**, 39-44.

Eisma, D., 1981. Supply and deposition of suspended matter in the North Sea. *Spec. Publ. Int. Ass. Sedim.* **5**, 41 5-428.

Eisma, D., 1986, Flocculation and de-flocculation of suspended matter in estuaries. *Neth. Jour.*

Sea Res., **20** (2/3), 183-199.

Eisma, D., S. J. van der Gaast, J. M. Martin, and A. J. Thomas, 1978. Suspended matter and bottom deposits of the Orinoco delta: turbidity, mineralogy and elementary composition. *Neth. Jour. Sea Res.* **12**, 224-251.

Eisma, D., G. C. Cadee, and R. Laane, 1982. Supply of suspended matter and particulate and dissolved organic carbon from the Rhine to the coastal North Sea. *Mitt. Geol. Palaont. Inst. Univ. Hamburg*, SCOPE/UNEP Sonderb. **52**, 483-505.

Eisma, D., J. Boon, R. Groenewegen, V. Ittekkot, J. Kalf,, and W. G. Mook, 1983. Observations on macro-aggregates, particle size and organic composition of suspended matter in the Ems estuary. *Mitt. Geol. Palaont. Inst. Univ. Hamburg*, SCOPE/UNEP Sonderb. **55**, 295-314.

Eisma, D., P. Bernard, J. J. Boon, R. van Grieken, J. Kalf, and W. G. Mook, 1984. Loss of particulate organic matter in estuaries as exemplified by the Ems and the Gironde estuaries. *Mitt. Geol. Palaont. Inst. Univ. Hamburg*, SCOPE/UNEP Sonderb. **58**, 397-412.

Eisma, D., and J. Kalf, 1984. Dispersal of Zaire river suspended matter in the estuary and the Angola Basin. *Neth. Jour. Sea Res.* **17**, 385-411.

Eisma, D., and J. Kalf, 1987, Dispersal, concentration and deposition of suspended matter in the North Sea. *Bull. Geol. Soc. London,* in press.

Elliot, A. J., 1978. Observations of the meteorologically induced circulation in the Potomac estuary. *Est. Coast. Mar. Sci.* **6**, 285-299.

Fairbridge, R. W., 1980. The estuary: its definition and geodynamic cycle, *in*. E. Olausson, and I. Cato eds., *Chemistry and Biogeochemistry of Estuaries*, Wiley and Sons, 1-35.

Ferentinos, O., and M. Collins, 1979. Tidally induced secondary circulations and their associated sedimentation processes. *Jour. Oceanogr. Soc. Japan* **35**, 65-74.

Festa, J. F., and D. V. Hansen, 1978. Turbidity maxima in partially mixed estuaries: a two-dimensional numerical model. *Est. Coast. Mar. Sci.* **7**, 347-359.

Feuillet, J. P., and P. Fleischer, 1980. Estuarine circulation: controlling factor of clay mineral distribution in James river estuary, Virginia. *Jour. Sed. Petrol.* **50**, 267-280.

Gallenne, B., 1974. Study of fine material in suspension in the estuary of the loire and its dynamic grading. *Est. Coast. Mar. Sci.* **2**, 261-272.

Greensmith, J. T., and E. V. Tucker, 1973. Holocene transgressions and regressions on the Essex coast, Outer Thames estuary. *Geol. Mijnb.* **52**, 193-202.

Gross, M. O., M. Karweit, W. B. Cronin, and J. R. Schubel, 1978. Suspended sediment discharge from the Susquehanna River to northern Chesapeake Bay, 1966-1976. *Estuaries* **1**, 106-110.

Hayes, M. O., 1978. Impact of hurricanes on sedimentation in estuaries, bays and lagoons, *in* M. L. Wiley ed., *Estuarine Interactions*, Acad. Press., 323-346.

Hayes, M. O., and J. Michel, 1982. Shoreline sedimentation within a forearc embayment, Lower Cook Inlet, Alaska. *Jour. Sed. Petrol.* **52**, 251-263.

Hunter, K. A., and P. S. Liss, 1982. Organic matter and the surface charge of suspended particles in estuarine waters. *Limnol. Oceanogr.* **27**, 322-335.

Jumars, P. A., A. R. M. Nowell, and R. F. L. Self, 1981. A simple model of flow-sediment-organism interaction. *Mar. Geol.* **42**, 155-172.

Kandiah, A., P. Loganathan, and R. B. Krone, 1975. Pore and eroding fluid in fluences on surface erosion of soil. *Jour. Geotechn. Eng. Div. ASCE,* **101**, GTI: 51-65.

Kirby, R., and W. R. Parker, 1982. A suspended sediment front in the Severn Estuary. *Nature,* **295**, 396-399.

Koppelman, L. E., P. K. Weyl, M. Grant Gross, and DeW. S. Davies, 1976. *The Urban Sea: Long Island Sound.* Praeger Publ., N. Y., 223 p.

Kranck, K., 1984. The role of flocculation in the filtering of particulate matter in estuaries, *in* V. S. Kennedy ed., *The Estuary as a Filter.* Acad. Press Inc., 159-175.

Kranck, K., and T. Milligan, 1980. Macroflocs: production of marine snow in the laboratory. *Mar. Ecol. Progr. Ser.* **3**, 19-24.

McCave, I. N., 1972. Transport and escape of fine-grained sediment from shelf areas, *in* D. J. P. Swift, D. B. Duane, and O. H. Pilkey eds., *Shelf Sediment Transport,* Dowden,

Hutchinson and Ross, 225-248.

McCave, I. H., 1984. Size spectra and aggregation of suspended particles in the deep ocean. *Deep-Sea Res.* **31**, 329-352.

Meade, R. H., 1972. Transport and deposition of sediments in estuaries. *in* B. W. Nelson ed., Environmental Framework of Coastal Plain Estuaries, *Geol. Soc. Amer. Mem.* **133**, 91-120.

Mehta, A. J., and E. Partheniades, 1973. *Depositional behavior of cohesive sediments.* Coast. Oceanogr. Eng. Lab. Univ. Florida Techn. Rep. **16**, 274 p.

Meybeck, M., 1976. Total mineral dissolved transport by world major rivers. *Hydrol. Sc. Bull.* **XXI**, 265-284. Erratum XXI, 651.

Meybeck, M., 1981. River transport of organic carbon to the ocean. Rep. Workshop U. S. Dept. Energy and W.H.O.I., Sept. 21-25, no. 016, CONF 80-09-140, 219-269.

Milliman, J. D., C. P. Summerhayes, and H. T. Barretto, 197S. Oceanography and suspended matter off the Amazon river, February-March 1973. *Jour. Sed. Petrol.* **45**, 189-206.

Milliman, J. D., J. Butenko, J. P. Barbot, and J. Hedberg, 1982. Depositional patterns of modern Orinoco/Amazon muds on the northern Venezuelan shelf. *Jour. Mar. Res.* **40**, 643-657.

Milliman, J. D., and R. H. Meade, 1983. World-wide delivery of river sediment to the oceans. *Jour. Geol.* **91**, 1-21.

NEDECO, 1965. *A study on the siltation of the Bangkok Port Channel. II. The Field Investigation.* Den Haag, 474 p.

Nowell, A. R. M., P. A. Jumars, and J. E. Eckman, 1981. Effects of biological activity on the entrainment of marine sediments. *Mar. Geol.* **42**, 133-153.

Officer, Ch. B., 1976. *Physical Oceanography of Estuaries.* Wiley Sons, 465 p.

Officer, Ch. B., and M. M. Nichols, 1980. Box model application to a study of suspended sediment distributions and fluxes in partially mixed estuaries, *in* V. S. Kennedy ed., *Estuarine Perspectives.* Acad. Press, 329-340.

Pocklington, R., and J. D. Leonard, 1979. Terrigenous organic matter in sediments of the St. Lawrence estuary and the Saguenay Fjord. *Jour. Fish. Res. Board Canada* **36**, 1250-1255.

Postma, H., 1980. Sediment transport and sedimentation, *in* E. Olausson, and I. Cato eds., *Chemistry and Biogeochemistry of Estuaries*, Wiley and Sons, 153-186.

Rumeau, J. L., and J. R. Varney, 1968. Elements traces de vases marines du plateau continental Atlantique au sud-ouest du massif armoricain (Orande Vasiere). *Bull. Centre Rech. Pau-SNPA* **2**, 69-81.

Salomons, W., 1975. Chemical and isotopic composition of carbonates in recent sediments and soils from Western Europe. *Jour. Sed. Petrol.* **45**, 440-449.

Salomons, W., P. Hofman, R. Boelens, and W. G. Mook, 1975. The oxygen isotopic composition of the fraction less than 2 microns (clay fraction) in Recent sediments from western Europe. *Mar. Geol.* **18**, M23-M28.

Schubel, J. R., 1972. Distribution and transportation of suspended sediment in Upper Chesapeake Bay, *in* B. W. Nelson ed., Environmental Framework of Coastal Plain Estuaries. *Geol. Soc. Amer. Mem.* **133**, 151-168.

Sholkovitz, E. R, 1976. Flocculation of dissolved organic and in organic matter during the mixing of river water and seawater. *Geochim. Cosmochim. Acta* **40**, 831-845.

Sholkovitz, E. R., E. A. Boyle, and N. B. Price, 1978. The removal of dissolved humic acids and iron during estuarine mixing. *Earth Plan. Sci. Lett.* **40**, 130-136.

Simpson, J. H., C. M. Allen, and N. C. O. Morris, 1978. Fronts on the continental shelf. *Jour. Geophys. Res.* **83** (C9), 4607-4614.

Syvitski, J. P. M., and J. W. Murray, 1981. Particle interaction in fjord suspended sediment. *Mar. Geol.* **39**, 215-242.

Tan, F. C., and P. M. Strain, 1979. Organic carbon isotope ratios in Recent sediments in the St. Lawrence estuary and the Gulf of St. Lawrence. *Est. Coast. Mar. Sci.* **8**, 213-225.

Teleki, P. O., 1972. Wave boundary layers and their relation to sediment transport, *in* D. J. P. Swift, D. B. Duane, and O. H. Pilkey, eds. *Shelf Sediment Transport: Process and Pattern*, Dowden, Hutchinson and Ross, 21-59.

Terwindt, J. H. J., 1977. Deposition, transportation and erosion of mud. *in* H. L. Golterman ed., *Interactions Between Sediments and Fresh Water*. Proc. Int. Symp. Amsterdam 6-10 Sept. 1976. Junk BV, 19-24.

Tipping, E., and D. Cooke, 1982. The effects of adsorbed humic substances on the surface charge of goethite (-FeOOH) in freshwaters. *Geochim. Cosmochim. Acta* **46**, 75-80.

Trent, J. D., A. L. Shanks, and M. W. Silver, 1978. *In situ* and laboratory measurements on macroscopic aggregates in Monterey Bay, California. *Limnol. Oceanogr.* **23**, 626-635.

de Vooys, C. O. N., 1979. Primary production in aquatic environments, *in* B. Bolin, E. T. Degens, S. Kempe, and P. Ketner eds., *The global carbon cycle*. SCOPE vol. 13, Wiley and Sons, 259-292.

Wells, J. T., 1983. Dynamics of coastal fluid muds in low-, moderate- and high-tide environments. *Can. J. Fish. Aquat. Sci.* **40** (suppl. 1), 130-142.

Whitehouse, U. O., L. M. Jeffrey, and J. D. Debrecht, 1960. Differential settling tendencies of clay minerals in saline waters, *in* A. Swineford ed., *Clays and Clay Minerals*, Proc. 7th Natl. Conf., Pergamon Press, 1-79.

Wright, L. D. , 1978. River deltas, *in* R. A. Davis Jr. ed., *Coastal Sedimentary Environments*. Springer, Ch. 1, 5-68.

Wollast, R., and G. Billen, 1981. The fate of terrestrial organic carbon in the coastal sea. Rep. Workshop U. S. Dept. Energy and W.H.O.I., Sept. 21-25, no. 016, CONF 80-09-140, 331-359.

RIVERBORNE MATERIALS AND THE CONTINENT-OCEAN INTERFACE

J. D. Burton
Department of Oceanography
The University
Southampton SO9 5NH
U.K.

1. INTRODUCTION

Rivers form the dominant route for the transport to the ocean of material derived from the weathering of the continents. Although inputs through the atmosphere and by glacial transport can be important in some oceanic regions, these routes, together with direct coastal inputs through erosion and ground water movement, account globally for only about 10% of the total transport (Garrels and Mackenzie, 1971). Consequently the major part of the continentally derived products of weathering passes through estuaries. The pronounced changes in physicochemical characteristics that occur during the mixing of river water and seawater and, commonly, the close coupling between estuarine processes in the water column and in sediments, lead to a variety of reactions of geochemical significance. These factors give the estuarine zone an importance for the major sedimentary cycle which is disproportionate to its geographical extent.

Definitions of the seaward limit of an estuary in terms of topographic features or specific boundaries determined by oceanographic parameters such as salinity are often artificial from the standpoint of geochemical processes as there is a continuum of processes in estuaries and adjacent coastal waters. Certain geochemical phenomena are, however, clearly associated with the earlier stages of mixing of river waters and seawater and it is with these that the present account is mainly concerned.

The study of estuarine chemistry is challenging because of the complexity of the relevant geochemical and hydrodynamic processes and their interactions. It has been stimulated particularly by interest in the role of estuaries in modifying the riverine fluxes of material to the shelf seas and oceans, and by the need to understand the fate of anthropogenic inputs to rivers and estuaries. Several extensive accounts of the subject are available (Burton and Liss, 1976; Aston, 1978; Olausson and Cato, 1980; Martin et al., 1981; Head, 1985); the last of these particularly emphasizes practical considerations. The papers in Olausson and Cato (1980) include background accounts of physical and sedimentological processes in estuaries. Fuller accounts of these aspects are given by Dyer (1973, 1986) and Officer (1976).

A. Lerman and M. Meybeck (eds.), Physical and Chemical Weathering in Geochemical Cycles, 299–321.

2. FLUXES AND COMPOSITIONS OF DISSOLVED AND PARTICULATE MATERIALS ENTERING ESTUARIES

Recent work on the transport of total suspended sediment in rivers (Milliman and Meade, 1983) suggests that the total delivery to estuaries is 1.3×10^{16} g yr^{-1} (see also Meade and Meybeck, this volume). These workers estimate that a further 1-2×10^{15} g yr^{-1} may be transported as bedload and flood discharge. The estimation of suspended sediment discharge is based upon a critical evaluation of data from many sources but Milliman and Meade (1983) emphasize the many uncertainties in its derivation. These include the inadequacy of the data base for many important rivers, uncertainties in the contribution made by severe floods, and inappropriate choices of sampling stations. The considerable errors that can be introduced by interpolation and extrapolation procedures used to estimate suspended material loads from non-continuous records of concentration and discharge have been discussed in detail by Walling and Webb (1981).

Some 70% of the total suspended sediment discharge is derived from southeastern Asia and large western Pacific islands. The difference between the estimate by Milliman and Meade (1983) of total flux to estuaries and the earlier one by Holeman (1968) of 1.8×10^{16} g yr^{-1} is in part a consequence of Holeman's extrapolation of high sediment yields for Himalayan and Chinese rivers to the whole of Asia, a procedure which is not justified in the light of present data.

The most recent assessment of the dissolved loads of rivers is that made by Meybeck (1979, 1981; see also this volume) who estimated the mean load as 33×10^6 g km^{-2} yr^{-1}, corresponding to a flux of dissolved material from the continents of 3.7×10^{15} g yr^{-1}. An examination by Walling and Webb (1983) of data for 490 rivers gave a global mean load of 39×10^6 g km^{-2} yr^{-1}. The range for individual rivers was 0.34 to 311×10^6 g km^{-2} yr^{-1}, with most values in the range of about 5-100×10^6 g km^{-2} yr^{-1}. Errors in the estimation of annual dissolved loads for these rivers may be no greater than ±20% (Walling and Webb, 1983) but, as with the estimates of delivery of particulate material, the data base has deficiencies in coverage, especially for large rivers in undeveloped areas.

Estimates of the world average concentrations of the major dissolved constituents in river water are given in Table I, together with values for five regions or systems. The dispersion of concentration values differs considerably for individual constituents, with the variability decreasing in the following order (Meybeck, 1981):

$$Cl^- > SO_4^{2-} > Ca^{2+} = Na^+ > Mg^{2+} > HCO_3^- > SiO_2 > K^+$$

It is not clear to what extent this order is influenced by physicochemical controls on concentrations in solution, as well as by source variability, although the existence of such a mechanism for silicon, possibly depending upon sorption processes, has been argued (Edwards and Liss, 1973).

Some of the processes which give rise to differences in the proportions of major constituents in river waters were considered by Gibbs (1970) as the basis for a rationalization of the major types. He showed that there was a general tendency for the ratio $Na^+/(Na^+ + Ca^{2+})$ or $Cl^-/(Cl^- + HCO_3^-)$ to decrease with increasing concentration of total dissolved material up to concentrations in the range of about 100-1000 mg l^{-1}, above which the ratios tended to increase. These features were interpreted as corresponding to two series. In the first, influence of salts in rainwater decreased relative to that of rock weathering, while in the second, removal of calcium carbonate was occurring with increase in total salt content. Meybeck (1979) has estimated that the type dominated by rock weathering accounts for some 98% of surface waters.

Table I. Average concentrations of dissolved constituents in river waters.

	Concentration (mg l^{-1})								
	Ca^{2+}	Mg^{2+}	Na^+	K^+	Cl^-	SO_4^{2-}	HCO_3^-	SiO_2	Total*
Global average [(a)]	15	4.1	6.3	2.3	7.8	11.2	58.4	13.1	118
Global average [(b)]	13.4	3.3	5.1	1.3	5.7	8.3	52	10.4	100
N.W. Territories	3.3	0.7	0.6	0.4	1.9	1.9	10.1	0.4	19
Guyana	2.6	1.1	2.5	0.7	3.9	2.0	12.2	10.9	36
Philippines	31	6.6	10.4	1.7	3.9	13.6	131	30	228
Mackenzie	33	10.4	7.0	1.1	8.9	36	111	3.0	210
Colorado	83	24	95	5.0	82	270	135	9.3	703

* Total = total dissolved material.

[(a)] Estimated by Livingstone (1963).

[(b)] All other estimates are from Meybeck (1979). Compilation from Meybeck (1981).

Detailed studies on the Amazon Basin (Stallard and Edmond, 1983; Stallard, this volume), like those on the Mackenzie system (Reeder et at., 1972), show clear relationships of both total cation concentrations and the proportions of major constituents with the lithological and pedological characteristics of the catchment areas. In the Amazon Basin these features, rather than the composition of rainwater, appear generally to control the composition even of rivers with low contents of total dissolved material.

The data discussed above suggest that on a world-wide basis rivers transport about four times as much material in total particulate forms as in solution. The proportions of dissolved and particulate material vary widely among individual rivers, and in some systems the load of dissolved material exceeds that of suspended sediment as shown, for example, by the data presented by Walling and Webb (1983). These workers have discussed the disagreement in some of the earlier literature as to whether the relationship between mean dissolved and particulate loads for individual rivers is inverse or positive. They examined the relationship between these loads for 302 rivers and found no clear trend, although there was some indication of a positive relationship.

For individual elements the fractions carried in solution or in suspension vary widely (Martin and Meybeck, 1979). The dissolved transport index (i.e., the ratio of dissolved to total transport) ranges from high values for elements which readily enter into or remain in solution during weathering and transport (e.g., about 90% for bromine and 60% for sodium and calcium) to values of $<1\%$ for strongly particle-associated elements such as aluminium, iron and the rare earth elements. This index can vary considerably between individual rivers.

From the present standpoint, the results of these studies indicate the great diversity of compositions of river water with respect to concentrations of particles, total dissolved material and individual major dissolved constituents. They emphasise that estuaries are highly individual systems in respect of the chemical character of the river input. Within individual systems wide temporal variability in input fluxes may also be found. In particular there is a marked tendency for concentrations of total dissolved material to be inversely related to discharge (Walling and Webb, 1983).

Accurate measurements of trace metals in rivers are few, particularly for the major systems, and meaningful estimates of global average concentrations cannot generally be made. A discussion of several of these elements is given in this volume by Meybeck who considers also dissolved organic carbon (DOC) and the micronutrient elements, nitrogen and phosphorus, for which a large influence of man is apparent in many systems. Other accounts concerning global average fluxes include those by van Bennekom (1981) and Wollast (1983) for micronutrients and by Mantoura and Woodward (1983) for DOC.

3. MAJOR FACTORS INFLUENCING GEOCHEMICAL BEHAVIOUR IN ESTUARIES

3.1. General Features

The principal processes affecting the distribution of chemical species in estuaries may be summarized in terms of the generalized picture given in Figure 1. The mixing of river water and seawater gives rise to gradients in ionic strength, composition and pH that are dominantly horizontal, although stratification occurs to a greater or lesser extent, so that pronounced vertical gradients are also commonly found. Changes in concentration of dissolved constituents and particles are brought about by mixing processes. For particulate material, however, such changes can also occur by the gravitational settling of particles (which may be accelerated by aggregation, including aggregation in faecal pellets) and by resuspension of settled sediment (Eisma, this volume).

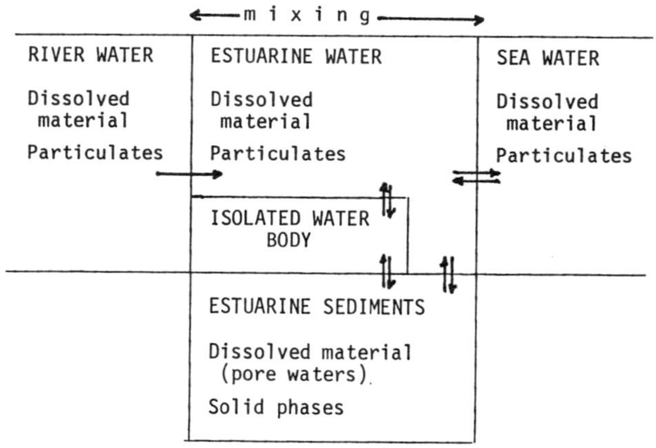

Figure 1. Schematic representation of some important processes in estuarine chemistry.

In some systems, exemplified by fjords, a body of water in the estuary may become isolated from exchange either seasonally or over a longer period. The flux of metabolisable organic material from surface waters to deep waters, as a result of *in situ* productivity in the euphotic zone and the settling of allogenic particles containing organic material, typically leads to the depletion of oxygen within sediments or within water bodies isolated to a degree which leads to removal of free oxygen at a rate greater than its renewal. Important geochemical interactions occur when oxic and anoxic waters mix, because the resulting change in oxidation states for certain elements may be associated with transfers from dissolved to particulate forms, or *vice versa.* Such reactions are of major importance also where reducing sediments are resuspended in overlying waters containing free oxygen. Coupling of chemical reactions in sediments with processes in overlying waters may also occur in the absence of resuspension, by diffusional fluxes across the benthic interface.

The extent to which specific geochemical processes occur in an estuary is to a high degree determined by hydrodynamic and topographic factors and their interaction. Thus isolation of bottom waters is favoured by restriction of lateral mixing by a sill and of vertical mixing by density stratification. Different circulation patterns in estuaries lead to differences in the distribution of particles (Postma, 1980; Eisma, this volume). Particularly important in the present context is the formation of a turbidity maximum in meso- or macrotidal estuaries. The common occurrence of this feature within the zone of early mixing of river water with saline water is an additional factor making this zone of crucial importance in many estuarine chemical processes (Morris et al., 1978).

The different characteristics of water circulation and mixing in estuaries also mean that the history of a parcel of water of a given salinity can be quite different in different estuaries and at different periods in the same estuary. The degree to which a particular estuarine chemical reaction occurs in an estuary or a particular compartment of an estuary is not determined simply by the availability of the reactants in the form of solutions and particles but may often be markedly influenced by the residence time of the reactants in the estuary or compartment.

3.2. Salinity and Compositional Variations

Concentrations of total dissolved material in river waters vary widely, as already discussed, but for many rivers lie in the range of 20-400 mg l^{-1}. At the seaward end of most estuaries, as defined topographically, the total salt concentration will usually be above 30 g l^{-1}, although in areas subject to seasonal flooding, as under monsoon regimes, water at the estuary mouth may at times be almost fresh. The use of salinity as an index of total salt concentration over the range encountered in estuaries, while acceptable for almost all practical purposes, is not rigorous, since salinity is measured using a practical scale which is based on an assumption of near constancy of the proportions of individual ions in waters of different salinities. This condition clearly is not fulfilled when seawater is diluted with river water. Chlorinity or chlorosity is sometimes used as an alternative index of the extent of estuarine mixing of seawater and river water. For consideration of many physicochemical properties of estuarine waters, ionic strength is an appropriate parameter, reflecting both the concentrations and charges of the ionic species present.

From the standpoint of solution chemistry, an estuary can be considered in the simplest case as a mixing series with one freshwater and one saline end-member. Coastal seawater can be regarded for present purposes as having essentially constant proportions of the major constituents. The minor constituents in seawater include a group of elements (e.g., Li, Cs, V, Mo, U) which behave more or less conservatively in the ocean and which consequently can be expected to vary in shelf waters rather narrowly and predictably, and a group, including many transition metals, which show greater spatial and temporal variability in these waters. It is primarily, however, the wide variability of both the major and minor constituents in river water that makes generalizations about estuarine mixing series difficult. As a basis for a general examination of the major chemical variables in estuaries the situation may be considered in which river water of globally average composition mixes with seawater in a model estuary.

The compositions of global average river water and seawater, in terms of major dissolved constituents, are compared in Table II. In Table III, the variations in molality (moles per 1 kg H_2O) of the major constituents with salinity are presented. The values given for ionic strength (I) refer to the nominal ionic strength; the effective ionic strength is affected by the associations between major ions and is somewhat lower than the nominal. Large variations in the proportions of the major constituents, with the exception of the bicarbonate ion, occur only in the early stages of mixing, up to a salinity of about 5‰. The compositional variations in estuaries are accompanied by changes in chemical speciation that are related to differences in the activity coefficients of free ions and complexes with ionic strength and variations in the concentrations of complexing ligands. Information on this topic can be found in papers by Turner et al. (1981) and Mantoura and Morris (1983) and in references cited by these authors.

Table II. Comparison of the compositions of seawater and the global average river water supply to the ocean

Constituent	River Water Concentration mg l^{-1}	River Water % of Total*	Seawater (salinity 35 g/kg) Concentration mg l^{-1}	Seawater (salinity 35 g/kg) % of Salinity
Sodium	5.1	5	10765	31
Potassium	1.3	1	399	1
Magnesium	3.3	3	1294	4
Calcium	13.4	13	412	1
Chloride	5.7	6	19353	55
Sulphate	8.3	8	2712	8
Bicarbonate	52	52	142[a]	0.4
Silicon (as SiO_2)	10.4	10	<0.1–>10[b]	-

[a]Inorganic carbon as bicarbonate.

[b]Variable with location and depth.

*% of Total = % of Total Dissolved Material

3.3. Variations in pH

Variations in pH in estuaries are important in relation to both the equilibria between dissolved species and the distribution of species between solution and particulate phases. The pH of oceanic water in equilibrium with the atmospheric partial pressure of carbon dioxide is about 8.1. River waters can be of highly variable pH, ranging from acid drainage from mineralized catchments, and drainage in areas with high organic and low mineral inputs, to waters of high alkalinity.

The estuarine distribution of pH cannot be treated as a simple function of salinity. Mook and Koene (1975) modelled estuarine pH distributions assuming the system to be closed with respect to both titration alkalinity and total inorganic carbon. Measured distributions in the western Scheldt Estuary showed fair agreement with the model predictions. Continuous measurements of pH made on the Tamar Estuary

Table III. Variations in molality of some major constituents and ionic strength (I) with salinity in mixtures of surface oceanic seawater and global average river water (RW). Data from Kester et al. (1975).

Constituent	RW	Salinity 4 o/oo	10 o/oo	20 o/oo	35 o/oo
Na^+	0.00027	0.05426	0.13731	0.27574	0.48614
Mg^{2+}	0.00017	0.00623	0.01556	0.03110	0.05473
Ca^{2+}	0.00037	0.00151	0.00327	0.00620	0.01065
K^+	0.00006	0.00123	0.00303	0.00627	0.01060
Cl^-	0.00022	0.06306	0.15973	0.32086	0.56577
SO_4^{2-}	0.00011	0.00335	0.00833	0.01664	0.02926
HCO_3^-	0.00096	0.00113	0.00139	0.00182	0.00247
F^-	0.00001	0.00002	0.00003	0.00004	0.00007
I	0.00210	0.08200	0.20510	0.41020	0.72180

by Morris (1978) did not, however, confirm the validity of the model and in many circumstances considerable departures from equilibrium predictions evidently occur, particularly as a result of biological cycling of carbon.

3.4. Variations in Redox Conditions

The redox conditions in estuarine waters range widely from those in water saturated with free oxygen to those in the sulphide-containing zones that occur in some deeper waters of restricted circulation and more widely in reducing sediments. These variations are created primarily by the decomposition of organic material in a sequence of reactions involving different terminal electron acceptors, the most important of these stages being the utilization of free oxygen, of nitrate (denitrification reaction) and its initial reduction products nitrite and nitrous oxide, and of sulphate.

Equilibrium calculations of thermodynamically stable oxidation states for trace elements in natural waters have been of importance because metals can differ greatly in their geochemical behaviour, and in their bioavailability and toxicity, according to their oxidation state. Predictions of the oxidation states of metals and other elements in oxygenated waters have generally been made by assigning a pE value of 12.5 as determined by the oxygen/water couple. From the electron free-energy diagram given in Figure 2 it is evident that if the pE were determined by the oxygen/water couple, the elements shown would be present in their higher oxidation states. Experimental data show, however, that lower oxidation states often occur to a significant extent in estuarine and other natural waters for many elements. The role of Mn(II) in estuaries is discussed in detail subsequently. Recent reports on other elements with variable oxidation states include work on arsenic (Howard et al., 1984), selenium (Takayanagi and Wong, 1984) and chromium (Cranston and Murray, 1980).

The use of a pE of 12.5 in predicting redox equilibrium speciation in oxygenated waters must exaggerate the extent of disequilibrium in many situations. It assumes the validity of the concept of a unique value determining all redox equilibria and it neglects the role of the oxygen/hydrogen peroxide couple. It is clear, nevertheless, that elements in lower oxidation states are often present as transient species, introduced by transport from external sources such as the products of weathering, by transfer from compartments where conditions are reducing, or by *in situ* biological or photochemical processes. In these circumstances kinetic factors become of major

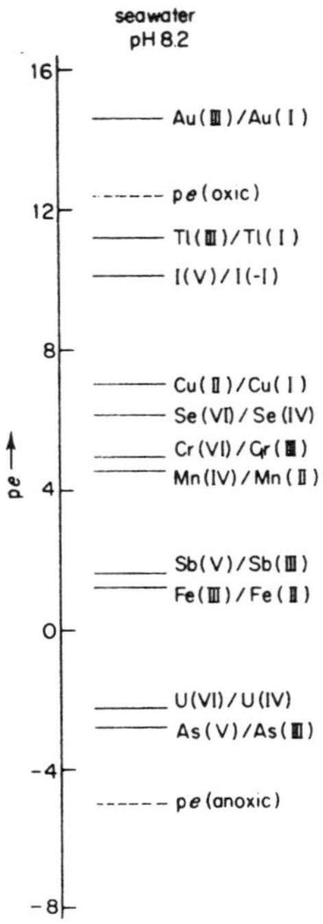

Figure 2. Electron free-energy diagram for some redox couples in sea water (from Turner et al., 1981).

importance in determining speciation.

3.5. Particulate Material

Although very high concentrations of suspended particulate material (SPM) enter some estuaries by transport in turbid rivers, the concentrations of SPM in many estuaries are determined not so much by the concentrations in the riverine and marine end-members as by the resuspension of particles from sediments. Concentrations in turbidity maxima in estuaries with low tidal ranges are commonly in the range of 100-200 mg l^{-1}, but in those with high tidal ranges, concentrations of 1-10 $g \cdot l^{-1}$ are found (Dyer, 1986). The concentrations are temporally variable, for example, over a spring-neap tide cycle. Particles of different sizes have different life times in suspension in the turbidity maximum and at any one time the material reflects in a complex way the effects of erosion, suspension and settling. Particularly because of the differences in redox conditions between the bottom sediments and the overlying water column, the material introduced by resuspension is potentially reactive, and turbidity maxima are important regions for the redistribution of material between particles and solution.

As well as modified particles introduced in this way, new particles are produced in the estuarine water column, principally by the destabilization and aggregation of microcolloidal material and by biological activity. Suspended particulate material thus consists of a complex mixture of inorganic and organic substances of continental, marine and *in situ* origins. The majority of constituents do not occur in discrete phases but are distributed between the bulk components formed by more abundant elements. They may be incorporated in lattice structures of lithogenic minerals (mainly aluminosilicates), associated with the hydrous oxide phases formed, notably by iron, in aquatic environments during weathering and transport, bound to organic matter or held at particle surfaces by adsorption or ion-exchange. The factors that determine the aggregation of particles and their role as sites for geochemical reactions are poorly understood in detail but some information is available on certain relevant characteristics such as surface area and charge.

Martin et al. (1985) have shown that the specific surface area of suspended sediment from the Gironde and Loire Estuaries is greatest in the high turbidity region around the salt intrusion. It is determined largely by amorphous oxide coatings on particles and is higher for particles from which organic matter has been removed. In contrast, the cation exchange capacity of the particles was increased by the presence of organic coatings, in agreement with earlier findings.

Micro-electrophoretic measurements by Hunter and Liss (1982) on suspended particles from four estuaries showed a rather narrow range of negative values for electrophoretic mobility over the salinity range from 5 to >30‰, with a slight decrease in the values with increase in salinity. At lower salinities the variations in electrophoretic mobility depended upon the concentration of dissolved cations, a pronounced increase in mobility for particles in the low salinity region being observed only when the river water was low in cations, particularly Ca^{2+}. In all the systems no positively charged particles were observed. The general uniformity of charge suggested that the surface properties of particles of widely differing compositions were controlled by surface coatings. Recent work (Hunter, 1980; Loder and Liss, 1985) has confirmed earlier indications of the importance of organic coatings. A variety of inorganic and organic solid phases become coated in seawater with organic matter which then largely determines their electrophoretic properties. The major functional groups in this ubiquitous material are carboxylic or phenolic in character.

4. APPROACHES TO THE STUDY OF THE BEHAVIOUR OF CONSTITUENTS DURING ESTUARINE MIXING

The extent of transfer of material between solution and solid phases in estuaries determines whether the flux of a dissolved constituent leaving the estuary differs significantly from that entering. The identification of source and sink functions for dissolved constituents has thus been a primary concern in estuarine chemistry. The emphasis on dissolved constituents reflects the difficulties in interpreting data on particulate material because of the complexity of both its composition and transport pathways, and also the fact that in some cases mass balance considerations make the dissolved pool a more sensitive indicator of transfer processes.

A basic tool for the study of environmental phase transfer processes has been the estuarine mixing diagram in which the concentration of a dissolved constituent is plotted against a conservative indicator of mixing, usually salinity. The basis of this approach is shown in Figure 3 and the more complex relationships that are often encountered in estuaries are illustrated in Figure 4. Deviation from the linear relationship corresponding to conservative behaviour (that is, mixing in which no transfer between phases takes place) is indicative of removal or addition of the dissolved constituent provided that the system is constrained as a mixing series between two end-members of constant composition. The basic data for this approach are obtained by collecting samples over the whole range of riverine and estuarine compositions, and it may therefore be termed the survey approach. It has a number of well-recognised limitations.

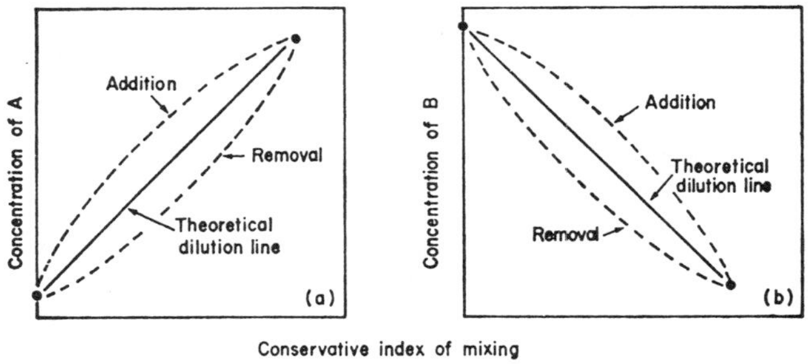

Figure 3. Simple model relationships between the concentration of a dissolved constituent and a conservative index of mixing in an estuary under steady state, with single sources of river and sea water (a) for a component (A) with a concentration higher in sea water than in river water; (b) for a component (B) with a concentration lower in seawater than in river water (from Liss, 1976).

First, so far as the fate of riverborne material is concerned, it is in practice informative only for those constituents which show a marked inverse relationship with salinity. A constituent which is enriched in the marine end-member may show conservative or non-conservative behaviour in the estuary, but it is difficult to follow processes affecting a low concentration riverine input against a marked gradient of increasing concentration with increase in salinity. Indeed, for a conservative constituent with a concentration in seawater much higher than that in river waters, data for various systems may conform very closely to a single dilution line representing in effect a global estuary. This is the case for methylated germanium species, particularly monomethylgermanium, which is remarkably conservative in the ocean (Lewis et al., 1984). Another example is boron in the Zaire and Magdalena Estuaries (Fanning and Maynard, 1978), although removal of this element has been reported to occur in the much smaller estuary of the Alde (Liss and Pointon, 1973).

Secondly, there are various hydrographic complications. Some estuaries have several inputs of fresh water with differing compositions and such tributary effects can simulate non-conservative behaviour. The identification of the marine end-member appropriate for an estuarine mixing series may also not be straightforward. Temporal variations in end-member concentrations, particularly in rivers, may be a particularly important factor. In interpreting dilution curves, such as those in Figures 3 and 4, in terms of source and sink functions, it is implicitly assumed that the estuary is a steady-state system. Changes in river water concentrations can give rise to estuarine mixing curves that simulate removal or addition, the precise effects depending upon the magnitude and periodicity of the changes and the integrating time in the estuarine reservoir (Loder and Reichard, 1981).

While erroneous conclusions can be reached if the survey approach is used uncritically, and its limitations, together with sampling and analytical errors, render it always difficult to establish small extents of removal or addition, it nevertheless

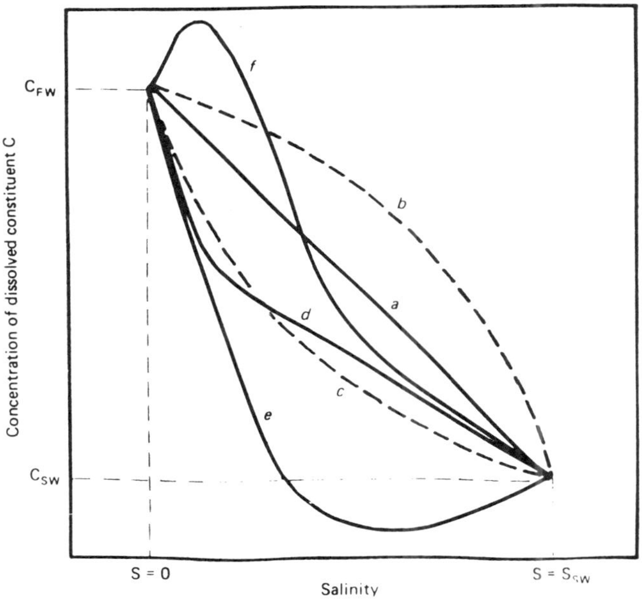

Figure 4. Model relationships corresponding to those shown in Fig. 3b, but showing a wider range of behaviour. Constituent C has a concentration of C_{FW} in river water (S=O) and of C_{SW} in seawater ($S=S_{SW}$). Line (a) and curves (b) and (c) correspond to those shown in Fig. 3. Curve (d) shows removal occurring in the upper estuary only; curve (e) shows removal within the estuary that exceeds the riverine input; curve (f) shows addition to the upper estuary, followed by removal in the lower estuary (from Morris, 1985).

remains a valuable tool and it has been developed to provide more critical quantitative treatments.

Boyle et al. (1974) used a mathematical model to demonstrate that continuous removal would generate a curve in the mixing relationship and pointed out that many data sets conform rather to two or more linear segments, which indicate either removal within a narrow salinity range or tributary effects. Removal has often been calculated by extrapolating the best linear fit in the high salinity region back to zero salinity to give an effective zero salinity end-member and comparing its concentration with that in the river water entering the system. Officer (1979) has examined this procedure in terms of hydrodynamic considerations and has shown it to be valid given the assumptions of a steady state, one dimensional, tidally averaged model. Factors influencing the distributions of nutrients in an estuarine system have been described mathematically in the model of Kaul and Froelich (1984), which incorporates functions for conservative behaviour, biological productivity (first order removal) and regeneration (parabolic input).

The survey approach has been further developed operationally in work on the Tamar Estuary. The use of sampling from a vessel equipped for continuous measurements of salinity has enabled changes in parameters in the low salinity range, below 1‰, to be followed in detail revealing the importance of processes in the very early stages of mixing (Morris et al., 1978). It has also been shown that a clearer picture of behaviour is obtainable by repeated surveys which enable time-averaged characteristics of the system to be examined (Morris et al., 1981, 1982a). Knox et al. (1981) have described the statistical analysis of estuarine survey data for two constituents that are likely to have a common source; this approach has the advantage that interpretation is less dependent upon knowledge of end-member concentrations.

An important approach, which is complementary to those involving analysis of survey data, has been the laboratory simulation of estuarine mixing processes. This approach was pioneered by Sholkovitz (1976) who carried out experiments in which filtered river waters and seawater were mixed batchwise to produce samples of intermediate salinities and the new particles produced then separated and analyzed. In this form the method had several limitations. First, the use of filtered waters meant that potentially important reaction sites on existing particles were removed and releases from such particles could not be identified; in some subsequent experiments unfiltered samples were also used. Secondly, the mixing procedures did not accurately simulate the continuous gradient of salinity actually encountered by the river input in estuarine mixing. Thirdly, such experiments were not designed to examine biological removal processes which can be important for some constituents. The advantages have been major, however, in enabling certain processes, particularly the destabilization and aggregation of colloidal material, to be studied under defined conditions and in providing evidence concerning the mechanisms and rates of such processes. Apparatus for simulating continuous mixing more closely has been subsequently developed (Bale and Morris, 1981).

5. ESTUARINE CHEMICAL BEHAVIOUR AND THE FATE OF RIVERBORNE MATERIAL IN ESTUARIES

5.1. Particulate Material

The deposition of sediment in the deep ocean and on the continental shelf appears to account for only some 10%, at most, of the amount supplied by rivers, implying large storage globally in estuaries and in the submarine fans formed by some major sediment-transporting systems (Postma, 1980). Individual estuaries, however, have trapping efficiencies that range essentially from 0 to 100%. Estuaries and coastal marshes on the eastern seaboard of the United States retain virtually all the sediment delivered by rivers and may accumulate additional sediment from the shelf (Bokuniewicz and Gordon, 1980; Meade, 1982). The capacity of an estuary such as Long Island Sound continues to increase, as the rate of rise of sea level exceeds the accumulation rate of the deposits. In other regimes storage capacity may be exceeded and all the sediment delivered by rivers be exported.

Coarse particles entering an estuary retain their individual sizes and are governed in their distribution mainly by hydrodynamic factors, but the depositional behaviour of particles with sizes in the clay (<2 μm) and colloidal ranges is affected by coagulation. The data on charges on particles as discussed in a previous section suggest that marked reduction in the negative charges on particles with increasing salinity may occur only in the low salinity region of certain estuaries. It seems therefore probable that compression of the electrical double layer as a result of the increased ionic strength of the solution may be more important than neutralization of surface charge by specific adsorption of cationic species. The modification of surface properties by coatings may in part explain the fact that the distributions of clay minerals in estuaries do not show a general conformity to that expected on the basis of the greater stability of montmorillonite suspensions, as compared to suspensions of kaolinite and illite (Hahn and Stumm, 1970).

5.2. Dissolved Material

Removal of dissolved constituents takes place in estuaries dominantly by scavenging on particle surfaces or incorporation into biologically produced material. The scavenging processes include simple exchange of ions at charged surfaces and the more important specific adsorption (Duinker, 1980). True precipitation is not a common process in the water column. The "dissolved" fraction as conventionally defined by filtration using membrane filters with average pore diameters usually of 0.4-0.5 μm can, however, include a significant amount of macromolecular and colloidal material, and the destabilization and aggregation of colloids to form flocculated particulate material is a significant process in many estuaries. Addition of dissolved constituents in the water column can occur by desorption from particles and by dissolution or decomposition of solid phases. Transfer of dissolved constituents across the benthic interface is important in the mass balance of some constituents.

In this section these processes are illustrated mainly by reference to the behaviour in various estuaries of some widely-studied constituents, namely silicon, manganese, iron and organic carbon.

5.2.1. Silicon. Two types of removal have been considered in the interpretation of estuarine distributions of silicon, namely reactions with mineral particles and uptake by estuarine organisms, principally diatoms, which utilise the element for skeletal formation. Early work on the behaviour of silicon in estuaries has been reviewed by Liss (1976). Distributions conforming to removal curves were reported for some estuaries, particularly those with high concentrations of dissolved silicon. It was concluded that significant removal could occur even under conditions of low biological activity. Hydrographic information on the systems was not generally sufficient to make this conclusion unequivocal, however, and for many systems which were studied with careful attention to end-member concentrations, removal could not be clearly demonstrated (e.g., Fanning and Pilson, 1973).

Uptake of dissolved silicon by both natural particles and pure silicate minerals, in the presence of sea salts, has been demonstrated in laboratory experiments (reviewed by Liss, 1976). These experiments have, however, used high concentrations of solid phases and the processes have not been specifically demonstrated to occur to a significant extent in the environment. Oxy-hydroxide precipitates, formed particularly by iron, have also been considered to provide potential sites for silicon scavenging. Boyle et al. (1974), however, found no evidence for removal of silicon in the Merrimack Estuary where substantial removal of "dissolved" iron occurred, and laboratory mixing experiments by Sholkovitz (1976) showed that the uptake of dissolved silicon by the iron-rich flocs formed on mixing filtered sea and river waters was less than about 10%. Removal which did not appear to be due to biological processes has been recorded by Morris et al. (1981) in the low salinity region of the Tamar Estuary, the extent of removal ranging from about 10-30%.

The potential importance of biological uptake was strikingly shown by the data of Wollast and de Broeu (1971) for the Scheldt Estuary. Uptake by diatoms was intense in a region of intermediate salinity, because of high concentrations of phosphate and combined nitrogen due to waste inputs upstream. This activity depleted the concentrations of dissolved silicon, creating an intermediate end-member which mixed essentially conservatively with the river waters over the low salinity range in the estuary. Silicon consumption giving almost total depletion by a salinity of 10-12‰ has been reported also in Charlotte Harbor (Froelich et al., 1985) where phosphate is highly enriched from industrial sources. In the same study it was shown that inorganic germanium was also completely depleted by uptake into the biogenic silica produced.

Uptake of dissolved silicon by diatoms under non-eutrophic conditions was found by Milliman and Boyle (1975) in the Amazon Estuary at salinities less than 5‰. DeMaster et al. (1983) found that removal by diatoms in the plume of the Amazon became significant at a lower salinity in the southeastern sections than in the northeastern, uptake being dependent on turbidity, turbulence and availability of nutrients. Turbidity is important in determining whether or not the nutrients supplied in river water lead to enhanced primary productivity. In the Zaire Estuary concentrations of SPM decrease less markedly with increasing salinity than in the Amazon and algal activity is not favored until dilution has reduced nutrient concentrations to levels which are too low to produce large algal blooms (Cadée, 1978). In this situation, dissolved silicon behaves conservatively up to salinities of 25‰ or greater (van Bennekom et al., 1978).

A relationship between biological removal of dissolved silicon and hydrographic factors was demonstrated by Peterson et al. (1978) who applied a two-dimensional steady-state numerical model to interpret data from the San Francisco Bay system, where the northern sector has the characteristics of a partially mixed estuary for much of the year. Nearly conservative behaviour was observed with a river flow of 550 $m^3 s^{-1}$ but increasingly marked removal at flows of 350 and 260 $m^3 s^{-1}$. The authors consider that water circulation and biological uptake act as independent competing processes. Their model simulations show how a given uniform silicon removal rate can produce a variable degree of non-conservative behaviour depending upon the advective water-replacement time which is a function of river flow. Similar findings were obtained by applying analytical solutions, using steady-state one-dimensional conservative equations, to the same problem (Rattray and Officer, 1979).

Several other studies point to similar conclusions. Callaway and Specht (1982) showed that in the Yaquina Estuary non-conservative behaviour of dissolved silicon is related to combined conditions of high insolation, and increased residence time resulting from low river flow. In this system removal appears to be greatest within a relatively narrow range of salinity so that the data fitted well to two linear dilution relationships with different slopes. Results for Tokyo Bay (Kamatani and Takano, 1984) show a change from conservative behaviour in winter to non-conservative behaviour during spring and mid-summer.

Given a constant removal rate of dissolved silicon, changes in residence times related to differences in river discharge will affect the extent of removal in a predictable way. In temperate latitudes, however, periods of low flow are often coincident with enhanced phytoplankton productivity. Since variations in flow can affect the phytoplankton populations themselves, particularly through effects on stratification of the water column, it is clear that there can be a complex interaction of hydrodynamic and biological factors.

Anderson (1986) also found that in the estuaries of several rivers entering Chesapeake Bay, conservative behaviour of dissolved silicon was associated with winter or high river discharge conditions. Non-conservative behaviour which was observed to occur under certain other conditions showed different characteristics than those already discussed. Diatom blooms were responsible for the removal of dissolved silicon just upstream of the salt intrusion limit, with almost total removal sometimes occurring. Recycling of silicon into solution was observed in the early stages of mixing with seawater, up to a salinity of about 5‰, followed by conservative behaviour at higher salinities. These data emphasize the importance of enhanced phytoplankton growth in the interfacial zone of the estuary. A generalized model of the distribution of silicon in these estuaries, in relation to phytoplankton biomass and salinity is shown in Figure 5. It was suggested by Anderson (1986) that the distribution of dissolved silicon in the Hudson Estuary, reported by Simpson et al. (1975), with depletion in the freshwater zone and increase in the low salinity region, may be explained by this model, rather than by a waste input source for the element as suggested in the original paper.

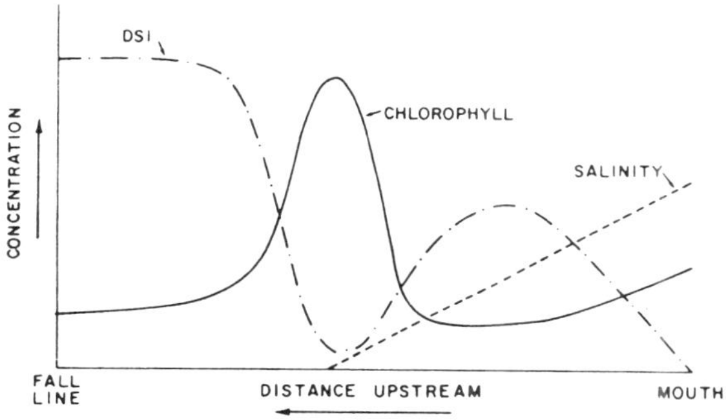

Figure 5. Generalized relationships between concentrations of dissolved silicon (DSi) and chlorophyll, and salinity, in an estuary where diatom blooms remove dissolved silicon above the limit of the salt intrusion (from Anderson, 1986).

Thus, while there is no compelling evidence that riverine fluxes of dissolved silicon undergo widespread modification by inorganic reactions in estuaries, considerable removal can occur when there is high biological uptake in systems with relatively long residence times. There is insufficient information to generalize about the fate of the biogenic silica which is produced. Milliman and Boyle (1975) concluded that in the low salinity region of the Amazon Estuary, much of it was transported landward. In the higher salinity region over the Amazon continental shelf, dissolution in deeper waters appears to be dominant (DeMaster et al., 1983). Kaul and Froelich (1984) report that the biogenic silica produced in Ochlockonee Bay undergoes dissolution within the system.

5.2.2. Manganese. The behaviour of manganese in estuaries, as in other aquatic environments, is dominantly controlled by its oxidation-reduction chemistry. Under reducing conditions, Mn (II) is the thermodynamically stable oxidation state and relatively high dissolved concentrations can exist in equilibrium with solid phases. In oxidizing environments, oxidation to give Mn(IV), which has a low solubility, is predicted thermodynamically (Figure 2). The rate of oxidation is, however, often slow enough to permit manganese to be transported in the dissolved form to a significant extent under conditions where Mn(II) is thermodynamically unstable.

The consequences of these changes in oxidation state are shown classically by the distribution of manganese in waters where there is an interface between anoxic and oxic zones. A recent example is provided by data on Balstadfjord (Hamilton-Taylor and Price, 1983). The vertical profile shown in Figure 6, for a station in the inner basin, shows a maximum in excess particulate manganese (that is, manganese in excess of that calculated from the aluminium content and the Mn/Al ratio characteristic of detrital aluminosilicate) above the interface. The high concentrations of dissolved manganese in the anoxic waters reflect the presence of Mn(II). In the lower part of the oxic zone Mn(II) enters by diffusion, and perhaps advection, from the anoxic zone. Its oxidation produces high particulate manganese concentrations, the particles consisting of brown flocs, attributable to the formation of iron and manganese oxyhydroxides. The significant increases in dissolved manganese maintained in the oxic waters above the interface are explicable in terms of relatively slow oxidation of Mn(II).

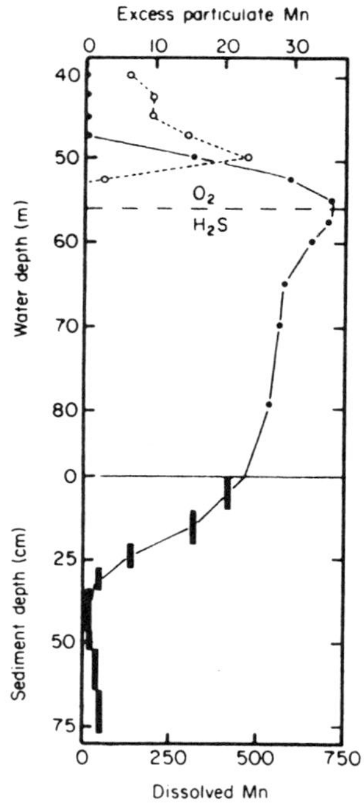

Figure 6. Vertical profiles of dissolved (continuous line) and excess particulate (dashed line) manganese ($\mu g\ l^{-1}$) at a station in Bolstadfjord (from Hamilton-Taylor and Price, 1983).

In many estuarine sediment cores with a thin oxidized layer overlying reducing sediment, a maximum in concentration of manganese occurs within the oxidized layer, reflecting the trapping of manganese mobilized at lower pE. Manganese(II) may, however, escape through such a layer by the mixing of pore waters and overlying waters. The exchange of manganese between pore waters and the estuarine water column can be greatly increased by resuspension of bottom deposits. Such an effect can arise through bioturbation, but it is dominantly associated with turbulent resuspension of material, especially in the turbidity maximum. Extensive mobilization of manganese from surface sediments can also occur in systems which undergo seasonal changes from oxygenated to anoxic conditions in bottom waters (Eaton, 1979).

For the Beaulieu Estuary (Holliday and Liss, 1976; Moore et al., 1979) and a Delaware salt marsh creek (Eastman and Church, 1984) essentially conservative behaviour of dissolved manganese has been observed. The majority of estuaries, however, have an input of manganese from reducing sediments producing a maximum concentration at low to intermediate salinity. Behaviour of this kind has been shown, for example, in the Newport River Estuary (Evans et al., 1977) and the Tamar Estuary (Knox et al., 1981; Morris et al., 1982b). In the latter estuary highly variable concentrations were observed in the region of the interface between fresh and saline waters, coincident with the turbidity maximum zone. Under neap to spring tidal conditions, leading to maximum mobilization of sediment into suspension, a pronounced maximum in dissolved manganese occurred in this zone but during the subsequent period of net sediment accretion concentrations decreased, as a result of scavenging and advection, to values lower than those upstream or downstream. In such a system there is therefore an intensive internal cycling of manganese caused by the combination of chemical reactions involving phase transfer, with the hydrodynamic processes responsible for movements of sediment and water.

The same redox phenomena govern the distribution of manganese in the Scheldt Estuary, but the regime in the low salinity region is very different there, as there is a zone in the upper estuary where reducing conditions prevail in the water column. The distribution of manganese here conforms reasonably closely to that predicted by a thermodynamic model, with $MnCO_3$ as the solid phase in equilibrium with Mn(II) (Wollast et al., 1979). In the lower estuary, in oxygenated waters, oxidative scavenging on particles takes place, reflected in a non-conservative distribution (Duinker et al., 1979). Upstream movement of bottom sediment containing scavenged manganese transports the metal back into the reducing zone where remobilization occurs either through the low pH and pE in the water column or through bottom sediments.

The overall consequences of the cycling of manganese in respect of the mass balance of the element in estuaries have generally not been evaluated. For the upper St. Lawrence Estuary, Bewers and Yeats (1978) calculated that the output of dissolved manganese was greater than the inputs from river water and inflowing seawater, implying overall addition from particles within the estuary.

There is a substantial literature on the mechanisms of manganese scavenging in fresh and saline waters and the nature of the solid phases produced by oxidation of Mn(II). Only the principal points can be summarized here. The role of microorganisms in uptake and oxidation has been shown by electron micrographic studies of particles, effects of bacterial inhibitors on removal rates, and the temperature dependence of removal rates (for example, Emerson et al., 1982; Vojak et al., 1985). Grill (1982) has described the removal of Mn(II) above an anoxic-oxic interface in terms of two parallel processes, bacterial oxidation and inorganic autocatalytic oxidation. The reaction does not involve a direct conversion to MnO_2. Solid phases commonly occur with oxidation numbers intermediate between +2 and +4 (Emerson et al., 1982; Grill, 1982) and it seems probable that forms of MnOOH are important as initial products.

5.2.3. Iron and organic carbon. Questions concerning the removal of iron and organic carbon are closely linked and it is convenient to consider them together.

Large-scale removal of iron has been observed in numerous estuaries (Boyle et al., 1974; Duinker, 1980). Most investigators have considered the behaviour of the "dissolved" fraction, operationally defined as that passing a membrane filter of specified average pore diameter, usually about 0.5 μm. Studies using ultracentrifugation or ultrafiltration methods have shown that much of this fraction is microcolloidal

(Boyle et al., 1977; Figuères et al., 1978; Sholkovitz et al., 1978; Moore et al., 1979) and it is primarily this material which undergoes removal. The data in Figure 7 show the non-conservative behaviour of "dissolved" iron and the reduction in its concentration in river and estuarine waters with decrease in the nominal pore size of filter, this effect being most marked at lower salinities.

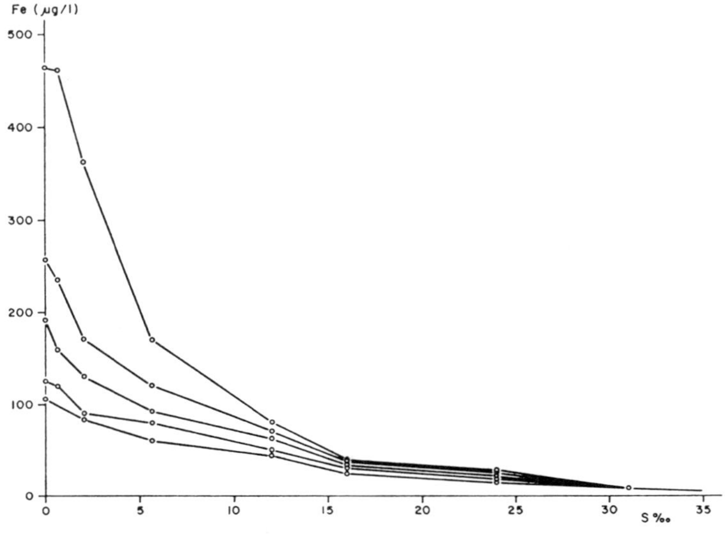

Figure 7. Relationship between salinity and the concentrations of iron in filtrates passing filters of various pore sizes for the Zaire Estuary. The curves are in a sequence of reducing concentrations of iron with decreasing filter diameters of 1.2, 0.45, 0.22, 0.05 and 0.025 μm (from Figuères et al., 1978).

The transport of iron in rivers at concentrations greatly above those predicted by thermodynamic equilibria is generally considered to reflect the presence of microcolloidal material stabilized by association with organic material, although the importance of complexation by organic molecules has also been argued by some workers (e.g., Perdue et al., 1976). According to the former view, the estuarine removal process involves destabilization and aggregation of the microcolloids to form larger-size flocculated material. This process has been extensively studied by laboratory mixing experiments (Sholkovitz, 1976, 1978; Sholkovitz and Copland, 1981; Bale and Morris, 1981). The efficiency of cations as flocculating agents increases in the order $Na^+ < Mg^{2+} < Ca^{2+}$ in accordance with classical ideas about colloidal stability based on the electrical double-layer theory, although specific chemical interactions as well as electrostatic interactions are involved. The kinetics of the process have been modelled in terms of a second order reaction with respect to iron (see, for example, Fox and Wofsy, 1983), an interpretation based upon theoretical expectations that the rate of aggregation of particles of the sizes concerned should be dominated by the frequency of collisions between particles brought about by Brownian motion.

In determining the behaviour of iron in the oxygenated waters of estuaries, the reactions of colloidal forms of Fe(III) are more important than redox reactions analogous to those discussed for manganese, even though these latter reactions also occur.

Hong and Kester (1985) reported that about 20% of the dissolved iron in the Connecticut River Estuary occurs as Fe(II) and that the oxidation of Fe(II) was not a significant factor in the removal of dissolved iron.

Much of the dissolved organic matter (DOM) in natural waters has not been characterized at the molecular level. It is apparent, however, that in simplistic terms it consists of a pool of labile compounds which are utilized relatively rapidly by micro-organisms and a larger pool of more refractory humic material. Because of the difficulty in characterising this latter material, the behaviour of organic material has been generally considered in terms of the total dissolved organic carbon (DOC). The relationships between concentrations of DOC and salinity in many estuaries show that the bulk of the DOC behaves conservatively (Sholkovitz et al., 1978; Moore et al., 1979; Hunter and Liss, 1982; Mantoura and Woodward, 1983). Removal of humic acids (the acid-insoluble fraction of DOM) has been shown to occur (Sholkovitz et al., 1978; Fox, 1983) but these constitute a relatively small part of the DOC. Addition of microbiologically labile DOC in the low salinity region of the Tamar Estuary has been observed (Morris et al., 1978) but generally the cycling of such material within estuaries does not produce strong signals relative to the concentrations of more microbially inert compounds.

The removal of a fraction of DOC during estuarine mixing has been most clearly demonstrated by laboratory mixing experiments (Sholkovitz, 1976; Sholkovitz et al., 1978; Sholkovitz and Copland, 1981), which show that about 10% of the DOC is removed in iron-rich flocs. These findings, together with observations that the bulk of the "dissolved" iron and a relatively small fraction of DOC occur in high molecular weight "dissolved" fractions, have been interpreted as indicating a close association of iron and humic acids in the colloids entering estuaries in river water. This interpretation has, however, been challenged by Fox (1984) who found that the kinetics of removal of dissolved iron and humic acids were significantly different and that it was not consistent with changes in the ratio of dissolved iron to humic acids along axial sections in some estuaries.

The character of the organic compounds that stabilize colloidal iron in river water has not been established. Moore et al. (1979) showed that in Beaulieu River water, most of the iron passing a 0.45 μm membrane filter has a nominal molecular weight greater than 10^5 and that some 10% of the DOC occurred in the same fraction. The DOC of lower nominal molecular weight was, however, effective in stabilizing iron as a colloid in river water. Cameron and Liss (1984) showed that humic acid, like tannic acid, certain surfactants, and the inorganic species, silicate and phosphate, could stabilize iron in forms passing an 0.45 μm filter. Fox (1984) found, however, that humic acids were not able to stabilize ferric hydroxide. This difference in findings may reflect differences in the composition of the DOM described as humic acid. Thus, while the role of organic matter in stabilizing iron in colloidal forms is clearly important the details of the process remain to be elucidated.

As with other solid phases formed in estuaries, the fate of the flocculates formed by destabilization of microcolloids has been little considered, but it seems possible that considerable amounts are transported in suspension into adjacent waters.

6. CONCLUDING COMMENTS

The examples discussed here are only illustrative of the range and complexity of interactions occurring in the estuarine region. Information is also available on many other constituents although they have mostly been studied less intensively from a geochemical standpoint. It is apparent that estuaries trap globally an important fraction of the particulate material entering in rivers. It is also evident that some dissolved and microcolloidal constituents are transferred to solid phases in estuaries. The implications of these transfers for the fluxes of constituents from estuaries to shelf and oceanic regions remain to be adequately and quantitatively elucidated.

7. REFERENCES

Anderson, G.F., 1986, Silica, diatoms and a freshwater productivity maximum in Atlantic coastal plain estuaries, Chesapeake Bay, *Estuarine Coastal Shelf Sci.*, **22**, 183-197.

Aston, S.R., 1978, Estuarine Chemistry, *in* J.P. Riley and R. Chester (eds.) *Chemical Oceanography,* 2nd Ed., Vol. 7, pp. 361-440. Academic Press, London.

Bale, A.J. and A.W. Morris, 1981, Laboratory simulation of chemical processes induced by estuarine mixing: The behaviour of iron and phosphate in estuaries, *Estuarine Coastal Shelf Sci.*, **13**, 1-10.

Bewers, J.M. and P.A. Yeats, 1978, Trace metals in the waters of a partially mixed estuary, *Estuarine Coastal Mar. Sci.*, **7**, 147-162.

Bokuniewicz, H.J. and R.B. Gordon, 1980, Sediment transport and deposition in Long Island Sound, *Adv. Geophys.*, **22**, 69-106.

Boyle, E.A., Collier, R., Dengler, A.T., Edmond, J.M., Ng, A.C. and R.F. Stallard, 1974, On the chemical mass balance in estuaries, *Geochim. Cosmochim. Acta,* **38**, 1719-1728.

Boyle, E.A., Edmond, J.M. and E.R. Sholkovitz, 1977, The mechanism of iron removal in estuaries, *Geochim. Cosmochim. Acta,* **41**, 1313-1324.

Burton, J.D. and P.S. Liss (eds.), 1976, *Estuarine Chemistry,* Academic Press, London, 229 pp.

Cadée, G.C., 1978, Primary production and chlorophyll in the Zaire river, estuary and plume, *Neth. J. Sea Res.*, **12**, 368-381.

Callaway, R.J. and D.T. Specht, 1982, Dissolved silicon in the Yaquina estuary, Oregon, *Estuarine Coastal Shelf Sci.*, **15**, 561-567.

Cameron, A.J. and P.S. Liss, 1984, The stabilization of "dissolved" iron in freshwaters, *Water Res.*, **18**, 179-185.

Cranston, R.E. and J.W. Murray, 1980, Chromium species in the Columbia river and estuary, *Limnol. Oceanogr.*, **25**, 1104-1112.

DeMaster, D.J., Knapp, G.B. and C.A. Nittrouer, 1983, Biological uptake and accumulation of silica on the Amazon continental shelf, *Geochim. Cosmochim. Acta,* **47**, 1713-1723.

Duinker, J.C., 1980, Suspended matter in estuaries: Adsorption and desorption processes, *in* E. Olausson and I. Cato (eds.) *Chemistry and Biogeochemistry of Estuaries,* pp. 122-151, Wiley, Chichester.

Duinker, J. C., Wollast, R. and G. Billen, 1979, Behaviour of manganese in the Rhine and Scheldt estuaries II. Geochemical cycling, *Estuarine Coastal Mar. Sci.*, **9**, 727-738.

Dyer, K.R., 1973, *Estuaries: A Physical Introduction.* Wiley, London, 140pp.

Dyer, K.R., 1986, *Coastal and Estuarine Sediment Dynamics*, Wiley, Chichester, 342pp.

Eastman, K.W. and T.M. Church, 1984, Behaviour of iron, manganese, phosphate and humic acid during mixing in a Delaware salt marsh creek, *Estuarine Coastal Shelf Sci.*, **18**, 447-458.

Eaton, A., 1979, The impact of anoxia on Mn fluxes in the Chesapeake Bay, *Geochim. Cosmochim. Acta,* **43**, 429-432.

Edwards, A.M.C. and P.S. Liss, 1973, Evidence for buffering of dissolved silicon in fresh waters, *Nature,* **243**, 341-342.

Emerson, S., Kalhorn, S., Jacobs, L., Tebo, B.M., Nealson, K.H. and R.A. Rosson, 1982, Environmental oxidation rate of manganese (II): Bacterial catalysis, *Geochim. Cosmochim. Acta,* **46**, 1073-1079.

Evans, D.W., Cutshall, N.H., Cross, F.A. and D.A. Wolfe, 1977, Manganese recycling in the Newport River estuary, North Carolina, *Estuarine Coastal Mar. Sci.*, **5**, 71-80.

Fanning, K.A. and V.I. Maynard, 1978, Dissolved boron and nutrients in the mixing plumes of major tropical rivers, *Neth. J. Sea Res.*, **12**, 345-354.

Fanning, K.A. and M.E.Q. Pilson, 1973, The lack of inorganic removal of dissolved silica during river-ocean mixing, *Geochim. Cosmochim. Acta,* **37**, 2405-2415.

Figuères, G., Martin, J.-M. and M. Meybeck, 1978, Iron behaviour in the Zaire estuary, *Neth. J. Sea Res.*, **12**, 329-337.

Fox, L.E., 1983, The removal of dissolved-humic acid during estuarine mixing, *Estuarine Coastal Shelf Sci.*, **16**, 431-440.

Fox, L.E., 1984, The relationship between dissolved humic acids and soluble iron in estuaries, *Geochim. Cosmochim. Acta,* **48**, 879-884.

Fox, L.E. and S.C. Wofsy, 1983, Kinetics of removal of iron colloids from estuaries, *Geochim. Cosmochim. Acta,* **47**, 211-216.

Froelich, P.N., Kaul, L.W., Byrd, J.T., Andreae, M.O. and K.K. Roe, 1985, Arsenic, barium, germanium, tin, dimethylsulfide and nutrient biogeochemistry in Charlotte Harbor, Florida, a phosphorus-enriched estuary, *Estuarine Coastal Shelf Sci.*, **20**, 239-264.

Garrels, R.M. and F.T. Mackenzie, 1971, *Evolution of Sedimentary Rocks*, Norton, New York, 397 pp.

Gibbs, R.J., 1970, Mechanisms controlling world water chemistry, *Science,* **170**, 1088-1090.

Grill, E.V., 1982, Kinetic and thermodynamic factors controlling manganese concentrations in oceanic waters, *Geochim. Cosmochim. Acta,* **46**, 2435-2446.

Hahn, H.H. and W. Stumm, 1970, The role of coagulation in natural waters, *Amer. J. Sci.*, **268**, 354-368.

Hamilton-Taylor, J. and N.B. Price, 1983, The geochemistry of iron and manganese in the waters and sediments of Bolstadfjord, S.W. Norway, *Estuarine Coastal Shelf Sci.*, **17**, 1-19.

Head, P.C. (ed.), 1985, *Practical Estuarine Chemistry,* Cambridge University Press, Cambridge, 337 pp.

Holeman, J.N., 1968, The sediment yield of major rivers of the world, *Water Resources Res.*, **4**, 737-747.

Holliday, L.M. and P.S. Liss, 1976, The behaviour of dissolved iron, manganese and zinc in the Beaulieu estuary, *Estuarine Coastal Mar. Sci.*, **4**, 349-353.

Hong, H. and D.R. Kester, 1985, Chemical forms of iron in the Connecticut River estuary, *Estuarine Coastal Shelf Sci.*, **21**, 449-459.

Howard, A.G., Arbab-Zavar, M.H. and S. Apte, 1984, The behaviour of dissolved arsenic in the estuary of the River Beaulieu, *Estuarine Coastal Shelf Sci.*, **19**, 493-504.

Hunter, K.A., 1980, Microelectrophoretic properties of natural surface-active organic matter in coastal seawater, *Limnol. Oceanogr.*, **25**, 807-872.

Hunter, K.A. and P.S. Liss, 1982, Organic matter and the surface charge of suspended particles in estuarine water, *Limnol. Oceanogr.*, **27**, 322-335.

Kamatani, A. and M. Takano, 1984, The behaviour of dissolved silica during the mixing of river and sea waters in Tokyo Bay, *Estuarine Coastal Shelf Sci.*, **19**, 505-512.

Kaul, L.W. and P.N. Froelich, Jr., 1984, Modeling estuarine nutrient geochemistry in a simple system, *Geochim. Cosmochim. Acta,* **48**, 1417-1433.

Kester, D.R., Byrne, R.H., Jr. and Y.-J. Liang, 1975, Redox reactions and solution complexes of iron in marine systems, *in* T.M. Church (ed.) *Marine Chemistry in the Coastal Environment,* pp. 56-79, American Chemical Society, Washington, D.C.

Knox, S., Turner, D.R., Dickson, A.G., Liddicoat, M.I., Whitfield, M. and E.I. Butler, 1981, Statistical analysis of estuarine profiles, application to manganese and ammonium in the Tamar Estuary, *Estuarine Coastal Shelf Sci.*, **13**, 357-371.

Lewis, B.L., Froelich, P.N. and M.O. Andreae, 1984, Methylgermanium in natural waters, *Nature,* **313**, 303-305.

Liss, P.S., 1976, Conservative and non-conservative behaviour of dissolved constituents during estuarine mixing, *in* J.D. Burton and P.S. Liss (eds.) *Estuarine Chemistry,* pp. 93-130, Academic Press, London.

Liss, P.S. and M.J. Pointon, 1973, Removal of dissolved boron and silicon during estuarine mixing of sea and river waters, *Geochim. Cosmochim. Acta,* **37**, 1493-1498.

Livingstone, D.A., 1963, Chemical composition of rivers and lakes, *U.S. Geol. Surv. Prof.*

Paper **440-G**, 64 pp.

Loder, T.C. and P.S. Liss, 1985, Control by organic coatings of the surface charge of estuarine sediment particles, *Limnol. Oceanogr.*, **30**, 418-421.

Loder, T.C. and R.P. Reichard, 1981, The dynamics of conservative mixing in estuaries, *Estuaries*, **4**, 64-69.

Mantoura, R.F.C. and A.W. Morris, 1983, Measurement of chemical distributions and processes, *in* A.W. Morris (ed.) *Practical Procedures for Estuarine Studies*, pp. 55-100, Natural Environment Research Council.

Mantoura, R.F.C. and E.M.S. Woodward, 1983, Conservative behaviour of riverine dissolved organic carbon in the Severn Estuary: Chemical and geochemical implications, *Geochim. Cosmochim. Acta*, **47**, 1293-1309.

Martin, J.-M. and M. Meybeck, 1979, Elemental mass balance of material carried by world major rivers, *Mar. Chem.*, **7**, 173-206.

Martin, J.-M., Burton, J.D. and D. Eisma (eds), 1981, *River Inputs to Ocean Systems*, United Nations Press, 384 pp.

Martin, J.-M., Mouchel, J.M. and P. Nirel, 1985, Some recent developments in the characterization of estuarine particulates, *Water Sci. Technol.*, in press.

Meade, R.H., 1982, Sources, sinks, and storage of river sediment in the Atlantic drainage of the United States, *J. Geol.*, **90**, 235-252.

Meybeck, M., 1979, Concentrations des eaux fluviales en éléments majeurs et apports en solution aux océans, *Revue Géol. Dyn. Géogr. Phys.*, **21**, 215-246.

Meybeck, M., 1981, Pathways of major elements from land to ocean through rivers, *in* J.-M. Martin, J.D. Burton and D. Eisma (eds.) *River Inputs to Ocean Systems*, pp. 18-30, United Nations Press.

Milliman, J.D. and E. Boyle, 1975, Biological uptake of dissolved silica in the Amazon estuary, *Science*, **189**, 995-997.

Milliman, J.D. and R.H. Meade, 1983, World-wide delivery of river sediment to the oceans, *J. Geol.*, **91**, 1-21.

Mook, W.G. and B.K.S. Koene, 1975, Chemistry of dissolved inorganic carbon in estuarine and coastal brackish waters, *Estuarine Coastal Mar. Sci.*, **3**, 325-336.

Moore, R.M., Burton, J.D., Williams, P.J.leB. and M.L. Young, 1979, The behaviour of dissolved organic material, iron and manganese in estuarine mixing, *Geochim. Cosmochim. Acta*, **43**, 919-926.

Morris, A.W., 1978, Chemical processes in estuaries: The importance of pH and its variability, *in* W. E. Krumbein (ed.) *The Aquatic Environment*, pp. 179-187, Ann Arbor Science, Ann Arbor.

Morris, A.W., 1985, Estuarine chemistry and general survey strategy, *in* P.C. Head (ed.) *Practical Estuarine Chemistry*, pp. 1-60, Cambridge University Press, Cambridge.

Morris, A.W., Mantoura, R.F.C., Bale, A.J. and R.J.M. Howland, 1978, Very low salinity regions of estuaries: Important sites for chemical and biological reactions, *Nature*, **274**, 678-680.

Morris, A.W., Bale, A.J. and R.J.M. Howland, 1981, Nutrient distributions in an estuary: Evidence of chemical precipitation of dissolved silicate and phosphate, *Estuarine Coastal Shelf Sci.*, **12**, 205-216.

Morris, A.W., Bale, A.J. and R.J.M. Howland, 1982a, Chemical variability in the Tamar Estuary, southwest England, *Estuarine Coastal Shelf Sci.*, **14**, 649-661.

Morris, A.W., Bale, A.J. and R.J.M. Howland, 1982b, The dynamics of estuarine manganese cycling, *Estuarine Coastal Shelf Sci.*, **14**, 175-192.

Officer, C.B., 1976, *Physical Oceanography of Estuaries (and Associated Coastal Waters)*, Wiley-Interscience, New York, 465 pp.

Officer, C.B., 1979, Discussion of the behaviour of nonconservative dissolved constituents in estuaries, *Estuarine Coastal Mar. Sci.*, **9**, 91-94.

Olausson, E. and I. Cato (eds.), 1980, *Chemistry and Biogeochemistry of Estuaries*, Wiley, Chichester, 452 pp.

Perdue, E.M., Beck, K.C. and J.H. Reuter, 1976, Organic complexes of iron and aluminium in natural waters, *Nature*, **260**, 418-420.

Peterson, D.H., Festa, J.F. and T.J. Conomos, 1978, Numerical simulation of dissolved silica in the San Francisco Bay, *Estuarine Coastal Mar. Sci.*, **7**, 99-116.

Postma, H., 1980, Sediment transport and sedimentation, *in* E. Olausson and I. Cato (eds.) *Chemistry and Biogeochemistry of Estuaries*, pp. 153-186, Wiley, Chichester.

Rattray, M., Jr. and C.B. Officer, 1979, Distribution of a non-conservative constituent in an estuary with application to the numerical simulation of dissolved silicon in the San Francisco Bay, *Estuarine Coastal Mar. Sci.*, **8**, 489-494.

Reeder, S.W., Hitchon, B. and A.A. Levinson, 1972, Hydrochemistry of the surface waters of the Mackenzie River drainage basin, Canada, I. Factors controlling the inorganic composition, *Geochim. Cosmochim. Acta*, **36**, 181-192.

Sholkovitz, E.R., 1976, Flocculation of dissolved organic and inorganic matter during the mixing of river water and seawater, *Geochim. Cosmochim. Acta*, **40**, 831-845.

Sholkovitz, E.R., 1978, The flocculation of dissolved Fe, Mn, Al, Cu, Ni, Co and Cd during estuarine mixing, *Earth Planet. Sci. Lett.*, **41**, 77-86.

Sholkovitz, E.R. and D. Copland, 1981, The coagulation, solubility and adsorption properties of Fe, Mn, Cu, Ni, Cd, Co and humic acids in a river water, *Geochim. Cosmochim. Acta*, **45**, 181-189.

Sholkovitz, E.R., Boyle, E.A. and N.B. Price, 1978, The removal of dissolved humic acids and iron during estuarine mixing, *Earth Planet. Sci. Lett.*, **40**, 130-136.

Simpson, H.J., Hammond, D.E., Deck, B.L. and S.C. Williams, 1975, Nutrient budgets in the Hudson River Estuary, *in* T. M. Church (ed.) *Marine Chemistry in the Coastal Environment*, pp. 618-635, American Chemical Society, Washington, D.C.

Stallard, R.F. and J.M. Edmond, 1983, Geochemistry of the Amazon 2. The influence of geology and weathering environment on the dissolved load, *J. Geophys. Res.*, **88**, 9671-9688.

Takayanagi, K. and G.T.F. Wong, 1984, Total selenium and selenium(IV) in the James River Estuary and Southern Chesapeake Bay, *Estuarine Coastal Shelf Sci.*, **18**, 113-119.

Turner, D.R., Whitfield, M. and A.G. Dickson, 1981, The equilibrium speciation of dissolved components in freshwater and seawater at 25°C and 1 atm pressure, *Geochim. Cosmochim. Acta*, **45**, 855-881.

van Bennekom, A.J. and W. Salomons, 1981, Pathways of nutrients and organic matter from land to ocean through rivers, *in* J.-M. Martin, J.D. Burton and D. Eisma (eds.) *River Inputs to Ocean Systems*, pp. 33-51, United Nations Press.

van Bennekom, A.J., Berger, G.W., Helder, W. and R.T.P. de Vries, 1978, Nutrient distribution in the Zaire estuary and river plume, *Neth. J. Sea. Res.*, **12**, 296-323.

Vojak, P.W.L., Edwards, C. and M.V. Jones, 1985, Evidence for microbiological manganese oxidation in the River Tamar Estuary, South West England, *Estuarine Coastal Shelf Sci.*, **20**, 661-671.

Walling, D.E. and B.W. Webb, 1981, The reliability of suspended sediment load data, *in Erosion and Sediment Transport Measurement*, International Association of Hydrological Sciences Publication No. **133**, 177-194.

Walling, D.E. and B.W. Webb, 1983, The dissolved loads of rivers: A global overview, *in Dissolved Loads of Rivers and Surface Water Quantity/Quality Relationships*, International Association of Hydrological Sciences Publication No. **141**, 3-20.

Wollast, R., 1983, Interactions in estuaries and coastal waters, *in* B. Bolin and R. B. Cook (eds.) *The Major Biogeochemical Cycles and Their Interactions*, pp. 385-407, Wiley, Chichester.

Wollast, R. and F. De Broeu, 1971, Study of the behaviour of dissolved silica in the estuary of the Scheldt, *Geochim. Cosmochim. Acta*, **35**, 613-620.

Wollast, R., Billen, G. and J.C. Duinker, 1979, Behaviour of manganese in the Rhine and Scheldt Estuaries. I. Physicochemical behaviour, *Estuarine Coastal Mar. Sci.*, **9**, 161-169.

STRONTIUM STORAGE AND RELEASE DURING DEPOSITION AND DIAGENESIS OF MARINE CARBONATES RELATED TO SEA-LEVEL VARIATIONS

S.O. Schlanger
Department of Geological Sciences
Northwestern University
Evanston, Illinois, 60201
U.S.A.

ABSTRACT

By virtue of their high content of aragonite, that contains $\sim 5{,}000 \times 10^{-6}$ g/g strontium, biogenic carbonates that accumulate as shallow-water reef and shelf limestones serve as a major sink and source of strontium in the ocean. In contrast, deep-sea pelagic carbonates are almost entirely calcite that contains $\sim 1400 \times 10^{-6}$ g/g strontium. The strontium sink represented by present-day aragonitic shallow-water carbonate deposition sequesters $\sim 3.3 \times 10^{12}$ g/yr whereas the deep-sea pelagic carbonate strontium sink accounts for $\sim 2.0 \times 10^{12}$ g/yr; a total of 5.3×10^{-12} g/yr of strontium is being stored now in all marine carbonate sediments. The riverine input of strontium from the continents amounts to $\sim 2.2 \times 10^{12}$ g/yr. This present-day annual deficit of $\sim 3.1 \times 10^{12}$ g/yr of strontium in the continent-oceanic carbonate sediment system is a short-term one because reef and shelf limestones store strontium during periods of rising sea level, such as the modern interglacial episode. During periods of falling sea level, reef and shelf limestones are exposed to fresh-water diagenesis which results in the dissolution of strontium-rich aragonite and the consequent return of strontium to the sea; thus short-term strontium deficits that accumulate while sea level is rising are partially paid back during periods of falling sea level at the rate of $\sim 1.8 \times 10^{12}$ g/yr from shallow-water carbonate sediments. Medium-term dissolution of pelagic carbonates at the sea floor returns strontium to the sea water at a rate of $\sim 1.0 \times 10^{12}$ g/yr and long-term burial diagenesis of pelagic carbonates returns $\sim 0.5 \times 10^{12}$ g/yr. Therefore, the long-term storage rate of strontium in reef, shelf and pelagic carbonates of $\sim 2.0 \times 10^{12}$ g/yr derived in this analysis is roughly in balance with the present-day riverine strontium input of $\sim 2.2 \times 10^{12}$ g/yr determined by Brass (1976) and by Martin and Meybeck (1979). Prolonged periods of reef and shelf carbonate accumulation during high sea-level stands result in significant changes in the Sr/Ca ratio in the oceanic reservoir which are detectable in the geological record as discussed by Graham et al. (1982).

1. INTRODUCTION

Studies of the geochemical cycles of individual elements are based on establishing a presumably complete set of geologically reasonable reservoirs, determining the amount of the element of interest in each reservoir and estimating the fluxes of the element between reservoirs. Holland (1978) reviewed the approaches taken to such studies,

A. Lerman and M. Meybeck (eds.), Physical and Chemical Weathering in Geochemical Cycles, 323–339.

and summarized the concepts useful in describing the behavior of reservoirs and analyzing the fluxes between reservoirs. His (1984) treatment of the major aspects of the geochemical cycle of strontium is shown here, somewhat simplified by the deletion of $^{87}Sr/^{86}Sr$ fluxes, as Figure 1. The major components of Holland's model are the large riverine flux of strontium from the continents to the ocean and a large, but unspecified, flux of strontium both into and out of the oceanic sea water reservoir due to the deposition and diagenesis of oceanic carbonate sediments. The flux of strontium from the continents to the ocean has been estimated by Holland (1984) at 3.24×10^{12} g/yr, by Brass (1976) at 2.2×10^{12} g/yr and by Martin and Meybeck (1979) at 2.24×10^{12} g/yr; the estimates of Holland (1984) and Brass (1976) differ because two different estimations of global stream flow were used. This difference nicely illustrates one of the problems in devising budgetary schemes for elements. The oceanic reservoir of strontium, based on an oceanic water mass of 1400×10^{21} g and a strontium content of 8×10^{-6} g/g is taken in this paper as 1.12×10^{19} g. For the purposes of this discussion, following Holland (1984), the fluxes of strontium into and out of ocean-floor basalts are not considered, and the removal of strontium from the ocean due to the formation of marine clays is also considered trivial (Turekian, 1964). The remaining reservoirs and fluxes to be considered are those involved in the deposition,

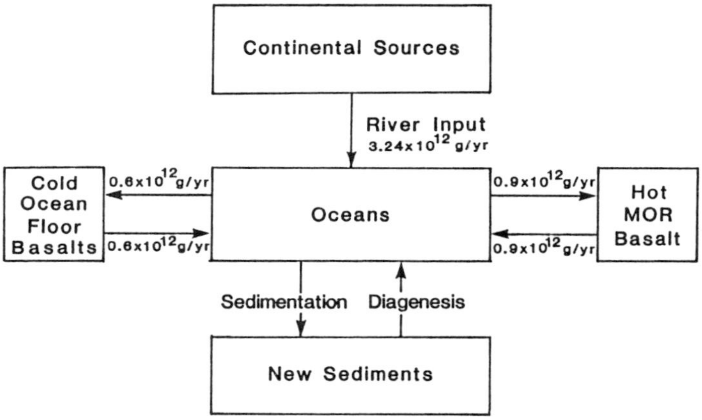

Figure 1. The geochemical cycle of strontium (modified from Holland, 1984). The ocean-carbonate sediment system is shown in detail on Figure 2.

diagenesis and dissolution of deep-sea pelagic carbonates, shallow-water shelf carbonates and reef complexes. Turekian (1964) took the point of view that the major sink for strontium in the ocean is deep-sea pelagic carbonate deposition and that the strontium budget could be balanced since the calcium carbonate sediment deposited as sediment in the deep sea realm is nearly equal to the equivalent calcium carbonate supplied in solution from the continents. He concluded (1964, p. 1495) that: "At present the deep sea appears to be the major area for the removal of both calcium carbonate and associated strontium. Using stream and deep-sea carbonate composition data, the material balance for strontium at present appears satisfied." Holland (1984) however, recognized that the rapid mass accumulation rates of shallow-water carbonates, such as those forming on the Grand Bahama Banks, make global carbonate budgeting difficult if only deep-sea carbonate sediments are taken into account. This recognition of the importance of shallow-water carbonate deposition in the global calcium carbonate and, by implication, the global strontium budget is taken as the theme of this paper which is an attempt to demonstrate by simple accounting procedures that reef and shallow-water shelf limestones are an important factor in the strontium geochemical cycle. Further, because these limestones serve as strontium sinks as they are deposited during periods of rising sea level and serve as strontium sources as they undergo dissolution during periods of falling sea level, their formation and dissolution affect the Sr/Ca ratio of sea water. The work of Graham et al. (1982), for example, has shown that the Sr/Ca ratio in sea water has varied markedly over Cenozoic time and that one of the mechanisms responsible for this variation could have been the storage of strontium in biogenic aragonite which is abundant in reef and shallow-shelf sediments but lacking in deep-sea sediment.

2. CONTINENTS AND MARINE CARBONATE SEDIMENTS AS STRONTIUM SOURCES AND SINK

A discussion of strontium fluxes between continents and oceanic carbonate sediments must be viewed in the light of a general $CaCO_3$ budget, although as shown below, estimates of the strontium flux from the land may be derived independently (Brass, 1976). A recent comprehensive analysis of the marine $CaCO_3$ budget in the light of new data from the Deep Sea Drilling Project on oceanic sedimentation rates is that of Davies and Worsley (1981). They base their analysis, and their attempt to balance the carbonate budget, on the following considerations:

1) The best available estimate of the global carbonate river input of Ca^{2+} is that of Garrels and Mackenzie (1971) who calculated it to be 4.88×10^{14} g/yr, which is equivalent to 12.2×10^{14} g/yr of $CaCO_3$ available for deposition in the oceans. It should be noted here that Holland (1984) pointed out that estimates of the calcium input to the global ocean may contain uncertainties of up to 50%; he used a figure of 7.0×10^{14} g/yr of Ca^{2+} of river input to arrive at a figure of 18×10^{14} g/yr as the equivalent $CaCO_3$ available in the oceans.

2) The current estimates of the long-term global accumulation rate of marine carbonate sediments of Hay and Southam (1977) of 17.8×10^{14} g/yr, and of Broeker (1974) of 18×10^{14} g/yr, are reasonable. I note here that these estimates include 6.4×10^{14} g/yr of limestone deposited in the form of reefs (0-20 m water depth) and on shelves (20-200 m water depth) as estimated by Milliman (1974).

3) The excess of the global marine carbonate sediment accumulation rate of $\sim 18 \times 10^{14}$ g/yr (Hay and Southam, 1977; Broeker, 1979) over the $\sim 12 \times 10^{14}$ g/yr of $CaCO_3$ available (Garrels and Mackenzie, 1971), amounting to $\sim 6 \times 10^{14}$ g/yr, leads to the conclusion that there is a source of carbonate that has not been included or that the long-term accumulation rate of carbonate sediment in the deep-sea is overestimated (Hay and Southam, 1977).

4) Davies and Worsley (1981) balance out this excess by asserting that their estimate of the oceanic deposition rate of $\sim 13 \times 10^{14}$ g/yr of carbonate sediment calculated from Deep Sea Drilling Project data is accurate but that the contention of Moore and Heath (1977) that only one-half to one-third of the carbonate sediment deposited in the deep sea is incorporated into the long-term stratigraphic record is also correct (see also van Andel et al., 1975). Thus, as Davies and Worsley (1981) point out, only $4.3\text{-}6.5 \times 10^{14}$ g/yr of deep-ocean carbonate

sediment is incorporated into the oceanic stratigraphic record. This amount, taken together with the 6.4×10^{14} g/yr of limestone deposited as reef and shelf deposits approximates the equivalent of 12.2×10^{14} g/yr of $CaCO_3$ supplied by rivers as calculated by Garrels and Mackenzie (1971). The long-term calcium carbonate budget is then considered by them to be balanced. The strontium fluxes associated with the above described carbonate budget is shown on Figure 2. These fluxes shown there have been arrived at in the following manner.

a) The riverine strontium input, according to Brass (1976), based on a global river flow of 3.23×10^{16} l/yr and a strontium content of 68.5 μg/l amounts to 2.2×10^{12} g/yr; Martin and Meybeck (1979) arrived at a figure of 2.24×10^{14} g/yr for the riverine strontium input to the ocean. Again it should be stated that Holland (1984) used a global river flow of 4.6×10^{16} l/yr and a strontium content of 70 μg/l to arrive at a strontium flux of 3.24×10^{12} g/yr as shown on Figure 1. For the purposes of this discussion the riverine strontium inputs estimates of Brass (1976) and Martin and Meybeck (1979) of 2.2×10^{12} g/yr are used as shown in Figure 2.

b) The strontium fluxes into and out of the deep-sea pelagic carbonate reservoir must be considered in the light of three time scales; short-, medium- and long-term. (In this paper short-, medium- and long-term refer to periods of 10^3 to 10^6 yr, 10^6-10^7 yrs and 10^7-10^8 yrs, respectively.) According to Davies and Worsley (1981), from 0-3 My ago the global carbonate deposition rate in the deep ocean averaged $\sim14.6\times10^{14}$ g/yr. At a strontium content of $\sim1400\times10^{-6}$ g/g determined from analyses of planktonic foraminifera and coccoliths which make up this sediment, this short-term strontium flux into these deep sea pelagic carbonates amounts to $\sim2.0\times10^{12}$ g/yr. A medium-term flux of strontium back into the ocean reservoir takes place as approximately one third to one half of deep-sea pelagic carbonate is dissolved during periods when oceanic bottom water is particularly corrosive. This dissolution is related to paleoceanographic factors operating on time scales of 10^6-10^7 years. The average Cenozoic accumulation rate of strontium can be taken as equal to approximately 1.0×10^{12} g/yr. On a long-term time scale of 10^7-10^8 years strontium is recycled back into the ocean through the mechanism of deep burial diagenesis. As shown in Figure 3, data on the strontium content of pelagic carbonates of Recent to Cretaceous age reveals that deep-sea limestones 120 million years old contains only about 400×10^{-6} g/g strontium compared with the $\sim1400\times10^{-6}$ g/g in modern pelagic calcite. The loss of strontium during burial diagenesis results from the dissolution of relatively strontium-rich foraminifera tests and coccoliths and the local reprecipitation of the carbonate as strontium-poor cement (Schlanger and Douglas, 1974). Thus the long-term storage rate of strontium in pelagic carbonates is probably equal to 0.5×10^{12} g/yr. For the purposes of this discussion, which is primarily to examine the short-term aspects of the strontium budget, the deep-sea carbonate accumulation rate over the past 3 My of about 14.6×10^{14} g/yr is considered as the controlling factor in the present-day strontium budget.

Other reservoirs and fluxes in the short-term strontium budget shown in Figure 2 are those associated with the deposition of reef and shallow-water limestones. Milliman (1974) defines reef limestones as those deposited in water depths of from 0-20 meters and shelf limestones as those deposited in water depths of from 20-200 meters. The global reef limestone reservoir for strontium which is discussed in detail in a subsequent section, forms at a rate of 500×10^{12} g/yr; at a strontium content of

~5,000×10^{-6} g/g the reef limestone sink absorbs ~2.5×10^{12} g/yr. The shelf limestone reservoir forms at the rate of 140×10^{12} g/yr and absorbs ~0.8×10^{12} g/yr of strontium (limestone formation rates from Milliman, 1974).

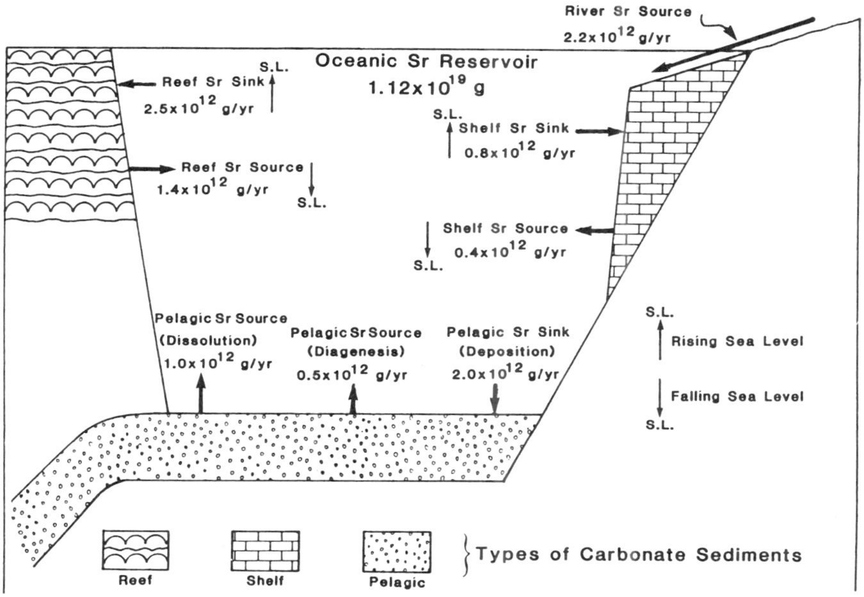

Figure 2. Strontium sources and sinks in the continent-ocean carbonate sediment system. During periods of rising sea level, reef and shelf carbonate sediments are strontium sinks. During periods of falling sea level these sediments become strontium sources.

In summary then, and in terms of present-day fluxes, the riverine input of strontium amounts to 2.2×10^{12} g/yr, the deep-sea pelagic carbonate sink takes in 2.0×10^{12} g/yr, the reef limestone sink accounts for 2.5×10^{12} g/yr and the shallow-shelf sink accounts for another 0.8×10^{12} g/yr of strontium. At the present time then, ~5.3×10^{12} g/yr of strontium is being removed from the ocean and about 2.2 ×10^{12} g/yr are entering the ocean from river flow. The implications of this short-term net loss of strontium from the oceanic reservoir is discussed below.

3. REEF AND SHELF LIMESTONES AS STRONTIUM SINKS

The calcium carbonate sediments that make up reef complexes and shallow-water shelf limestones consist largely of biologically mediated precipitates that accumulate in and around the body tissues of a wide variety of organisms. The most common organisms that contribute to these limestone accumulations are: corals (aragonite), calcareous green algae, such as *Halimeda* (aragonite), calcareous red algae (calcite), foraminifera (calcite and aragonite), molluscs (calcite and aragonite); echinoids and bryozoans which are calcitic are also present in small amounts. In some areas, such as the Grand Bahama Banks, ooids (aragonite) and fine-grained lime mud (aragonite)

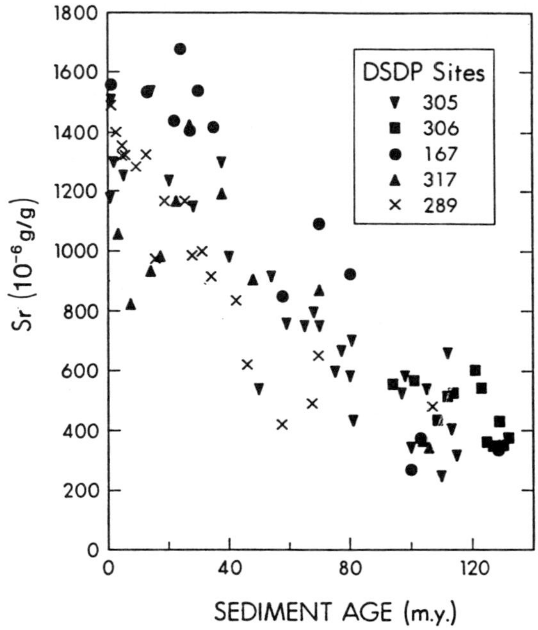

Figure 3. Strontium content related to age of deep-sea pelagic carbonate sediments from DSDP sites in the central Pacific. The linear relationship where Sr (10^{-6} g/g) = 1356 - (8.3 $\times$ age in Ma) is taken as evidence that long-term burial diagenesis results in the recycling to the oceanic reservoir of much of the original strontium contained in the calcite tests of foraminifera and coccoliths that make up these sediments (from Manghnani, Schlanger and Milholland, 1980).

make up a major fraction of the sediment. A synthesis of the sedimentology of reefs and shallow-water shelves is beyond the scope of this chapter, and for a detailed discussion of modern carbonate environments the following references are particularly valuable: Bathurst (1971), Milliman (1974), and Scholle et al. (1983) for general reviews; Hopley (1982) on the Great Barrier Reef; Ginsburg (1972) on the South Florida region; Purdy (1963) on the Grand Bahama Bank; and Emery, Tracey, and Ladd (1954) on Pacific atolls.

The importance of reef and shallow-water carbonates as strontium sinks lies in the fact that these limestones are largely aragonite as contrasted to deep-sea pelagic carbonates that are entirely calcite. As reviewed by Graham et al. (1982), the strontium distribution coefficient expressed as:

$$D_{Sr} = \left[(Sr/Ca)_{CaCO_3}/(Sr/Ca)_{seawater}\right]$$

for most biogenic calcite is about 0.16 $\pm$ 0.02. Graham et al. (1982) point out that this empirically determined D_{Sr} is close to experimentally determined values (Lorens, 1981). The most abundant constituents of deep-sea pelagic carbonates, planktonic foraminifera and nannoplankton contain about 1400×10^{-6} g/g strontium. However, reef and shelf carbonate is dominated by the skeletal remains of aragonitic coral and algae and ooids and mud that contained as much as $9{,}000 \times 10^{-6}$ g/g of strontium; the D_{Sr} for these carbonates is close to 1.0, a value about 6 times as great as the D_{Sr} for pelagic carbonate.

The strontium content of a typical reef carbonate accumulation can be derived from data collected during field and laboratory studies by many workers in various regions of the world. As an example, the composition of a typical Pacific atoll lagoon sediment is:

Foraminifera (calcite) - 5%
Fine debris (calcite and aragonite) - 43%
Halimeda (aragonite) - 36%
Coral (aragonite) - 16%

Strontium analyses of these sediments described by Emery, Tracey and Ladd (1954) show strontium contents of 1500×10^{-6} g/g for the Foraminifera, $8{,}000 \times 10^{-6}$ g/g for the fine debris, $8{,}000 \times 10^{-6}$ g/g for the coral and $7{,}500 \times 10^{-6}$ g/g for the green algae, *Halimeda*. The bulk sediment has an aragonite content of about 75% and a calcite content of about 25% which translates into a bulk strontium content of about $6{,}700 \times 10^{-6}$ g/g. That this aragonite/calcite ratio and strontium content is typical of present-day shallow-water carbonate deposits is confirmed by the data of Roselle (1985) reproduced here as Figure 4 and Table I. As shown, the average aragonite content of such sediments deposited in water depths of less than 200 m is about 70% on a global basis and the average strontium content is about $5{,}000 \times 10^{-6}$ g/g. The figure of $5{,}000 \times 10^{-6}$ g/g is taken here as representative of reef carbonates.

Table I. Aragonite and strontium contents of shallow-water modern carbonate sediments from various regions including the Bahamas, the Persian Gulf, Florida, Mauritania and British Honduras. (From Roselle, 1985.)

	Aragonite (%)		Strontium ($\times 10^{-6}$ g/g)	
Latitude	Range	Mean	Range	Mean
26 °N	43-46	54	2260-5640	3840
24-25 °N	80-98	93	7050-9710	8700
24-25 °N	28-100	69	2960-6080	3810
24-25 °N	40-100	74	2000-8000	5175
24-25 °N	42-90	67	2700-6500	4621
20-21 °N	10-84	53	1350-3210	2060
17 °N	0-100	69	1410-9080	5730

Extrapolation of these data to an estimate of the total strontium incorporated into reef complexes annually can be accomplished by using data on the annual amount of carbonate deposited as reefs developed by Milliman (1974). Milliman considered that the area occupied by reefs forming in water depths of 20 m or less is approximated by the area of twice the world's atolls plus the area of the Great Barrier Reef, The Grand Bahama Bank and the Campeche Bank; this area totals 1.4×10^6 km^2. Further, the rate of deposition in these carbonate environments was calculated by Milliman (1974) to be about 35 g/cm^2/10^3 yrs. Thus present-day reefs are accumulating 500×10^{12} g of $CaCO_3$/yr. At an average strontium content of $5{,}000 \times 10^{-6}$ g/g reefs are accumulating about 2.5×10^{12} g of strontium per year.

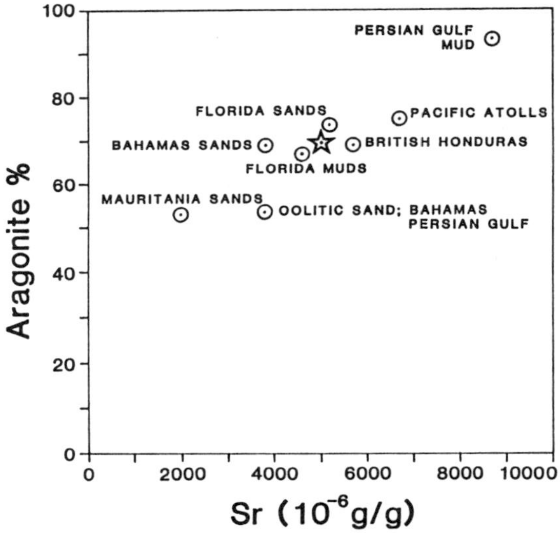

Figure 4. Strontium and aragonite contents of modern marine carbonate sediments from shallow-water environments. The points plotted are the means of aragonite percentages and strontium contents from Table I with Pacific atoll data added (from Roselle, 1985). The star indicates the strontium content used for calculations in this paper.

Similarly, using Milliman's (1974) estimate of the accumulation rate of 143×10^{12} g of $CaCO_3$/yr on shelves between depths of 20 to 200 m, a figure of 0.8×10^{12} g of Sr/yr is arrived at for the carbonate shelf sink.

This annual accumulation of 3.3×10^{12} g Sr/yr in reef complexes and shelf limestones is significant in light of the estimates of the strontium flux from the continents to the ocean of 2.2×10^{12} g/yr and the estimated flux of strontium into deep-sea pelagic carbonates of about 2.0×10^{12} g/yr (Fig. 2).

4. REEF AND SHELF LIMESTONES AS STRONTIUM SOURCES

The role of reefs as strontium sinks during periods of rising sea level is reversed during periods of falling sea level when the reefs become strontium sources due to the diagenesis of the aragonite-rich sediments under the influence of fresh water. The importance of exposure to fresh water in the diagenesis of reefs was shown by the drilling results from Bikini and Enewetak Atolls in the mid-Pacific (Emery, Tracey and Ladd, 1963). The sub-surface stratigraphy of the atolls is marked by alternating sequences of aragonite-rich carbonate, in which corals and *Halimeda* still retain their original mineralogy and completely calcitized limestone from which the aragonite has been removed by dissolution and calcite cement precipitated. The contact between the leached, aragonite-free section and the overlying, aragonite-bearing section has been termed a "solution unconformity"; it is a buried karst surface (Schlanger, 1963). The evolution of a solution unconformity as it relates to sea-level change is shown as

Figure 5. The major control on the diagenesis of these limestones is the introduction of fresh water into the limestone-ocean hydrologic system as shown in Figure 6. A carbonate atoll or bank that emerges above sea level becomes the site of a Ghyben-Herzberg lens of fresh water. Rain water percolating down through the permeable reef sediments comes to rest on the surface of the sea water that fills the limestone column still below sea level. Because of the difference in density between the fresh rain water (ρ=1) and the sea water (ρ=1.025) an isostatic equilibrium situation develops so that a biconvex lens of fresh water with a height above sea level to height below sea level ratio of 1:40 forms. The water in the lens is in constant motion and returns to the ocean; replenishment depends on the rate of rainfall. The sediments above the lens are subjected to constant dissolution by infiltrating fresh water and the sediments in the lens are subjected to constant flushing by fresh water. The diagenetic reactions of the mixture of calcite and aragonite in these hydrologic regimes is summarized on Figure 7. The aragonite dissolves releasing calcium, strontium and other trace cations to the pore waters. Much of the dissolved carbonate is reprecipitated as calcite cement. Note that the strontium loss is marked in the vadose and phreatic zones; in carbonates that remain in a sea-water environment the strontium content is maintained. The overall diagenetic trends developed during diagenesis of a reef complex are shown on Figure 8. The original mixed assemblage of biogenic aragonite and calcite is replaced by an assemblage of calcite cement and low-magnesium calcite relic from original skeletal material.

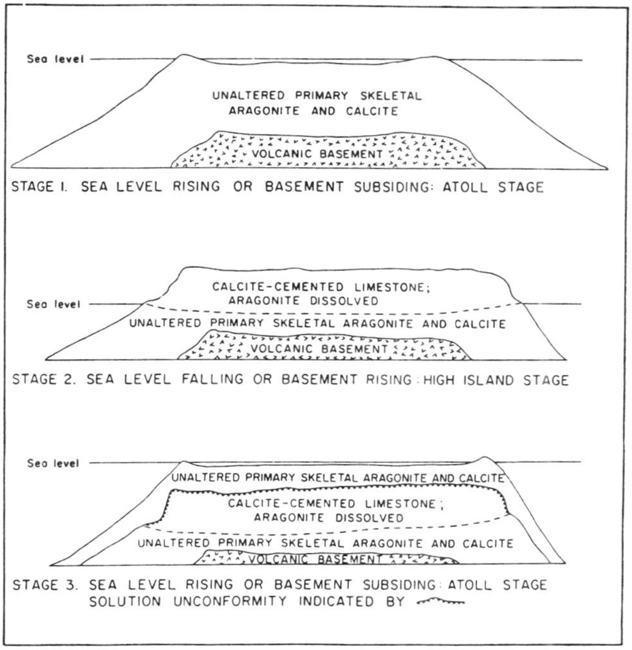

Figure 5. Schematic representation of the development of a solution unconformity below an atoll (modified from Schlanger, 1963).

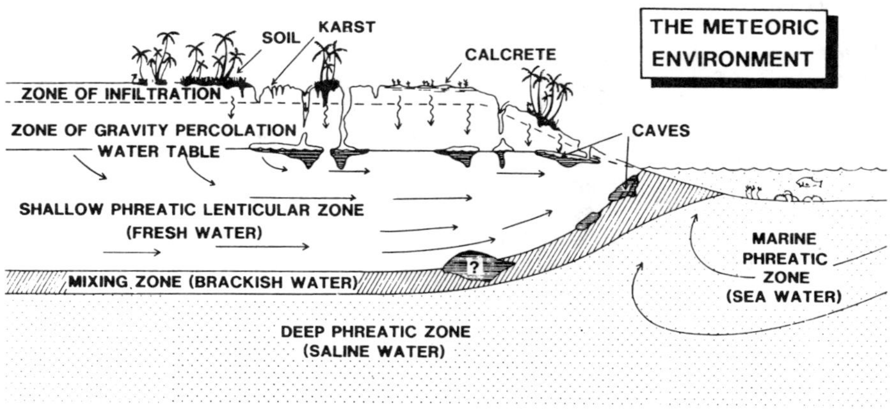

Figure 6. Hydrologic zones of the meteoric environment. These various zones would develop in both shelf carbonate sequences and atoll structures upon emergence above sea level (from James and Choquette, 1984).

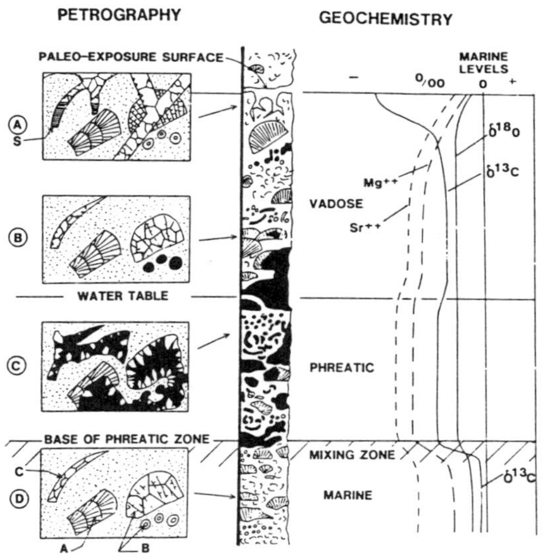

Figure 7. Petrographic and geochemical characteristics of a hypothetical sequence of shallow-water reef limestone that has been exposed to meteoric diagenesis under humid climatic conditions. Loss of strontium is most marked in the phreatic zone where aragonite is dissolved and calcite cement is precipitated (from James and Choquette, 1984).

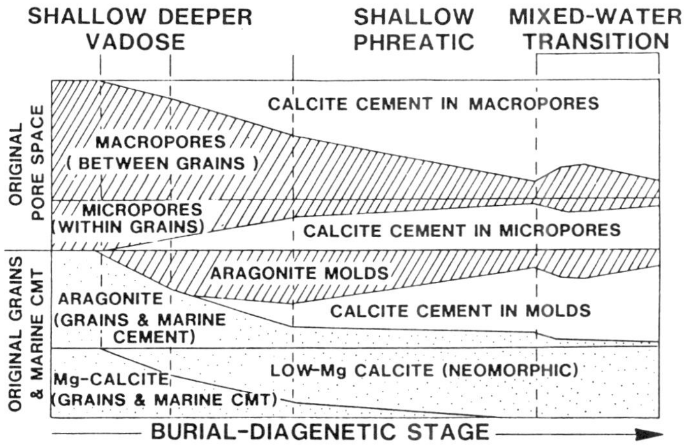

Figure 8. Diagenetic evolution of a sediment originally composed of biogenic aragonite and magnesium-rich calcite as it is affected by passage through meteoric diagenetic zones. The major changes are the elimination of aragonite and the transformation of the sediment to a limestone made up of low magnesium calcite (from Jones and Choquette, 1984).

The extent to which strontium is released from these carbonates can be estimated from strontium analyses of originally aragonitic limestone compared with analyses of diagenetically altered limestone. Such analyses, carried out on core material from Bikini Atoll (Emery, Tracey and Ladd, 1963), led to the following conclusions:

1. Coral and detrital sands that are only slightly altered or unaltered contain 0.6 to 0.9% strontium (6,000 to 9,000 $\times 10^{-6}$ g/g).
2. Coral and detrital sands that are broken down to microgranular aragonite still have high percentages of strontium, about 0.6%, but are, in general, less strontium-rich than unaltered material.
3. Corals and detrital sands that are broken down and recrystallized have intermediate amounts of strontium, about 0.4%.
4. Corals and detritus that are completely recrystallized and replaced by crystalline calcite have low percentages of strontium, about 0.03 to 0.1%.

Analyses of similar material from Enewetak Atoll (Schlanger, 1963) show similar results. Dolomite from Enewetak shows extremely low strontium contents of 0.003% indicating that dolomitization is a particularly effective diagenetic mechanism for removing strontium from a reef complex. In detail the reaction of various mineral species present as skeletal elements to diagenesis is a function of both the skeletal microstructure of the calcite and aragonite and the degree of saturation of the pore water with respect to calcite and aragonite. Walter (1985) has conducted experiments on dissolution rates of reef-forming organisms. A major result of these experiments is shown on Table II. In calcite undersaturated solutions, such as would be present in the fresh water lens, the dissolving phase is always more strontium-rich than the bulk sediment and the aragonite weight-percent in the dissolving phase is always significantly greater than in the bulk sediment.

Table II. Bulk sediment composition related to the composition of the dissolving phase. Data from dissolution experiments conducted below calcite saturation on biogenic carbonates by Walter (1985).

Sediment type	Bulk Sr ($\times 10^{-6}$ g/g) (sediment)	wt% Aragonite (sediment)	Sr^{2+} ($\times 10^{-6}$ g/g) (dissolving phase)	wt% Aragonite (dissolving phase)
beach	4560	55.0	6412	77
lagoon	6075	64.5	8400	87
reef	5410	71.0	7660	100

The amount of strontium released to the sea from reefs during a sea-level fall can be estimated by combining data on the global area occupied by reefs in the world ocean and the limestone removal rate from karst terranes due to dissolution. As shown on Figure 9, Jennings (1983) has compiled data on limestone removal rates from present-day karst regions. Limestone removal rates on Jamaica, which may be taken as typical, range from 40 to 95 $m^3/km^2/yr$; 70 $m^3/km^2/yr$ is taken here as representative. Milliman (1974), as discussed above, has estimated the area of reefs to be 1.4×10^6 km^2. Thus the strontium release rate from reefs can be calculated by considering that (1) the limestone volume removed $= 70 \times 10^6$ $cm^3/km^2/yr \times 1.4 \times 10^6$ km^2 or 98×10^{12} cm^3/yr, (2) the limestone mass removed $= 274 \times 10^{12}$ g/yr (at $\rho = 2.8$), (3) the strontium content of the limestone $= 5{,}000 \times 10^{-6}$ g/g, which gives (4) a strontium release rate $= 1.4 \times 10^{12}$ g/yr. Applying the same line of argument to the release of strontium from exposed shelf limestones during sea-level falls yields the figure of 0.4×10^{12} g Sr/yr for the shelf source.

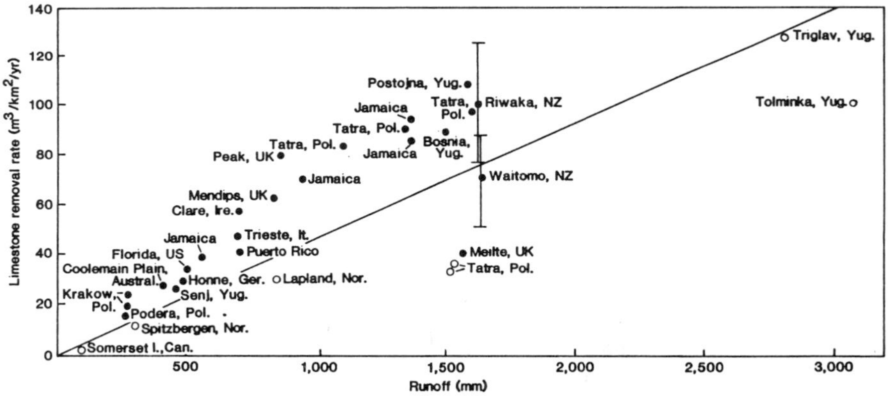

Figure 9. Limestone removal rate plotted against runoff in karst terrains including data from both tropical and non-tropical regions; high latitude or high altitude karsts where bare rock dominates are shown by open circles (from Jennings, 1983).

The implications of the difference between the rate of strontium storage in reef complexes and shelf limestones during periods of rising sea level at 3.3×10^{12} g/yr, and the release of strontium from reef complexes and shelf limestones during periods of falling sea level at a rate of 1.8×10^{12} g/yr are discussed below.

5. THE EFFECT OF SEA LEVEL VARIATIONS ON THE STRONTIUM CYCLE

Frequent major changes in relative sea level have been accepted to explain hiatuses, unconformities and vertical facies changes seen in the stratigraphic record (see Vail et al., 1977 for review). Carbonate sediments deposited on the continental shelves, in interior seaways, and in atoll and bank settings in the open ocean have therefore been subjected to repeated cycles of submergence and emergence. For the purposes of this discussion, sea-level changes during the Cenozoic Era over the past 65 million years are considered as an example of the influence of sea-level changes on the global strontium cycle and budget. In order to illustrate the frequency and magnitude of these sea-level changes, the sea-level curve of Vail and Hardenbohl (1979) is used and shown here as Figure 10.

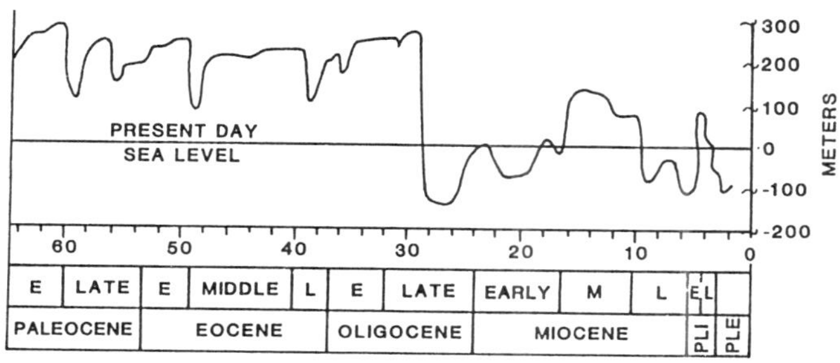

Figure 10. Tertiary sea-level curve related to present-day sea level and the geological time scale (from Vail and Hardenbohl, 1979).

The storage of strontium in actively forming reef and shallow shelf carbonates, as discussed above, indicates that such deposits serve as a strontium sink during periods of rising sea level and as strontium sources during periods of falling sea level. As shown in Figure 2, the present-day strontium sinks which have been forming during the recent period of rising sea level include reef carbonates that store strontium at a rate of 2.5×10^{12} g/yr, shelf carbonates storing strontium at a rate of 0.8×10^{12} g/yr, and pelagic carbonates storing 2.0×10^{12} g/yr of strontium. This total of 5.3×10^{12} g/yr of strontium shows, considering the riverine input of 2.2×10^{12} g/yr of strontium, that there is a net present-day annual loss of strontium of 3.1×10^{12} g/yr from the oceanic reservoir, that contains 1.12×10^{19} g. As a simple example then, a steady condition of rising sea level over a period of 10^6 yrs would lead to the sequestering of 3.1×10^{18} g of strontium, roughly equal to ~28% of the oceanic reservoir. During a period of falling sea level, reef carbonates become sources and release 1.4×10^{12} g/yr of strontium. Assuming that shelf limestones would also be affected and would release strontium at a rate of 0.4×10^{12} g/yr, i.e., in the same proportion that reef limestones do, we can arrive at the conclusion that ~1.8×10^{12} g/yr of strontium would be released into the oceanic reservoir from both kinds of shallow-water limestones during sea-level falls. During periods of relatively high but fluctuating sea level, such as seen on Figure 11 from 65 to 29 Ma, reef and shelf limestones would act as net sinks for strontium because the release rate of 1.8×10^{12} g/yr is less than the 3.3×10^{12} g/yr stored during periods of rising sea level. Further, due to tectonic subsidence not all of the shallow-water carbonates are subjected to fresh-water diagenesis, as seen in the significant stratigraphic intervals below atolls that retain their original aragonite content. Over the Cenozoic Era shallow-water carbonates then were sinks that accounted for a long-term average of $\sim 1.5 \times 10^{12}$ g/yr of strontium that was not recycled back into the oceanic reservoir. The deep-sea pelagic carbonate strontium sink, based on the carbonate sedimentation rate data of Davies and Worsley (1981) for the past 3 My accounts for $\sim 2.0 \times 10^{12}$ g/yr. However, the long-term pelagic strontium storage rate is less than the 0-3 My rate due to two recycling paths. First, the stratigraphic record in the deep sea reveals hiatuses and periods of intense dissolution of pelagic carbonate so that, as pointed out by Moore and Heath (1977), only one-half to one-third of the pelagic carbonate is incorporated into the medium-term record. Further, long-term diagenesis results in the release of much of the strontium originally present in the foraminifera and coccoliths that make up the pelagic carbonate (Figure 3). Probably only about one-quarter of the originally precipitated deep-sea pelagic strontium is retained so that the long-term pelagic deep-sea sink accounts for $\sim 0.5 \times 10^{12}$ g/yr. Thus, the long-term strontium storage rate for the oceanic carbonate sediment system would equal the net of $\sim 1.5 \times 10^{12}$ g/yr retained in the reef and shelf carbonates plus the 0.5×10^{12} g/yr retained in the pelagic carbonate. This long-term storage rate of $\sim 2.0 \times 10^{12}$ g/yr in the total oceanic carbonate record is roughly in balance with the present-day riverine input of $\sim 2.2 \times 10^{12}$ g/yr determined by Brass (1976) and Martin and Meybeck (1979).

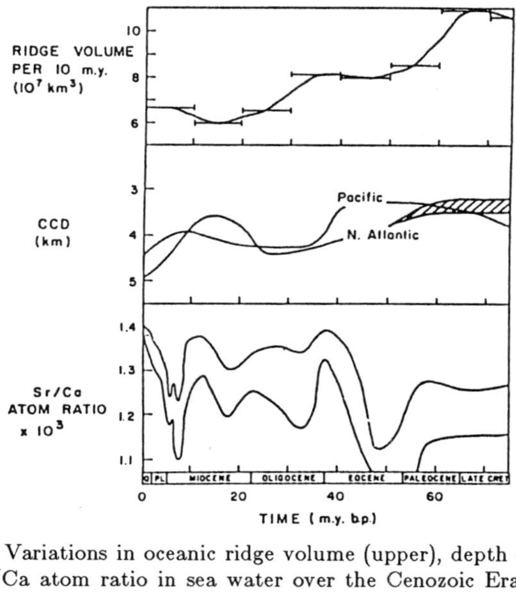

Figure 11. Variations in oceanic ridge volume (upper), depth of the CCD (middle) and the Sr/Ca atom ratio in sea water over the Cenozoic Era (lower); the 2 solid lines in the lower figure represent the 80% confidence limit of the mean Sr/Ca ratios determined from analyses of the tests of planktonic foraminifera (from Graham et al., 1982).

6. DEPOSITION OF REEF AND SHELF LIMESTONES RELATED TO CHANGES IN THE Sr/Ca RATIO IN SEA WATER

The arguments presented above show that reef and shelf limestones are important sinks and sources for strontium. Graham et al. (1982) undertook an analysis of the

Cenozoic history of the Sr/Ca ratios in sea water; the following discussion and Figure 11 are in part drawn from the work of these authors. Graham et al. (1982) measured the Sr/Ca ratio in tests of planktonic foraminifera and inferred the Sr/Ca ratio in sea water throughout the Cenozoic Era, based on the assumption that the ratio in the biogenic calcite reflects the ratio in sea water. As shown in Figure 11, the curves depicting the Sr/Ca ratio in the biogenic calcite which makes up the pelagic carbonate record show the following characteristics:

1) The ratio has been within ~10% of 1.25×10^{-3} over the long-term, that is, the 65 My of the Cenozoic Era. This conclusion is in accord with the conclusion reached in the present paper that the strontium budget of the continent-ocean system has been in balance over the past 65 My.
2) There is a long-term increase towards the present in the average ratio over the past 65 My.
3) There are distinct minima in the Sr/Ca ratio at 50 Ma and 4-9 Ma.

Graham et al. (1982) point out that if only rivers are considered as sources and if only biogenic calcite is considered as a sink (assuming a constant composition of river water) a steady state situation would develop where the Sr/Ca ratio in the biogenic calcite would tend to become equal to the Sr/Ca ratio in the river input. However, there are other sources and sinks as these authors point out. Ca is added to the sea water reservoir through hydrothermal activity linked to spreading at submarine ridge crests. Graham et al. (1982) suggest that the long term increase in the Sr/Ca ratio seen in their data is due to decreasing sea floor spreading rates over the past 65 My, expressed as a decrease in ridge volume seen in Figure 11. The Sr/Ca minimum at 4-9 Ma is ascribed to an increase in the hydrothermal supply of Ca due to an increase in spreading rates during that time interval. The strong minimum at 50 Ma is attributed by Graham et al. (1982) to an increase in aragonite deposition on shallow shelves and reefs.

As discussed above, data and interpretations developed in the present paper show that medium-term, 10^6-10^7 year, perturbations in the Sr/Ca ratio of sea water such as shown by Graham et al. (1982), can be accounted for by the growth and dissolution of reef complexes; in 10^6 years the present-day strontium deficit would cause a reduction in the oceanic sea water reservoir of strontium of some 28%. However, as Graham et al. (1982) have shown, the actual Sr/Ca ratio in sea water is a function of several factors including mid-ocean ridge crest spreading rates, the level of the Calcite Compensation Depth (CCD) (a deep CCD results in increased dissolution of pelagic carbonates and a shallow CCD promotes preservation of these carbonates) and the rate of reef and shelf carbonate deposition. Given our still imprecise knowledge of the exact timing, duration and magnitude of Cenozoic sea-level changes, it is difficult to precisely assign each Sr/Ca ratio perturbation to a particular aragonite deposition period. At this time it can only be pointed out that the effect of reef and shelf carbonate deposition and sub-aerial and fresh-water diagenesis on the oceanic strontium budget is considerable.

7. CONCLUSIONS

Recognizing the uncertainties inherent in the kinds of data used in developing a strontium balance sheet for the ocean system, the following points appear to be clear:

1) The argument that the deep sea is the major area of removal of strontium from the oceanic reservoir, as expressed by Turekian (1964), needs to be modified. Reef and shallow shelf carbonate deposits are probably more important than deep-sea pelagic carbonates as strontium sinks and sources.
2) Over the short term (10^3-10^6 years), such as represented by the recent period of rising sea level, there is a marked strontium deficit in the budget due to the high rate of storage of strontium in reef and shelf carbonates, and in pelagic deep-sea carbonates of 5.3×10^{12} g/yr, compared to the riverine flux of $\sim 2.2 \times 10^{12}$ g/yr.
3) Medium-term (10^6-10^7 years) deficits in the strontium budget due to the storage of strontium in reef and shelf carbonates during prolonged periods of rising sea level cause observable changes in the Sr/Ca of seawater. These medium-term deficits are only partially "paid back" during periods of falling sea level.
4) Over the long-term (10^7-10^8 years) the continent-ocean strontium budget is in balance due to the recycling of strontium from both reef and shallow-water shelf carbonates and deep-sea pelagic carbonates back into the oceanic reservoir.

8. REFERENCES

Bathurst, R.G.C., 1971, *Carbonate Sediments and Their Diagenesis: Developments in Sedimentology,* v. 12, New York, Elsevier, 620 p.

Brass, G.W., 1976, The variation of the marine $^{87}Sr/^{86}Sr$ ratio during Phanerozoic time: interpretation using a flux model, *Geochim. Cosmochim. Acta,* **40**, 721-730.

Broeker, W.S., 1974, *Chemical Oceanography,* New York, Harcourt, Brace & Jovanovich, 214 p.

Davies, T.A. and Worsley, T.R., 1981, Paleoenvironmental implications of oceanic carbonate sedimentation rates, *in* Warme, J.E., Douglas, R.G. and Winterer, E.L. (eds.) The Deep Sea Drilling Project: A Decade of Progress, *Soc. Econ. Paleo. Mineral. Spec. Pub.* **32**, 169-179.

Emery, K.O., Tracey, J.I. and Ladd, H.S., 1954, Geology of Bikini and nearby atolls, *U.S. Geol. Survey Prof. Paper* **260-A**, 1-265.

Garrels, R.M. and Mackenzie, F.T., 1971, *Evolution of Sedimentary Rocks*, New York, W.W. Norton and Co., 397 p.

Ginsburg, R.N., 1972, *South Florida Carbonate Sediments,* The University of Miami, Miami, Florida, 72 p.

Graham, D.W., Bender, M.L., Williams, D.F. and Keigwin, L.D., 1982, Strontium-calcium ratios in Cenozoic planktonic foraminifera, *Geochim. Cosmochim. Acta,* **46**, 1281-1292.

Hay W.W. and Southam, J.R., 1977, Modulation of marine sedimentation by the continental shelves, *in,* Anderson, N.R. and Malahoff, A. (eds.) *The Fate of Fossil Fuel* CO_2 in the Ocean, pp. 569-604, New York, Plenum Publishing Co.

Holland, H.D., 1978, *The Chemistry of the Atmosphere and Oceans,* New York, John Wiley & Sons, 351 p.

Holland, H.D., 1984, *The Chemical Evolution of the Atmosphere and Oceans,* New Jersey, Princeton Univ. Press, 582 p.

Hopley, D., 1982, *The Geomorphology of the Great Barrier Reef: Quarternary Development of Coral Reefs,* John Wiley and Sons, 453 p.

James, N.P. and Choquette, P.W., 1984, Diagenesis 9: Limestones—the meteoric environment, *Geosciences Canada,* **11**, 161-194.

Jennings, J.N., 1983, Karst landforms, *Amer. Scientist,* **71**, 578-585.

Lorens, R.B., 1981, Sr, Cd, Mn, and Co distribution coefficients in calcite as a function of calcite precipitation rate, *Geochim. Cosmochim. Acta,* **45**, 553-561.

Manghnani, M.H., Schlanger, S.O. and Milholland, P.D., 1980, Elastic properties related to depth of burial, strontium content and age, and diagenetic stage in pelagic carbonate sediments, *in*, Kuperman, W.A. and Jensen, F.B. (eds.) *Bottom-Interacting Ocean Acoustics*, pp. 45-51, New York, Plenum Publishing Co.

Martin, J.-M. and Meybeck, M., 1979, Elemental mass balance of material carried by major world rivers, *Marine Chemistry,* **7**, 173-206.

Milliman, J.D., 1974, Marine carbonates, *in*, Milliman, J.D., Muller, G. and Forstner, U., (eds.) *Recent Sedimentary Carbonates*, New York, Springer-Verlag, 375 p.

Moore, T.C. and Heath, G.R., 1977, Survival of deep-sea sedimentary sections, *Earth Planet. Sci. Lett.,* **37**, 71-80.

Purdy, E.G., 1963, Recent calcium carbonate facies of the Great Bahama Bank, *Jour. Geol.,* **71**, 331-355; 472-497.

Roselle, P., 1985, Modern cool-water beach sands of southwest England—a discussion, *Jour. Sed. Pet.,* **55**, 778-783.

Schlanger, S.O., 1963, Subsurface geology of Enewetak Atoll, *U.S. Geol. Survey Prof. Paper* **260-BB**, 991-1065.

Schlanger, S.O. and Douglas, R.G., 1974, The pelagic ooze-chalk-limestone transition and its implications for marine stratigraphy, *in* Hsu, K.J. and Jenkyns, H.C. (eds.) *Pelagic Sediments on Land and Under the Sea,* pp. 117-148, Special Pub. no. 1, Internat'l. Assoc. Sedimentologists.

Scholle, P.A., Bebout, D.G. and Moore, C.H. (eds.) 1983, Carbonate Depositional Environments, *Amer. Assoc. Pet. Geol. Mem.* **33**, 708 p.

Turekian, K. T., 1964, The marine geochemistry of strontium, *Geochim. Cosmochim. Acta,* **28**, 1479-1496.

Vail, P.R. and Hardenbohl, J., 1979, Sea-level changes during the Tertiary, *Oceanus,* **22**, 71-79.

Vail, P.R., Mitchum, R.M., Jr., and Thompson, S., III, 1977, Seismic stratigraphy and global changes of sea level, Part 4: Global cycles of relative changes in sea level, *in* Payton, C.E. (ed.) Seismic Stratigraphy—Applications to Hydrocarbon Exploration, *Amer. Assoc. Petroleum Geologists Mem.* **26**, 83-97.

Walter, L.M., 1985, Relative reactivity of skeletal carbonate during dissolution :implications for diagenesis, *in* Schneidermann, N. and Harris, P. (eds.) Carbonate Cements, *Soc. Econ. Min. Paleo. Spec. Pub.* **36**, 3-16.

van Andel, Tj. H., Heath, G.R. and Moore, T.C., 1975, Cenozoic History and Paleooceanography of the Central Equatorial Pacific Ocean, *Geol. Soc Amer. Mem.* **143**, 134 p.

SEDIMENT CYCLING DURING EARTH HISTORY

R. M. Garrels
Department of Marine Science
University of South Florida
St. Petersburg, Florida 33701
U.S.A.

ABSTRACT

The mass-age relations of sedimentary rocks are roughly consistent with a model in which the rate of sedimentation has been approximately constant through time, averaged over very long time intervals. The total mass of sedimentary rocks is assumed to be constant through time, and the probability of erosion of sedimentary rocks of a given period is assumed to be proportional to the mass/year of that age interval remaining. The model is not entirely cannibalistic; subduction of sediments, return of their volatiles to the atmosphere, and production of new sediments by reaction of these volatiles with crystalline rocks is an important process.

This model, with multiple turnovers of the sedimentary mass, without striking changes in its gross composition, has led to the concept of an average steady state among the reservoirs of chemical sediments. Although there seem to be few important secular elemental trends in the total sedimentary mass, there are important reciprocal changes among the mineral reservoirs, notably reciprocal changes between sulfide and sulfate, balanced by changes between carbonate and organic carbon.

The earth surface environment, so dependent on oxygen and carbon dioxide levels in the ocean and atmosphere, is dependent, in the long term, on the transfers of carbon and sulfur from one reservoir to another. The ocean-atmosphere contents of oxygen and carbon dioxide are small relative to the amounts of carbon and sulfur that have been transferred back and forth between the mineral reservoirs. The ocean and atmosphere, over geologic time, are media of transfer and not reservoirs. The above relations have led to mathematical models of sediment cycling which, initially, assumed constancy of ocean-atmosphere. The models, using the isotopic composition of the carbon of carbonates and organic carbon on the one hand, and that of sulfide and sulfate on the other as "ground truth" have yielded results in terms of transfer rates between sulfide-sulfate-carbonate-organic carbon reservoirs that probably are accurate within an order of magnitude, and indicate general trends through Phanerozoic time. The models have recently been expanded to include transfers among all the major chemical sediment reservoirs, and functional relations have been developed on transfer rates as functions of reservoir sizes, temperatures, land area, sea floor spreading rates, oceanic circulation through ridges, and metamorphic and volcanic processes in general.

A. Lerman and M. Meybeck (eds.), Physical and Chemical Weathering in Geochemical Cycles, 341–355.

1. INTRODUCTION

It is generally held that as the earth accreted, and developed core, mantle, and crust, that volatile compounds, especially H_2O, CO_2, HCl and S gases were released. The H_2O made the oceans; the acid gases attacked the original crustal rocks to make limestones, sulfate compounds, clay minerals, and the salt in the oceans. Thus most of the original sedimentary rocks and the ocean itself were created early in earth history. Most of the mass of sedimentary rocks was created long ago, and over Phanerozoic time (the last 600 million years) it certainly has not changed more than a few percent. Detailed analysis of sediment cycling indicates that about 6 times the volume of currently existing sediments have been deposited over the last 3 billion years.

The first sedimentary rocks were formed from primordial crystalline crust through reactions with water and other volatile substances. Volatile compounds, especially water and acid gases of carbon, sulfur, and chlorine, were released during the crystallization of molten rocks at high temperatures. They back-reacted with the igneous materials at low temperature. The water condensed to form the oceans and reacted with primary silicates to form hydrated mineral phases. Some typical reactions of igneous rock minerals with water and/or acids to give sedimentary mineral species are as follows:

$$\underset{\text{pyroxene}}{CaSiO_3} + CO_2 = \underset{\text{calcite}}{CaCO_3} + \underset{\text{quartz}}{SiO_2}$$

$$\underset{\text{feldspar}}{CaAl_2Si_2O_8} + CO_2 + 2H_2O = \underset{\text{calcite}}{CaCO_3} + \underset{\text{kaolinite}}{Al_2Si_2O_5(OH)_4}$$

$$\underset{\text{feldspar}}{2NaAlSi_3O_8} + 2HCl = \underset{\substack{\text{in ocean}\\ \text{or evaporite}}}{2NaCl} + \underset{\text{montmorillonite}}{Al_2Si_4O_{10}(OH)_2} + \underset{\text{quartz}}{2SiO_2}$$

$$\underset{\text{pyroxene}}{CaSiO_3} + H_2SO_4 + H_2O = \underset{\text{gypsum}}{CaSO_4{\cdot}2H_2O} + \underset{\text{quartz}}{SiO_2}$$

The budget for sedimentary mineral formation has been calculated in detail by Li (1972) and is shown in Figure 1. Rubey (1951) was first to discuss in detail the role of volatiles in the genesis of sedimentary rocks.

If the CO_2 released from the formation of the crust had remained in the atmosphere it would have exerted a pressure of about 50 atmospheres, probably enough to have caused a "runaway greenhouse" like that of Venus, which has surface temperatures near 450°C and a CO_2 pressure of 90+ atmospheres (Walker, 1977). Similarly the pressure of SO_2 gas would have been 2.5 atm, and that of HCl about 300 atm. The CO_2 is now sequestered in carbonate rocks or has been reduced to organic carbon in a mole ratio of about 4 to 1 inorganic to organic carbon. The Cl of HCl is in the ocean or in evaporite deposits of NaCl (1/1), the H^+ has reacted with crystalline rocks forming silicates to make chiefly clay minerals. Original H_2S or SO_2 gases have been partly oxidized to sulfuric acid and stored as sulfate in the oceans as the minerals gypsum ($CaSO_4{\cdot}2H_2O$) or anhydrite ($CaSO_4$) in sedimentary rocks, or stored in reduced form as pyrite. The molar proportions are ocean sulfate 1, gypsum-anhydrite 5, pyrite 5.

The total mass of sedimentary rocks made up of shale, sandstone, and limestone is estimated at about $25{,}000 \times 10^{20}$ g. Slates and schists are included in the shale category, just as quartzite is included with sandstone, and marble with limestone.

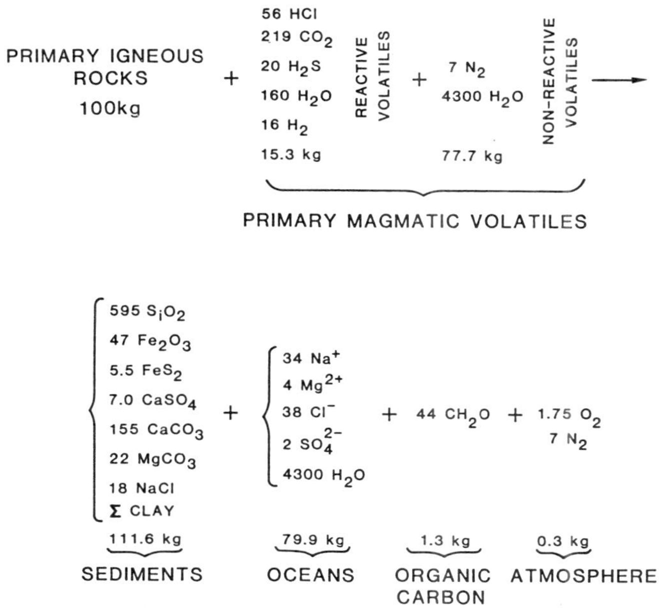

Figure 1. Mass balance for the origin of sedimentary rocks, oceans, atmosphere, and buried organic carbon. Reaction coefficients are in relative moles. (Li, 1972)

The mass ratios are about 65% shale, 18% limestone, 15% sandstones, and 2% evaporites ($CaSO_4$ and NaCl). Thus the volume of existing sedimentary rocks of all ages, with a density of about 2.5 g/cm^3 is about $10{,}000\times10^{20}$ cm^3, or somewhat less than that of the ocean, $14{,}000\times10^{20}$ cm^3 (Garrels and MacKenzie, 1971a).

The distribution of sedimentary mass as a function of sedimentary rock age is not well known, but most plots of mass per unit time existing today versus rock age show a rough exponential relation (Figure 2). Such a distribution is consistent with a model in which the total mass of sedimentary rock has been constant for the last 2.5×10^9 years, the rate of deposition has been constant, and the probability of destruction (erosion and redeposition) has been proportional to the existing mass (Garrels and Mackenzie, 1971a). Such a model obviously is but a crude first approximation, and has been modified (Garrels and MacKenzie, 1971b) and elaborated (Veizer and Jansen, 1979). However, the basic aspects of the model—a constant sedimentary mass and a constant rate of destruction and deposition of sediments (if averaged over long time intervals)—appear adequate as a basis for modification as more information becomes available. Current estimates indicate that 5 or 6 times as much sedimentary rock as now exists has been deposited over the last 2.5 to 3.0 billion years, and that whereas large fluctuations of depositional rates with time have taken place, there are no marked continuous trends with time. Instead a model with fluctuations about a mean rate of deposition seems valid.

Figure 3 shows mass distribution starting with an original sedimentary mass equal to that of today. Every 500 million years a mass equal to 1/2 the original mass is deposited, and is derived by erosion of the earlier sediments. After about 2.5 billion years the original mass is almost destroyed, and the mass-age distribution clearly

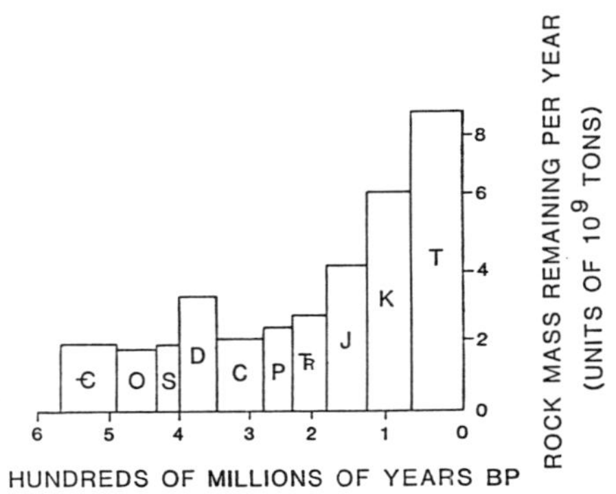

Figure 2. Mass-age distribution of sedimentary rocks for Phanerozoic time. Note the roughly exponential increase in surviving mass over the last 600 million years (adapted from Gregor, 1985).

becomes exponential. Each volume of sediment of a given age is diminished by one half each 500 million years in this illustrative model. Each volume of sediment decays according to the same rate law.

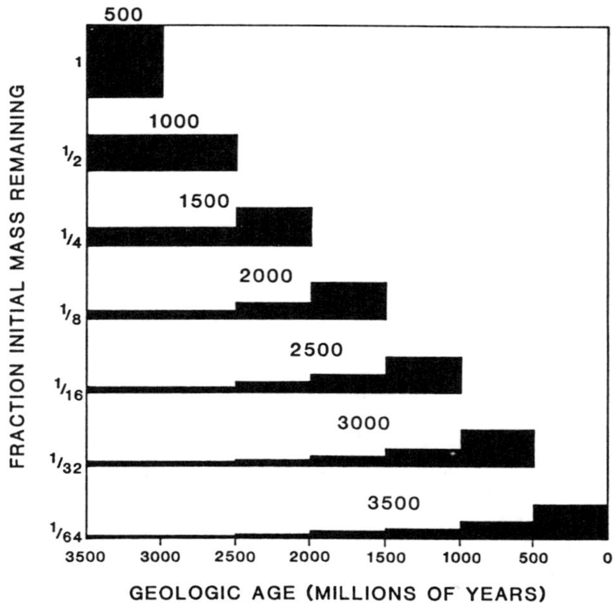

Figure 3. Constant mass model, 1/2 mass cycled every 500 million years, 3 × original mass cycled. Erosion proportional to mass present; all sediments have equal opportunity for erosion.

The chief alternative to this constant mass model is one in which volatiles are emitted at a constant rate through time and hence the sedimentary mass grows linearly with time. It has been shown (Garrels and Mackenzie, 1971a) that the constant mass model and the linear growth model converge with time, and become indistinguishable at about 500 million years before the present.

Igneous rocks are being weathered today and converted into sediments at the expense of volatiles, chiefly CO_2. However, sedimentary rocks are also being destroyed by metamorphic processes, which release the volatiles required to make them in the first place. These volatiles can then react to form new sediments to replace those lost. This hot-cold system, probably dominated by the heating and volatile production accompanying subduction of ocean sediments, is an essentially closed system, but it involves the crust and upper mantle, and is more realistic than a simple cannibalistic model in which sediments feed only upon themselves.

2. SEDIMENTARY ROCK CHEMISTRY

The major element chemistry of sedimentary rocks has changed little over the last 3.5 billion years. Table I shows the average composition of some 400 Archean shales, 400 early Proterozoic shales, and the average Phanerozoic shale (Cameron and Garrels, 1980). The similarities of the major oxide percentages are obvious. In fact, the differences between Phanerozoic shales (0-600 million years) and early Proterozoic shales (1800-2500 million years) are generally greater than those between Phanerozoic and Archean shales (2500 million-3800 million years). The differences may not even be significant but a result of limited sampling. Of great interest are the percentages of organic carbon, which are of the order of 1 weight percent for all three sets of shale analyses. The presence of abundant organic carbon shows that CO_2 was being reduced by organisms 3.5 or more billion years ago, which requires oxidation of other chemical species to complete the redox couple. It is a matter of continuing controversy as to which species were oxidized, but the presence of abundant ferric oxides in Archean sediments need no longer be interpreted as post-depositional change.

Table I. Comparison of the major constituents of Archean, early Proterozoic (Aphebian), and Phanerozoic shales (modified after Cameron and Garrels, 1980, Table III).

	Archean	Aphebian	Phanerozoic (carbonate-free)
SiO_2	60.52	65.00	63.11
TiO_2	0.76	0.76	0.89
Al_2O_3	17.63	16.22	18.80
Fe_2O_3	7.91	6.37	7.79
MgO	3.47	2.51	3.11
CaO	2.16	0.52	0.07
Na_2O	2.62	1.44	1.27
K_2O	2.84	4.83	3.69
C	0.77	1.68	0.87
S	0.66	0.65	0.38
$\delta^{34}S$ (‰)	$+1.31 \pm 0.32$	$+11.79 \pm 4.37$	-12.
Total	99.34	99.98	99.98

3. SEDIMENTARY ROCK MINERALOGY

The absence of significant secular change in the overall chemical composition of sedimentary rocks has been overshadowed in scientific publications by reports of mineralogic changes with time. The changes in the distribution of elements after deposition are known as diagenetic change. New sediments are typically derived from a variety of sources, and they are a mixture of many minerals. As time goes on there is a tendency, on the one hand, to reaction and recrystallization in the original mineral mass, and on the other, to transport of elements into and out of the sediments. Furthermore, when sediments are isolated from oxygenated water by burial there is almost always enough reducing material, chiefly organic carbon, to convert initially oxidized compounds, such as sulfate and ferric oxides, to reduced species such as pyrite (FeS_2).

Most carbonate rocks initially contain calcium carbonate as nearly pure calcite, as its polymorph aragonite, and as magnesium calcite with up to about 18 mole% $MgCO_3$. With time aragonite and high-magnesium calcite disappear, usually recrystallizing to the more stable nearly pure calcite, or are simply selectively removed from the system by aggressive percolating waters that dissolve them more rapidly than the nearly pure calcite. The time scale of such changes has a tremendous range. Under ideal conditions of burial, in which the system is closed and temperature and pressure remain low, aragonite may persist for 2×10^8 years or more (Sandberg, 1975).

Other polymorphous mineral species also tend to recrystallize to the least soluble form. Amorphous SiO_2 and the SiO_2 minerals cristobalite and tridymite tend to convert to quartz. The time interval required ranges from very short times to about the 2×10^8 years observed for carbonate mineral species. The aluminum oxide trihydrate gibbsite, which is the chief Al species found in commercial Al deposits, tends to lose its H_2O molecules after burial. However, the process is seldom complete, and in deposits as much as 10^8 years old, one finds a mixture of monohydrates with the trihydrate (Valeton, 1972).

The layer silicates (clay minerals) that make up the bulk of shaly rocks exhibit many mineral species in fresh sediments. Figure 4 shows the compositions of a number of such species. As pointed out before, the bulk composition of shales shows little variation with time. The average compositions of Archean, Proterozoic, and Phanerozoic shales are plotted in the Figure and are indicated with x's. Under these nearly identical compositional relations, only three minerals make up the stable association reached during diagenetic change—the clay minerals illite and chlorite, and quartz. Shales older than about 2×10^8 years tend to have a monotonous stable mineralogy. One of the major chemical reactions that occurs in clay diagenesis is the conversion of highly hydrous clays to illite, a micaceous mineral with no interlayer water, by reaction of the clay minerals with the potassium in detrital feldspars (Burst, 1957; Perry and Hower, 1970).

The diagenetic changes described briefly here could be supplemented by many published studies of other mineral relations, showing that diagenesis obscures original mineralogy, and causes marked decrease in the information available from studies of sediments. Age differences in the mineralogy of sediments can usually be ascribed to post-depositional change. Only in a few instances can real primary secular differences be recognized. One striking example is that of the Banded Iron Formations (BIF) of Archean to early Proterozoic age (3.8 to 1.9×10^9 yrs). These chemically precipitated rocks, consisting almost entirely of iron minerals and silica, and exhibiting striking banded structures, have only a few unimportant Late Proterozoic or Phanerozoic analogs (Gole and Klein, 1981).

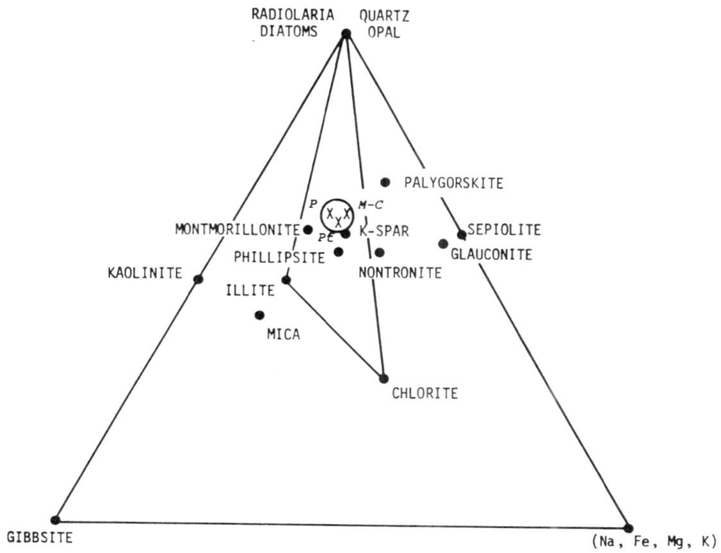

Figure 4. Schematic diagram to show that a mixture of original minerals in a shale with bulk composition lying in the circle could react to yield a simple three phase system—quartz, illite, and chlorite. The three average compositions in the circle are PC=Precambrian shales; P=Paleozoic shales; M-C=Mesozoic-Cenozoic shales (adapted from Garrels and Mackenzie, 1974.)

4. STABLE ISOTOPES AS MONITORS OF THE PAST

The stable isotope ratios of $^{34}S/^{32}S$ and of $^{13}C/^{12}C$ have emerged recently as two reliable indicators of conditions in the environments of the ancient past. Values for stable isotope ratios are usually expressed in parts per mil (parts per thousand, ‰) difference between the ratio in a given sample and the ratio in a standard, i.e., $\delta^{34}S = +15$ ‰ . The ratio of ^{34}S to ^{32}S in unaltered sulfate evaporite deposits records the ratio in the water from which they were deposited, because there is only little isotope fractionation during deposition. The marine origin for a given evaporite can often be established by determining the ratio of chloride to iodide or bromide in Cl-containing minerals (but see Hardie, 1984). Also, because of their high solubility, gypsum or anhydrite deposits tend either to be found with their original composition, or to be dissolved away completely. The problem of the formation of secondary minerals is far less severe than, for example, it is for the silicates.

The ratio of ^{13}C to ^{12}C in carbonate minerals also provides a measure of the ratio in the seawater from which the carbonate was precipitated. Although there is greater uncertainty than in the case of $\delta^{34}S$ in assigning a measured $\delta^{13}C$ of a sample to that of the ocean from which it came, the measured $\delta^{13}C$ at least seems to be close to the original value of the seawater. There is great potential in the use of other stable isotope ratios (such as ^{87}Sr to ^{86}Sr), and many studies have already been made. Here, however, the discussion will be limited to $\delta^{13}C$ and $\delta^{34}S$.

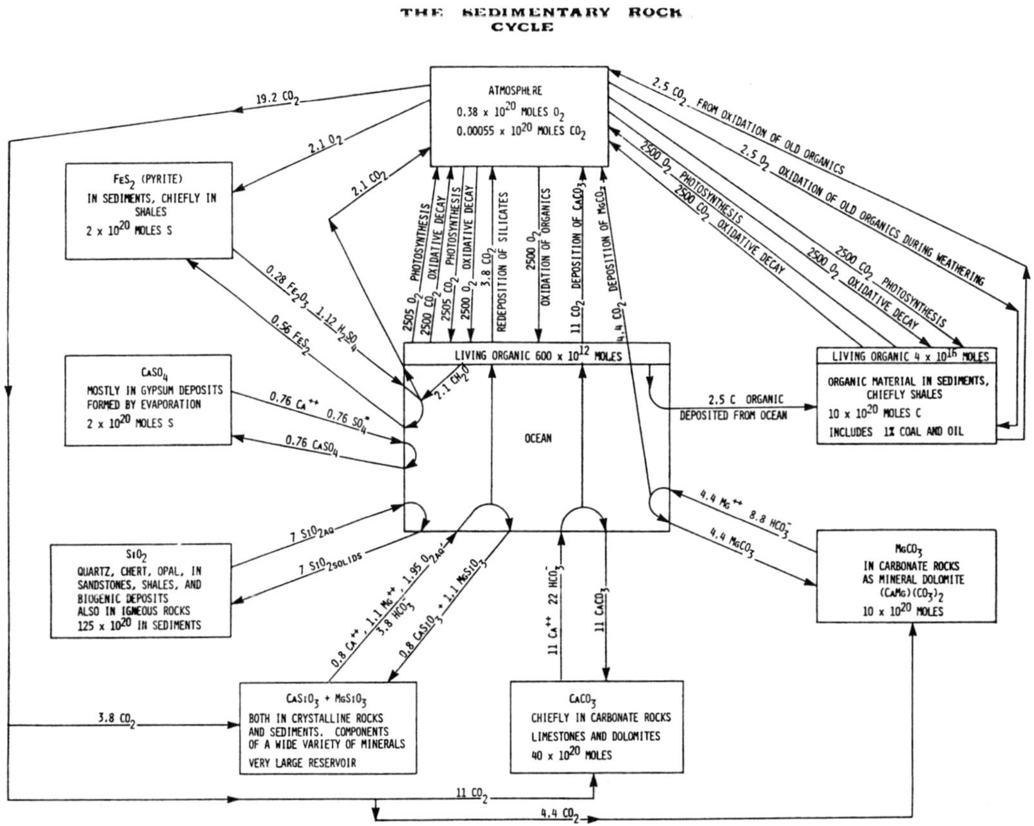

Figure 5. Schematic steady system suggesting average relations over the past few hundreds of millions of years. Not all important reservoirs are shown. Transfer of solid materials from various reservoirs to the ocean is not shown; emphasis is on transfer of dissolved materials to ocean, and on the reversal of the solution reactions and restoration of the atmosphere. Fluxes are in units of 10^{12} moles/yr. (Garrels, Mackenzie, and Hunt, 1973.)

5. COUPLING OF SEDIMENTARY RESERVOIRS

One of the important areas of progress in the last 20 years has been the development of quantitative models that interrelate mass relations among the reservoirs of chemical sediments (for example, Holser and Kaplan, 1966; Holland, 1973; Rees, 1970; Garrels and Perry, 1974; Schidlowski et al., 1977). These models were based on the concept of a constant total mass of sediments with nearly constant mean elemental composition (see section on sedimentary rock chemistry) for the last 600 or 700 million years of sedimentary cycling. The range of deviation from this assumption is not yet known, and it is still a highly controversial topic.

The sedimentary reservoirs of many chemical species are far greater than their amounts in the ocean. The ocean residence time (time required to flux into the ocean an amount of a species equal to that currently in the ocean) of most elements is less than 10 million years, whereas the residence times of elements stored in sediments characteristically is hundreds of millions of years. This relation can be restated to say that transfers of materials from one sedimentary reservoir to another over long periods of geologic time are characteristically many times the amount of those materials stored in the ocean today, as well as many times the probable variations of oceanic content through time (Holland, 1973, 1984).

The ocean, considered on a time scale of hundreds of millions of years, is a medium of transfer rather than one of element storage. This point will emerge clearly from the models described here. Figure 5 shows one set of estimated values for the reservoir sizes and the fluxes to and from them for a number of chemical sedimentary species (Garrels, Mackenzie and Hunt, 1973). As a first approximation, the system is shown in steady state, in which there is no net transfer of materials from one reservoir to another. The next step toward modeling the system is to keep ocean and atmosphere constant, and to assess the changes in reservoirs required to increase the gypsum reservoir by an arbitrary amount (Garrels and Perry, 1974). Figure 6 shows the responses required of other reservoirs in order to increase gypsum storage. It is worthwhile to follow qualitatively the reasons for the changes shown.

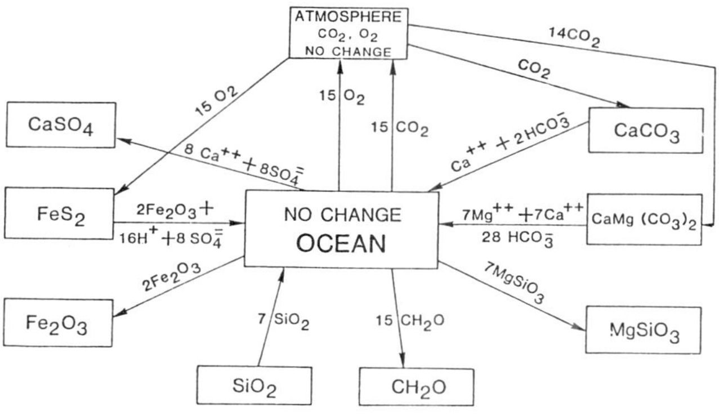

Figure 6. Changes in sedimentary reservoirs necessary to increase the size of the $CaSO_4$ reservoir by 8 units, with no change of ocean or atmosphere (all units in terms of moles; adapted from Garrels and Perry, 1974).

The deposition of 8 units of gypsum (all discussion on a molar basis) requires 8 units of Ca, 8 units of S, and 16 units of O_2. The 8 units of Ca come from the $CaCO_3$

reservoir, the 8 units of S from the pyrite reservoir, and the 16 units of O_2 from H_2O and from reduction of carbon dioxide to organic matter. The oxidation of 8 units of S and 4 units of Fe in pyrite requires reduction of 15 units of oxidized carbon (from +4 to 0 valence). Eight of these are available from the $CaCO_3$ broken up to obtain the Ca for gypsum, the 7 others are taken from $MgCO_3$ (in nature, from the $MgCO_3$ of dolomite). The 7 Mg's released are removed from the system as clay minerals (here stylized as $MgSiO_3$), which demands 7 SiO_2 from the SiO_2 reservoir. The 15 carbons in carbonate that are reduced to obtain the oxygen for oxidation of pyrite are added to the organic carbon (CH_2O) reservoir. The overall reaction for all these changes is:

$$7MgCO_3+15H_2O+4FeS_2+8CaCO_3+7SiO_2 = 2Fe_2O_3+15CH_2O+8CaSO_4+7MgSiO_3$$

Details on the individual reactions that are added to give this overall reaction are available in Garrels and Perry (1974). So far as is known now and in the absence of a significant participation by other minerals (e.g., Fe-silicates), the reaction shown, with its coefficients, is a unique solution for gypsum storage without change in ocean and atmosphere, with the single restraint that $MgCO_3$ must be used to provide 7 of the required oxidized carbons. The oxidized carbon is not taken from additional $CaCO_3$ because deposition of $CaSiO_3$ minerals from the ocean is not important, and the Ca in excess of that going into gypsum would have to be removed as a calcium silicate. However, Mg silicates do form from seawater.

This tight coupling fixes the masses transferred in and out of 8 different reservoirs, so that in principle the predicted changes could be investigated through increases or decreases in any of the reservoirs. On the other hand, any change in any of the reservoirs could be regarded as the "forcing function," and whatever was responsible for an increase or decrease in any reservoir would require the stipulated changes in all others. Greatest interest seems to have involved the reciprocal relations between the sulfide-sulfate and carbonate-organic carbon oxidation-reduction pairs, and a number of investigators have examined the system during the last 20 years (Holser and Kaplan, 1966; Rees, 1970; Holland, 1973; Garrels and Perry, 1974; Schidlowski et al., 1977, as well as many later studies).

Garrels and Lerman (1981, improved model 1984) proposed modeling the coupled sulfur-carbon systems by using the changes with time of sulfur isotopes in evaporite sulfates to drive the coupled system. They then calculated the carbon isotope changes of the carbonates of the carbonate-organic carbon system as well as the changes in size of the various sedimentary reservoirs. Using one isotope time curve to predict the other provided a check on the validity of their model.

6. THE CARBON-SULFUR MODEL

Garrels and Lerman (1981, 1984) isolated the coupled C-S system from the other reservoirs involved, and set up the system shown in Figure 7. Their article can be consulted for details of the reasons for choosing reservoir sizes, fluxes, isotope values, and other parameters. They assumed reservoir sizes and general conditions corresponding to the present. The sulfur isotope curve for evaporites published by Claypool et al. (1980) was used to drive the model backward in time, by forcing the flux and reservoir relations to reproduce an ocean (and hence gypsum deposits) with the sulfur isotope composition of sulfate deposits of known ages. They assumed that the flux of sulfate from gypsum, or from sulfide oxidized during weathering into the ocean, was simply proportional to the size of the appropriate source reservoir. The outputs from the ocean were manipulated so as to duplicate the required sulfur isotope changes of the ocean.

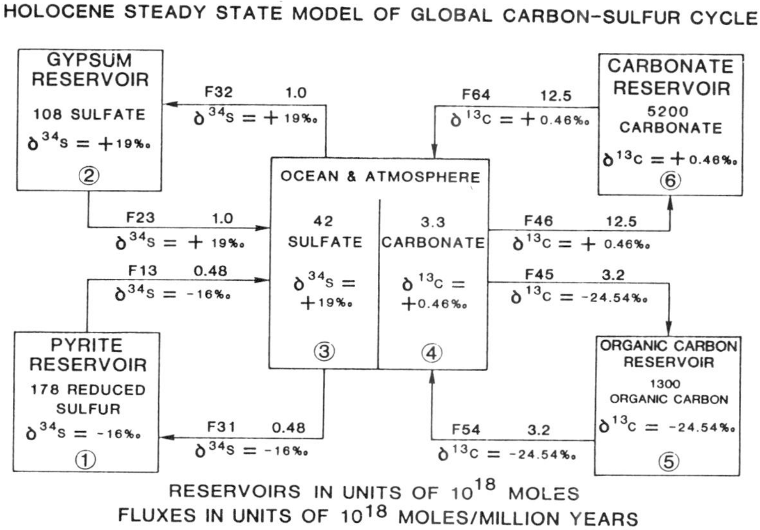

Figure 7. Some reservoirs, fluxes, and isotopic compositions in the global sulfur-carbon system (Garrels and Lerman, 1984).

Changes in the fluxes of today's models yield results strictly applicable only to the future. One technique that is used to obviate this difficulty is to assume a set of conditions for a time in the past, then run the model forward in time to the present, manipulating past conditions until present conditions are achieved by the model. Another, used by Garrels and Lerman (1984) is to run the model backward in time by the simple expedient of changing the signs of the fluxes. In other words, the flux of sulfate from evaporites to the ocean, which takes place forward in time, becomes a flux of sulfate back from the oceans to the sulfate reservoir. Streams run back to their sources, deposited gypsum is dissolved back into the ocean, and so on. The simple expedient of changing signs works for simple linear systems; the model, run backward and then forward in time, returns to all original conditions.

The general calculation procedure for running into the future is straightforward. Assume that the $\delta^{34}S$ of the oceans will be $+22‰$ 5 million years in the future, as compared to $+20‰$ today. To make the ocean isotopically heavier, it is necessary to remove light sulfur selectively. Evaporites have the same $\delta^{34}S$ as the ocean from which they came; sulfides formed by reduction of oceanic sulfate typically are about $35‰$ lighter than the oceanic sulfate. To achieve a heavier ocean 5 million years hence, the amount of sulfur going into the ocean from sulfide weathering is exceeded by the sulfide output; the sulfate coming in is correspondingly more than the sulfate removed. The ocean becomes heavier isotopically, the sulfide reservoir increases in mass, the sedimentary sulfate reservoir decreases. The sizes of the fluxes of sulfur into the ocean are maintained proportional to these masses; the fluxes of sulfur out of the ocean are adjusted to give the required isotopic change of the ocean.

As a result of their modeling, Garrels and Lerman obtained values for the past 700 million years of the fluxes of sulfate into and out of the ocean; they also produced numbers for the sizes of the sulfide and sulfate reservoirs. Then, using the relation from reservoir coupling that each mole of increase of sulfate in the gypsum reservoir corresponds to an increase of 15/8 moles of organic material in the organic carbon reservoir (and a corresponding decrease of carbonate in the carbonate reservoir), they calculated a curve for the $\delta^{13}C$ values for oceanic carbonate for Phanerozoic time and compared it with the observed curve.

Figure 8 represents a variant of this general technique. The size of the gypsum reservoir during Phanerozoic time was calculated separately from the sulfur isotope curve and from the carbon isotope curve. The qualitative trends are similar, and give credence to the conclusion that material transfers between reservoirs can be obtained from the coupling of sediment reservoirs.

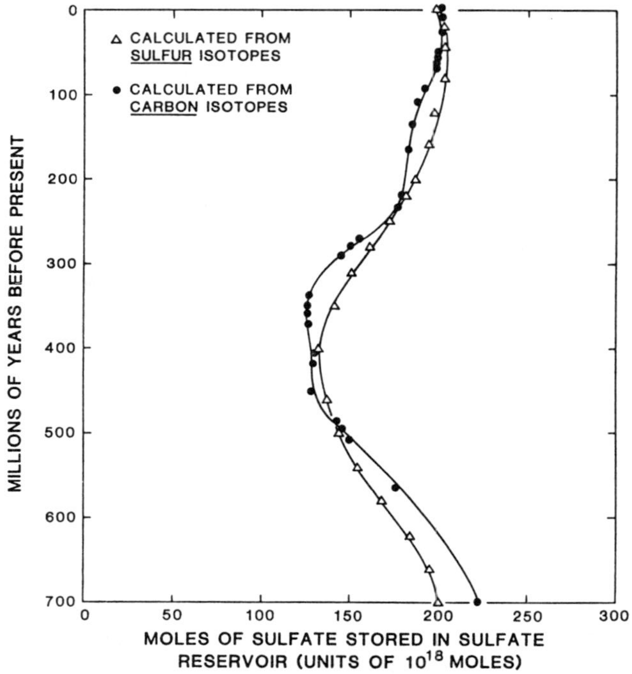

Figure 8. Comparison of size of sulfate reservoir through time calculated from carbon isotopes to that calculated from sulfur isotopes (Garrels and Lerman, 1984).

7. DISCUSSION

The success of the carbon-sulfur model, if only qualitative, has some far-reaching consequences. Even though it was necessary to average flux rates over very long time intervals (about 20×10^6 yrs), the model seems to be one of the first geochemical demonstrations of the principle of Uniformitarianism. To obtain the carbon and sulfur isotope curves it is not necessary to postulate remarkable deviations in flux rates of carbon and sulfur species into and out of the oceans. The system seems closely constrained by various feedbacks. The success of the model also bolsters the conclusion that carefully chosen values of $\delta^{34}S$ and $\delta^{13}C$ for $CaSO_4$ and $CaCO_3$ deposits indeed reveal oceanic conditions of the past, and that these isotope ratios can be used as "ground truth" checks on models.

The good response of the model also supports the idea that atmospheric oxygen has been well buffered through time. The model assumed a constant ocean and atmosphere. Any excess oxygen generated by an increase of the burial of organic carbon was assumed to be immediately consumed by oxidative weathering of sulfide and of fossil carbon. In more recent modeling (Kump and Garrels, 1986) an increase in oxygen production rate is accompanied by an increase in atmospheric oxygen levels, as well as by an increase in the rate of sulfide oxidation. The close coupling of the sulfur-carbon system limits the levels to which oxygen can rise.

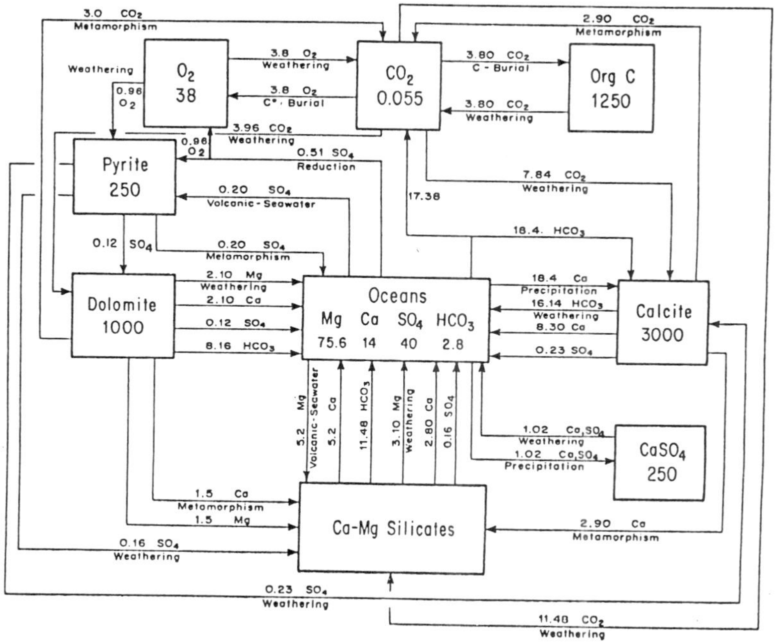

Figure 9. Box model used by Lasaga et al. (1985) to study atmospheric CO_2 changes with time. Reservoirs are in units of 10^{18} moles, fluxes in 10^{18} moles per million years. Comparison with Figure 5, constructed 12 years before, gives an estimate of the uncertainties in the numbers used.

8. WORK IN PROGRESS

A great deal of paleoclimate reconstruction is currently in progress. General interest has stimulated biologists to increase their efforts to reconstruct climates from the fossil record of the biota. Classical climate modeling, as practiced by atmospheric scientists, is being applied more and more to predicting both future and past (can one predict the past, or does one postdict it?) climates. Geochemical modeling, like that described here, has quickly become complex. Lasaga et al. (1985) have developed a model that takes into consideration atmospheric CO_2 levels, temperature, sea floor spreading rates, flow of seawater through ridge crests, CO_2 production in soils, volcanic and metamorphic activity. This model (Figure 9) is being expanded to include the carbon-sulfur system in detail, and attempts are being made to model all of Phanerozoic time instead of just the last 100 million years. Many advances will be made in our understanding of the earth's paleoenvironment as more and better data accumulate on sedimentary rock mass-age relations, stable isotope-age curves, isotope fractionation factors, diagenetic changes versus primary sediment mineralogy, and composition.

9. ACKNOWLEDGMENTS

Various aspects of this work have been supported by NSF grant 83-06390, and by a grant from NASA NGWA-339 to Northwestern University. As revealed by the references in the text, most of the material presented came from collaboration with F. Mackenzie, A. Lerman, and the late E.A. Perry, Jr. In addition, I am especially indebted to R.A. Berner and A. Lasaga, who have not only reviewed and corrected various early manuscripts, but who also have collaborated with me in recent work. My research assistant, Terri Woods, and my graduate assistant, Lee Kump, also have contributed ideas and sweat.

10. REFERENCES

Burst, J. F., 1959, Post-diagenetic clay mineral-environmental relationships in the Gulf Coast Eocene in clays and clay minerals, *Clays Clay Min.* **6**, 327-341.

Cameron, E. M. and Garrels, R. M, 1980, Geochemical compositions of some Precambrian shales from the Canadian Shield: *Chem. Geol.*, **28**, 181-197.

Claypool, G. E., Holser, W. T., Kaplan, I. R., and Zak, I., 1980, The age curves of sulfur and oxygen isotopes in marine sulfate and their mutual interpretations: *Chem. Geol.*, **28**, 199-260.

Garrels, R. M. and MacKenzie, F. T., 1971a, *Evolution of Sedimentary Rocks*: Norton and Co., New York, 500 pp.

Garrels, R. M. and MacKenzie, F. T., 1971b, Gregor's denudation of the continents: *Nature*, **231**, No. 5302, 382-383.

Garrels, R. M., MacKenzie, F. T. and Hunt, C., 1973, *Chemical Cycles and the Global Environment*, William Kaufmann, Inc., Los Altos, CA, 206 pp.

Garrels, R. M. and MacKenzie, F. T., 1974, Chemical history of the oceans deduced from post-depositional changes in sedimentary rocks, *in* Hay, W. W., ed., Studies in Paleo-Oceanography, *Soc. Econ. Paleo. Mineral. Spec. Pub.* **20**, p. 193-204,

Garrels, R. M. and Perry, E. A., Jr., 1974, Cycling of carbon, sulfur, and oxygen through geologic time, *in* Goldberg, E. D., ed., *The Sea, 5*, pp. 303-336, John Wiley and Sons, New York.

Garrels, R. M. and Lerman, A., 1981, Phanerozoic cycles of sedimentary carbon and sulfur, *Proc. Natl. Acad. Sci.*, **78**, 4652-4656.

Garrels, R. M. and Lerman, A., 1985, Coupling of the sedimentary sulfur and carbon cycles—an improved model, *Amer. Jour. Sci.*, **284**, 989-1007.

Cole, M. J. and Klein, C., 1981, Banded iron formations through much of Precambrian time, *J. Geol.* **89**, 169-183.

Gregor, C. B., 1985, The mass-age distribution of Phanerozoic sediments: *in* Snelling, N. J., ed., *The Chronology of the Geological Record*, pp. 284-289, Blackwell, Oxford.

Hardie, L. A., 1984, Evaporites: marine or non-marine?, *Amer. Jour. Sci.* **284**, 193-240.

Holland, H. D., 1973, Systematics of the isotopic composition of sulfur in the oceans during the

Phanerozoic and its implications for atmospheric oxygen, *Geochem. Cosmochim. Acta,* **37**, 2605-2616.

Holland, H. D., 1984, *The Chemical Evolution of the Atmosphere and Oceans*, Princeton Univ. Press, Princeton, N. J., 582 pp.

Holser, W. T. and Kaplan, I. R., 1966, Isotope geochemistry of sedimentary sulfates, *Chem. Geol.,* **6**, 93-135.

Kump, L. R. and Garrels, R. M., 1986, Modeling atmospheric O_2 in the global sedimentary redox cycle, *Amer. Jour. Sci.* **286**, 337-360.

Lasaga, A. C., Berner, R. A. and Garrels, R. M., 1985, An improved geochemical model of atmospheric CO_2 fluctuations over the past 100 million years, *in* Sundquist, E. T. and Broecker, W. S., eds., The Carbon Cycle and Atmospheric CO_2: Natural Variations Archean to Present, *Geophysical Monograph* **52**, Amer. Geophys. Union, Washington, D. C., p. 397-411.

Li, Y. H., 1972, Geochemical mass balance among lithosphere, hydrosphere, and atmosphere: *Am. Jour. Sci,* **272**, 119-137.

Perry, E. and Hower, J., 1970, Burial diagenesis in Gulf Coast pelitic sediments, *Clays Clay Min.* **18**, 165-177.

Rees, C. E., 1970, The sulfur isotope balance of the ocean: an improved model: *Earth Planet. Sci. Lett.,* **7**, 336-370.

Rubey, W. W., 1951, Geologic history of sea water: an attempt to state the problem: *Geol. Soc. Amer. Bull.,* **62**, 1111-1147.

Sandberg, P. A., 1975, New interpretations of Great Salt Lake ooids and of ancient non-skeletal carbonate mineralogy: *Sedimentology,* **22**, 497-537.

Schidlowski, M., Junge, C. E. and Pietrik, N., 1977, Sulfur isotope variations in marine sulfate evaporites and the Phanerozoic oxygen budget, *J. Geophys. Res.,* **82**, 2557-2565.

Valeton, I., 1972, *Bauxites*, Elsevier Pub. Co., 226 pp.

Veizer, J. and Jansen, S. L., 1979, Basement and sedimentary recycling and continental evolution, *Jour. Geol.,* **87**, 341-370.

Walker, J. C. G., 1977, *Evolution of the Atmosphere*, Macmillan Pub. Co., New York, 318 pp.

SOLID EARTH AS A RECYCLING SYSTEM: TEMPORAL DIMENSIONS OF GLOBAL TECTONICS*

Ján Veizer
Derry Laboratory
Department of Geology
University of Ottawa
Ottawa, Ontario K1N 6N5
Canada

ABSTRACT

Plots of mass (volume, area, tonnage) against time for geologic entities—for example, continental and oceanic crust, sediments, and mineral resources—display an exponential or a power law relationship, with entity per unit time increasing toward the present. This relationship is analogous to that of population dynamics, where the recycling of an entity goes through continuous birth and death cycles. The discussed approach is based on the plate tectonic theory, and it considers area-age or mass-age distributions of crystalline basement and sediments for major global tectonic realms. Each tectonic realm is characterized by a specific life-span (τ_{MAX}), which is an inverse function of its recycling rate. The estimated average half-area or half-mass ages (τ_{50}) are the following: basins of active margins ~30 Ma, oceanic sediments ~40 Ma, oceanic crust ~55 Ma, basins of passive margins ~80 Ma, immature orogenic belts ~100 Ma, mature orogenic belts ~380 Ma, and platforms ~380 Ma. The τ_{50}'s for continental crust are ~690 Ma for K/Ar, ~1200 Ma for Rb/Sr and U-Th/Pb, and <1800 Ma for Sm/Nd dating pairs. The estimated maximal survival (or oblivion) ages τ_{MAX} for these tectonic realms are ~3.0-3.5 times longer than their τ_{50}. Tectonic diversity preserved in the geologic record is therefore a function of time, with oceanic tectonic realms, because of their rapid recycling, underrepresented in the rocks older than ~300 Ma. Similarly, the relative proportions of lithological types of sediments in crustal segments of different ages are, in the first instance, a reflection of this tectonic recycling because of the preferential association of specific facies assemblages with a given tectonic realm. The rate of generation or destruction of global oil reserves indicates τ_{50} of ~85 Ma, and the τ_{50} for post-Paleozoic coals is ~250 Ma. Reserves of metals reflect the fact that the genetic types of deposits, and thus their tectonic associations, evolved during geologic history. The preserved age distribution patterns are complex. The apparent rates of geological processes are inversely proportional to the duration of the utilized time intervals. Therefore the "excessively high" rates of modern erosional processes, attributed usually to anthropogenic causes, may be only a reflection of the methods of the estimation.

---

*Publication 18-85 of the Ottawa-Carleton Centre for Geoscience Studies.

A. Lerman and M. Meybeck (eds.), Physical and Chemical Weathering in Geochemical Cycles, 357–372.

1. INTRODUCTION

The dynamics of release, transport, and sequestering of an element or isotope in surficial terrestrial environments, that is its geochemical cycle, is controlled by the rates of physical and chemical weathering of rocks. Weathering rates are, in turn, a function of the exposure of rocks to surficial weathering agents, and it is mechanical weathering that dominates erosional processes. If the present-day ratio of suspended to dissolved load of world rivers is taken as a guide to the relative importance of physical vs chemical weathering, the former accounts for ~80% of the total (Martin and Meybeck, 1979; Milliman and Meade, 1983). The integrated rate of mechanical erosion, and ultimately the suspended load of rivers, is directly proportional to the average height of continents (Garrels and Mackenzie, 1971a; Southam and Hay, 1981). Thus, uplift of potential source areas is the rate limiting step of mechanical denudation. It is therefore essential to understand the periodicities and the rates of mountain building (orogenic) cycles responsible for the formation and destruction of elevated structural features. This chapter deals with the temporal dynamics of the global tectonic framework, with the consequences of this approach for differentiation of cyclic from evolutionary phenomena, and with the signatures preserved in the available rock record. The theoretical concept underlying all subsequent discussion is that of population dynamics.

2. AN INTRODUCTION TO POPULATION DYNAMICS

A natural population, characterized by continuous birth and death cycles, is usually typified by an age distribution similar to that shown in Figure 1. The proportion of progressively older constituent units, such as human individuals, decreases exponentially, because mortality is usually a first-order function of the size of a given age group. The internal age distribution of the population can be expressed either as a histogram or a cumulative curve. The latter is the preferred representation for our purposes because it defines all the necessary parameters for a given population. These are its size A_0, half-life τ_{50}, mean age τ_{MEAN}, and oblivion age τ_{MAX} (Figure 1). Mathematically (e.g., Lerman, 1979)

$$A_t = A_0 e^{-kt} \tag{1}$$

where A_t is the cumulative fraction of the surviving population older than t, t is age (not time), A_0 equals unity, and k is the rate constant for the net birth and death process. In the subsequent text, the birth and death rate is discussed in the form of a recycling proportionality constant b, which is related to the above formalism as

$$b = 1 - e^{-kT}$$

where T is the total duration of recycling (see Veizer and Jansen, 1979, 1985, for details). In general, the larger the b, that is the higher the mortality rate, the steeper the slope of the cumulative curve and the shorter the τ_{50}, τ_{MEAN} and τ_{MAX} of the population. In other words, the chance of survival for older individuals diminishes with increasing mortality rate. For a steady-state population, natality per unit time equals combined mortality for all age groups during the same time interval. Consequently, the cumulative slope remains the same but propagates into the future (Figure 1, bottom).

The above exponential relationship is a general expression of population dynamics for ideal populations. In nature, populations of many species have large numbers

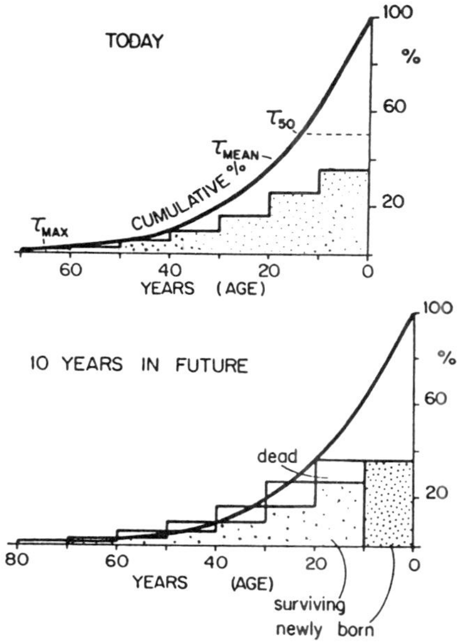

Figure 1. Simplified internal age distribution for a steady-state living system. τ_{50}, τ_{MEAN}, τ_{MAX} as in the text. In this case, the natality/mortality rate is 35% of the total population for a 10 year interval ($b = 35 \times 10^{-3}$ year^{-1}). τ_{MAX} is defined as the 5th percentile. Reproduced from Veizer and Jansen (1985).

of young individuals (e.g., planktonic larval stages) because their mortality is very high. The chances for survival improve considerably only with the attainment of an advanced growth stage. Such populations, if compared to the ideal ones, are characterized by shorter τ_{50} and τ_{MEAN}, but their τ_{MAX} may be comparable (Figure 2). Mathematical formalism for such populations is a power law function

$$A_t = A_0(1 + kt)^{-z} \qquad (2)$$

where the power exponent z increases for populations with progressively larger mortality rates of young individuals. Another common exception are populations with smaller than usual mortality of young individuals, such as in the advanced industrial nations. In these instances the death rates increase rapidly as the life-span τ_{MAX} is approached (Figure 2). Mathematical expression for this relationship is

$$A_t = A_0 \left[1 - (kt)^z\right] \qquad (3)$$

where the power exponent z increases for populations with progressively larger mortality rates of old individuals.

The above relationships are valid for steady-state populations, but not necessarily for the growing or declining ones. For populations characterized by overall rates of growth or decline much slower than the rates of mortality and natality of its constituents, the steady-state solutions may still be applicable. This is not so for populations where the rates of overall growth and decline approach those of the internal recycling. In such a case, it is permissible to interpret the relationship expressed by

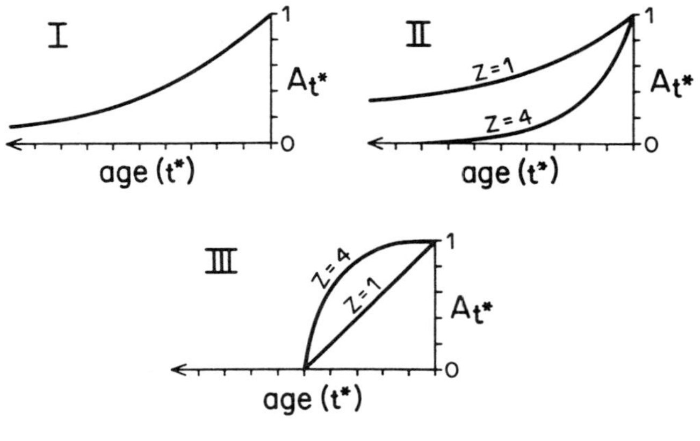

Figure 2. Cumulative age distribution functions for various populations. I=exponential distribution (eq. 1), II=power law distribution approaching zero assymptotically (eq. 2), III=age distribution attaining zero at finite time. For explanation of z see the text. Modified from Lerman (1979); Veizer (in press b).

equation (2) either as indicating a high mortality of young individuals within a steady-state population, or as representing a population which expanded its overall size by excessive births during its latest history. Similarly, the relationship approximated by equation (3) may be a consequence of greater mortality of old constituents or of the fact that the population was characterized by fast growth rates in its early evolutionary stages. The distinction between these two limiting alternatives, and their combinations, can be made only with the knowledge of additional constraining parameters.

The principles of population dynamics are well established for biological systems (e.g., Deevey, 1947; Slobodkin, 1962; MacArthur and Connel, 1966; Odum, 1971; Hallam, 1972; Valentine, 1973; Hutchinson, 1978; Dodd and Stanton, 1981; Raup, 1985) and the concept in its formalistic expression is widely, frequently unconsciously, utilized in such diverse fields as astronomy (birth and death of stars), insurance industry (survivorship tables), religion (reincarnation of souls into "new" bodies), or banking (populations = reserves; constituent units = units of currency, and birth and death rates = credit and debit, or cash flow). Because of its fundamental validity, the mathematical formalism of population dynamics is applicable also to the major units of the earth, including its atmosphere and hydrosphere (Veizer, in press b) and the solid earth (Veizer and Jansen, 1985).

3. GLOBAL TECTONIC REALMS AND THEIR RECYCLING RATES

The concept of global plate tectonics (Hess, 1962; Dietz, 1961; Morgan, 1968; Le Pichon et al., 1973) combined the earlier ideas of continental drift and seafloor spreading into a unified theory of earth dynamics. It introduced the notion of continual generation and destruction of oceanic crust and implied similar consequences for other tectonic realms. One possible classification of global tectonic realms, reflecting to some extent the restrictions of the available data base, encompasses the following tectonic entities (Figure 3):

a) oceanic crust and its overlying abyssal and pelagic sediments,
b) active margin basins,
c) passive margin basins,
d) immature orogenic belts (young mountain ranges),
e) mature, worn-down, orogenic belts,
f) platforms, and
g) cratons (here taken as continental crystalline and metamorphic basement).

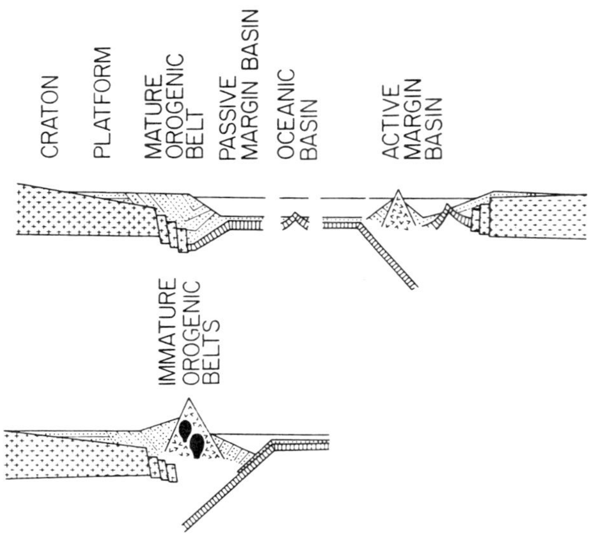

Figure 3. Schematic presentations of major tectonic realms in the context of global plate tectonics.

Age distributions of rocks in these tectonic realms (Figure 4) correspond well with the patterns espoused in Figures 1 and 2, suggesting that we are indeed dealing with populations characterized by a quasi-continuous generation and destruction of the constituent rock units. Assuming, as a first order approximation, that all deviations from strictly exponential relationships of equation (1) were due to unsatisfactory data base, Veizer and Jansen (1985) calculated the relevant recycling proportionality constants (b's) and estimated the half-lives (τ_{50}'s) for the studied tectonic realms. These results are summarized in Figure 5 and Table I. They are consistent with the existence of two fundamental tectonic domains, the continental and the oceanic one, with a transition at b $\sim 40 \times 10^{-10}$ a^{-1} and τ_{50} $\sim$225 Ma. The oceanic domain, encompassing tectonic realms (a) to (d), represents a relatively fast cycling system with τ_{50} in 10^7-10^8 years time range. In contrast, tectonic realms of the continental domain, (e) to (g), are being recycled on time scales of 10^8-10^9 years. The calculated parameters also indicate that mountain building (orogenesis), although a result of interaction of these two domains, is a response to the operation of the oceanic cycle. If so, seafloor spreading causes orogenesis and subsequent erosion of mountains back into the ocean. The overall consistency of the available data with the concept of population dynamics also leads to prediction of the probable oblivion age for each tectonic realm. Empirically, τ_{MAX} is $\sim$3.0-3.5$\times\tau_{50}$. Consequently, the degree of tectonic diversity observed in the geologic record is a function of time. This diversity diminishes as the given segment of solid earth ages and the rate of memory loss is inversely proportional to recycling rates (b's) of the constituent tectonic realms.

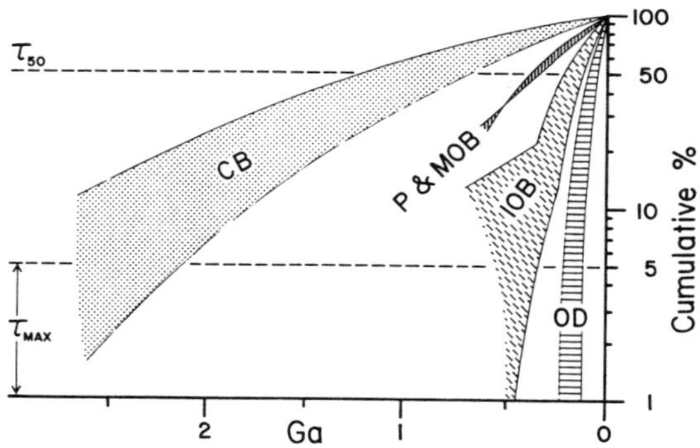

Figure 4. Observed mass-age cumulative distributions of sediments, and area-age distributions of oceanic and continental crusts, within tectonic realms as represented in Figure 3. Note that in this, and all subsequent figures, cumulative percentages are plotted on a $\log_{10}$ scale. Consequently, an ideal population (eq. 1) will plot as a straight line. Explanations: CB—continental basement, P & MOB—platforms and mature orogenic belts, IOB—immature orogenic belts, OD—oceanic domain encompassing, from left to right, passive margin basins, oceanic crust, oceanic sediments, and active margin basins. Based on the data summarized in Veizer and Jansen (1985).

If one accepts that the quality of data is sufficient for demonstration of nuances in internal age distributions for these tectonic realms, it appears that almost all distributions (Figure 4) conform to the mathematical formalism (3) rather than (1). The sole possible exception may be platforms. This may suggest enhanced rates of destruction ("mortality") as the constituent rock units age. Such an interpretation is in accord with our understanding of the mode of destruction (subduction and/or obduction) of the fast cycling oceanic crust and its overlying sediments. This destruction process is localized preferentially at old leading, and not young trailing, edges of the oceanic plates. For slowly cycling systems, however, the observed age distributions more likely indicate higher growth ("natality") rates during the youthful stages of the evolution of these populations. Such may be the explanation for the continental crust, which may have been generated at higher rates during the early history of the earth, and particularly during the interval of $\sim$3-2 Ga ago (Veizer and Jansen, 1979; Allègre, 1982; Taylor and McLennan, 1985). The geologic meaning of the observed second order patterns for the remaining tectonic realms, if real, is not yet understood.

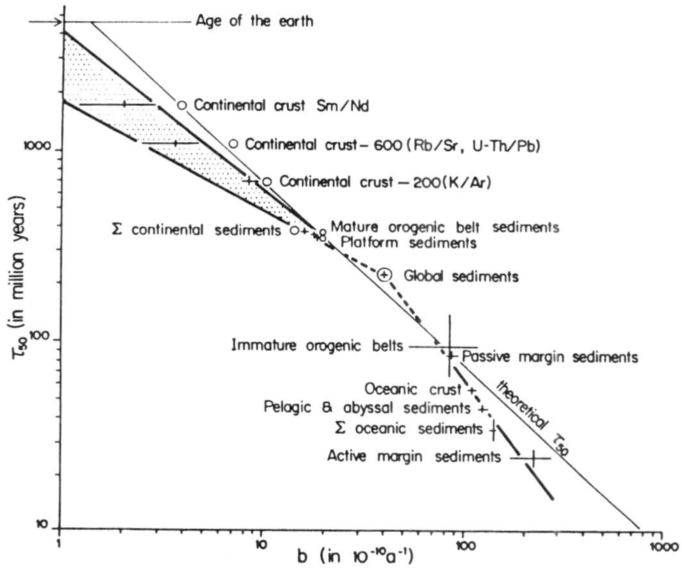

Figure 5. Plot of the observed half-mass and half-area ages (τ_{50}) and of recycling rates (b) for major global tectonic realms. Crosses represent solutions for the "preferred" model of continental growth, where it is assumed that continents started to grow (or started to be preserved) at ~4000 Ma ago. This was followed, ~3000-2000 Ma ago, by a fast growth rate and subsequently the continents have been of a near present-day size. The circles give results for the steady-state alternative. The two limiting solutions are identical at $b > 40\times10^{-10}$ For definition of b see the text and Figure 1. Continental crust -200, -600 represent solutions for the bottom and the top enveloping curves in Figure 4. The Sm/Nd curve is not shown here. Reproduced from Veizer and Jansen (1985).

4. PRESENT-DAY RATES OF CONTINENTAL EROSION

The present-day particular and dissolved loads for the world rivers are 16×10^{15} a^{-1} and 5.6×10^{15} g a^{-1}, respectively (Martin and Meybeck, 1979; Milliman and Meade, 1983). The value for the dissolved load is only that derived from rock weathering. This total flux of $\sim22\times10^{15}$ g a^{-1} exceeds considerably the geological average calculated for recycling of continental sediments ($3.0\pm0.3\times10^{15}$ g a^{-1}; Table I). Some of this discrepancy may be caused by the fact that igneous and metamorphic rocks supply additional material into weathering cycle. This contribution, however, must be relatively minor, because the rates of weathering of igneous and metamorphic rocks are much slower than those for sediments and because about 2/3 of the present-day continental area is covered by sediments (Blatt and Jones, 1975). Furthermore, metamorphism and anatexis of sediments will counteract the above process. In order to explain the high present-day erosion, it has been proposed that it reflects: (a) anthropogenically induced faster rates of erosion and transport (Judson, 1968; Garrels and Mackenzie, 1971a, b; Gregor, 1980), or (b) exceptionally high rates of Cenozoic

Table I. Sizes of reservoirs, their recycling rates, fluxes and half-lives (from Veizer and Jansen, 1985)

Reservoir	Size ($\times 10^{21}$ g)	"Preferred" recycling constant b ($\times 10^{-10}$ a^{-1})	Absolute quantity recycled ($\times 10^{15}$ g a^{-1})	Observed τ_{50}* (Ma)
Oceanic crust	6118	110 ± 5	67.3 ± 3.1	~55
Continental crust	20900 [200 × 10^6 km^2]			
I recycled via mantle (Sm/Nd)		2.1 ± 1.0	.04 ± .02 km^2 a^{-1}	≤1800
II involved in high-grade metamorphism (≥600°C) + I		3.7 ± 1.3	.07 ± .03 km^2 a^{-1}	~1200
III involved in low-grade metamorphism (≥200°C) + I + II		8.5 ± .7	.17 ± .1 km^2 a^{-1}	~690
Sediments				
global sediments	2701	40 ± 3	10.8 ± .8	~235
continental sediments	1882	16.2 ± 1.1	3.0 ± .3	~375
oceanic sediments (minus passive margins)	376	142	5.3	~45
Global tectonic realms (sediments)				
abyssal and pelagic	217	126	2.7	~40
active margin basins	144	223 ± 51	3.2 ± .9	~30
deep sea fans	15	230 ± 55	0.3 ± .1	~25
passive margin basins	337	88 ± 8	3.0 ± .2	~80
immature orogenic belts		85 ± 35		~100 (50–190)
mature orogenic belts	≤1350	18.4 ± .5	≤2.5	~380
platforms	540	18.6 ± .5	1.0	~380

weathering, as reflected in the unusually thick accumulations of deep sea sediments for at least the last 10^7 years (Davies et al., 1977; Southam and Hay, 1981). In this alternative, the high erosional rates clearly precede the emergence of technological societies.

While not discounting the above explanations, I would like to point out that the apparent discrepancy may be a consequence of methodology. In a fully or partially cannibalistic recycling system (such as represented by continental sediments) the rate of recycling, and thus of erosion, is inversely proportional to the length of the time interval under consideration (Veizer and Jansen, 1985; see also Garrels, this volume). Taking the example of human population, a census at 10 year intervals will not even register all individuals who were born and died before their first census. A biennial or yearly census will miss progressively fewer individuals. Consequently, the rate of birth or death per annum will increase for shorter time intervals. In the geologic context, the river flux calculated, for example, from the quantity of sediments deposited over 10^7 years time interval measures only the net flux. Yet, in the course of such long time increment, some of the sediments could have been repeatedly deposited in the sea, uplifted, eroded, and transported again by rivers. The island of Taiwan is a good example of such cannibalism (see also R.F. Stallard, this volume). Due to compression from the Philippine plate, the eastern coast is rapidly and continually uplifted and the erosion products channeled westward into the Taiwan Strait. However, the uplift itself migrates westward, thus cannibalising its former erosional products. Consequently, an instantaneous river flux must exceed considerably the long-term net flux, because the latter does not account for recycling.

The above situation represents a quasi continuous birth and death cycle. For discrete events, typical of the bulk of the geologic record, this inverse relationship is even more pronounced. As pointed out already by Barrell (1917), sedimentary process at any given time is typified by deposition, erosion, and non-deposition. Erosion and non-deposition account for progressively larger portions of the total time span for longer time intervals. This again is analogous to random walk in human population. Consider a nomadic hunter and his mode of life. He moves forward (deposition), rests (non-deposition), and frequently retraces his steps (erosion). On an active morning, his forward rate of movement is high, but as the day advances he rests more and more, culminating with daily sleep. Over months, he may even stay put in a camp for days, or move backwards. Thus, knowing only his initial and final position and the length of time required to cover this distance, the calculated average net rate of movement will be inversely proportional to time resolution. Rest and retrograde movement account for progressively larger portion of the total time interval. This inverse relationship reaches a plateau at a time when the hunter completes the cycle of yearly migration and returns to his starting camp. Stated in other words, the time span under consideration now exceeds the periodicity of the event. Consequently, observation over 2, 3 or 10 years yields comparable net rates of movement per year. Direct geologic confirmation of this relationship has been furnished by Sadler (1981).

In summary, sedimentary and, hence, erosional rates (calculated usually as total thickness of sediments divided by the duration of the time interval they represent) decrease with increase in the length of the observational interval (Gingerich, 1983; Veizer, 1984). For example, a total thickness of sediments for 5 million years divided by 5 is not the same as the thickness for 50 million years divided by 50. For the same population, the calculated mean rate of sedimentation must be slower in the latter case. At this stage, the possibility that the factor of 7 faster present-day ("instantaneous") rates of continental erosion, if compared to these of the geological past, are not simply a consequence of a 10^6-10^7 years shorter observational time interval should not be discounted (see also R.H. Meade, this volume).

5. GEOLOGICAL CONSEQUENCES

The concept of differential tectonic recycling introduced in the preceding text has direct consequences on age distribution patterns of associated rocks, mineral resources, paleontological record, and other related variables. Here, only the general features of the first two topics will be reviewed.

5.1. First-Order Age Distribution of Sedimentary Lithological Types

From extensive studies of the Russian Platform and its marginal geosynclines, as well as from complementary data from other continents, the Russian geologists (Ronov, 1964, 1980; Strakhov, 1964, 1969) were able to demonstrate that the present-day relative proportions of sedimentary lithologies vary with geological age. The existing compilations, although somewhat schematic and generalized, show that progressively older sequences have increasingly higher ratios of clastic:carbonate:evaporite rocks. This was interpreted by one school as reflecting a secular evolution of the outer spheres of the Earth (Ronov, 1964; Strakhov, 1964, 1969; Cloud, 1969). An opposing view was championed by Garrels and Mackenzie (1971a), who suggested that this apparent evolutionary trend is a reflection of differential recycling rates, with evaporites being recycled faster than carbonates and carbonates faster than clastics. This increasing survival rate was attributed to decreasing susceptibility of such rocks to chemical (and to some degree mechanical) weathering. An intermediate position was taken by Veizer (1973, in press a), who attributed the observed age distribution of lithologies to a combination of primary and secondary factors. The discussion of the topic is, despite the imperfections of the data, more instructive with a somewhat finer lithological subdivisions (Figure 6).

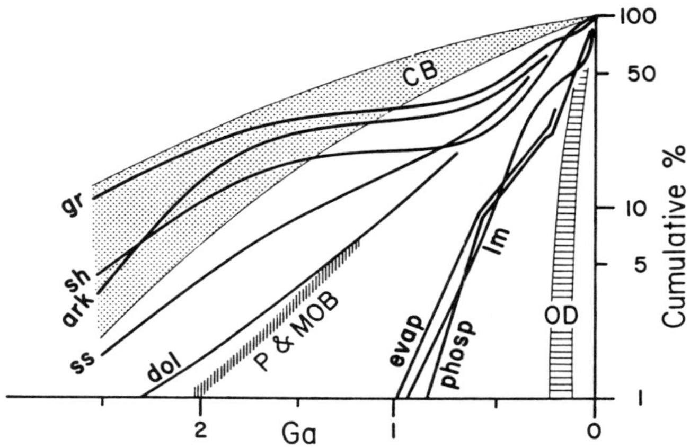

Figure 6. Observed cumulative mass-age distributions of major sedimentary lithological types. Carbonates and clastics after Ronov (1964, fig. 1), evaporites from Holser (1984), and phosphorites after Cook and McElhinny (1979). Explanations: gr=graywackes, sh=shales, ark=arkoses, ss=sandstones, dol=dolostones, evap=evaporites, lm=limestones, phosp=phosphorites. CB, P & MOB, and OD as in Figure 4.

The rate of continental erosion is essentially controlled by relief, which is a function of the size of continents (Garrels and Mackenzie, 1971a; Southam and Hay, 1977). Mechanical erosion is therefore an order of magnitude more important weathering process than is chemical erosion (Meybeck, 1979). Such mechanical disintegration of sediments usually leads to removal of all lithologies en masse. In a layer-cake stratigraphy, any selective removal of carbonate and/or evaporite strata must result also in the collapse and removal of the interlayered clastics. Salts perhaps can be squeezed out selectively through diapirism during deep burial and this may be a partial reason for their apparently rapid recycling (Figure 6). However, limestones and phosphorites are not subject to diapirism, yet their age slopes are comparable to that for evaporites. Thus, if recycling were a cause of the observed age distribution of lithologies, such recycling must have been related to the preferential association of sediment types with specific tectonic realms.

Evaporites, limestones and phosphorites indicate τ_{50} of ~150 Ma (Figure 6). This rate is about 2.5 times faster than the rate of recycling for platforms and mature (worn down) orogenic belts and ~2.5 times slower than the recycling rates characteristic of the oceanic domain, including its passive and active margin basins. Such present-day age distribution pattern may be a consequence of a possible 50:50 split of evaporites, limestones and phosphorites between intracontinental and oceanic (marginal basins) depositional environments. In contrast, dolostones appear to have been associated mostly with intracontinental (platforms, worn down orogenic belts) tectonic realms. The age curves for clastic lithologies, however imperfect the data, indicate a type of distribution expressed mathematically by equation (2), that is of enhanced mortality for young sediments. During the Phanerozoic, they appear to have been associated mostly with intracratonic settings, while older sediments display shallower slopes, comparable to those of the continental basement. Since the expected maximal life-span τ_{MAX} of intracratonic tectonic realms is ~1300 Ma (Veizer and Jansen, 1985), these sediments can be preserved beyond such time limits only if incorporated into the basement. Following such incorporation, their rate of recycling is similar to that of their host tectonic realm.

The previous discussion suggests that the first-order age distribution of sedimentary lithological types is indeed a consequence of recycling, but the rates of recycling are controlled by tectonism and not by lithologies. It must be, however, emphasized that the acceptance of the recycling concept should not be construed as a proof for steady-state and equated with dogmatic uniformitarianism. The recycling concept does not preclude simultaneous evolution of first and higher order significance, such as the role of evolving biota in the formation of different carbonate facies (Wilson, 1975; James, 1979). Such evolutionary phenomena may be responsible, for example, for the observed age distribution patterns of clastic rocks. Their chemical, mineralogical and tectonic stability increases from greywackes and arkoses to shales and sandstones, yet their apparent preservation potential is a reverse of this order (Figure 6). This reversal is understandable if the overall tectonic evolution of the earth is taken into account (Windley, 1984). The Archean earth was dominated by tectonic environments of oceanic domain (greenstone belts) and this stage was followed by a pronounced interval of Late Archean to Early Proterozoic cratonization (Cloud, 1976; Engel et al., 1975; Veizer, 1983). The early, pre-2.5 Ga, tectonic environments were therefore dominated by the immature greywacke-argillite facies, while the post-Archean ones were typified by the mature sandstone-shale as well as by the immature arkose-rich assemblages derived from granitoids. Because of the high original abundance of the different facies subsequent recycling did not yet obliterate the inherited differences. As a consequence, slowly cycling systems, such as the continental crust, still preserve the vestiges of the earth's early terrestrial record. It must be pointed out, however, that the discussed data were derived mostly from exposed continents. Inclusion of rocks from modern marine shelves may shift most curves somewhat to the right, that is towards faster recycling rates.

5.2. Age Distribution of Hydrocarbons

The present-day cumulative age distributions of oil, gas, and coal (Tissot, 1979; Bois et al., 1982) are summarized in Figure 7. Excepting the Cretaceous anomaly, the reserves of oil fit well the exponential distribution pattern, with τ_{50} of ~85 Ma. This is similar to that of basins of passive margins (Table I) and, on geological time scales, it could have been the rate of initiation, compaction, and destruction of such basins that controlled the rate of generation and dissipation of oil. However, intracontinental tectonic realms account for substantial portion of global oil reserves (Bois et al., 1982). Since the continental recycling rates are slow, the factors which control oil dissipation must be therefore related to some second-order process, such as, for example, the hydrodynamic characteristics of sedimentary basins.

In contrast to oil, the age distribution of coal conforms clearly with the equation (3) population structure. Since vascular plants did not colonize continents until mid-Paleozoic, the observed pattern must reflect the exceptionally high rates of coal

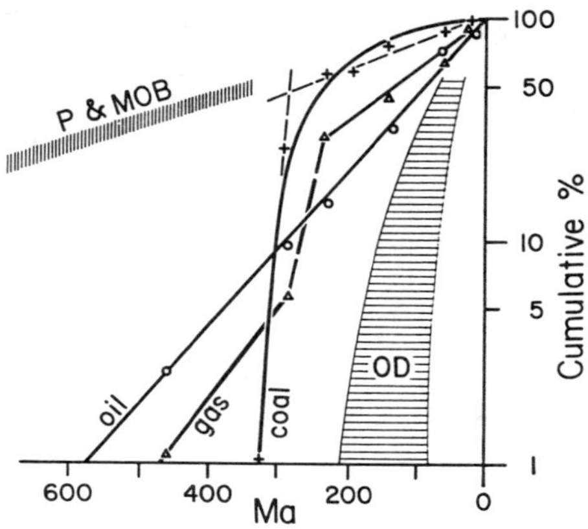

Figure 7. Present-day cumulative reserves-age distributions for oil, gas, and coal. Based on compilations of Bois et al. (1983). OD and P & MOB as in Figure 4.

formation in the Late Paleozoic and not a preferential destruction of pre-Mesozoic coals. Following the Paleozoic, the rate of formation and destruction of coal deposits may have attained a steady-state, with τ_{50} of ~250 Ma. Comparison with Table I suggests that this recycling rate may be a reflection of a predominant association of coal swamps with worn-down orogenic belts (molasse) and only to a lesser degree with the shorelines of marginal marine basins.

The age distribution of gas reserves shows a clear Late Paleozoic brake in the slope of the age curve (Figure 7). The older population follows the slope similar to that for oil, while the Mesozoic-Cenozoic one has a slope intermediate between that of oil and coal. In my view, this reflects generation and dissipation of gas from marine sources prior to the advanced Late Paleozoic colonization of continents by plants and from both marine and terrestrial sources during the subsequent geologic history.

5.3. Age Distribution of Metallic Ore Reserves

The present-day cumulative age distributions of metallic ore reserves are summarized in Figure 8 (Laznicka, 1973; Veizer and Jansen, 1979). Although the age slopes, and hence the rates of recycling, span almost the entire range, the common feature to all curves is that they reflect equation (2) populations, that is populations with higher "mortalities" for young deposits. In contrast to sedimentary lithological packages, which were associated predominantly with specific tectonic realms, the case for metallic ores is more complicated. Some deposits, such as mercury, are associated mostly with the shallow near-surface levels of immature orogenic belts. Their age distribution patterns deviate therefore only marginally from those of the oceanic domain. In other instances, such as gold, the dominant types of deposits have been the greenstone-belt associated Au veins in the Archean (Boyle, 1976), the Witwatersrand type paleoplacers in the Late Archean-Early Proterozoic (Pretorius, 1981), subvolcanic hydrothermal vein deposits of orogenic belts in the Phanerozoic, and placers in the Quaternary. Thus, the age curve for total reserves incorporates deposit types from variable tectonic settings (Meyer, 1985). All these genetic types, and their hosting tectonic realms, have variable rates of recycling and must therefore be considered independently. Overall, the concave upward shapes of cumulative age curves suggest that genetic types associated with fast cycling tectonic realms dominate the young portion of the terrestrial record. This does not necessarily mean that these transient genetic types of ores have not been represented at all in the geologic past, but only that the old deposits have been dispersed. Only the deposit types which have been associated with slowly cycling realms, or those of the transient types which were subsequently incorporated into such realms by, for example, metamorphism, are preserved in old segments of the present-day continents.

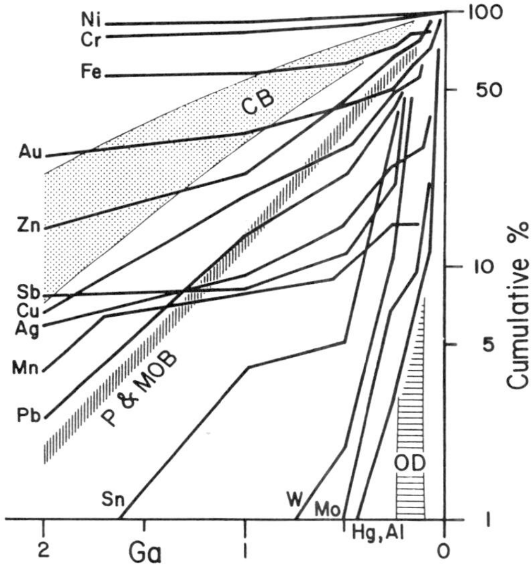

Figure 8. Present-day cumulative reserves-age distributions of metallic ores. Background data are summarized in Laznicka (1973) and Veizer and Jansen (1979). The curves were terminated at 2 Ga ago, because the data base for the older crustal segments is poor. Most curves show large inflections in Archean time, but the exact timing of these inflections is not easily resolvable with the available data. For this reason the pre-2.0 Ga history of mineralization is not discussed in the text. OD, CB, and P & MOB as in Figure 4.

In analogy to clastic sediments, the age-distribution patterns of the deposit of Ni, Cr, Fe, Au, and possibly Zn (Figure 8) suggest that the present-day distributions of their reserves are not solely a consequence of post-formational recycling. Their Precambrian age slopes are shallower than those of the slowest cycling tectonic realm, that is continental basement. As pointed out by Veizer and Jansen (1979), this is a logical contradiction, because destruction (recycling) of the host environment must lead also to the dispersal of the associated deposits. Consequently, the minimal recycling rates for the ores must have been at least those of their hosting tectonic realms. The present-day "excess" reserves of old Zn, Au, Fe, Cr, and Ni ores must be therefore a consequence of exceptionally high original abundance of such deposits during early terrestrial history. The subsequent recycling did not yet succeed in obliteration of these anomalies. If so, these data are a reflection of evolutionary trends in the composition of the crust and/or the atmosphere-hydrosphere-biosphere system (for further discussion see, for example, Veizer, 1976; Veizer and Jansen, 1979; Windley, 1984) during geologic history.

5.4. Chemical Evolution of Sediments

The systematics of recycling processes, discussed in this chapter have direct bearing on the nature of chemical and isotopic signals observed in Recent and ancient sediments. For more details on this special subject the reader is referred to Veizer and Jansen (1979), Veizer (1984), Taylor and McLennan (1985), Garrels and Mackenzie (1971a), Holland (1978, 1984), and particularly, Veizer (in press b).

6. CONCLUDING REMARKS

The geological evolution of the earth is characterized by a combination of unidirectional (evolutionary) and repetitive (cyclic) phenomena. Our perception of their relative importance depends on the time framework of our observations. In particular, it depends on the total duration of the (contemplated) record in comparison to the life-span τ_{MAX} of a given system or "population". For slowly cycling systems, such as the continental crust, the whole terrestrial record represents only 1-2 life-spans. This is barely sufficient to produce the exponential or power law age distributions, and the record of shorter-term and episodic events, visible as a scatter ("peaks" and "valleys") around the overall trend, is still clearly discernible. In contrast, for fast cycling systems, the same terrestrial record represents many complete cycles (e.g., ~25 τ_{MAX} for oceanic crust) and the exponential or power law relationship is well developed. Furthermore, beyond ~2-3 half-lives (τ_{50}) the residual "peaks" and "valleys" around the overall cumulative curves are indistinct, which leads to the loss of an interpretable record for shorter-term rates and episodic events. As a result, we tend to emphasize evolutionary features in the case of the continental crust, and recycling phenomena in the case of the oceanic crust. However, to a student of the oceanic environment, where record extends to only ~180 Ma, shorter-term rates and episodic phenomena again loom large, despite the fact that, on terrestrial time scales, he is dealing with a fast cycling system and a short time span. The appreciation of the importance of time to our perception of the geologic evolution of the earth is of paramount importance. Understanding of this variable will help us to reconcile uniformitarianism with evolution and to formulate a unified theory of terrestrial development. Quantification of this history may eventually be based on the concept of population dynamics encompassing not only the solid earth but also its complementary exogenic spheres.

7. ACKNOWLEDGMENTS

I thank Natural Sciences and Engineering Research Council of Canada for financial support and J. Hayes for typing the manuscript.

8. REFERENCES

Allègre, C.J., 1982, Chemical geodynamics, *Tectonophysics* **81**, 109-132.

Barrell, J., 1917, Rhythms and the measurement of geologic time, *Geol. Soc. Amer. Bull.* **28**, 745-904.

Blatt, H. and Jones, R.L., 1975, Proportions of exposed igneous, metamorphic, and sedimentary rocks, *Geol. Soc. Amer. Bull.* **86**, 1085-1088.

Bois, C., Bouche, P. and Pelet, R., 1982, Global geologic history and distribution of hydrocarbon reserves, *Amer. Assoc. Petrol. Geol. Bull.* **66**, 1248-1270.

Boyle, R.W., 1976, Mineralization processes in Archean greenstone and sedimentary belts, *Geol. Survey Canada Paper* **75-15**, 1-45.

Cloud, P., 1969, Pre-Paleozoic sediments and their significance for organic geochemistry, *in* G. Eglington and M.T.J. Murphy (eds.) *Organic Geochemistry*, Springer, Berlin, p. 772-786.

Cloud, P., 1976, Major features of crustal evolution, *Geol. Soc. South Africa* LXXIX, 1-32.

Cook, P.J. and McElhinny, M.W., 1979, A reevaluation of the spatial and temporal distribution of sedimentary phosphate deposits in the light of plate tectonics, *Econ. Geol.* **74**, 315-330.

Davies, T.A., Hay, W.W., Southam, J.R. and Worsley, T.R., 1977, Estimates of Cenozoic oceanic sedimentation rates, *Science,* **197**, 53-55.

Deevey, E.S., 1947, Life tables for natural populations of animals, *Quat. Rev. Biol.* **22**, 283-314.

Dietz, R.S., 1961, Continent and ocean basin evolution by spreading of the sea floor, *Nature* **190**, 854-857.

Dodd, J.R. and Stanton, R.J., 1981, *Paleoecology, Concepts and Applications*, Wiley, New York, 559 pp.

Engel, A.E.J., Itson, S.P., Engel, C.G., Stickney, D.M. and Cray, E.J., (1975) Crustal evolution and global tectonics: a petrogenic view, *Geol. Soc. Amer. Bull* **85**, 843-858.

Garrels, R.M. and Mackenzie, F.T., 1971a, *The Evolution of Sedimentary Rocks*, Norton, New York, 397 pp.

Garrels, R.M. and Mackenzie, F.T., 1971b, Gregor's denudation of the continents, *Nature,* **231**, 382-383.

Gingerich, P.D., 1983, Rates of evolution: effects of time and temporal scaling, *Science* **222**, 159-161.

Gregor, C.B., 1980, Weathering rates of sedimentary and crystalline rocks, *Proc. Koninklijke Nederlandse Akad. Wetenschappen,* **B 83**, 173-181.

Hallam, A., 1972, Models involving population dynamics, *in* T.J.M. Shopf (ed.), *Models in Paleobiology*, Freeman, Cooper, and Co., San Francisco, p. 62-80.

Hess, H.H., 1962, History of ocean basins, *in* A.E.J. Engel, H.L. James and B.F. Leonard (eds.) *Petrologic Studies,* Geological Society of America, p. 599-620.

Holland, H.D., 1978, *The Chemistry of the Atmosphere and Oceans,* Wiley-Interscience, New York, 351 pp.

Holland, H.D., 1984, *The Chemical Evolution of the Atmosphere and Oceans,* Princeton University Press, Princeton, 582 pp.

Holser, W.T., 1984, Gradual and abrupt shifts in ocean chemistry during Phanerozoic time, *in* H.D. Holland and A.F. Trendall (eds.) *Patterns of Change in Earth Evolution,* Springer, Berlin, p. 123-143.

Hutchinson, G.E., 1978, *An Introduction to Population Ecology,* Yale University Press, New Haven, 260 pp.

James, N.P., 1979, Facies models II: Reefs, *Geoscience Canada* **5**, 121-133.

Judson, S., 1968, Erosion of the land—or what's happening to our continents, *American Scientist* **56**, 356-374.

Laznicka, P., 1973, Development of non-ferrous metal deposits in geological time, *Can. J. Earth Sci.* **10**, 18-25.

LePichon, X., Francheteau, J. and Bonnin, J., 1978, *Plate tectonics,* Elsevier, Amsterdam, 300 pp.

Lerman, A., 1979, *Geochemical Processes,* Wiley-Interscience, New York, 385 pp.

MacArthur, R.H. and Connel, J.H., 1966, *The Biology of Populations*, Wiley, N.Y., 200 pp.
Martin, J.M. and Meybeck, M., 1979, Elemental mass balance of material carried by major world rivers, *Marine Chem.* **7**, 173-206.
Meybeck, M., 1979, Concentrations des eaux fluviales en éléments majeurs et apports en solution aux océans, *Rev. Géol. Dynam. Géogr. Phys.* **21**, 215-246.
Meyer, C., 1985, Ore metals through geologic history, *Science* **227**, 1421-1428.
Milliman, J.D. and Meade, R.H., 1983, World-wide delivery of river sediment to the oceans, *J. Geol.* **91**, 1-21.
Morgan, W.J., 1968, Rises, trenches, great faults, and crustal blocks, *J. Geophys. Res.* **73**, 1959-1982.
Odum, E.P., 1971, *Fundamentals of Geology*, W.B. Saunders Co., Philadelphia, 574 pp.
Pretorius, D.A., 1981, Gold and uranium in quartz-pebble conglomerates, *Econ. Geol., 75th Anniv. Volume*, 117-138.
Raup, D.M., 1985, Mathematical models of cladogenesis, *Paleobiology* **11**, 42-52.
Ronov, A.B., 1964, Common tendencies in the chemical evolution of the earth's crust, ocean, and atmosphere, *Geochem. Internat.* **1**, 713-737.
Ronov, A.B., 1980, *The Sedimentary Layer of the Earth*, Nauka, Moscow, 131 pp. (in Russian).
Sadler, P.M., 1981, Sediment accumulation rates and the completeness of stratigraphic sections, *J. Geol.* **89**, 569-584.
Slobodkin, C.B., 1962, *Growth and Regulation of Animal Populations*, Holt, Rinehart, and Winston, New York, 184 pp.
Southam, J.R. and Hay, W.W., 1977, The scales and dynamic models of deep-sea sedimentation, *J. Geophys. Res.* **82**, 3825-3842.
Southam, J.R. and Hay, W.W., 1981, Global sedimentary mass balance and sea level changes, *in* C. Emiliani (ed.) *The Sea*, v. 7, Wiley-Interscience, p. 1917-1984.
Strakhov, N.M., 1964, States and development of the external geospheres and formation of sedimentary rocks in the history of the earth, *Intern. Geol. Rev.* **6**, 1466-1482.
Strakhov, N.M., 1969, *Principles of Lithogenesis*, v. 2, Oliver-Boyd, Edinburgh, 609 pp.
Taylor, S.R. and McLennan, S.M., 1985, *The Continental Crust: Its Composition and Evolution*, Blackwell, Oxford, 312 pp.
Tissot, B., 1979, La répartition mondiale des combustibles fossiles, *La Recherche* **104**, 984-991.
Valentine, J.W., 1973, *Evolutionary Paleoecology of the Marine Biosphere*, Prentice-Hall, Englewood Cliffs, 511 pp.
Veizer, J., 1973, Sedimentation in geologic history: recycling vs. evolution or recycling with evolution, *Contrib. Mineral. Petrol.* **38**, 261-278.
Veizer, J., 1976, Evolution of ores of sedimentary affiliation during geologic history; relations to the general tendencies in the evolution of the crust, hydrosphere, atmosphere and biosphere, *in* K.H. Wolf (ed.) *Handbook of Strata-Bound and Stratiform Ore Deposits*, v. 3, Elsevier, Amsterdam, p. 1-41.
Veizer, J., 1983, Geologic evolution of the Archean-Early Proterozoic earth, *in* J.W. Schopf (ed.) *The Earth's Earliest Biosphere: Its Origin and Evolution*, Princeton Univ. Press, New Jersey, p. 240-259.
Veizer, J., 1984, Recycling on the evolving earth: geochemical record in sediments, *Proc. 27th Intern. Geol. Congr.*, v. 11, VNU Science Press, Utrecht, p. 325-345.
Veizer, J. (in press a), 1988, The evolving exogenic cycle, *in* C.B. Gregor and others (eds.) *Geochemical cycles in the evolution of the earth*, Wiley, New York, p. 175-220.
Veizer, J. (in press b), 1987, The earth and its life: geologic record of interactions and controls, *in* J. M. Robson (ed.) *Origin and Evolution of the Universe: Evidence for Design?*, Royal Soc. Canada, McGill-Queen's University Press, Montreal, p. 167-194.
Veizer, J. and Jansen, S.L., 1979, Basement and sedimentary recycling and continental evolution, *J. Geol.* **87**, 341-370.
Veizer, J. and Jansen, S.L., 1985, Basement and sedimentary recycling—2: Time dimension to global tectonics, *J. Geol.* **93**, 625-643.
Wilson, J.L., 1975, *Carbonate Facies in Geologic History*, Springer, New York, 471 pp.
Windley, B.F., 1984, *The Evolving Continents*, Wiley, New York, 399 pp.

SUBJECT INDEX

Page numbers in **bold face** (*e. g.*, **181**) indicate a major discussion of the subject.